MW00845297

A Course in Convexity

Alexander Barvinok

Graduate Studies in Mathematics
Volume 54

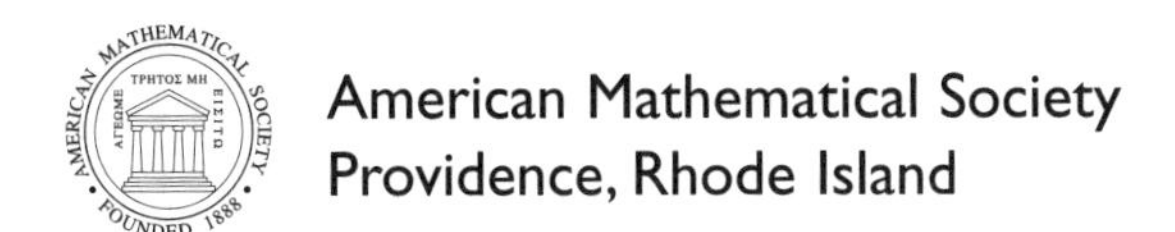

2000 *Mathematics Subject Classification.* Primary 52–01, 52–02, 52B45, 52C07, 46A20, 46N10, 90C05, 90C08, 90C22, 49N15.

Library of Congress Cataloging-in-Publication Data

Barvinok, Alexander, 1963–
A course in convexity / Alexander Barvinok.
p. cm. — (Graduate studies in mathematics, ISSN 1065-7339 ; v. 54)
Includes bibliographical references and index.
ISBN 0-8218-2968-8 (alk. paper)
1. Convex geometry. 2. Functional analysis. 3. Programming (Mathematics) I. Title. II. Series.

QA639.5.B37 2002

2002028208

Printed in the United States of America.

∞ The paper used in this book is acid-free and falls within the guidelines
established to ensure permanence and durability.
Visit the AMS home page at http://www.ams.org/

10 9 8 7 6 5 4 3 2 1   07 06 05 04 03 02

Contents

Preface

Convexity is very easy to define, to visualize and to get an intuition about. A set is called convex if for every two points a and b in the set, the straight line interval $[a, b]$ is also in the set. Thus the main building block of convexity theory is a straight line interval.

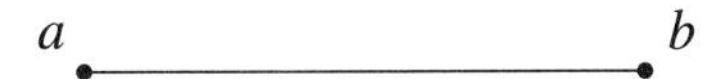

Convexity is more intuitive than, say, linear algebra. In linear algebra, the interval is replaced by the whole straight line. We have some difficulty visualizing a straight line because it runs unchecked in both directions.

On the other hand, the structure of convexity is richer than that of linear algebra. It is already evident in the fact that all points on the line are alike whereas the interval has two points, a and b, which clearly stand out.

Indeed, convexity has an immensely rich structure and numerous applications. On the other hand, almost every "convex" idea can be explained by a two-dimensional picture. There must be some reason for that apart from the tautological one that all our pictures are two-dimensional. One possible explanation is that since the definition of a convex set involves only three points (the two points a and b and a typical point x of the interval) and every three points lie in some plane, whenever we invoke a convexity argument in our reasoning, it can be properly pictured (moreover, since our three points a, b and x lie on the same line, we have room for a fourth point which often plays the role of the origin). Simplicity, intuitive appeal and universality of applications make teaching convexity (and writing a book on convexity) a rather gratifying experience.

About this book. This book grew out of sets of lecture notes for graduate courses that I taught at the University of Michigan in Ann Arbor since 1994. Consequently, this is a graduate textbook. The textbook covers several directions, which, although not independent, provide enough material for several one-semester three-credit courses.

One possibility is to follow discrete and combinatorial aspects of convexity: combinatorial properties of convex sets (Chapter I) – the structure of some interesting polytopes and polyhedra (the first part of Chapter II, some results of Chapter IV and Chapter VI) – lattice points and convex bodies (Chapter VII) – lattice points and polyhedra (Chapter VIII).

Another possibility is to follow the analytic line: basic properties of convex sets (Chapter I) – the structure of some interesting non-polyhedral convex sets, such as the moment cone, the cone of non-negative polynomials and the cone of positive semidefinite matrices (Chapter II and some results of Chapter IV) – metric properties of convex bodies (Chapter V).

Yet another possibility is to follow infinite-dimensional and dimension-free applications of convexity: basic properties of convex sets in a vector space (Chapter I) – separation theorems and the structure of some interesting infinite-dimensional convex sets (Chapter III) – linear inequalities and linear programming in an abstract setting (Chapter IV).

The main focus of the book is on applications of convexity rather than on studying convexity for its own sake. Consequently, mathematical applications range from analysis and probability to algebra to combinatorics to number theory. Finite- and infinite-dimensional optimization problems, such as the Transportation Problem, the Diet Problem, problems of optimal control, statistics and approximation are discussed as well.

The choice of topics covered in the book is entirely subjective. It is probably impossible to write a textbook that covers "all" convexity just as it is impossible to write a textbook that covers all mathematics. I don't even presume to claim to cover all "essential" or "important" aspects of convexity, although I believe that many of the topics discussed in the book belong to both categories.

The audience. The book is intended for graduate students in mathematics and other fields such as operations research, electrical engineering and computer science. That was the typical audience for the courses that I taught. This is, of course, reflected in the selection of topics covered in the book. Also, a significant portion of the material is suitable for undergraduates.

Prerequisites. The main prerequisite is linear algebra, especially the coordinate-free linear algebra. Knowledge of basic linear algebra should be sufficient for understanding the main convexity results (called "Theorems") and solving problems which address convex properties per se.

In many places, knowledge of some basic analysis and topology is needed. In most cases, some general understanding coupled with basic computational skills will be sufficient. For example, when it comes to the topology of Euclidean space, it suffices to know that a set in Euclidean space is compact if and only if it is closed and bounded and that a linear functional attains its maximum and minimum on such a set. Whenever the book says "Lebesgue integral" or "Borel set", it does so for the sake of brevity and means, roughly, "the integral makes sense" and "the set is nice and behaves predictably". For the most part, the only properties of the integral that the book uses are linearity (the integral of a linear combination of two functions is the linear combination of the integrals of the functions) and monotonicity (the integral of a non-negative function is non-negative). The relative abundance of integrals in a textbook on convexity is explained by the fact that the most natural way to define a linear functional is by using an integral of some sort. A few exercises openly require some additional skills (knowledge of functional analysis or representation theory).

When it comes to applications (often called "Propositions"), the reader is expected to have some knowledge in the general area which concerns the application.

Style. The numbering in each chapter is consecutive: for example, Theorem 2.1 is followed by Definition 2.2 which is followed by Theorem 2.3. When a reference is made to another chapter, a roman numeral is included: for example, if Theorem 2.1 of Chapter I is referenced in Chapter III, it will be referred to as Theorem I.2.1. Definitions, theorems and other numbered objects in the text (except figures) are usually followed by a set of problems (exercises). For example, Problem 5 following Definition 2.6 in Chapter II will be referred to as Problem 5 of Section 2.6 from within Chapter II and as Problem 5 of Section II.2.6 from everywhere else in the book. Figures are numbered consecutively throughout the book. There is a certain difference between "Theorems" and "Propositions". Theorems state some general and fundamental convex properties or, in some cases, are called "Theorems" historically. Propositions describe properties of particular convex sets or refer to an application.

Problems. There are three kinds of problems in the text. The problems marked by * are deemed difficult (they may be so marked simply because the author is unaware of an easy solution). Problems with straightforward solutions are marked by °. Solving a problem marked by ° is essential for understanding the material and its result may be used in the future. Some problems are not marked at all. There are no solutions at the end of the book and there is no accompanying solution manual (that I am aware of) , which, in my opinion, makes the book rather convenient for use in courses where grades are given. On the other hand, many of the difficult and some of the easy problems used later in the text are supplied with a hint to a solution.

Acknowledgment. My greatest intellectual debt is to my teacher A.M. Vershik. As a student, I took his courses on convexity and linear programming. Later, we discussed various topics in convex analysis and geometry and he shared his notes on the subject with me. We planned to write a book on convexity together and actually started to write one (in Russian), but the project was effectively terminated by my relocation to the United States. The overall plan, structure and scope of the book have changed since then, although much has remained the same. All unfortunate choices, mistakes, typos, blunders and other slips in the text are my own. A.M. Vershik always insisted on a "dimension-free" approach to convexity, whenever possible, which simplifies and makes transparent many fundamental facts, and on stressing the idea of duality in the broadest sense. In particular, I learned the algebraic approach to the Hahn-Banach Theorem (Sections II.1, III.1–3) and the general view of infinite-dimensional linear programming (Chapter IV) from him. This approach makes the exposition rather simple and elegant. It makes it possible to deduce a variety of strong duality results from a single simple theorem (Theorem IV.7.2). My interest in quadratic convexity (Section II.14) and other "hidden convexity" results (Section III.7) was inspired by him. He also encouraged my preoccupation with lattice points (Chapter VIII) and various peculiar polytopes (Sections II.5–7).

On various stages of the project I received encouragement from A. Björner, L. Billera, R. Pollack, V. Klee, J.E. Goodman, G. Kalai, A. Frieze, L. Lovász, W.T. Gowers and I. Bárány.

I am grateful to my colleagues in the Department of Mathematics at the University of Michigan in Ann Arbor, especially to P. Hanlon, B.A. Taylor, J. Stembridge and S. Fomin with whose blessings I promoted convexity within the Michigan combinatorics curriculum. I thank the students who took Math 669 convexity classes in 1994–2001. Special thanks to G. Blekherman who contributed some of his interesting results on the metric structure of the set of non-negative multivariate polynomials (Problems 8 and 9 of Section V.2.4).

Since the draft of this book was posted on the web, I received very useful and detailed comments from R. Connelly, N. Ivanov, J. Lawrence, L. Lovász, G. Ziegler and A.M. Vershik. I am particularly grateful to J. Lawrence who suggested a number of essential improvements, among them are the greater generality of the "polarity as a valuation" theorem (Theorem IV.1.5), a simplified proof of the Euler-Poincaré Formula (Corollary VI.3.2) and an elegant proof of Gram's relations (Problem 1 of Section VIII.4.4) and many mathematical, stylistic and bibliographical corrections.

I thank A. Yong for reading the whole manuscript carefully and suggesting numerous mathematical and stylistic corrections. I thank M. Wendt for catching a mistake and alerting me by e-mail.

I thank S. Gelfand (AMS) for insisting over a number of years that I write the book and for believing that I was able to finish it.

I am grateful to the National Science Foundation for its support.

Ann Arbor, 2002
Alexander Barvinok

Chapter I

Convex Sets at Large

We define convex sets and explore some of their fundamental properties. In this chapter, we are interested in the "global" properties of convex sets as opposed to the "local" properties studied in the next chapter. Namely, we are interested in what a convex set looks like as a whole, how convex sets may intersect and how they behave with respect to linear transformations. In contrast, in the next chapter, we will discuss what a convex set looks like in a neighborhood of a point. The landmark results of this chapter are classical theorems of Carathéodory, Radon and Helly and the geometric construction of the Euler characteristic. We apply our results to study positive multivariate polynomials, the problem of uniform (Chebyshev) approximation and some interesting valuations on convex sets, such as the intrinsic volumes. Exercises address some other applications (such as the Gauss-Lucas Theorem), discuss various ramifications of the main results (such as the Fractional Helly Theorem or the Colored Carathéodory Theorem) and preview some of the results of the next chapters (such as the Brickman Theorem, the Schur-Horn Theorem and the Birkhoff-von Neumann Theorem). We introduce two important classes of convex sets, polytopes and polyhedra, discussed throughout the book.

1. Convex Sets. Main Definitions, Some Interesting Examples and Problems

First, we set the stage where the action is taking place. Much of the action, though definitely not all, happens in Euclidean space $\mathbb{R}^d$.

(1.1) Euclidean space. The d-dimensional Euclidean space $\mathbb{R}^d$ consists of all d-tuples $x = (\xi_1, \ldots, \xi_d)$ of real numbers. We call an element of $\mathbb{R}^d$ a *vector* or (more often) a *point*. We can add points: we say that

$$z = x + y \quad \text{for} \quad x = (\xi_1, \ldots, \xi_d), \quad y = (\eta_1, \ldots, \eta_d) \quad \text{and} \quad z = (\zeta_1, \ldots, \zeta_d),$$

provided

$$\zeta_i = \xi_i + \eta_i \quad \text{for} \quad i = 1, \ldots, d.$$

We can multiply a point by a real number:

$$\text{if} \quad x = (\xi_1, \ldots, \xi_d) \quad \text{and} \quad \alpha \quad \text{is a real number,}$$

then

$$\alpha x = (\alpha\xi_1, \ldots, \alpha\xi_d)$$

is a point from $\mathbb{R}^d$. We consider the *scalar product* in $\mathbb{R}^d$:

$$\langle x, y \rangle = \xi_1\eta_1 + \ldots + \xi_d\eta_d, \qquad \text{where} \qquad x = (\xi_1, \ldots, \xi_d) \quad \text{and} \quad y = (\eta_1, \ldots, \eta_d).$$

We define the (Euclidean) *norm*

$$\|x\| = \sqrt{\xi_1^2 + \ldots + \xi_d^2}$$

of a point $x = (\xi_1, \ldots, \xi_d)$ and the *distance* between two points x and y:

$$\operatorname{dist}(x, y) = \|x - y\| \quad \text{for} \quad x, y \in \mathbb{R}^d.$$

Later in the text we will need *volume*. We do not define volume formally (that would lead us too far away from the main direction of this book). Nevertheless, we assume that the reader is familiar with elementary properties of the volume (cf. Section 8.3). The volume of a set $A \subset \mathbb{R}^d$ is denoted $\operatorname{vol} A$ or $\operatorname{vol}_d A$.

Let us introduce the central concept of the book.

(1.2) Convex sets, convex combinations and convex hulls.
Let $\{x_1, \ldots, x_m\}$ be a finite set of points from $\mathbb{R}^d$. A point

$$x = \sum_{i=1}^m \alpha_i x_i, \qquad \text{where} \quad \sum_{i=1}^m \alpha_i = 1 \quad \text{and} \quad \alpha_i \geq 0 \quad \text{for} \quad i = 1, \ldots, m$$

is called a *convex combination* of $x_1, \ldots, x_m$. Given two distinct points $x, y \in \mathbb{R}^d$, the set

$$[x, y] = \Big\{ \alpha x + (1 - \alpha) y : \ 0 \leq \alpha \leq 1 \Big\}$$

of all convex combinations of x and y is called the *interval* with endpoints x and y. A set $A \subset \mathbb{R}^d$ is called *convex*, provided $[x, y] \subset A$ for any two $x, y \in A$, or in words: a set is convex if and only if for every two points it contains the interval that connects them. We agree that the empty set $\emptyset$ is convex. For $A \subset \mathbb{R}^d$, the set of all convex combinations of points from A is called the *convex hull* of A and denoted $\operatorname{conv}(A)$. We will see that $\operatorname{conv}(A)$ is the smallest convex set containing A (Theorem 2.1).

(1.3) Some interesting examples. Sometimes, it is very easy to see whether the set is convex or not (see Figure 1).

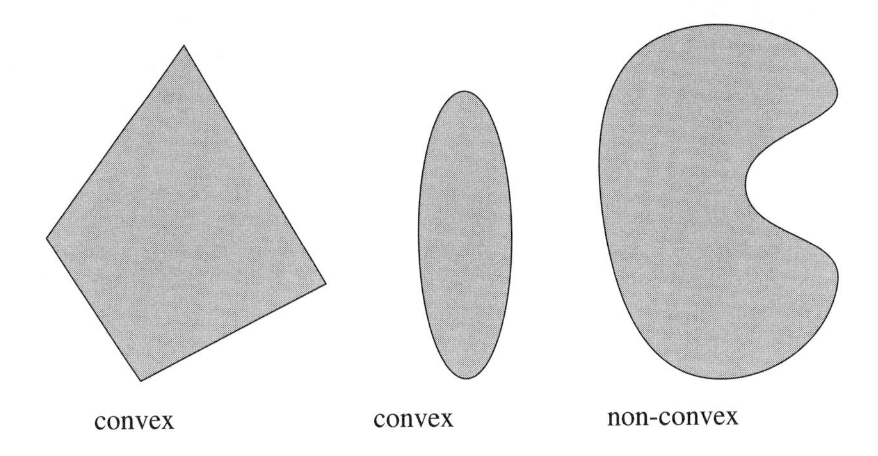

Figure 1. Two convex and one non-convex set

Sometimes, however, this is not so easy to see (cf. Problems 3, 4 and 5 below), or a convex set may have a number of equivalent descriptions and their equivalence may be not obvious (cf. Problems 6 and 7 below).

PROBLEMS.

We will encounter many of the harder problems later in the text.

1°. Prove that the convex hull of a set is a convex set.

2°. Let $c_1, \ldots, c_m$ be vectors from $\mathbb{R}^d$ and let $\beta_1, \ldots, \beta_m$ be numbers. The set

$$A = \Big\{ x \in \mathbb{R}^d : \quad \langle c_i, x \rangle \leq \beta_i \quad \text{for} \quad i = 1, \ldots, m \Big\}$$

is called a *polyhedron.* Prove that a polyhedron is a convex set.

3. Let $v_1, \ldots, v_m \in \mathbb{R}^d$ be points. Let us fix positive numbers $\rho_1, \ldots, \rho_m$ and let us define a map: $H : \mathbb{R}^d \longrightarrow \mathbb{R}^d$ by

$$H(x) = \frac{1}{f(x)} \sum_{i=1}^{m} \rho_i \exp\big\{\langle x, v_i \rangle\big\} v_i, \quad \text{where} \quad f(x) = \sum_{i=1}^{m} \rho_i \exp\big\{\langle x, v_i \rangle\big\}.$$

a°) Prove that the image of H lies in the convex hull of $v_1, \ldots, v_m$.

b*) Prove that the image of H is convex and that for any $\epsilon > 0$ and for any $y \in \operatorname{conv}\big(v_1, \ldots, v_m\big)$ there exists an $x \in \mathbb{R}^d$ such that $\operatorname{dist}\big(H(x), y\big) < \epsilon$.

c*) Assume that one cannot find a non-zero vector $c \in \mathbb{R}^d$ and a number α such that $\langle c, v_i \rangle = \alpha$ for $i = 1, \ldots, m$. Prove that H is injective.

Remark: The map H is an example of a *moment map*; see Chapter 4 of [**F93**].

4* (The Brickman Theorem). Let $q_1, q_2 : \mathbb{R}^n \longrightarrow \mathbb{R}$ be quadratic forms and let $\mathbb{S}^{n-1} = \{x \in \mathbb{R}^n : \ \|x\| = 1\}$ be the unit sphere. Consider the map $T : \mathbb{R}^n \longrightarrow \mathbb{R}^2$, $T(x) = \big(q_1(x), q_2(x)\big)$. Prove that the image $T(\mathbb{S}^{n-1})$ of the sphere is a convex set in $\mathbb{R}^2$, provided $n > 2$.

Remark: We prove this in Chapter II (see Theorem II.14.1).

5* (The Schur-Horn Theorem). For an $n \times n$ real symmetric matrix $A = (\alpha_{ij})$, let $\operatorname{diag}(A) = (\alpha_{11}, \ldots, \alpha_{nn})$ be the diagonal of A, considered as a vector from $\mathbb{R}^n$. Let us fix real numbers $\lambda_1, \ldots, \lambda_n$. Consider the set $X \subset \mathbb{R}^n$ of all diagonals of $n \times n$ real symmetric matrices with the eigenvalues $\lambda_1, \ldots, \lambda_n$. Prove that X is a convex set. Furthermore, let $l = (\lambda_1, \ldots, \lambda_n)$ be the vector of eigenvalues, so $l \in \mathbb{R}^n$. For a permutation σ of the set $\{1, \ldots, n\}$, let $l^\sigma = (\lambda_{\sigma(1)}, \ldots, \lambda_{\sigma(n)})$ be the vector with the permuted coordinates. Prove that

$$X = \operatorname{conv}\Big(l^\sigma : \sigma \quad \text{ranges over all } n! \text{ permutations of the set } \ \{1, \ldots, n\}\Big).$$

Remark: See Theorem II.6.2.

6 (The Birkhoff - von Neumann Theorem). For a permutation σ of the set $\{1, \ldots, n\}$, let us define the $n \times n$ permutation matrix $X^\sigma = (\xi_{ij}^\sigma)$ as

$$\xi_{ij}^\sigma = \begin{cases} 1 & \text{if } \ \sigma(j) = i, \\ 0 & \text{if } \ \sigma(j) \neq i. \end{cases}$$

Prove that the convex hull of all $n!$ permutation matrices X^σ is the set of all $n \times n$ doubly stochastic matrices, that is, matrices $X = (\xi_{ij})$, where

$$\sum_{i=1}^n \xi_{ij} = 1 \quad \text{for all} \quad j, \qquad \sum_{j=1}^n \xi_{ij} = 1 \quad \text{for all} \quad i \qquad \text{and}$$
$$\xi_{ij} \geq 0 \quad \text{for all} \quad i, j.$$

We consider an $n \times n$ matrix X as a point in $\mathbb{R}^{n^2}$.

Remark: We prove this in Chapter II (see Theorem II.5.2).

7. Let us fix an even number $n = 2m$ and let us interpret $\mathbb{R}^{n+1}$ as the space of all polynomials $p(\tau) = \alpha_0 + \alpha_1 \tau + \ldots + \alpha_n \tau^n$ of degree at most n in one variable τ. Let

$$K = \Big\{p \in \mathbb{R}^{n+1} : \quad p(\tau) \geq 0 \quad \text{for all} \quad \tau \in \mathbb{R}\Big\}$$

be the set of all non-negative polynomials. Prove that K is a convex set and that K is the set of all polynomials that are representable as sums of squares of polynomials of degree at most m:

$$K = \Big\{\sum_{i=1}^k q_i^2, \quad \text{where} \quad \deg q_i \leq m\Big\}.$$

Remark: See Section II.11 and, especially, Problem 3 of Section II.11.3.

Often, we consider convex sets in a more general setting.

(1.4) Convex sets in vector spaces. We recall that a set V with the operations "+" (addition): $V \times V \longrightarrow V$ and "·" (scalar multiplication): $\mathbb{R} \times V \longrightarrow V$ is called a (real) *vector space* provided the following eight axioms are satisfied:

(1) $u + v = v + u$ for any two $u, v \in V$;

(2) $u + (v + w) = (u + v) + w$ for any three $u, v, w \in V$;

(3) $(\alpha\beta)v = \alpha(\beta v)$ for any $v \in V$ and any $\alpha, \beta \in \mathbb{R}$;

(4) $1v = v$ for any $v \in V$;

(5) $(\alpha + \beta)v = \alpha v + \beta v$ for any $v \in V$ and any $\alpha, \beta \in \mathbb{R}$;

(6) $\alpha(v + u) = \alpha v + \alpha u$ for any $\alpha \in \mathbb{R}$ and any $u, v \in V$;

(7) there exists a zero vector $\mathbf{0} \in V$ such that $v + \mathbf{0} = v$ for each $v \in V$ and

(8) for each $v \in V$ there exists a vector $-v \in V$ such that $v + (-v) = \mathbf{0}$.

We often say "points" instead of "vectors", especially when we have no particular reason to consider $\mathbf{0}$ (which we often denote just by 0) to be significantly different from any other point (vector) in V.

A set $A \subset V$ is called *convex*, provided for all $x, y \in A$ the *interval*

$$[x, y] = \big\{\alpha x + (1 - \alpha)y : \ 0 \leq \alpha \leq 1\big\}$$

is contained in A. Again, we agree that the empty set is convex. A *convex combination* of a finite set of points in V and a *convex hull* $\operatorname{conv}(A)$ of a set $A \subset V$ are defined just as in the case of Euclidean space.

PROBLEM.

1°. Let V be the space of all continuous real-valued functions $f : [0, 1] \longrightarrow \mathbb{R}$. Prove that the sets

$$B = \Big\{f \in V : \quad |f(\tau)| \leq 1 \quad \text{for all} \quad \tau \in [0, 1]\Big\} \qquad \text{and}$$
$$K = \Big\{f \in V : \quad f(\tau) \leq 0 \quad \text{for all} \quad \tau \in [0, 1]\Big\}$$

are convex.

(1.5) Operations with convex sets. Let V be a vector space and let $A, B \subset V$ be (convex) sets. The *Minkowski sum* $A + B$ is a subset in V defined by

$$A + B = \Big\{x + y : \quad x \in A, \ y \in B\Big\}.$$

In particular, if $B = \{b\}$ is a point, the set

$$A + b = \Big\{x + b : x \in A\Big\}$$

is a *translation* of A. For a number α and a subset $X \subset V$, the set

$$\alpha X = \Big\{\alpha x : x \in X\Big\}$$

is called a *scaling* of X (for $\alpha > 0$, the set αX is also called a *dilation* of X). Some properties of convex sets are obvious, some are not so obvious, and some are quite surprising.

PROBLEMS.

We will encounter some of the harder problems below later in the text.

1°. Prove that the intersection $\bigcap_{i \in I} A_i$ of convex sets is convex.

2°. Let $A \subset V$ be a convex set and let $T : V \longrightarrow W$ be a linear transformation. Prove that the image $T(A)$ is a convex set in W.

3. Let $A \subset \mathbb{R}^n$ be a polyhedron (see Problem 2, Section 1.3) and let $T : \mathbb{R}^n \longrightarrow \mathbb{R}^m$ be a linear transformation. Prove that the image $T(A)$ is a polyhedron in $\mathbb{R}^m$.

Remark: We prove this in Section 9; see Theorem 9.2.

4. Prove that $A + B$ is a convex set provided A and B are convex. Prove that for a convex set A and non-negative numbers α and β one has $(\alpha + \beta)A = \alpha A + \beta A$. Show that the identity does not hold if A is not convex or if α or β are allowed to be negative.

5*. For a set $A \subset \mathbb{R}^d$, let $[A] : \mathbb{R}^d \longrightarrow \mathbb{R}$ be the *indicator function* of A:

$$[A](x) = \begin{cases} 1 & \text{if } x \in A, \\ 0 & \text{if } x \notin A. \end{cases}$$

Let $A_1, \ldots, A_k$ be compact convex sets in $\mathbb{R}^n$ and let $T : \mathbb{R}^n \longrightarrow \mathbb{R}^m$ be a linear transformation. Let $B_i = T(A_i)$ be the image of A_i. Suppose that $\sum_{i=1}^k \alpha_i [A_i] = 0$ for some numbers α_i. Prove that $\sum_{i=1}^k \alpha_i [B_i] = 0$. Show that this is no longer true if the A_i are not convex.

Remark: We prove this in Section 8; see Corollary 8.2.

6. Let $A \subset \mathbb{R}^d$ be a compact convex set and $B = (-1/d)A$. Prove that there exists a vector $b \in \mathbb{R}^d$ such that $b + B \subset A$.

Remark: Figure 2 illustrates the statement for $d = 2$. We go back to this problem in Section 5 when we discuss Helly's Theorem (see Problem 1 of Section 5.2 and the hint thereafter).

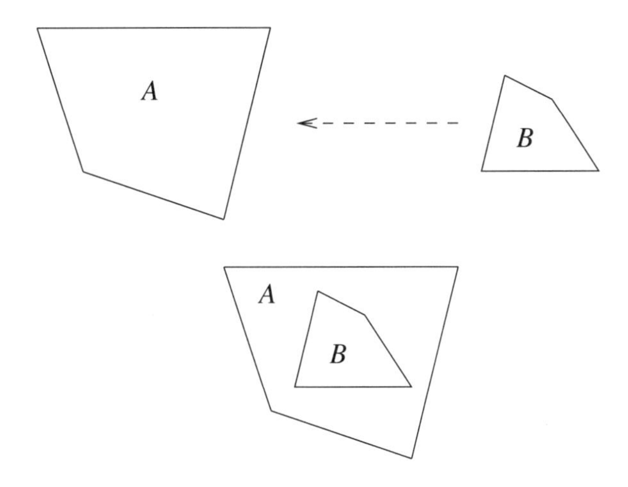

Figure 2. Example: the polygon $B = (-1/2)A$ can be translated inside A.

2. Properties of the Convex Hull. Carathéodory's Theorem

Recall from (1.2) that the convex hull $\operatorname{conv}(S)$ of a set S is the set of all convex combinations of points from S. Here is our first result.

(2.1) Theorem. *Let V be a vector space and let $S \subset V$ be a set. Then the convex hull of S is a convex set and any convex set containing S also contains* $\operatorname{conv}(S)$. *In other words,* $\operatorname{conv}(S)$ *is the smallest convex set containing S.*

Proof. First, we prove that $\operatorname{conv}(S)$ is a convex set (cf. Problem 1, Section 1.3). Indeed, let us choose two convex combinations $u = \alpha_1 u_1 + \ldots + \alpha_m u_m$ and $v = \beta_1 v_1 + \ldots + \beta_n v_n$ of points from S. The interval $[u, v]$ consists of the points $\gamma u + (1-\gamma)v$ for $0 \leq \gamma \leq 1$. Each such point $\gamma\alpha_1 u_1 + \ldots + \gamma\alpha_m u_m + (1-\gamma)\beta_1 v_1 + \ldots + (1-\gamma)\beta_n v_n$ is a convex combination of points $u_1, \ldots, u_m, v_1, \ldots, v_n$ from S since

$$\sum_{i=1}^{m} \gamma\alpha_i + \sum_{i=1}^{n} (1-\gamma)\beta_i = \gamma \sum_{i=1}^{m} \alpha_i + (1-\gamma) \sum_{i=1}^{n} \beta_i = \gamma + (1-\gamma) = 1.$$

Therefore, $\operatorname{conv}(S)$ is convex.

Now we prove that for any convex set A such that $S \subset A$, we have $\operatorname{conv}(S) \subset A$. Let us choose a convex combination

$$u = \alpha_1 u_1 + \ldots + \alpha_m u_m$$

of points $u_1, \ldots, u_m$ from S. We must prove that $u \in A$. Without loss of generality, we may assume that $\alpha_i > 0$ for $i = 1, \ldots, m$. We proceed by induction on m. If

$m = 1$, then $u = u_1$ and $u \in A$ since $S \subset A$. Suppose that $m > 1$. Then $\alpha_m < 1$ and we may write

$$u = (1 - \alpha_m)w + \alpha_m u_m, \quad \text{where} \quad w = \frac{\alpha_1}{1 - \alpha_m}u_1 + \ldots + \frac{\alpha_{m-1}}{1 - \alpha_m}u_{m-1}.$$

Now, w is a convex combination of $u_1, \ldots, u_{m-1}$ because

$$\sum_{i=1}^{m-1} \frac{\alpha_i}{1 - \alpha_m} = \frac{1}{1 - \alpha_m} \sum_{i=1}^{m-1} \alpha_i = \frac{1 - \alpha_m}{1 - \alpha_m} = 1.$$

Therefore, by the induction hypothesis, we have $w \in A$. Since A is convex, $[w, u_m] \subset A$, so $u \in A$. □

PROBLEMS.

1°. Prove that $\operatorname{conv}\big(\operatorname{conv}(S)\big) = \operatorname{conv}(S)$ for any $S \subset V$.

2°. Prove that if $A \subset B$, then $\operatorname{conv}(A) \subset \operatorname{conv}(B)$.

3°. Prove that $\Big(\operatorname{conv}(A) \cup \operatorname{conv}(B)\Big) \subset \operatorname{conv}(A \cup B)$.

4. Let $S \subset V$ be a set and let $u, v \in V$ be points such that $u \notin \operatorname{conv}(S)$ and $v \notin \operatorname{conv}(S)$. Prove that if $u \in \operatorname{conv}\big(S \cup \{v\}\big)$ and $v \in \operatorname{conv}\big(S \cup \{u\}\big)$, then $u = v$.

5 (Gauss-Lucas Theorem). Let $f(z)$ be a non-constant polynomial in one complex variable z and let $z_1, \ldots, z_m$ be the roots of f (that is, the set of all solutions to the equation $f(z) = 0$). Let us interpret a complex number $z = x + iy$ as a point $(x, y) \in \mathbb{R}^2$. Prove that each root of the derivative $f'(z)$ lies in the convex hull $\operatorname{conv}(z_1, \ldots, z_m)$.

Hint: Without loss of generality we may suppose that $f(z) = (z - z_1) \cdots (z - z_m)$. If w is a root of $f'(z)$, then $\sum_{i=1}^m \prod_{j \neq i}(w - z_j) = 0$, and, therefore, $\sum_{i=1}^m \prod_{j \neq i}(\overline{w - z_j}) = 0$, where $\overline{z}$ is the complex conjugate of z. Multiply both sides of the last identity by $(w - z_1) \cdots (w - z_n)$ and express w as a convex combination of $z_1, \ldots, z_m$.

Next, we introduce two important classes of convex sets.

(2.2) Definitions. The convex hull of a finite set of points in $\mathbb{R}^d$ is called a *polytope.*

Let $c_1, \ldots, c_m$ be vectors from $\mathbb{R}^d$ and let $\beta_1, \ldots, \beta_m$ be numbers. The set

$$P = \Big\{x \in \mathbb{R}^d : \langle c_i, x \rangle \leq \beta_i \quad \text{for} \quad i = 1, \ldots, m\Big\}$$

is called a *polyhedron* (see Problem 2 of Section 1.3).

PROBLEMS.

1. Prove that the set

$$\Delta = \Big\{(\xi_1, \ldots, \xi_{d+1}) \in \mathbb{R}^{d+1}: \quad \xi_1 + \ldots + \xi_{d+1} = 1 \quad \text{and} \quad \xi_i \geq 0 \quad \text{for} \quad i = 1, \ldots, d+1\Big\}$$

is a polytope in $\mathbb{R}^{d+1}$. This polytope is called the *standard d-dimensional simplex.*

2. Prove that the set

$$I = \Big\{(\xi_1, \ldots, \xi_d) \in \mathbb{R}^d: \quad 0 \leq \xi_i \leq 1 \quad \text{for} \quad i = 1, \ldots, d\Big\}$$

is a polytope. This polytope is called a *d-dimensional cube.*

3. Prove that the set

$$O = \Big\{(\xi_1, \ldots, \xi_d) \in \mathbb{R}^d: \quad |\xi_1| + \ldots + |\xi_d| \leq 1\Big\}$$

is a polytope. This polytope is called a *(hyper)octahedron* or *crosspolytope.*

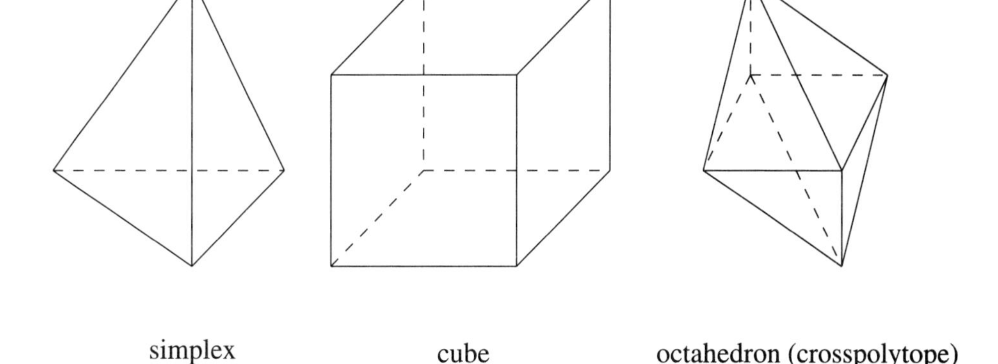

Figure 3. Some 3-dimensional polytopes: simplex (tetrahedron), cube and octahedron

4. Prove that the disc $B = \{(\xi_1, \xi_2) \in \mathbb{R}^2: \ \xi_1^2 + \xi_2^2 \leq 1\}$ is not a polytope.

5. Let $V = C[0,1]$ be the space of all real-valued continuous functions on the interval $[0,1]$ and let $A = \Big\{f \in V: \ 0 \leq f(\tau) \leq 1 \text{ for all } \tau \in [0,1]\Big\}$. Prove that A is not a polytope.

The following two problems constitute the Weyl-Minkowski Theorem.

6*. Prove that a polytope $P \subset \mathbb{R}^d$ is also a polyhedron.

Remark: We prove this in Chapter II; see Corollary II.4.3.

7*. Prove that a bounded polyhedron $P \subset \mathbb{R}^d$ is also a polytope.

Remark: We prove this in Chapter IV; see Corollary IV.1.3.

It seems intuitively obvious that in the space of a small dimension, to represent a given point x from the convex hull of a set A as a convex combination, we would need to use only a few points of A, although their choice will, of course, depend on x. For example, in the plane, to represent x as a convex combination, we need to use only three points; see Figure 4.

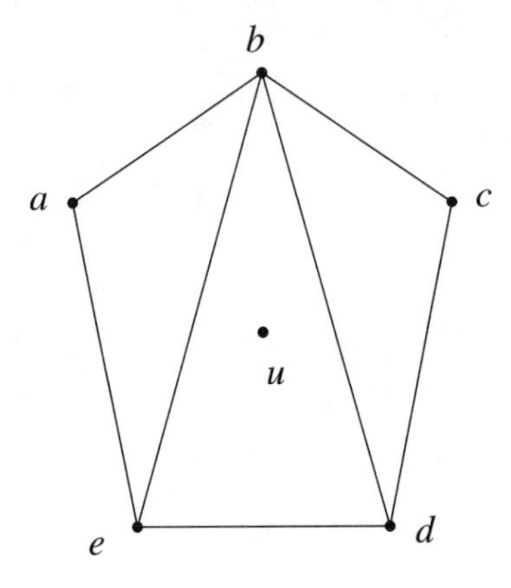

Figure 4. Example: to represent u as a convex combination of a, b, c, d and e, we need only three points, for instance b, e and d.

The general fact is known as Carathéodory's Theorem, which was proved by C. Carathéodory around 1907.

(2.3) Carathéodory's Theorem. *Let $S \subset \mathbb{R}^d$ be a set. Then every point $x \in \operatorname{conv}(S)$ can be represented as a convex combination of $d+1$ points from S:*

$$x = \alpha_1 y_1 + \ldots + \alpha_{d+1} y_{d+1}, \quad \text{where} \quad \sum_{i=1}^{d+1} \alpha_i = 1, \quad \alpha_i \geq 0$$

and $y_i \in S$ for $i = 1, \ldots, d+1$.

Proof. Every point $x \in \operatorname{conv}(S)$ can be written as a convex combination

$$x = \alpha_1 y_1 + \ldots + \alpha_m y_m$$

of some points $y_1, \ldots, y_m \in S$. We can assume that $\alpha_i > 0$ for all $i = 1, \ldots, m$. If $m < d+1$, we can add terms $0 y_1$, say, to get a convex combination with $d+1$ terms. Suppose that $m > d+1$. Let us show that we can construct a convex combination with fewer terms. Let us consider a system of linear homogeneous equations in m real variables $\gamma_1, \dots, \gamma_m$:

$$\gamma_1 y_1 + \dots + \gamma_m y_m = 0 \quad \text{and} \quad \gamma_1 + \dots + \gamma_m = 0.$$

The first vector equation reads as d real linear equations

$$\gamma_1 \eta_{1j} + \dots + \gamma_m \eta_{mj} = 0 : \quad j = 1, \dots, d$$

in the coordinates η_{ij} of y_i: $y_i = (\eta_{i1}, \dots, \eta_{id})$. Altogether, we have $d+1$ linear homogeneous equations in m variables $\gamma_1, \dots, \gamma_m$. Since $m > d+1$, there must be a non-trivial solution $\gamma_1, \dots, \gamma_m$. Since $\gamma_1 + \dots + \gamma_m = 0$, some γ_i are strictly positive and some are strictly negative. Let

$$\tau = \min\{\alpha_i / \gamma_i : \quad \gamma_i > 0\} = \alpha_{i_0} / \gamma_{i_0}.$$

Let $\widetilde{\alpha}_i = \alpha_i - \tau \gamma_i$ for $i = 1, \dots, m$. Then $\widetilde{\alpha}_i \geq 0$ for all $i = 1, \dots, m$ and $\alpha_{i_0} = 0$. Furthermore,

$$\widetilde{\alpha}_1 + \dots + \widetilde{\alpha}_m = (\alpha_1 + \dots + \alpha_m) - \tau(\gamma_1 + \dots + \gamma_m) = 1$$

and

$$\widetilde{\alpha}_1 y_1 + \dots + \widetilde{\alpha}_m y_m = \alpha_1 y_1 + \dots + \alpha_m y_m - \tau(\gamma_1 y_1 + \dots + \gamma_m y_m) = x.$$

Therefore, we represented x as a convex combination

$$x = \sum_{i \neq i_0} \widetilde{\alpha}_i y_i$$

of $m-1$ points $y_1, \dots, \widehat{y_{i_0}}, \dots, y_m$ (y_{i_0} omitted).

So, if x is a convex combination of $m > d+1$ points, it can be written as a convex combination of fewer points. Iterating this procedure, we get x as a convex combination of $d+1$ (or fewer) points from S. $\square$

PROBLEMS.

1°. Show by an example that the constant $d+1$ in Carathéodory's Theorem cannot be improved to d.

2* (I. Bárány). Let $S_1, \dots, S_{d+1}$ be subsets of $\mathbb{R}^d$. Prove that if $u \in \operatorname{conv}(S_i)$ for each S_i, then there exist points $v_i \in S_i$ such that $u \in \operatorname{conv}(v_1, \dots, v_{d+1})$.

Hint: Choose points $v_i \in S_i$ in such a way that the distance from u to $\operatorname{conv}(v_1, \dots, v_{d+1})$ is the smallest possible. Prove that if $u \notin \operatorname{conv}(v_1, \dots, v_{d+1})$, the distance could have been decreased further. This result is known as the "Colored Carathéodory Theorem"; see [**Bar82**].

3*. Let $S \subset \mathbb{R}^d$ be a set and let u be a point *in the interior* of $\operatorname{conv}(S)$. Prove that one can choose $2d$ points $v_1, \dots, v_{2d} \in S$ such that u lies *in the interior* of $\operatorname{conv}(v_1, \dots, v_{2d})$.

4. Suppose that $S \subset \mathbb{R}^d$ is a set such that every two points in S can be connected by a continuous path in S or a union of at most d such sets. Prove that every point $u \in \operatorname{conv}(S)$ is a convex combination of some d points of S.

Here is a useful corollary relating convexity and topology.

(2.4) Corollary. *If $S \subset \mathbb{R}^d$ is a compact set, then* $\operatorname{conv}(S)$ *is a compact set.*

Proof. Let $\Delta \subset \mathbb{R}^{d+1}$ be the standard d-dimensional simplex; see Problem 1 of Section 2.2:

$$\Delta = \Big\{(\alpha_1, \dots, \alpha_{d+1}) : \sum_{i=1}^{d+1} \alpha_i = 1 \quad \text{and} \quad \alpha_i \geq 0 \quad \text{for} \quad i = 1, \dots, d+1\Big\}.$$

Then Δ is compact and so is the direct product

$$S^{d+1} \times \Delta = \Big\{\big(u_1, \dots, u_{d+1}; \alpha_1, \dots, \alpha_{d+1}\big) : \quad u_i \in S \quad \text{and} \quad (\alpha_1, \dots, \alpha_{d+1}) \in \Delta\Big\}.$$

Let us consider the map $\Phi : S^{d+1} \times \Delta \longrightarrow \mathbb{R}^d$,

$$\Phi(u_1, \dots, u_{d+1}; \alpha_1, \dots, \alpha_{d+1}) = \alpha_1 u_1 + \ldots + \alpha_{d+1} u_{d+1}.$$

Theorem 2.3 implies that the image of Φ is $\operatorname{conv}(S)$. Since Φ is continuous, the image of Φ is compact, which completes the proof. □

PROBLEMS.

1. Give an example of a closed set in $\mathbb{R}^2$ whose convex hull is not closed.
2. Prove that the convex hull of an open set in $\mathbb{R}^d$ is open.

3. An Application: Positive Polynomials

In this section, we demonstrate a somewhat unexpected application of Carathéodory's Theorem (Theorem 2.3). We will use Carathéodory's Theorem in the space of (homogeneous) polynomials.

Let us fix positive integers k and n and let $H_{2k,n}$ be the real vector space of all homogeneous polynomials $p(x)$ of degree $2k$ in n real variables $x = (\xi_1, \dots, \xi_n)$. We choose a basis of $H_{2k,n}$ consisting of the monomials

$$e_a = \xi_1^{\alpha_1} \cdots \xi_n^{\alpha_n} \quad \text{for} \quad a = (\alpha_1, \dots, \alpha_n) \quad \text{where} \quad \alpha_1 + \ldots + \alpha_n = 2k.$$

Hence $\dim H_{2k,n} = \binom{n+2k-1}{2k}$. At this point, we are not particularly concerned with choosing the "correct" scalar product in $H_{2k,n}$. Instead, we declare $\{e_a\}$ the orthonormal basis of $H_{2k,n}$, hence identifying $H_{2k,n} = \mathbb{R}^d$ with $d = \binom{n+2k-1}{2k}$.

We can change variables in polynomials.

(3.1) Definition. Let $U : \mathbb{R}^n \longrightarrow \mathbb{R}^n$ be an orthogonal transformation and let $p \in H_{2k,n}$ be a polynomial. We define $q = U(p)$ by

$$q(x) = p(U^{-1}x) \quad \text{for} \quad x = (\xi_1, \dots, \xi_n).$$

Clearly, q is a homogeneous polynomial of degree $2k$ in $\xi_1, \dots, \xi_n$.

PROBLEMS.

1°. Check that $(U_1U_2)(p) = U_1(U_2(p))$.

2°. Let

$$p(x) = \|x\|^{2k} = \left(\xi_1^2 + \ldots + \xi_n^2\right)^k.$$

Prove that $U(p) = p$ for any orthogonal transformation U.

It turns out that the polynomial of Problem 2, Section 3.1, up to a scalar multiple, is the only polynomial that stays invariant under any orthogonal transformation.

(3.2) Lemma. *Let $p \in H_{2k,n}$ be a polynomial such that $U(p) = p$ for every orthogonal transformation U. Then*

$$p(x) = \gamma\|x\|^{2k} = \gamma\left(\xi_1^2 + \ldots + \xi_n^2\right)^k \quad \text{for some} \quad \gamma \in \mathbb{R}.$$

Proof. Let us choose a point $y \in \mathbb{R}^d$ such that $\|y\| = 1$ and let $\gamma = p(y)$. Let us consider

$$q(x) = p(x) - \gamma\|x\|^{2k}.$$

Thus q is a homogeneous polynomial of degree $2k$ and $q(Ux) = q(x)$ for any orthogonal transformation U and any vector x. Moreover, $q(y) = 0$. Since for every vector x such that $\|x\| = 1$ there is an orthogonal transformation U_x such that $U_x y = x$, we have $q(x) = q\big(U_x y\big) = q(y) = 0$ and hence $q(x) = 0$ for all x such that $\|x\| = 1$. Since q is a homogeneous polynomial, we have $q(x) = 0$ for all $x \in \mathbb{R}^n$. Therefore, $p(x) = \gamma\|x\|^{2k}$ as claimed. □

We are going to use Theorem 2.3 to deduce the existence of an interesting identity.

(3.3) Proposition. *Let k and n be positive integers. Then there exist vectors $c_1, \ldots, c_m \in \mathbb{R}^n$ such that*

$$\|x\|^{2k} = \sum_{i=1}^{m} \langle c_i, x\rangle^{2k} \quad \text{for all} \quad x \in \mathbb{R}^n.$$

In words: the k-th power of the sum of squares of n real variables is a sum of $2k$-th powers of linear forms in the variables.

Proof. We are going to apply Carathéodory's Theorem in the space $H_{2k,n}$.

Let

$$\mathbb{S}^{n-1} = \Big\{c \in \mathbb{R}^n : \quad \|c\| = 1\Big\}$$

be the unit sphere in $\mathbb{R}^n$. For a $c \in \mathbb{S}^{n-1}$, let

$$p_c(x) = \langle c, x\rangle^{2k} \quad \text{where} \quad x = (\xi_1, \ldots, \xi_n).$$

Hence we have $p_c \in H_{2k,n}$. Let

$$K = \operatorname{conv}\Big(p_c : \quad c \in \mathbb{S}^{n-1}\Big)$$

be the convex hull of all polynomials p_c. Since the sphere $\mathbb{S}^{n-1}$ is compact and the map $c \longmapsto p_c$ is continuous, the set $\big\{p_c : \ c \in \mathbb{S}^{n-1}\big\}$ is a compact subset of $H_{2k,n}$. Therefore, by Corollary 2.4, we conclude that K is compact.

Let us prove that $\gamma\|x\|^{2k} \in K$ for some $\gamma > 0$. The idea is to average the polynomials p_c over all possible vectors $c \in \mathbb{S}^{n-1}$. To this end, let dc be the rotation invariant probability measure on $\mathbb{S}^{n-1}$ and let

$$p(x) = \int_{\mathbb{S}^{n-1}} p_c(x)\ dc = \int_{\mathbb{S}^{n-1}} \langle c, x\rangle^{2k}\ dc \tag{3.3.1}$$

be the average of all polynomials p_c. We observe that $p \in H_{2k,n}$. Moreover, since dc is a rotation invariant measure, we have $U(p) = p$ for any orthogonal transformation U of $\mathbb{R}^n$ and hence by Lemma 3.2, we must have

$$p(x) = \gamma\|x\|^{2k} \quad \text{for some} \quad \gamma \in \mathbb{R}.$$

We observe that $\gamma > 0$. Indeed, for any $x \neq 0$, we have $p_c(x) > 0$ for all $c \in \mathbb{S}^{n-1}$ except from a set of measure 0 and hence $p(x) > 0$.

The integral (3.3.1) can be approximated with arbitrary precision by a finite Riemann sum:

$$p(x) \approx \frac{1}{N}\sum_{i=1}^{N} p_{c_i}(x) \quad \text{for some} \quad c_i \in \mathbb{S}^{n-1}.$$

Therefore, p lies in the closure of K. Since K is closed, $p \in K$. By Theorem 2.3, we can write $p(x) = \gamma\|x\|^{2k}$ as a convex combination of some $\binom{n+2k-1}{2k} + 1$ polynomials $p_{c_i}(x) = \langle c_i, x\rangle^{2k}$. Dividing by γ, we complete the proof. □

It is not always easy to come up with a particular choice of c_i in the identity of Proposition 3.3.

PROBLEMS.

1. Prove Liouville's identity:

$$(\xi_1^2 + \xi_2^2 + \xi_3^2 + \xi_4^2)^2 = \frac{1}{6}\sum_{1 \leq i < j \leq 4} (\xi_i + \xi_j)^4 + \frac{1}{6}\sum_{1 \leq i < j \leq 4} (\xi_i - \xi_j)^4.$$

2. Prove Fleck's identity:

$$(\xi_1^2+\xi_2^2+\xi_3^2+\xi_4^2)^3 = \frac{1}{60}\sum_{1 \leq i < j < k \leq 4} (\xi_i \pm \xi_j \pm \xi_k)^6 + \frac{1}{30}\sum_{1 \leq i < j \leq 4} (\xi_i \pm \xi_j)^6 + \frac{3}{5}\sum_{1 \leq i \leq 4} \xi_i^6,$$

where the sums containing $\pm$ signs are taken over all possible independent choices of pluses and minuses.

3. Prove that one can choose $m \leq \binom{n+2k-1}{2k}$ in Proposition 3.3.

Remark: In his solution of Waring's problem, for all positive integers k and n, D. Hilbert constructed *integer* vectors c_i and *rational* numbers γ_i such that

$$\|x\|^{2k} = \sum_{i=1}^{m} \gamma_i \langle c_i, x\rangle^{2k} \quad \text{for all} \quad x \in \mathbb{R}^n;$$

see, for example, Chapter 3 of [**N96**].

We apply Proposition 3.3 to study positive polynomials.

(3.4) Definition. Let $p \in H_{2k,n}$ be a polynomial. We say that p is *positive* provided $p(x) > 0$ for all $x \neq 0$. Equivalently, $p \in H_{2k,n}$ is positive provided $p(x) > 0$ for all $x \in \mathbb{S}^{n-1}$. Similarly, a polynomial $p \in H_{2k,n}$ is *non-negative* if $p(x) \geq 0$ for all x.

PROBLEM.

1°. Prove that the set of all positive polynomials is a non-empty open convex set in $H_{2k,n}$ and that the set of all non-negative polynomials is a non-empty closed convex set in $H_{2k,n}$.

We apply Proposition 3.3 to prove that a homogeneous polynomial is positive if and only if it can be multiplied by a sufficiently high power of $\|x\|^2$ to produce a sum of even powers of linear functions. The proof below is due to B. Reznick [**R95**] and [**R00**].

(3.5) Proposition. *Let $p \in H_{2k,n}$ be a positive polynomial. Then there exist a positive integer s and vectors $c_1, \ldots, c_m \in \mathbb{R}^n$ such that*

$$\|x\|^{2s-2k} p(x) = \sum_{i=1}^{m} \langle c_i, x\rangle^{2s} \quad \textit{for all} \quad x \in \mathbb{R}^n.$$

Sketch of Proof. For a polynomial $f \in H_{2k,n}$,

$$f(x) = \sum_{a=(\alpha_1,\ldots,\alpha_n)} \lambda_a \xi_1^{\alpha_1} \cdots \xi_n^{\alpha_n},$$

let us formally define the differential operator

$$f(\partial) = \sum_{a=(\alpha_1,\ldots,\alpha_n)} \lambda_a \frac{\partial^{\alpha_1}}{\partial \xi_1^{\alpha_1}} \cdots \frac{\partial^{\alpha_n}}{\partial \xi_n^{\alpha_n}}.$$

Let us choose a positive integer $s > 2k$ and the corresponding identity of Proposition 3.3:

$$\|x\|^{2s} = \sum_{i=1}^{m} \langle c_i, x\rangle^{2s}. \tag{3.5.1}$$

Let us see what happens if we apply $f(\partial)$ to both sides of the identity.

It is not very hard to see that

$$f(\partial)\Big(\langle c,x\rangle^{2s}\Big) = \frac{(2s)!}{(2s-2k)!} f(c)\cdot \langle c,x\rangle^{2s-2k}. \tag{3.5.2}$$

It suffices to check the identity when f is a monomial and then it is straightforward.

One can also see that

$$f(\partial)\Big(\|x\|^{2s}\Big) = \frac{2^{2k}s!}{(s-2k)!} g(x)\cdot \|x\|^{2s-2k} \quad \text{for some} \quad g \in H_{2k,n}. \tag{3.5.3}$$

The correspondence $f \longmapsto g$ defines a linear transformation

$$\Phi_s : H_{2k,n} \longrightarrow H_{2k,n}$$

and the crucial observation is that Φ_s converges to the identity operator I as s grows. Again, it suffices to check this when f is a monomial, in which case $\Phi_s(f) = f + O(1/s)$ by the repeated application of the chain rule.

Since $I^{-1} = I$, for all sufficiently large s the operator Φ_s is invertible and Φ_s^{-1} converges to the identity operator I as s grows. Now we note that the set of positive polynomials is open; see Problem 1 of Section 3.4. Therefore, for a sufficiently large s the polynomial $q = \Phi_s^{-1}(p)$ lies in a sufficiently small neighborhood of $p = I(p)$ and hence is positive. Applying $q(\partial)$ to both sides of (3.5.1), by (3.5.2) and (3.5.3) we get

$$\frac{2^{2k}s!}{(s-2k)!}\Phi_s(q)\cdot \|x\|^{2s-2k} = \frac{(2s)!}{(2s-2k)!}\sum_{i=1}^{m} q(c_i)\langle c_i,x\rangle^{2s-2k}.$$

Now $\Phi_s(q) = p$ and $q(c_i) > 0$ for $i = 1,\ldots,m$. Rescaling, we obtain a representation of $p\cdot\|x\|^{2s-2k}$ as a sum of powers of linear forms. □

PROBLEMS.

1°. Check formulas (3.5.2) and (3.5.3).

2°. Check that Φ_s indeed converges to the identity operator on $H_{2k,n}$ as s grows.

3. For polynomials $f, g \in H_{2k,n}$, let us define

$$\langle f, g\rangle = f(\partial)g.$$

Note that since $\deg f = \deg g$, we get a *number*. Prove that $\langle f,g\rangle$ is a scalar product in $H_{2k,n}$ and that

$$\langle U(f), U(g)\rangle = \langle f, g\rangle$$

for every orthogonal transformation of $\mathbb{R}^n$.

4. Construct an example of a non-negative polynomial $p \in H_{2k,n}$ for which the conclusion of Proposition 3.5 does not hold true.

5. Using Proposition 3.5, deduce *Polya's Theorem*:

Let p be a real homogeneous polynomial of degree k in n real variables $x = (\xi_1, \ldots, \xi_n)$ and let

$$\mathbb{R}^n_+ = \Big\{(\xi_1, \ldots, \xi_n) : \quad \xi_i \geq 0 \quad \text{for} \quad i = 1, \ldots, n\Big\}$$

be the non-negative orthant in $\mathbb{R}^n$. Suppose that

$$p(x) > 0 \quad \text{for all} \quad x \in \mathbb{R}^n_+ \setminus \{0\}.$$

Then there exists a positive integer s such that the coefficients $\{\lambda_a\}$ of the polynomial

$$(\xi_1 + \ldots + \xi_n)^s p(\xi_1, \ldots, \xi_n) = \sum_{\substack{a=(\alpha_1, \ldots, \alpha_n) \\ \alpha_1 + \ldots + \alpha_n = s+k}} \lambda_a \xi_1^{\alpha_1} \cdots \xi_n^{\alpha_n}$$

are non-negative.

We discuss the structure of the set of *non*-homogeneous non-negative *uni*variate polynomials in Chapter II; see Section II.11. The results there can be translated in a more or less straightforward way to homogeneous non-negative *bi*variate polynomials by applying the following "homogenization trick": if $p(t)$ is a non-homogeneous polynomial of degree d, let $q(x, y) = y^d p(x/y)$. Some interesting metric properties of the set of non-negative multivariate polynomials are discussed in exercises of Chapter V; see Problems 8 and 9 of Section V.2.4.

4. Theorems of Radon and Helly

The following very useful result was first stated in 1921 by J. Radon as a lemma.

(4.1) Radon's Theorem. *Let $S \subset \mathbb{R}^d$ be a set containing at least $d + 2$ points. Then there are two non-intersecting subsets $R \subset S$ ("red points") and $B \subset S$ ("blue points") such that*

$$\operatorname{conv}(R) \cap \operatorname{conv}(B) \neq \emptyset.$$

Proof. Let $v_1, \ldots, v_m$, $m \geq d+2$, be distinct points from S. Consider the following system of $d + 1$ homogeneous linear equations in variables $\gamma_1, \ldots, \gamma_m$:

$$\gamma_1 v_1 + \ldots + \gamma_m v_m = 0 \quad \text{and} \quad \gamma_1 + \ldots + \gamma_m = 0.$$

Since $m \geq d + 2$, there is a non-trivial solution to this system. Let

$$R = \big\{v_i : \ \gamma_i > 0\big\} \quad \text{and} \quad B = \big\{v_i : \ \gamma_i < 0\big\}.$$

Then $R \cap B = \emptyset$.

Let $\beta = \sum_{i:\gamma_i > 0} \gamma_i$. Then $\beta > 0$ and $\sum_{i:\gamma_i < 0} \gamma_i = -\beta$, since γ's sum up to zero. Since $\gamma_1 v_1 + \ldots + \gamma_m v_m = 0$, we have

$$\sum_{i:\gamma_i > 0} \gamma_i v_i = \sum_{i:\gamma_i < 0} (-\gamma_i) v_i.$$

Let

$$v = \sum_{i:\gamma_i>0} \frac{\gamma_i}{\beta} v_i = \sum_{i:\gamma_i<0} \frac{-\gamma_i}{\beta} v_i.$$

Hence v is a convex combination of points from R and a convex combination of points from B. In other words, $v \in \text{conv}(R)$ and $v \in \text{conv}(B)$. □

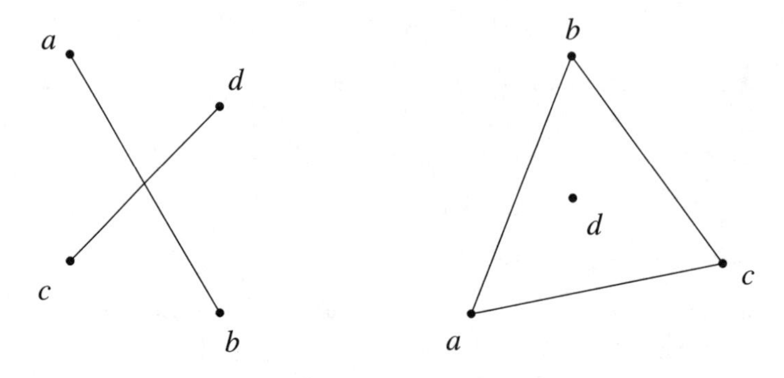

Figure 5. Example: for any set of four points in the plane, either one of the points lies within the convex hull of the other three, or the points can be split into two pairs whose convex hulls intersect.

PROBLEMS.

1°. Show by an example that the constant $d+2$ in Radon's Theorem cannot be improved to $d+1$.

2* (Tverberg's Theorem). Let $k \geq 2$. Prove that for any set S of $(k-1)(d+1)+1$ or more points in $\mathbb{R}^d$, one can find k pairwise non-intersecting subsets $A_1, \ldots, A_k \subset S$ such that the intersection

$$\text{conv}(A_1) \cap \text{conv}(A_2) \cap \ldots \cap \text{conv}(A_k)$$

is not empty. Show that for some sets of $(k-1)(d+1)$ points, such subsets $A_1, \ldots, A_k$ cannot be found.

Remark: See, for example, Chapter 8 of [**Mat02**].

The following result (one of the most famous results in convexity) was discovered by E. Helly in 1913. The proof below is due to Radon (1921).

(4.2) Helly's Theorem. *Let $A_1, \ldots, A_m$, $m \geq d+1$, be a finite family of convex sets in $\mathbb{R}^d$. Suppose that every $d+1$ of the sets have a common point:*

$$A_{i_1} \cap \ldots \cap A_{i_{d+1}} \neq \emptyset.$$

Then all the sets have a common point:

$$A_1 \cap \ldots \cap A_m \neq \emptyset.$$

Proof. The proof is by induction on m (starting with $m = d + 1$). Suppose that $m > d + 1$. Then, by the induction hypothesis, for every $i = 1, \ldots, m$ there is a point p_i in the intersection $A_1 \cap \ldots \cap A_{i-1} \cap A_{i+1} \cap \ldots \cap A_m$ (A_i is missing). Altogether, we have $m > d + 1$ points p_i, each of which belongs to all the sets, except perhaps A_i. If two of these points happened to coincide, we get a point which belongs to all the A_i's. Otherwise, by Radon's Theorem (Theorem 4.1) there are non-intersecting subsets $R = \{p_i : i \in I\}$ and $B = \{p_j : j \in J\}$ such that there is a point

$$p \in \operatorname{conv}(R) \cap \operatorname{conv}(B).$$

We claim that p is a common point of $A_1, \ldots, A_m$. Indeed, all the points $p_i : i \in I$ of R belong to the sets $A_i : i \notin I$. All the points $p_j : j \in J$ of B belong to the sets $A_j : j \notin J$. Since the sets A_i are convex, every point from $\operatorname{conv}(R)$ belongs to the sets $A_i : i \notin I$. Similarly, every point from $\operatorname{conv}(B)$ belongs to the sets $A_j : j \notin J$. Therefore,

$$p \in \bigcap_{i \notin I} A_i \quad \text{and} \quad p \in \bigcap_{j \notin J} A_j.$$

Since $I \cap J = \emptyset$, we have

$$p \in \bigcap_{i=1}^{m} A_i$$

and the proof follows. □

PROBLEMS.

1°. Show that the theorem does not hold for non-convex sets A_i.

2°. Construct an example of convex sets A_i in $\mathbb{R}^2$, such that every two sets have a common point, but there is no point which would belong to all the sets A_i.

3°. Give an example of an infinite family $\{A_i : i = 1, 2, \ldots\}$ of convex sets in $\mathbb{R}^d$ such that every $d+1$ sets have a common point but there are no points common to all the sets A_i.

The theorem can be extended to infinite families of compact convex sets.

(4.3) Corollary. *Let $\{A_i : i \in I\}$, $|I| \geq d + 1$ be a (possibly infinite) family of compact convex sets in $\mathbb{R}^d$ such that the intersection of any $d+1$ sets is not empty:*

$$A_{i_1} \cap \ldots \cap A_{i_{d+1}} \neq \emptyset.$$

Then the intersection of all the sets A_i is not empty:

$$\bigcap_{i \in I} A_i \neq \emptyset.$$

Proof. By Theorem 4.2, for any finite subfamily $J \subset I$ the intersection $\bigcap_{i \in J} A_i$ is not empty. Now we use the fact that if the intersection of a family of compact sets is empty, then the intersection of the sets from some finite subfamily is empty. $\square$

Helly's Theorem has numerous generalizations, extensions, ramifications, etc. To list all of them is impossible; here are just some.

PROBLEMS.

1. Let $A_1, \ldots, A_m$ be convex sets in $\mathbb{R}^d$ and let $k \leq d+1$. Prove that if every k of the sets have a common point, then for every $(d-k+1)$-dimensional subspace L in $\mathbb{R}^d$ there exists a translate $L+u: \ u \in \mathbb{R}^d$ which intersects every set $A_i: \ i = 1, \ldots, m$.

2. Let $A_1, \ldots, A_m$ and C be convex sets in $\mathbb{R}^d$. Suppose that for any $d+1$ sets $A_{i_1}, \ldots, A_{i_{d+1}}$ there is a translate $C+u: \ u \in \mathbb{R}^d$ of C which *intersects* all $A_{i_1}, \ldots, A_{i_{d+1}}$. Prove that there is a translate $C+u$ of C which *intersects* all sets $A_1, \ldots, A_m$.

3. In Problem 2, replace *intersects* by *contains.*

4. In Problem 2, replace *intersects* by *is contained in.*

5* (Fractional Helly's Theorem). Prove that for any $0 < \alpha < 1$ and any d there exists a $\beta = \beta(d, \alpha) > 0$ with the following property:

Suppose that $A_1, \ldots, A_m$, $m \geq d+1$, are convex sets in $\mathbb{R}^d$. Let f be the number of $(d+1)$-subfamilies $A_{i_1}, \ldots, A_{i_{d+1}}$ that have a common point. If $f \geq \alpha\binom{m}{d+1}$, then at least some βm sets A_i have a common point.

Prove that one can choose $\beta = 1 - (1-\alpha)^{1/(d+1)}$.

Remark: The bound $\beta = 1 - (1-\alpha)^{1/(d+1)}$ is best possible. A weaker bound $\beta \geq \alpha/(d+1)$ is much easier to prove; see Chapter 8 of [**Mat02**].

6* ("Piercing" Theorem). Prove that for every triple (p, q, d) such that $p \geq q \geq d+1$ there exists a positive integer $c(p, q, d)$ such that if $A_1, \ldots, A_m \subset \mathbb{R}^d$ are convex sets, $m \geq p$ and out of every p sets $A_{i_1}, \ldots, A_{i_p}$ some q sets have a common point, then some set X of $c(p, q, d)$ points in $\mathbb{R}^d$ intersects every set A_i.

Remark: This is a conjecture of H. Hadwiger and M. Debrunner, proved by N. Alon and D. Kleitman; see [**We97**] and references therein.

7* (Measure of the intersection). Prove that for every d there exists a constant $\gamma = \gamma(d) > 0$ with the following property:

Let $A_1, \ldots, A_m \subset \mathbb{R}^d$ be convex sets. Suppose that $m \geq 2d$ and that the intersection of every $2d$ sets $A_{i_1}, \ldots, A_{i_{2d}}$ has volume at least 1. Prove that the intersection $A_1 \cap \ldots \cap A_m$ has volume at least γ.

Prove that one can choose $\gamma(d) = d^{-2d^2}$ (it is conjectured that the constant d^{-2d^2} can be improved to $d^{-\alpha d}$ for some absolute constant $\alpha > 0$).

Remark: This is a result of I. Bárány, M. Katchalski and J. Pach; see [**E93**] and references therein.

8* (Colored Helly's Theorem). Let $\mathcal{A}_1, \ldots, \mathcal{A}_{d+1}$ be non-empty finite families of convex sets in $\mathbb{R}^d$. Suppose that for each choice $A_i \in \mathcal{A}_i$, $i = 1, \ldots, d+1$, we have $A_1 \cap \ldots \cap A_{d+1} \neq \emptyset$. Prove that for some i the intersection of the sets in the family $\mathcal{A}_i$ is non-empty.

Hint: This can be deduced from Problem 2 of Section 2.3; see also [**Bar82**] and Chapter 8 of [**Mat02**].

5. Applications of Helly's Theorem in Combinatorial Geometry

In the next two sections, we discuss various applications and appearances of Helly's Theorem. The proofs are almost immediate once we recognize the relationship of the problem to Helly's Theorem but may be quite non-trivial if we miss that connection.

(5.1) Separating points by a hyperplane. Suppose that there is a finite set R of red points in $\mathbb{R}^d$ and a finite set B of blue points in $\mathbb{R}^d$. A *hyperplane* $H \subset \mathbb{R}^d$ is the set described by a linear equation $H = \{x \in \mathbb{R}^d : \langle c, x \rangle = \alpha\}$, where $c \neq 0$ is a non-zero vector and α is a number. We say that the hyperplane $H \subset \mathbb{R}^d$ strictly separates red and blue points if $\langle c, x \rangle < \alpha$ for all $x \in R$ and $\langle c, x \rangle > \alpha$ for all $x \in B$. The following result, called Kirchberger's Theorem, was proved by P. Kirchberger in 1903, that is, *before* Helly's Theorem.

Proposition. *Suppose that for any set $S \subset \mathbb{R}^d$ of $d+2$ or fewer points there exists a hyperplane which strictly separates the sets $S \cap R$ and $S \cap B$ of red, resp. blue, points in S. Then there exists a hyperplane which strictly separates the sets R and B.*

Proof. A hyperplane $H = \{x \in \mathbb{R}^d : \langle c, x \rangle = \alpha\}$, where $c = (\gamma_1, \ldots, \gamma_d) \in \mathbb{R}^d$, can be encoded by a point $(c, \alpha) = (\gamma_1, \ldots, \gamma_d, \alpha) \in \mathbb{R}^{d+1}$. For every point $r \in R$ we define a set $A_r \subset \mathbb{R}^{d+1}$:

$$A_r = \Big\{(c, \alpha) \in \mathbb{R}^{d+1} : \langle c, r \rangle < \alpha\Big\}$$

and for every point $b \in B$ we define a set $A_b \subset \mathbb{R}^{d+1}$:

$$A_b = \Big\{(c, \alpha) \in \mathbb{R}^{d+1} : \langle c, b \rangle > \alpha\Big\}.$$

It is clear that A_b and A_r are convex sets in $\mathbb{R}^{d+1}$. Therefore, by Helly's Theorem the intersection

$$\Big(\bigcap_{r \in R} A_r\Big) \cap \Big(\bigcap_{b \in B} A_b\Big)$$

is non-empty, provided for any subset $S \subset R \cup B$ of at most $d+2$ points the intersection

$$\Big(\bigcap_{r \in S \cap R} A_r\Big) \cap \Big(\bigcap_{b \in S \cap B} A_b\Big)$$

is non-empty. Since the sets A_r and A_b are open, their intersection is an open set; therefore an intersection of sets A_r and A_b is non-empty if and only if it contains a point (c, α) with $c \neq 0$. Hence for any subset $S \subset R \cup B$, the intersection

$$\Big(\bigcap_{r \in S \cap R} A_r \Big) \cap \Big(\bigcap_{b \in S \cap B} A_b \Big)$$

is not empty if and only if there is a point $(c, \alpha) \in \mathbb{R}^{d+1}$ such that the hyperplane

$$H = \big\{ x \in \mathbb{R}^d : \quad \langle c, x \rangle = \alpha \big\}$$

strictly separates the sets $B \cap S$ and $R \cap S$. This completes the proof. □

For example, if two sets of points in the plain cannot be separated by a straight line, one of the three configurations of Figure 6 must occur.

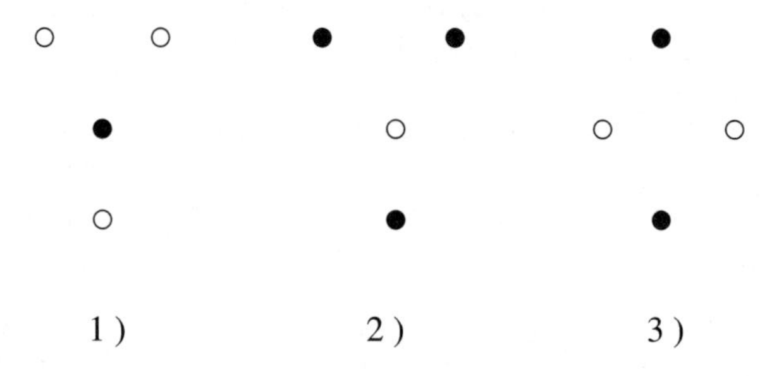

Figure 6. The three reasons points cannot be separated in the plane

PROBLEMS.

1. Prove that if a convex set is contained in the union of a finite family of halfspaces in $\mathbb{R}^d$ (sometimes we say *covered* by a finite family of halfspaces; see Section 5.2), then it is contained in the union of some $d+1$ (or fewer) halfspaces from the family (covered by some $d+1$ subspaces).

2. Let $I_1, \ldots, I_m$ be parallel line segments in $\mathbb{R}^2$, such that for every three $I_{i_1}, I_{i_2}, I_{i_3}$ there is a straight line that intersects all three. Prove that there is a straight line that intersects all the segments $I_1, \ldots, I_m$.

3. Let $A_i : i = 1, \ldots, m$ be convex sets in $\mathbb{R}^2$ such that for every two sets A_i and A_j there is a line parallel to the x-axis which intersects them both. Prove that there is a line parallel to the x-axis which intersects all the sets A_i.

(5.2) The center point. Let us fix a Borel probability measure μ on $\mathbb{R}^d$. This means, roughly speaking, that for any "reasonable" subset $A \subset \mathbb{R}^d$ a non-negative number $\mu(A)$ is assigned which satisfies some additivity and continuity properties and such that $\mu(\mathbb{R}^d) = 1$. We are not interested in rigorous definitions here; the following two examples are already of interest:

Counting measure. Suppose that there is a finite set $X \subset \mathbb{R}^d$ of $|X| = n$ points and $\mu(A) = |A \cap X|/n$ is the proportion of the points in X contained in A.

Integrable density. Suppose that there is an integrable function $f : \mathbb{R}^d \longrightarrow \mathbb{R}$ such that $f(x) \geq 0$ for all $x \in \mathbb{R}^d$ and such that $\displaystyle\int_{\mathbb{R}^d} f(x)\, dx = 1$, where dx is the Lebesgue measure. Let $\mu(A) = \displaystyle\int_A f(x)\, dx$ for all (Borel) measurable sets A.

With a hyperplane $H = \big\{x : \ \langle c, x \rangle = \alpha \big\}$ we associate two *open halfspaces*

$$H_+ = \big\{x \in \mathbb{R}^d : \quad \langle c, x \rangle > \alpha \big\} \quad \text{and} \quad H_- = \big\{x \in \mathbb{R}^d : \quad \langle c, x \rangle < \alpha \big\}$$

and two *closed halfspaces*

$$\overline{H_+} = \big\{x \in \mathbb{R}^d : \quad \langle c, x \rangle \geq \alpha \big\} \quad \text{and} \quad \overline{H_-} = \big\{x \in \mathbb{R}^d : \quad \langle c, x \rangle \leq \alpha \big\}.$$

Proposition. *Let μ be a Borel probability measure on $\mathbb{R}^d$. Then there exists a point $y \in \mathbb{R}^d$, called a center point, such that for any closed halfspace $\overline{H_\star}$ containing y one has*

$$\mu(\overline{H_\star}) \geq \frac{1}{d+1}.$$

Proof. For a closed halfspace $G \subset \mathbb{R}^d$, let $\widetilde{G} = \mathbb{R}^d \setminus G$ be the complementary open halfspace. Let $\mathcal{S}$ be the set of all closed halfspaces G such that $\mu\big(\widetilde{G}\big) < 1/(d+1)$. We observe that for any $d+1$ halfspaces $G_1, \ldots, G_{d+1}$ from $\mathcal{S}$ one has

$$\mu\big(\widetilde{G}_1 \cup \ldots \cup \widetilde{G}_{d+1}\big) < \frac{(d+1)}{(d+1)} = 1 \quad \text{and hence} \quad \widetilde{G}_1 \cup \ldots \cup \widetilde{G}_{d+1} \neq \mathbb{R}^d,$$

which implies that $G_1 \cap \ldots \cap G_{d+1} \neq \emptyset$. Helly's Theorem implies that any finite family $\{G_i\}$ of halfspaces from $\mathcal{S}$ has a non-empty intersection. Let us choose a finite number of halfspaces $G_1, \ldots, G_m \subset \mathbb{R}^d$ such that the intersection $B = G_1 \cap \ldots \cap G_m$ is bounded and hence compact. Enlarging the halfspaces by translations, if necessary, we can ensure that $G_1, \ldots, G_m$ are from $\mathcal{S}$. Thus $\{B \cap G : \ G \in \mathcal{S}\}$ is a family of compact sets such that every finite subfamily has a non-empty intersection. Hence there is a point y which belongs to all halfspaces $G \in \mathcal{S}$. If $H_\star$ is an open halfspace containing y, then the complementary closed halfspace does not contain y and hence does not belong to $\mathcal{S}$. Then we must have $\mu(H_\star) \geq 1/(d+1)$. Since μ is σ-additive and a closed halfspace can be represented as an intersection of countably many nested open halfspaces, the result follows. □

The above result was first obtained in 1916 by J. Radon. The above proof belongs to I.M. Yaglom and V.G. Boltyanskii (1956).

PROBLEMS.

1. Let $S \subset \mathbb{R}^d$ be a compact convex set. Prove that there is a point $u \in \mathbb{R}^d$ such that $(-1/d)S + u \subset S$.

Hint: For every point $x \in S$ consider the set $A_x = \{u : \ (-1/d)x + u \in S\}$. Use Helly's Theorem.

2* ("Ham Sandwich Theorem"). Let $\mu_1, \ldots, \mu_d$ be a set of Borel probability measures on $\mathbb{R}^d$. Prove that there exists a hyperplane $H \subset \mathbb{R}^d$ such that $\mu_i(\overline{H_+}) \geq 1/2$ and $\mu_i(\overline{H_-}) \geq 1/2$ for all $i = 1, \ldots, d$.

3* (Center Transversal Theorem). Let $\mu_1, \ldots, \mu_k$, $k \leq d$, be Borel probability measures on $\mathbb{R}^d$. Prove that there exists a $(k-1)$-dimensional affine subspace $L \subset \mathbb{R}^d$ such that for every closed halfspace $\overline{H_\star}$ containing L we have $\mu_i(\overline{H_\star}) \geq 1/(d-k+2)$.

Remark: For Problems 2 and 3, see [**Ž97**] and references therein.

4°. Prove that the theorem of Problem 3 implies both the result of Problem 2 above and the proposition of this section.

A couple of geometric problems.

5* (Krasnoselsky's Theorem). Let $X \subset \mathbb{R}^d$ be a set and let $a, b \in X$ be points. We say that b is *visible* from a if $[a, b] \subset X$. Suppose that $X \subset \mathbb{R}^d$ is an infinite compact set such that for any $d+1$ points of X there is a point from which all $d+1$ are visible. Prove that there is a point from which all points of X are visible.

6 (Jung's Theorem). For a compact set $X \subset \mathbb{R}^d$, let us call $\max_{y,z \in X} \|y - z\|$ the *diameter* of X. Prove that any compact set of diameter 2 is contained in a ball of radius $\sqrt{2d/(d+1)}$.

For Problems 5 and 6, see [**DG63**].

6. An Application to Approximation

We proceed to apply Helly's Theorem to an important problem of constructing the best approximation of a given function by a function from the required class. We will go back to this problem again in Section IV.13.

(6.1) Uniform approximations.

Let us fix some real-valued functions $f_i : T \longrightarrow \mathbb{R}$, $i = i, \ldots, m$, on some set T. Given a function $g : T \longrightarrow \mathbb{R}$ and a number $\epsilon \geq 0$, we want to construct a linear combination

$$f_x(\tau) = \sum_{i=1}^{m} \xi_i f_i(\tau), \quad x = (\xi_1, \ldots, \xi_m)$$

such that

$$|g(\tau) - f_x(\tau)| \leq \epsilon \quad \text{for all} \quad \tau \in T.$$

This is the problem of the *uniform* or *Chebyshev* approximation. Helly's Theorem implies that a uniform approximation exists if it exists on every reasonably small subset of T.

(6.2) Proposition. *Suppose that T is a finite set. Let us fix an $\epsilon \geq 0$. Suppose that for any $m+1$ points $\tau_1, \ldots, \tau_{m+1}$ from T one can construct a function $f_x(\tau)$ (with $x = (\xi_1, \ldots, \xi_m)$ depending on $\tau_1, \ldots, \tau_{m+1}$) such that*

$$|g(\tau) - f_x(\tau)| \leq \epsilon \quad \text{for } \tau = \tau_1, \ldots, \tau_{m+1}.$$

Then there exists a function $f_{\overline{x}}(\tau)$ such that

$$|g(\tau) - f_{\overline{x}}(\tau)| \leq \epsilon \quad \text{for all} \quad \tau \in T.$$

Proof. For a $\tau \in T$ let us define a set $A(\tau) \subset \mathbb{R}^m$:

$$A(\tau) = \Big\{(\xi_1, \ldots, \xi_m) : \ \big|g(\tau) - f_x(\tau)\big| \leq \epsilon\Big\}.$$

In other words, $A(\tau)$ is the set of functions f_x that approximate g within ϵ at the point τ. Now $A(\tau)$ are convex sets (see Problem 1 below), and

$$A(\tau_1) \cap \ldots \cap A(\tau_{m+1}) \neq \emptyset$$

for all possible choices of $m+1$ points $\tau_1, \ldots, \tau_{m+1}$ in T. Since T is finite, Helly's Theorem (Theorem 4.2) implies that the intersection of all sets $A(\tau) : \tau \in T$ is non-empty. It follows then that for a point

$$\overline{x} \in \bigcap_{\tau \in T} A(\tau)$$

we have $|g(\tau) - f_{\overline{x}}(\tau)| \leq \epsilon$ for all $\tau \in T$. □

PROBLEMS.

1°. Let $A(\tau)$ be as in the proof of Proposition 6.2. Prove that $A(\tau) \subset \mathbb{R}^{m+1}$ is a closed convex set.

2°. Show that for $m \geq 2$ the set $A(\tau)$ is *not* compact.

To prove a version of Proposition 6.2 for infinite sets T, we must assume some regularity of functions $f_1, \ldots, f_m$.

(6.3) Proposition. *Suppose there is a finite set of points $\sigma_1, \ldots, \sigma_n$ in T such that whenever $f_x = \xi_1 f_1 + \ldots + \xi_m f_m$ and $f_x(\sigma_1) = \ldots = f_x(\sigma_n) = 0$, then $\xi_1 = \ldots = \xi_m = 0$. Suppose further that for any set of $m+1$ points $\tau_1, \ldots, \tau_{m+1}$ in T one can construct a function f_x (with $x = (\xi_1, \ldots, \xi_m)$ depending on $\tau_1, \ldots, \tau_{m+1}$) such that*

$$|g(\tau) - f_x(\tau)| \leq \epsilon \quad \text{for} \quad \tau = \tau_1, \ldots, \tau_{m+1}.$$

Then there exists a function $f_{\overline{x}}(\tau)$ such that

$$|g(\tau) - f_{\overline{x}}(\tau)| \leq \epsilon \quad \text{for all} \quad \tau \in T.$$

Proof. Let $A(\tau) : \tau \in T$ be the sets defined in the proof of Proposition 6.2. Let

$$A = A(\sigma_1) \cap \ldots \cap A(\sigma_n).$$

First, we prove that A is compact. Indeed, by Problem 1, Section 6.2, the set A is closed. It remains to show that A is bounded. Let us define a function

$$N : \mathbb{R}^m \longrightarrow \mathbb{R}, \quad N(x) = \max\{|f_x(\sigma_i)| : \quad i = 1, \ldots, n\}.$$

Then $N(\lambda x) = |\lambda| N(x)$ for $\lambda \in \mathbb{R}$, $N(x) > 0$ for $x \neq 0$ and N is continuous (in fact, N is a norm in $\mathbb{R}^m$). Therefore,

$$\min\{N(x) : \quad \|x\| = 1\} = \delta > 0$$

and $N(x) > \delta\|x\|$.

Now, if $|g(\sigma_i) - f_x(\sigma_i)| \leq \epsilon$ for $i = 1, \ldots, n$, we have $|f_x(\sigma_i)| \leq |g(\sigma_i)| + \epsilon$ for $i = 1, \ldots, n$. Letting

$$R = \epsilon + \max\{|g(\sigma_i)| : i = 1, \ldots, n\},$$

we conclude that $N(x) \leq R$, and, therefore, $\|x\| \leq R/\delta$ for any $x \in A$. Thus A is compact.

For $\tau \in T$ let $\overline{A(\tau)} = A(\tau) \cap A$. Then each set $\overline{A(\tau)}$ is compact. Applying Helly's Theorem as in the proof of Proposition 6.2, we conclude that every intersection of a *finite* family of sets $\overline{A(\tau)}$ is non-empty. Therefore, every intersection $\overline{A(\tau_1)} \cap \ldots \cap \overline{A(\tau_{m+1})}$ is a non-empty compact convex set. Therefore, By Corollary 4.3, the intersection of all the sets $\overline{A(\tau)}$ is non-empty and so is the intersection of all the sets $A(\tau)$. A point

$$\overline{x} = (\xi_1, \ldots, \xi_m) \in \bigcap_{\tau \in T} A(\tau)$$

gives rise to a function

$$f_{\overline{x}} = \xi_1 f_1 + \ldots + \xi_m f_m,$$

which approximates g uniformly within the error ϵ. □

PROBLEMS.

In the problems below, $T = [0, 1]$ and $f_i(\tau) = \tau^i$, $i = 0, \ldots, m$ (note that we start with f_0).

1°. Prove that for any $m + 1$ distinct points $\tau_1, \tau_2, \ldots, \tau_{m+1}$ from $[0, 1]$ the intersection $A(\tau_1) \cap \ldots \cap A(\tau_{m+1})$ is compact.

2°. Let $g(\tau) = e^\tau$ for $\tau \in [0, 1]$. Let us choose $\epsilon = 0$. Check that each intersection $A(\tau_1) \cap \ldots \cap A(\tau_{m+1})$ is not empty for any choice of $\tau_1, \ldots, \tau_{m+1} \in [0, 1]$, but $\bigcap_{\tau \in [0,1]} A_\tau = \emptyset$. In other words, for every $m + 1$ points $\tau_1, \ldots, \tau_{m+1}$ there is a polynomial $p(\tau) = \xi_0 + \xi_1 \tau + \ldots + \xi_m \tau^m$ such that $p(\tau) = e^\tau$ for $\tau = \tau_1, \ldots, \tau_{m+1}$ but there is no polynomial $p(\tau)$ such that $e^\tau = p(\tau)$ for all $\tau \in [0, 1]$.

3*. Let $g : [0,1] \longrightarrow \mathbb{R}$ be any function. Prove that for any $m+2$ points $0 \leq \tau_1 < \tau_2 < \ldots < \tau_{m+2} \leq 1$ there is a unique polynomial $p(\tau) = \xi_0 + \xi_1 \tau + \ldots + \xi_m \tau^m$, such that

$$|g(\tau_1) - p(\tau_1)| = |g(\tau_2) - p(\tau_2)| = \ldots = |g(\tau_{m+2}) - p(\tau_{m+2})|$$

and the signs of the differences

$$g(\tau_1) - p(\tau_1), \quad g(\tau_2) - p(\tau_2), \quad \ldots, \quad g(\tau_{m+2}) - p(\tau_{m+2})$$

alternate. Prove that the polynomial p gives the unique best (that is, with the smallest ϵ) uniform approximation to g on the set of $m+2$ points $\tau_1, \ldots, \tau_{m+2}$. The error ϵ of this approximation can be found to be $\epsilon = |\eta|$, where $\xi_0, \ldots, \xi_m$ and η is the (necessarily unique) solution to the system of $m+2$ linear equations

$$g(\tau_1) - p(\tau_1) = \eta, \ g(\tau_2) - p(\tau_2) = -\eta, \ldots, g(\tau_{m+2}) - p(\tau_{m+2}) = (-1)^{m+1}\eta$$

in $m+2$ variables $(\xi_0, \ldots, \xi_m, \eta)$.

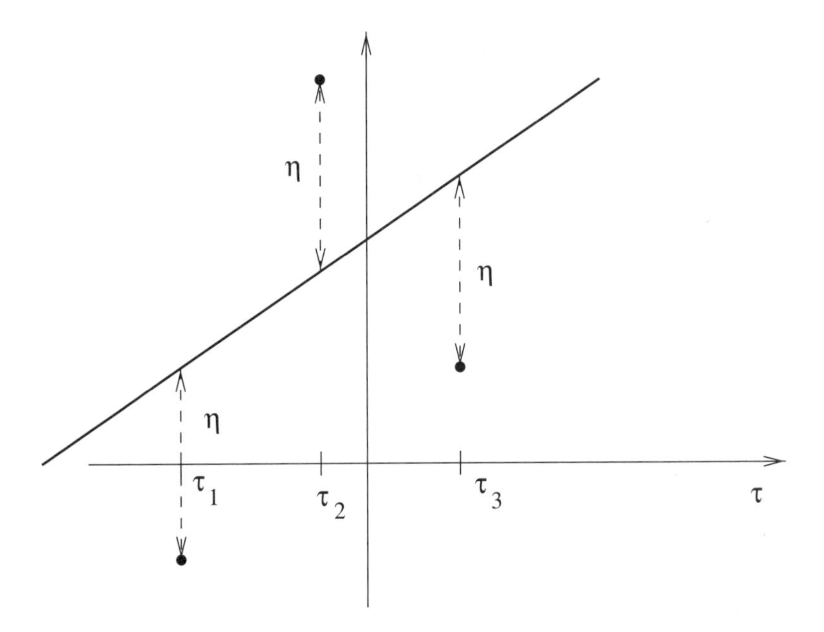

Figure 7. A linear function $p(\tau) = \xi_0 + \xi_1 \tau$ which provides the best uniform approximation for g at some three points τ_1, τ_2 and τ_2 and satisfies $p(\tau_1) - g(\tau_1) = -(p(\tau_2) - g(\tau_2)) = p(\tau_3) - g(\tau_3)$

7. The Euler Characteristic

Helly's Theorem tells us something special about how convex sets may intersect. Here we introduce another powerful tool to study intersection properties of convex sets.

(7.1) Definition. Let $A \subset \mathbb{R}^d$ be a subset. The *indicator function* $[A]$ of A is the function $[A] : \mathbb{R}^d \longrightarrow \mathbb{R}$ such that

$$[A](x) = \begin{cases} 1 & \text{if } x \in A, \\ 0 & \text{if } x \notin A. \end{cases}$$

PROBLEM.

1°. Prove that $[A] \cdot [B] = [A \cap B]$.

(7.2) Lemma (Inclusion-Exclusion Formula). *Let $A_1, \ldots, A_m \subset \mathbb{R}^d$ be sets. Then*

$$\begin{aligned} [A_1 \cup \ldots \cup A_m] &= 1 - (1 - [A_1]) \cdot (1 - [A_2]) \cdots (1 - [A_m]) \\ &= \sum_{k=1}^{m} (-1)^{k-1} \sum_{1 \leq i_1 < i_2 < \ldots < i_k \leq m} [A_{i_1} \cap \ldots \cap A_{i_k}]. \end{aligned}$$

In particular,

$$[A_1 \cup A_2] = [A_1] + [A_2] - [A_1 \cap A_2].$$

Proof. Let us choose an $x \in \mathbb{R}^d$. Then

$$[A_1 \cup \ldots \cup A_m](x) = \begin{cases} 1 & \text{if } x \in A_1 \cup \ldots \cup A_m, \\ 0 & \text{if } x \notin A_1 \cup \ldots \cup A_m. \end{cases}$$

On the other hand,

$$1 - [A_i](x) = \begin{cases} 1 & \text{if } x \notin A_i, \\ 0 & \text{if } x \in A_i. \end{cases}$$

Therefore,

$$\big(1 - [A_1](x)\big) \cdots \big(1 - [A_m](x)\big) = \begin{cases} 1 & \text{if } x \notin A_i \text{ for all } i, \\ 0 & \text{if } x \in A_i \text{ for some } i. \end{cases}$$

Hence $[A_1 \cup \ldots \cup A_m] = 1 - (1 - [A_1]) \cdot (1 - [A_2]) \cdots (1 - [A_m])$. Expanding the product, we complete the proof. □

PROBLEM.

1°. Researchers at a research institute speak French, Russian, and English. Among them, 20 people speak French, 15 speak Russian, and 10 speak English. Also, 8 people speak French and Russian, 5 people speak Russian and English and 7 speak French and English. Two people speak French, Russian and English. How many people work at the institute?

We are going to develop a technique which can be viewed as a combinatorial calculus of convex sets. First, we define the class of functions we will be dealing with.

(7.3) Definitions. The real vector space spanned by the functions $[A]$, where $A \subset \mathbb{R}^d$ is a compact convex set, is called the *algebra of compact convex sets* and is denoted $\mathcal{K}(\mathbb{R}^d)$. Thus a function $f \in \mathcal{K}(\mathbb{R}^d)$ is a linear combination

$$f = \sum_{i=1}^{m} \alpha_i [A_i],$$

where the $[A_i] \subset \mathbb{R}^d$ are compact convex sets and $\alpha_i \in \mathbb{R}$ are real numbers.

The real vector space spanned by the functions $[A]$, where $A \subset \mathbb{R}^d$ is a closed convex set, is called the *algebra of closed convex sets* and is denoted $\mathcal{C}(\mathbb{R}^d)$. Thus a typical function $f \in \mathcal{C}(\mathbb{R}^d)$ is a linear combination

$$f = \sum_{i=1}^{m} \alpha_i [A_i],$$

where $[A_i] \subset \mathbb{R}^d$ are closed convex sets and $\alpha_i \in \mathbb{R}$ are real numbers.

We use the term "algebra" since the spaces $\mathcal{K}(\mathbb{R}^d)$ and $\mathcal{C}(\mathbb{R}^d)$ are closed under multiplication of functions; see Problem 1 below.

A linear functional $\nu : \mathcal{K}(\mathbb{R}^d) \longrightarrow \mathbb{R}$, resp. $\nu : \mathcal{C}(\mathbb{R}^d) \longrightarrow \mathbb{R}$, is called a *valuation*. Thus $\nu(\alpha f + \beta g) = \alpha\nu(f) + \beta\nu(g)$ for any real α and β and any $f, g \in \mathcal{K}(\mathbb{R}^d)$, resp. $f, g \in \mathcal{C}(\mathbb{R}^d)$. More generally, we call a *valuation* any linear transformation $\mathcal{K}(\mathbb{R}^d), \mathcal{C}(\mathbb{R}^d) \longrightarrow V$, where V is a real vector space.

Valuations will emerge as analogues of "integrals" and "integral transforms" in our combinatorial calculus; see Sections 8 and IV.1 for some examples.

PROBLEMS.

1°. Prove that the product fg of functions $f, g \in \mathcal{K}(\mathbb{R}^d)$ is a function in $\mathcal{K}(\mathbb{R}^d)$ and that the product fg of functions $f, g \in \mathcal{C}(\mathbb{R}^d)$ is a function in $\mathcal{C}(\mathbb{R}^d)$.

2°. Do the functions $[A]$, where $A \subset \mathbb{R}^d$ is a non-empty compact convex set, form a basis of $\mathcal{K}(\mathbb{R}^d)$?

Now we prove the main result of this section.

(7.4) Theorem. *There exists a unique valuation $\chi : \mathcal{C}(\mathbb{R}^d) \longrightarrow \mathbb{R}$, called the Euler characteristic, such that $\chi([A]) = 1$ for every non-empty closed convex set $A \subset \mathbb{R}^d$.*

Proof. To show that χ must be unique, if it exists, is easy: let

$$f = \sum_{i=1}^{m} \alpha_i [A_i].$$

Then we must have

$$\chi(f) = \sum_{i: A_i \neq \emptyset} \alpha_i.$$

Let us prove that χ exists. First, we define χ on functions $f \in \mathcal{K}(\mathbb{R}^d)$.

We use induction on d. Suppose that $d = 0$. Then any function $f \in \mathcal{K}(\mathbb{R}^d)$ has the form $f = \alpha[0]$ for some $\alpha \in \mathbb{R}$ and we let $\chi(f) = \alpha$.

Suppose that $d > 0$.

For a point $x = (\xi_1, \ldots, \xi_d)$, let $\ell(x) = \xi_d$ be the last coordinate of x. For a $\tau \in \mathbb{R}$ let us consider the hyperplane

$$H_\tau = \{x \in \mathbb{R}^d : \quad \ell(x) = \tau\}.$$

The hyperplane H_τ can be identified with $\mathbb{R}^{d-1}$ and hence, by the induction hypothesis, there exists a valuation, say $\chi_\tau : \mathcal{K}(H_\tau) \longrightarrow \mathbb{R}$, which satisfies the required properties. For a function $f \in \mathcal{K}(\mathbb{R}^d)$, let f_τ be the restriction of f onto H_τ. Thus

$$\text{if} \quad f = \sum_{i=1}^m \alpha_i [A_i], \quad \text{then} \quad f_\tau = \sum_{i=1}^m \alpha_i [A_i \cap H_\tau]$$

and so $f_\tau \in \mathcal{K}(H_\tau)$ and we can define $\chi_\tau(f_\tau)$. Since $A_i \cap H_\tau$ are compact convex (possibly empty) sets, we must have

$$\chi_\tau(f_\tau) = \sum_{i: A_i \cap H_\tau \neq \emptyset} \alpha_i.$$

Let us consider the limit

$$\lim_{\epsilon \longrightarrow +0} \chi_{\tau-\epsilon}(f_{\tau-\epsilon}).$$

It may happen that the limit is equal to $\chi_\tau(f_\tau)$. This happens, for example, if for every i and small $\epsilon > 0$, we have $A_i \cap H_\tau \neq \emptyset \Longrightarrow A_i \cap H_{\tau-\epsilon} \neq \emptyset$ (see Figure 8).

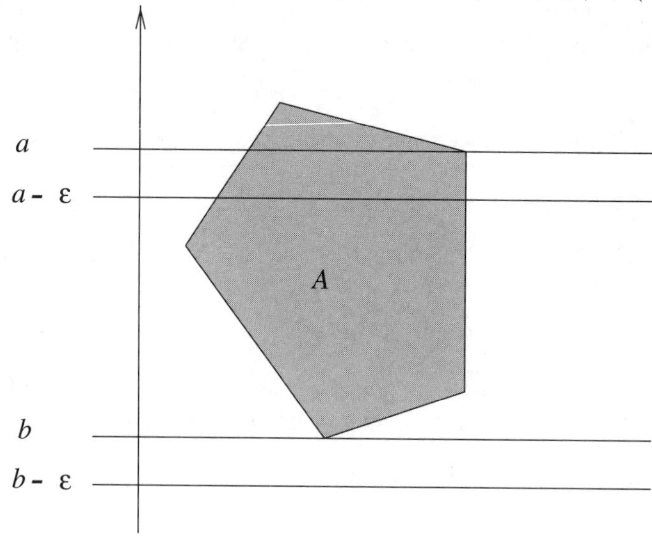

Figure 8. Example: for the function $f = [A]$, we have $\lim_{\epsilon \longrightarrow +0} \chi_{a-\epsilon}(f_{a-\epsilon}) = \chi_a(f_a) = 1$ but $0 = \lim_{\epsilon \longrightarrow +0} \chi_{b-\epsilon}(f_{b-\epsilon}) \neq \chi_b(f_b) = 1$.

In general, we conclude that $\lim_{\epsilon \longrightarrow +0} \chi_{\tau-\epsilon}(f_{\tau-\epsilon})$ is the sum of α_i such that $A_i \cap H_{\tau-\epsilon} \neq \emptyset$ for all sufficiently small $\epsilon > 0$. It follows then that

$$\chi_\tau(f_\tau) - \lim_{\epsilon \longrightarrow +0} \chi_{\tau-\epsilon}(f_{\tau-\epsilon}) = \sum_{i \in I} \alpha_i, \quad \text{where} \quad I = \big\{i : \min_{x \in A_i} \ell(x) = \tau\big\}.$$

In particular, $\lim_{\epsilon \longrightarrow +0} \chi_{\tau-\epsilon}(f_{\tau-\epsilon}) = \chi_\tau(f_\tau)$ unless τ is the minimum value of the linear function $\ell(x)$ on some set A_i.

Therefore, for a given function $f \in \mathcal{K}(\mathbb{R}^d)$ there are only finitely many τ's, where $\lim_{\epsilon \longrightarrow +0} \chi_{\tau-\epsilon}(f_{\tau-\epsilon}) \neq \chi_\tau(f_\tau)$. Now we define

$$\chi(f) = \sum_{\tau \in \mathbb{R}} \Big(\chi_\tau(f_\tau) - \lim_{\epsilon \longrightarrow +0} \chi_{\tau-\epsilon}(f_{\tau-\epsilon})\Big).$$

As we noted, the sum contains only finitely many non-zero summands, so it is well defined.

If $f, g \in \mathcal{K}(\mathbb{R}^d)$ are functions and $\alpha, \beta \in \mathbb{R}$ are numbers, then for every $\tau \in \mathbb{R}$ we have $(\alpha f + \beta g)_\tau = \alpha f_\tau + \beta g_\tau$. Since by the induction hypothesis χ_τ is a valuation and taking the limit is a linear operation, we conclude that $\chi(\alpha f + \beta g) = \alpha\chi(f) + \beta\chi(g)$, so χ is a valuation. Furthermore, if $A \subset \mathbb{R}^d$ is a compact convex set, then

$$\chi_\tau([A]_\tau) - \lim_{\epsilon \longrightarrow +0} \chi_{\tau-\epsilon}([A]_{\tau-\epsilon}) = \begin{cases} 1 & \text{if } \min_{x \in A} \ell(x) = \tau, \\ 0 & \text{otherwise.} \end{cases}$$

Since A is a non-empty compact convex set, there is unique minimum value of the linear function $\ell(x)$ on A. Therefore, $\chi([A]) = 1$.

Now we are ready to extend χ onto $\mathcal{C}(\mathbb{R}^d)$. Let $B(\rho) = \big\{x \in \mathbb{R}^d : \|x\| \leq \rho\big\}$ be the ball of radius ρ. For $f \in \mathcal{C}(\mathbb{R}^d)$ we define

$$\chi(f) = \lim_{\rho \longrightarrow +\infty} f \cdot [B(\rho)].$$

Clearly, χ satisfies the required properties. □

Theorem 7.4 and its proof belongs to H. Hadwiger.

If $A \subset \mathbb{R}^d$ is a set such that $[A] \in \mathcal{C}(\mathbb{R}^d)$, we often write $\chi(A)$ instead of $\chi([A])$ and call it the Euler characteristic of the set A. In the course of the proof of Theorem 7.4, we established the following useful fact, which will play the central role in our approach to the Euler-Poincaré Formula of Section VI.3.

(7.5) Lemma. *Let $A \subset \mathbb{R}^d$ be a set such that $[A] \in \mathcal{K}(\mathbb{R}^d)$. For $\tau \in \mathbb{R}$ let H_τ be the hyperplane consisting of the points $x = (\xi_1, \ldots, \xi_d)$ with $\xi_d = \tau$. Then $[A \cap H_\tau] \in \mathcal{K}(\mathbb{R}^d)$ and*

$$\chi(A) = \sum_{\tau \in \mathbb{R}} \Big(\chi(A \cap H_\tau) - \lim_{\epsilon \longrightarrow +0} \chi(A \cap H_{\tau-\epsilon})\Big).$$

□

Another useful result allows us to express the Euler characteristic of a union of sets in terms of the Euler characteristics of the intersections of the sets.

(7.6) Corollary. *Let $A_1, \ldots, A_m \subset \mathbb{R}^d$ be sets such that $[A_i] \in \mathcal{K}(\mathbb{R}^d)$ for all $i = 1, \ldots, m$. Then $[A_1 \cup \ldots \cup A_m] \in \mathcal{K}(\mathbb{R}^d)$ and*

$$\chi(A_1 \cup \ldots \cup A_m) = \sum_{k=1}^{m} (-1)^{k-1} \sum_{1 \leq i_1 < i_2 < \ldots < i_k \leq m} \chi(A_{i_1} \cap \ldots \cap A_{i_k}).$$

In particular,

$$\chi(A_1 \cup A_2) = \chi(A_1) + \chi(A_2) - \chi(A_1 \cap A_2).$$

Proof. Follows by Lemma 7.2 and Theorem 7.4. □

PROBLEMS.

1. Let $A_1, A_2, A_3 \subset \mathbb{R}^d$ be closed convex sets such that $A_1 \cap A_2 \neq \emptyset$, $A_1 \cap A_3 \neq \emptyset$, $A_2 \cap A_3 \neq \emptyset$ and $A_1 \cup A_2 \cup A_3$ is a convex set. Prove that $A_1 \cap A_2 \cap A_3 \neq \emptyset$.

2. Let $A_1, \ldots, A_m \subset \mathbb{R}^d$ be closed convex sets such that $A_1 \cup \ldots \cup A_m$ is a convex set. Suppose that the intersection of every k sets $A_{i_1}, \ldots, A_{i_k}$ is non-empty. Prove that there are $k+1$ sets $A_{i_1}, \ldots, A_{i_{k+1}}$ whose intersection is non-empty.

3. Let

$$\Delta = \Big\{(\xi_1, \ldots, \xi_d) : \quad \xi_1 + \ldots + \xi_d = 1, \quad \xi_i \geq 0 \quad \text{for} \quad i = 1, \ldots, d\Big\}$$

be the standard simplex in $\mathbb{R}^d$. Let $\Delta_i = \Big\{x \in \Delta : \ \xi_i = 0\Big\}$ be the i-th facet of Δ. Suppose that there are compact convex sets $K_1, \ldots, K_d \subset \mathbb{R}^d$, such that $\Delta \subset K_1 \cup \ldots \cup K_d$ and $K_i \cap \Delta_i = \emptyset$ for $i = 1, \ldots, d$. Prove that $K_1 \cap \ldots \cap K_d \neq \emptyset$.

Hint: Use induction on d and Problem 2.

4. Let $A_1, \ldots, A_m \subset \mathbb{R}^d$ be closed convex sets such that $A_1 \cap \ldots \cap A_m \neq \emptyset$. Prove that $\chi(A_1 \cup \ldots \cup A_m) = 1$.

5. Find the Euler characteristic of the "open square"

$$I_2 = \Big\{(\xi_1, \xi_2) : \ 0 < \xi_1, \xi_2 < 1\Big\}$$

and the "open cube"

$$I_3 = \Big\{(\xi_1, \xi_2, \xi_3) : \ 0 < \xi_1, \xi_2, \xi_3 < 1\Big\}.$$

6. Let $A_1, A_2, A_3, A_4 \subset \mathbb{R}^d$ be closed convex sets such that the union $A_1 \cup A_2 \cup A_3 \cup A_4$ is convex and all pairwise intersections $A_1 \cap A_2, A_1 \cap A_3, A_1 \cap A_4, A_2 \cap A_3, A_2 \cap A_4$ and $A_3 \cap A_4$ are non-empty. Prove that at least three of the four intersections $A_1 \cap A_2 \cap A_3, A_1 \cap A_2 \cap A_4, A_1 \cap A_3 \cap A_4$ and $A_2 \cap A_3 \cap A_4$ are non-empty and that if all the four intersections are non-empty, then the intersection $A_1 \cap A_2 \cap A_3 \cap A_4$ is non-empty. Construct an example where exactly three of the four intersections $A_1 \cap A_2 \cap A_3, A_1 \cap A_2 \cap A_4, A_1 \cap A_3 \cap A_4$ and $A_2 \cap A_3 \cap A_4$ are non-empty.

8. Application: Convex Sets and Linear Transformations

As an application of the Euler characteristic, we demonstrate an interesting behavior of collections of compact convex sets under linear transformations.

(8.1) Theorem. *Let $T : \mathbb{R}^n \longrightarrow \mathbb{R}^m$ be a linear transformation. Then there exists a linear transformation $\mathcal{T} : \mathcal{K}(\mathbb{R}^n) \longrightarrow \mathcal{K}(\mathbb{R}^m)$ such that $\mathcal{T}([A]) = [T(A)]$ for any compact convex set $A \subset \mathbb{R}^n$.*

Proof. Clearly, if $A \subset \mathbb{R}^n$ is a compact convex set, then $T(A) \subset \mathbb{R}^m$ is also a compact convex set. Let us define a function $G : \mathbb{R}^n \times \mathbb{R}^m \longrightarrow \mathbb{R}$, where

$$G(x,y) = \begin{cases} 1 & \text{if } T(x) = y, \\ 0 & \text{if } T(x) \neq y. \end{cases}$$

Let $f \in \mathcal{K}(\mathbb{R}^n)$ be a function. We claim that for every $y \in \mathbb{R}^m$ the function $g_y(x) = G(x,y)f(x)$ belongs to the space $\mathcal{K}(\mathbb{R}^n)$. Indeed, if

$$f = \sum_{i=1}^{k} \alpha_i [A_i], \tag{8.1.1}$$

where $\alpha_i \in \mathbb{R}$ and $A_i \subset \mathbb{R}^n$ are compact convex sets, then

$$g_y = \sum_{i=1}^{k} \alpha_i [A_i \cap T^{-1}(y)], \tag{8.1.2}$$

where $T^{-1}(y)$ is the affine subspace that is the inverse image of y. Hence $\chi(g_y)$ is well defined and we define $h = \mathcal{T}(f)$ by the formula $h(y) = \chi(g_y)$. We claim that $h \in \mathcal{K}(\mathbb{R}^m)$. Indeed, for f as in (8.1.1), the function g_y is given by (8.1.2) and

$$h(y) = \sum_{i \in I} \alpha_i, \quad \text{where} \quad I = \{i : \quad A_i \cap T^{-1}(y) \neq \emptyset\}.$$

However, $A_i \cap T^{-1}(y) \neq \emptyset$ if and only if $y \in T(A_i)$, so

$$h = \sum_{i \in I} \alpha_i [T(A_i)]. \tag{8.1.3}$$

Therefore, $h = \mathcal{T}(f) \in \mathcal{K}(\mathbb{R}^m)$ and the transformation $\mathcal{T}$ is well defined. We see that $\mathcal{T}$ is linear since for $f = \alpha_1 f_1 + \alpha_2 f_2$ we get

$$g_y(x) = \alpha_1 g_{1,y}(x) + \alpha_2 g_{2,y}(x),$$

where

$$g_y(x) = G(y,x)f(x), \quad g_{1,y} = G(y,x)f_1(x) \quad \text{and} \quad g_{2,y} = G(y,x)f_2(x).$$

Since χ is a linear functional (see Theorem 7.4), $h(y) = \alpha_1 h_1(y) + \alpha_2 h_2(y)$, where $h = \mathcal{T}(f)$, $h_1 = \mathcal{T}(f_1)$ and $h_2 = \mathcal{T}(f_2)$. It follows from (8.1.3) that $\mathcal{T}[A] = [T(A)]$. □

In particular, linear dependencies among the indicators of compact convex sets are preserved by linear transformations.

(8.2) Corollary. *Let $T : \mathbb{R}^n \longrightarrow \mathbb{R}^m$ be a linear transformation, let $A_1, \ldots, A_k$ be compact convex sets in $\mathbb{R}^n$ and let $\alpha_1, \ldots, \alpha_k$ be numbers such that*

$$\alpha_1[A_1] + \ldots + \alpha_k[A_k] = 0.$$

Then

$$\alpha_1[T(A_1)] + \ldots + \alpha_k[T(A_k)] = 0.$$

Proof. We apply the transformation T to both sides of the identity $\alpha_1[A_1] + \ldots + \alpha_k[A_k] = 0$. □

Corollary 8.2 is trivial for invertible linear transformations T but becomes much less obvious for projections; see Figure 9.

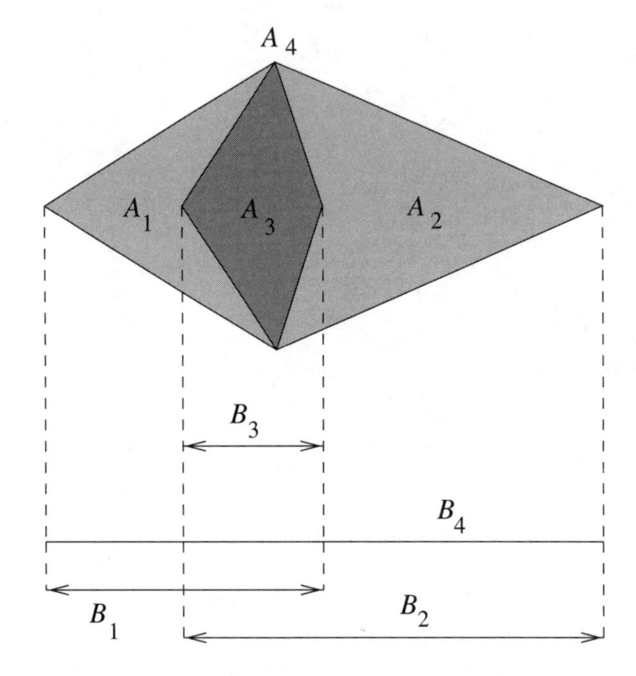

Figure 9. Four convex sets A_1, A_2, A_3, A_4 such that $[A_4] = [A_1] + [A_2] - [A_3]$ and their projections B_1, B_2, B_3, B_4. We observe that $[B_4] = [B_1] + [B_2] - [B_3]$.

PROBLEMS.

1. Prove that the Minkowski sum of compact convex sets is a compact convex set and that there exists a commutative and associative operation $f \star g$, called a *convolution*, for functions $f, g \in \mathcal{K}(\mathbb{R}^d)$ such that $(\alpha_1 f_1 + \alpha_2 f_2) \star g = \alpha_1(f_1 \star g) + \alpha_2(f_2 \star g)$ for any $f_1, f_2, g \in \mathcal{K}(\mathbb{R}^d)$ and any $\alpha_1, \alpha_2 \in \mathbb{R}$ and such that $[A_1] \star [A_2] = [A_1 + A_2]$ for any compact convex sets $A_1, A_2 \subset \mathbb{R}^d$.

2. Construct an example of compact non-convex sets $A_i \subset \mathbb{R}^n$ and real numbers α_i such that $\sum_{i=1}^k \alpha_i[A_i] = 0$ but $\sum_{i=1}^k \alpha_i[T(A_i)] \neq 0$ for some linear transformation $T : \mathbb{R}^n \longrightarrow \mathbb{R}^m$.

3. Construct an example of non-compact convex sets $A_i \subset \mathbb{R}^n$ and real numbers α_i such that $\sum_{i=1}^k \alpha_i[A_i] = 0$ but $\sum_{i=1}^k \alpha_i[T(A_i)] \neq 0$ for some linear transformation $T : \mathbb{R}^n \longrightarrow \mathbb{R}^m$.

(8.3) Some interesting valuations. Intrinsic volumes. Let $\operatorname{vol}_d(A)$ be the usual volume of a compact convex set $A \subset \mathbb{R}^d$. The function vol_d satisfies a number of useful properties:

(8.3.1) The volume is (finitely) additive: If $A_1, \ldots, A_m \subset \mathbb{R}^d$ are compact convex sets and if $\alpha_1, \ldots, \alpha_m$ are numbers such that $\alpha_1[A_1] + \ldots + \alpha_m[A_m] = 0$, then $\alpha_1 \operatorname{vol}_d(A_1) + \ldots + \alpha_m \operatorname{vol}_d(A_m) = 0$.

(8.3.2) The volume is invariant under *isometries* of $\mathbb{R}^d$, that is, orthogonal transformations and translations: $\operatorname{vol}_d\big(T(A)\big) = \operatorname{vol}_d(A)$ for any isometry $T : \mathbb{R}^d \longrightarrow \mathbb{R}^d$.

(8.3.3) The volume of a compact convex set $A \subset \mathbb{R}^d$ with a non-empty interior is positive.

(8.3.4) The volume in $\mathbb{R}^d$ is homogeneous of degree d: $\operatorname{vol}_d(\alpha A) = \alpha^d \operatorname{vol}_d(A)$ for $\alpha \geq 0$.

It turns out that for every $k = 0, \ldots, d$ there exists a measure w_k on compact convex sets in $\mathbb{R}^d$, which satisfies properties (8.3.1)–(8.3.3) and which is homogeneous of degree k: $w_k(\alpha A) = \alpha^k w_k(A)$ for $\alpha > 0$. These measures are called *intrinsic volumes.* For $k = d$ we get the usual volume and for $k = 0$ we get the Euler characteristic.

To construct the intrinsic volumes, we observe that the volume can be extended to a valuation $\omega_d : \mathcal{K}(\mathbb{R}^d) \longrightarrow \mathbb{R}$ such that $\omega_d([A]) = \operatorname{vol}_d(A)$ for any compact convex set A. Indeed, we define

$$\omega_d(f) = \int_{\mathbb{R}^d} f(x)\, dx \quad \text{for} \quad f \in \mathcal{K}(\mathbb{R}^d),$$

where dx is the usual Lebesgue measure on $\mathbb{R}^d$. Properties of the integral imply that $\omega_d(\alpha_1 f_1 + \alpha_2 f_2) = \alpha_1 \omega_d(f_1) + \alpha_2 \omega_d(f_2)$, so ω_d is a valuation.

Let $L \subset \mathbb{R}^d$ be a k-dimensional subspace and let P_L be the orthogonal projection $P_L : \mathbb{R}^d \longrightarrow L$. Using Theorem 8.1, let us construct a linear transformation $\mathcal{P}_L : \mathcal{K}(\mathbb{R}^d) \longrightarrow \mathcal{K}(L)$ and hence a valuation $\omega_{k,L} : \mathcal{K}(L) \longrightarrow \mathbb{R}$ by letting $\omega_{k,L}(f) = \omega_k\big(\mathcal{P}_L(f)\big)$. Thus, for a compact convex set $A \subset \mathbb{R}^d$, the value of $\omega_{k,L}([A])$ is the volume of the orthogonal projection of A onto $L \subset \mathbb{R}^d$.

The functional $\omega_{k,L}[A]$ satisfies (8.3.1) and (8.3.3), it is homogeneous of degree k, but it is not invariant under orthogonal transformations (although it is invariant under translations). To construct an invariant functional, we average $\omega_{k,L}$ over all k-dimensional subspaces $L \subset \mathbb{R}^d$. Let $G_k(\mathbb{R}^d)$ be the set of all k-dimensional subspaces $L \subset \mathbb{R}^d$. It is known that $G_k(\mathbb{R}^d)$ possesses a manifold structure (it is called the *Grassmannian*) and the rotationally invariant probability measure dL. Hence, for $f \in \mathcal{K}(\mathbb{R}^d)$ we let

$$\omega_k(f) = \int_{G_k(\mathbb{R}^d)} \omega_{k,L}(f)\, dL.$$

In other words, $\omega_k(f)$ is the average value of $\omega_{k,L}(f)$ over all k-dimensional subspaces $L \subset \mathbb{R}^d$.

Clearly, $\omega_k : \mathcal{K}(\mathbb{R}^d) \longrightarrow \mathbb{R}$ is a valuation. For a compact convex set $A \subset \mathbb{R}^d$ we define $w_k(A) := \omega_k([A])$.

Hence $w_k(A)$ is the average volume of projections of A onto k-dimensional subspaces in $\mathbb{R}^d$. The number $w_k(A)$ is called the k-th *intrinsic volume* of A. It satisfies properties (8.3.1)–(8.3.3) and it is homogeneous of degree k: $w_k(\alpha A) = \alpha^k w_k(A)$ for $\alpha \geq 0$. It is convenient to agree that $w_0(A) = \chi(A)$ and that $w_d(A) = \operatorname{vol}_d(A)$.

PROBLEMS.

1. Compute the intrinsic volumes of the unit ball $B = \{x \in \mathbb{R}^d : \ \|x\| \leq 1\}$.

2*. Let $A \subset \mathbb{R}^d$ be a compact convex set with non-empty interior. Prove that the *surface area* of A (*perimeter*, if $d = 2$) is equal to $c_d w_{d-1}(A)$, where c_d is a constant depending on d alone. Find c_d.

Here is another interesting valuation.

3. Let us fix a vector $c \in \mathbb{R}^d$. For a non-empty compact convex set $A \subset \mathbb{R}^d$, let

$$h(A; c) = \max_{x \in A} \langle c, x \rangle$$

(when A is fixed, the function $h(A, c) : \mathbb{R}^d \longrightarrow \mathbb{R}$ is called the *support function* of A). Prove that there exists a valuation $\nu_c : \mathcal{K}(\mathbb{R}^d) \longrightarrow \mathbb{R}$ such that $\nu_c([A]) = h(A; c)$ for every non-empty convex compact set $A \subset \mathbb{R}^d$.

Hint: If $c \neq 0$, let

$$\nu_c(f) = \sum_{\alpha \in \mathbb{R}} \alpha \Big(\chi(f_\alpha) - \lim_{\epsilon \longrightarrow +0} \chi(f_{\alpha+\epsilon}) \Big),$$

where f_α is the restriction of f onto the hyperplane $H = \{x : \ \langle c, x \rangle = \alpha\}$.

4*. Let $K_1, K_2 \subset \mathbb{R}^d$ be compact convex sets such that $K_1 \cup K_2$ is convex. Prove that $(K_1 \cup K_2) + (K_1 \cap K_2) = K_1 + K_2$.

Hint: Note that $[K_1 \cup K_2] + [K_1 \cap K_2] = [K_1] + [K_2]$ and use Problem 3 to conclude that $h(K_1 \cup K_2; c) + h(K_1 \cap K_2; c) = h(K_1; c) + h(K_2; c)$ for any $c \in \mathbb{R}^d$. Observe that $h(A + B; c) = h(A; c) + h(B; c)$.

9. Polyhedra and Linear Transformations

We would like to extend the results of Section 8 to certain unbounded sets, specifically to polyhedra. Recall (see Definition 2.2) that a polyhedron is a set of solutions to a finite system of linear inequalities. The main result of this section is that the image of a polyhedron under a linear transformation is a polyhedron. Hence the class of polyhedra is preserved by linear transformations. We will need this result in Section IV.8.

The proof is based on "going down one step at a time".

(9.1) Lemma. *Let $P \subset \mathbb{R}^d$ be a polyhedron and let $pr : \mathbb{R}^d \longrightarrow \mathbb{R}^{d-1}$ be the projection $pr(\xi_1, \dots, \xi_d) = (\xi_1, \dots, \xi_{d-1})$. Then the image $pr(P)$ is a polyhedron in $\mathbb{R}^{d-1}$.*

Proof. Suppose that P is defined by a system of linear inequalities for vectors $x = (\xi_1, \dots, \xi_d)$ in $\mathbb{R}^d$:

$$P = \Big\{x : \quad \sum_{j=1}^{d} \alpha_{ij}\xi_j \leq \beta_i \quad \text{for} \quad i = 1, \dots, m\Big\},$$

where α_{ij} and β_j are real numbers.

Let us define $I_+ = \{i : \ \alpha_{id} > 0\}$, $I_- = \{i : \ \alpha_{id} < 0\}$ and $I_0 = \{i : \ \alpha_{id} = 0\}$. Hence a point $(\xi_1, \dots, \xi_{d-1})$ belongs to the projection $pr(P)$ if and only if

$$\sum_{j=1}^{d-1} \alpha_{ij}\xi_j \leq \beta_i \quad \text{for} \quad i \in I_0,$$

and there exists a number ξ_d which satisfies the inequalities

$$\alpha_{id}\xi_d + \sum_{j=1}^{d-1} \alpha_{ij}\xi_j \leq \beta_i \quad \text{for all} \quad i \in I_+ \cup I_-.$$

The latter of these two conditions is equivalent to

$$\xi_d \leq \frac{\beta_i}{\alpha_{id}} - \sum_{j=1}^{d-1} \frac{\alpha_{ij}}{\alpha_{id}}\xi_j \quad \text{for all} \quad i \in I_+ \qquad \text{and}$$

$$\xi_d \geq \frac{\beta_i}{\alpha_{id}} - \sum_{j=1}^{d-1} \frac{\alpha_{ij}}{\alpha_{id}}\xi_j \quad \text{for all} \quad i \in I_-.$$

Such a number ξ_d exists if and only if for no pair of numbers consisting of one of the lower bound for ξ_d and one of the upper bound for ξ_d does the lower bound exceed the upper bound. Thus ξ_d exists if and only if

$$\frac{\beta_i}{\alpha_{id}} - \sum_{j=1}^{d-1} \frac{\alpha_{ij}}{\alpha_{id}}\xi_j \leq \frac{\beta_k}{\alpha_{kd}} - \sum_{j=1}^{d-1} \frac{\alpha_{kj}}{\alpha_{kd}}\xi_j \quad \text{for all pairs} \quad i \in I_- \quad \text{and} \quad k \in I_+.$$

Hence the projection $pr(P)$ is the polyhedron in $\mathbb{R}^{d-1}$ defined by the following linear inequalities for $(\xi_1, \ldots, \xi_{d-1})$:

$$\sum_{j=1}^{d-1} \alpha_{ij}\xi_j \leq \beta_i \quad \text{for all} \quad i \in I_0 \quad \text{and}$$

$$\frac{\beta_i}{\alpha_{id}} - \sum_{j=1}^{d-1} \frac{\alpha_{ij}}{\alpha_{id}}\xi_j \leq \frac{\beta_k}{\alpha_{kd}} - \sum_{j=1}^{d-1} \frac{\alpha_{kj}}{\alpha_{kd}}\xi_j \quad \text{for all pairs} \quad i \in I_- \quad \text{and} \quad k \in I_+.$$

If I_0 is empty, then there are no inequalities of the first kind, and if $I_-, I+$ or both are empty, then there are no inequalities of the second kind. □

PROBLEMS.

1°. Let $P \subset \mathbb{R}^d$ be a polyhedron defined by $m \geq 4$ linear inequalities, and let $Q = pr(P) \subset \mathbb{R}^{d-1}$ be its projection. Prove that Q can be defined by not more than $m^2/4$ linear inequalities.

2°. Let $P \subset \mathbb{R}^n$, $P = \{x : \langle a_i, x \rangle \leq \beta_i, i = 1, \ldots, m\}$ be a polyhedron and let $T : \mathbb{R}^n \longrightarrow \mathbb{R}^n$ be an invertible linear transformation. Prove that $Q = T(P)$ is a polyhedron defined by $Q = \{x : \langle c_i, x \rangle \leq \beta_i, i = 1, \ldots, m\}$, where $c_i = (T^*)^{-1}a_i$ and T^* is the conjugate linear transformation.

Now we can prove the result in full generality.

(9.2) Theorem. *Let $P \subset \mathbb{R}^n$ be a polyhedron and let $T : \mathbb{R}^n \longrightarrow \mathbb{R}^m$ be a linear transformation. Then $T(P)$ is a polyhedron in $\mathbb{R}^m$.*

Proof. If $n = m$ and T is invertible, the result follows by Problem 2 of Section 9.1. If $\ker T = \{0\}$, then the restriction $T : \mathbb{R}^n \longrightarrow \operatorname{im} T \subset \mathbb{R}^m$ is an invertible linear transformation and the result follows as above. For a general T, let us define a transformation $\widehat{T} : \mathbb{R}^n \longrightarrow \mathbb{R}^m \oplus \mathbb{R}^n = \mathbb{R}^{m+n}$ by $\widehat{T}(x) = \big(T(x), x\big)$. Then $\ker \widehat{T} = \{0\}$ and hence $\widehat{T}(P)$ is a polyhedron in $\mathbb{R}^{m+n}$. Now we observe that $T(P)$ is obtained from $\widehat{T}(P)$ by a series of n successive projections

$$\mathbb{R}^{m+n} \longrightarrow \mathbb{R}^{m+n-1} \longrightarrow \ldots \longrightarrow \mathbb{R}^m \quad \text{via}$$

$$(\xi_1, \ldots, \xi_{m+n}) \longmapsto (\xi_1, \ldots, \xi_{m+n-1}) \longmapsto \ldots \longmapsto (\xi_1, \ldots, \xi_m).$$

Applying Lemma 9.1 m times, we conclude that $T(P)$ is a polyhedron. □

The procedure of obtaining the description of $T(P)$ from the description of P which we employed in Lemma 9.1 and Theorem 9.2 is called the *Fourier-Motzkin Elimination.*

PROBLEMS.

1. Let $P \subset \mathbb{R}^n$ be a polyhedron defined by k linear inequalities and let $T : \mathbb{R}^n \longrightarrow \mathbb{R}^m$ be a linear transformation. Estimate the number of linear inequalities needed to define $T(P)$ using the construction of Theorem 9.2.

Remark: This number is way too big. In practice, after performing each one-step projection $\mathbb{R}^d \longrightarrow \mathbb{R}^{d-1}$, it is advisable to "clean" the list of obtained inequalities by removing those that can be removed without changing the image of the projection. Still, typically the number of inequalities needed to describe the projection is substantially larger than the number of inequalities needed to describe the original polyhedron.

2°. Prove that the Minkowski sum $P_1 + P_2$ of two polyhedra in Euclidean space is a polyhedron.

We define an important subalgebra of the algebra of closed convex sets from Definition 7.3.

(9.3) Definition. The real vector space spanned by the indicator functions $[P]$, where $P \subset \mathbb{R}^d$ is a polyhedron, is called the *algebra of polyhedra* and denoted $\mathcal{P}(\mathbb{R}^d)$.

PROBLEMS.

1. Let $T : \mathbb{R}^n \longrightarrow \mathbb{R}^m$ be a linear transformation. Prove that there exists a linear transformation $\mathcal{T} : \mathcal{P}(\mathbb{R}^n) \longrightarrow \mathcal{P}(\mathbb{R}^m)$ such that $\mathcal{T}[P] = [T(P)]$ for all polyhedra $P \subset \mathbb{R}^n$.

Hint: Cf. Theorem 8.1.

2. Prove that there exists a commutative and associative operation $f \star g$ for functions $f, g \in \mathcal{P}(\mathbb{R}^d)$ such that $(\alpha_1 f_1 + \alpha_2 f_2) \star g = \alpha_1 (f_1 \star g) + \alpha_2 (f_2 \star g)$ for any $f_1, f_2, g \in \mathcal{P}(\mathbb{R}^d)$ and such that $[P_1] \star [P_2] = [P_1 + P_2]$ for any two polyhedra $P_1, P_2 \subset \mathbb{R}^d$

Hint: Cf. Problem 1 of Section 8.2.

10. Remarks

A general reference in convexity is [**W94**]. Our discussion of positive polynomials in Section 3 follows [**R95**] and [**R00**] with some simplifications. A classical reference for Helly's Theorem and its numerous applications is [**DG63**]. More recent developments, including applications of topological methods, are surveyed in [**E93**], [**K95**], [**We97**] and [**Ž97**] (see also references therein). See also [**Bar82**] for a nice and elementary generalization of Radon's Theorem and Helly's Theorem and [**Mat02**] for further results in this direction. For the Euler characteristic and valuations, see [**Kl63**], [**Mc93a**] and [**MS83**]. Note that our definition of the relevant algebras (the algebra of compact convex sets, the algebra of closed convex sets and the algebra of polyhedra) may be different from those in [**Mc93a**], [**MS83**] and elsewhere. Often, an equivalence relation of some kind is imposed and the algebra is factored modulo that relation. The role of algebra multiplication is played by the convolution operation $\star$ (which we introduce in Problem 1 of Section 8.2 and Problem 2 of Section 9.3).

Intrinsic volumes in the context of the general theory of valuations are discussed in [**KR97**]. The Fourier-Motzkin elimination procedure is discussed in detail in [**Z95**].

Chapter II

Faces and Extreme Points

We take a closer look at convex sets. In this chapter, we are interested in local properties of closed convex sets in Euclidean space. A finite-dimensional closed convex set always has an interior when considered in a proper ambient space and, therefore, has a non-trivial boundary. We explore the structure of the boundary and define and study faces and extreme points. We look at the structure of some particular convex sets: the Birkhoff polytope, transportation polyhedra, the moment cone, the cone of non-negative univariate polynomials and the cone of positive semidefinite matrices. Our main tools are the Isolation Theorem in a general vector space and the Krein-Milman Theorem in Euclidean space. Applications include the Schur-Horn Theorem describing the set of possible diagonals of a symmetric matrix having prescribed eigenvalues, efficient formulas for numerical integration, a characterization of the polynomials that are non-negative on the interval and numerous quadratic convexity results, such as the Brickman Theorem, which describe various situations when the image of a quadratic map turns out to be convex. Quadratic convexity allows us to visualize often counterintuitive results about the facial structure of the cone of positive semidefinite matrices through the existence and rigidity properties of configurations of points in Euclidean space.

1. The Isolation Theorem

In this section, we develop one of the most useful and universal tools to explore the structure of a convex set, both in finite and infinite dimensions. We review some linear algebra first.

(1.1) Affine subspaces, affine hulls and linear functionals. Let V be a vector space and let $L \subset V$ be a subspace of V. The translation $A = L + u$ is called an *affine subspace* of V. The *dimension* of A is the dimension of L. We say that A is *parallel* to L. In particular, if $\dim A = 1$, the set A is called a *straight line.* A straight line can be written in the parametric form $A = \{u + \tau v : \tau \in \mathbb{R}\}$, where $u, v \in V$ are vectors and $v \neq 0$.

A *linear functional* is a map $f : V \longrightarrow \mathbb{R}$ such that

$$f(\alpha u + \beta v) = \alpha f(u) + \beta f(v)$$

for all $u, v \in V$ and all $\alpha, \beta \in \mathbb{R}$.

An important example of an affine subspace is an affine hyperplane.

Let $f : V \longrightarrow \mathbb{R}$ be a linear functional which is not identically 0 and let $\alpha \in \mathbb{R}$ be a number. The set

$$H = \{v \in V : \quad f(v) = \alpha\}, \quad \text{where } \alpha \in \mathbb{R},$$

is called an *affine hyperplane.* Often, we simply call it a *hyperplane.*

A linear combination

$$v = \alpha_1 v_1 + \ldots + \alpha_m v_m, \quad \text{where} \quad \alpha_1 + \ldots + \alpha_m = 1,$$

is called an *affine combination.* Similarly, points $v_1, \ldots, v_m \in V$ are said to be *affinely independent* if whenever

$$\alpha_1 v_1 + \ldots + \alpha_m v_m = 0 \quad \text{and} \quad \alpha_1 + \ldots + \alpha_m = 0,$$

we must have $\alpha_1 = \ldots = \alpha_m = 0$.

Given a set $X \subset V$, the set $\operatorname{aff}(X)$ of all affine combinations of points from X is called the *affine hull* of X.

PROBLEMS.

1°. Prove that an affine combination of vectors from an affine subspace is a vector from the subspace.

2°. Prove that the intersection of affine subspaces is an affine subspace.

3°. Prove that an affine hyperplane is an affine subspace.

4°. Prove that an affine subspace is a subspace if and only if it contains 0.

5°. Let $A \subset V$ be an affine subspace of dimension n. Prove that the maximum number of affinely independent points in A is $n + 1$.

6°. Let $L \subset V$ be a subspace and let $v, u \in V$ be vectors. Prove that $L + u = L + v$ if and only if $u - v \in L$.

7°. Let $X \subset V$ be a set of points in a vector space V. Prove that $\operatorname{aff}(X)$ is the smallest affine subspace containing X.

(1.2) Quotients, projections and codimension. Let V be a vector space and let $L \subset V$ be a subspace. We can form the *quotient space* V/L as follows: the points of V/L are the affine subspaces parallel to L. Addition in V/L is defined as follows:

$$A_1 + A_2 = A_3$$

provided

$$A_1 = L + v_1, \quad A_2 = L + v_2, \quad A_3 = L + v_3 \quad \text{and} \quad v_1 + v_2 = v_3$$

for some v_1, v_2 and v_3.

Scalar multiplication in V/L is defined as follows:

$$\alpha A_1 = A_2$$

provided

$$A_1 = L + v_1, \quad A_2 = L + v_2 \quad \text{and} \quad v_2 = \alpha v_1$$

for some $v_1, v_2 \in V$. Thus L itself is the 0 of the quotient V/L.

The dimension of V/L is called the *codimension* of L (denoted $\operatorname{codim} L$). If $A = L + u$ is an affine subspace, the codimension of A is defined to be the codimension of L. There is a linear transformation $pr : V \longrightarrow V/L$, where $pr(v) = L+v$, called the *projection*.

PROBLEMS.

1°. Prove that addition and scalar multiplication in V/L are well defined (do not depend on particular choices of v_1 and v_2).

For addition: let $L \subset V$ be a subspace. Suppose there are two vectors u_1, v_1 such that $L+u_1 = L+v_1 = A_1$ and two vectors u_2, v_2 such that $L+u_2 = L+v_2 = A_2$. Let $u_3 = u_1 + u_2$ and $v_3 = v_1 + v_2$. Prove that $L + v_3 = L + u_3 = A_3$.

For scalar multiplication: let $L \subset V$ be a subspace. Suppose there are two vectors u_1, v_1 such that $L + u_1 = L + v_1 = A_1$. For $\alpha \in \mathbb{R}$ let $v_2 = \alpha v_1$ and $u_2 = \alpha u_1$. Prove that $L + u_2 = L + v_2 = A_2$.

2°. Prove that the affine hyperplanes are exactly the affine subspaces of codimension 1.

3°. Prove that the projection $pr : V \longrightarrow V/L$ is indeed a linear transformation, that its image is the whole space V/L and that its kernel is L.

4°. Let $L \subset \mathbb{R}^d$ be a subspace. Prove that $\dim L + \operatorname{codim} L = d$.

Now, some convexity enters the picture.

(1.3) Halfspaces. Let V be a vector space and let $H \subset V$ be an affine hyperplane. Then the complement of H in V is the union of two convex sets, called *open halfspaces*: $V \setminus H = H_+ \cup H_-$. Indeed, suppose that $H = \{x \in \mathbb{R}^d : \ f(x) = \alpha\}$, where $f : V \longrightarrow \mathbb{R}$ is a linear functional and $\alpha \in \mathbb{R}$ is a number. We let

$$H_+ = \{x \in \mathbb{R}^d : \quad f(x) > \alpha\} \quad \text{and} \quad H_- = \{x \in \mathbb{R}^d : \quad f(x) < \alpha\}.$$

Of course, if we choose a different equation for H (say, $-f(x) = -\alpha$), then H_+ and H_- may be interchanged. The sets $\overline{H_+} = H \cup H_+$ and $\overline{H_-} = H \cup H_-$ are called *closed halfspaces*. We can write

$$\overline{H_+} = \{x \in \mathbb{R}^d : \quad f(x) \geq \alpha\} \quad \text{and} \quad \overline{H_-} = \{x \in \mathbb{R}^d : \quad f(x) \leq \alpha\}.$$

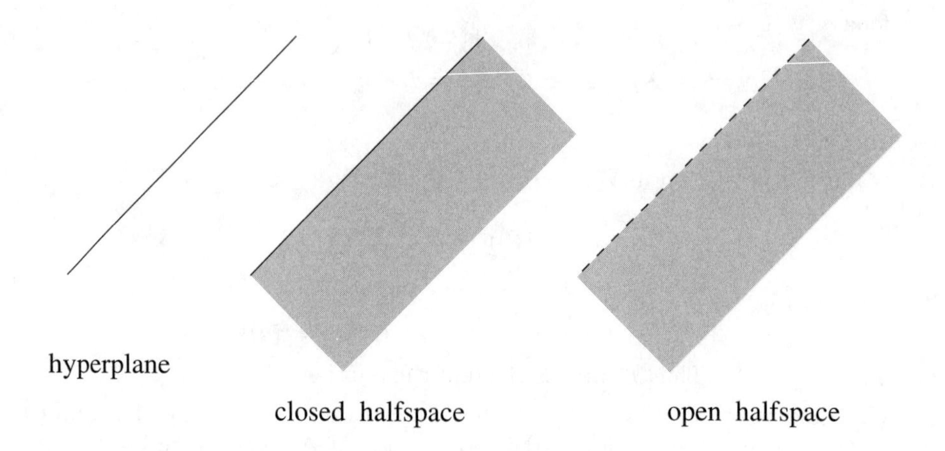

Figure 10. A hyperplane, a closed halfspace and an open halfspace

PROBLEMS.

1°. Prove that open halfspaces and closed halfspaces are convex.

2°. Prove that an open halfspace of $\mathbb{R}^d$ is an open subset of $\mathbb{R}^d$ and that a closed halfspace of $\mathbb{R}^d$ is a closed subset of $\mathbb{R}^d$.

Let us describe some basic cases of the relative position of an affine hyperplane and a (convex) set.

(1.4) Definitions. Let V be a vector space, let $A \subset V$ be a set and let $H \subset V$ be an affine hyperplane. We say that H *isolates* A if A is contained in one of the closed subspaces $\overline{H_-}$ or $\overline{H_+}$. We say that H *strictly isolates* A if A is contained in one of the open halfspaces H_- or H_+.

Let V be a vector space, let $A, B \subset V$ be sets and let $H \subset V$ be a hyperplane. We say that H *separates* A and B if A is contained in one closed halfspace and B is contained in the other. We say that H *strictly separates* A and B if A is contained in one open halfspace and B is contained in the other open halfspace.

PROBLEM.

1°. Prove that sets $A, B \subset V$ can be separated, respectively strictly separated, by an affine hyperplane if and only if there is a linear functional $f : V \longrightarrow \mathbb{R}^d$ and a number $\alpha \in \mathbb{R}$ such that $f(x) \leq \alpha \leq f(y)$, respectively $f(x) < \alpha < f(y)$, for all $x \in A$ and all $y \in B$.

It turns out that in infinite-dimensional spaces there exist remarkably "shallow" convex sets that consist of their own "boundary" alone. Such sets often demonstrate various kinds of pathological behavior; see Problem 2 of Section 1.6, Problem 1 of Section 2.5 and Section III.1.4. We would like to single out a class of reasonably "solid" convex sets, which behave much more predictably.

(1.5) Definition. Let V be a vector space and let $A \subset V$ be a convex set. The set A is called *algebraically open* if the intersection of A with every straight line in V is an open interval (possibly empty). Thus if $L = \{v + \tau u : \ \tau \in \mathbb{R}\}$ is a straight line in V, where $v, u \in V$, then

$$A \cap L = \{v + \tau u : \quad \alpha < \tau < \beta\}, \quad \text{where}$$

either $-\infty < \alpha < \beta < +\infty$ (the intersection is a non-empty open interval)

or $\alpha = -\infty < \beta < +\infty$ (the intersection is an open ray)

or $-\infty < \alpha < \beta = +\infty$ (the intersection is an open ray)

or $-\infty = \alpha < \beta = +\infty$ (the intersection is the whole straight line)

or $\alpha \geq \beta$ (the intersection is empty).

PROBLEMS.

1°. Prove that convex open sets in $\mathbb{R}^d$ are algebraically open.

2. Construct an example of a (non-convex) set $A \subset \mathbb{R}^2$ such that for every straight line $L \subset \mathbb{R}^2$ the intersection $A \cap L$ is an open subset in L but A is not open in $\mathbb{R}^2$.

3. Let $A \subset \mathbb{R}^d$ be a convex set. Prove that it is open if and only if it is algebraically open.

4. Prove that if an algebraically open set is isolated by an affine hyperplane, it is strictly isolated by the hyperplane.

5. Let V be a vector space and let $A, B \subset V$ be algebraically open sets. Prove that if A and B are separated by an affine hyperplane H, then A and B are strictly separated by H.

6°. Prove that the intersection of finitely many algebraically open sets is algebraically open.

7. Let V and W be vector spaces and let $T : V \longrightarrow W$ be a linear transformation such that $\operatorname{im}(T) = W$. Let $A \subset V$ be an algebraically open set in V. Prove that the image $T(A)$ is algebraically open in W.

8°. Let $A \subset V$ be an algebraically open set and let $L \subset V$ be a subspace. Prove that $A \cap L$ is algebraically open as a set in L.

We arrive at the main result of this section.

(1.6) The Isolation Theorem. *Let V be a vector space, let $A \subset V$ be an algebraically open convex set and let $u \notin A$ be a point. Then there exists an affine hyperplane H which contains u and strictly isolates A.*

Proof. Without loss of generality we may assume that $u = 0$ is the origin.

First, we prove the result in the case of $V = \mathbb{R}^2$. Let $S = \{x \in \mathbb{R}^2 : \|x\| = 1\}$ be the circle of radius 1 centered at the origin. Let us project A radially into S: $v \longmapsto v/\|v\|$. Since A is convex, it is connected, and, therefore, the image of this projection is a connected arc Γ of S. Furthermore, since A is algebraically open, Γ must be an open arc

$$\Gamma = \Big\{(\cos\phi,\ \sin\phi) : \quad \alpha < \phi < \beta\Big\}$$

of the circle S. Indeed, let $x \in \Gamma$ be a point. Then $x = v/\|v\|$ for some $v \in A$, so we can choose a straight line L through v parallel to the tangent line to S at x. Then the intersection $L \cap A$ will be an open interval containing v, so the radial projection of A will contain an open arc containing x.

Next, we observe that the length of Γ cannot be greater then π, because otherwise Γ would have contained two antipodal points x and $-x$ and 0 would have been in A, since A is convex. Now, let v be an endpoint of Γ (which is not in Γ, since Γ is open). The straight line through 0 and v is the desired hyperplane, containing 0 and strictly isolating A.

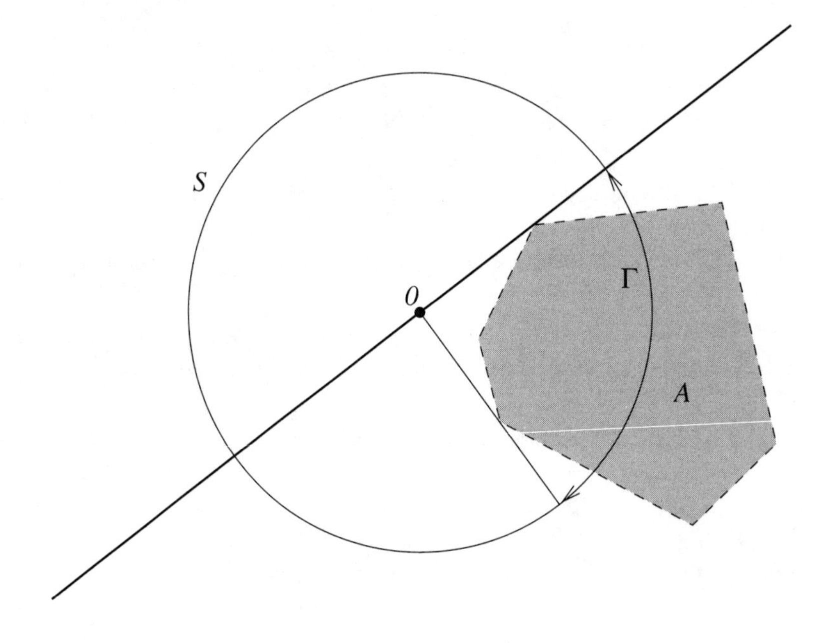

Figure 11. Constructing the isolating hyperplane when $d = 2$

Next, suppose that $\dim V \geq 2$. We prove that there is a straight line L such that $0 \in L$ and $L \cap A = \emptyset$. To prove this, let us consider any 2-dimensional plane P containing 0. The intersection $B = P \cap A$ is a convex algebraically open subset of P (possibly empty – see Problem 8 of Section 1.5) and as we proved, there is a line $L \subset P$ such that $0 \in L$ and $L \cap B = \emptyset$. Then L is the desired straight line.

Now, we prove the theorem. Let $H \subset V$ be the maximal affine subspace such that $0 \in H$ and $H \cap A = \emptyset$. By maximal we mean a subspace which has these properties and is not contained in a larger subspace with the same properties. If

V is finite-dimensional, we could choose H to be a subspace of the largest possible dimension such that $0 \in H$ and $H \cap A = \emptyset$. If V is arbitrary, the existence of such an H is ensured by Zorn's Lemma. We claim that H is a hyperplane. To prove this, consider the quotient V/H and let $pr : V \longrightarrow V/H$ be the projection. If H is not a hyperplane, then $\dim V/H \geq 2$ and $pr(A)$ is an algebraically open subset in V/H (see Problem 7, Section 1.5). Then, as we proved, there is a straight line $L \subset V/H$ such that $0 \in L$ and $L \cap pr(A) = \emptyset$. Then the preimage $G = pr^{-1}(L) = \{x : pr(x) \in L\}$ is a subspace in V, such that $0 \in G$, $G \cap A = \emptyset$, $H \subset G$ and G is strictly larger than H. This contradiction shows that H must be a hyperplane. □

PROBLEMS.

1°. Construct an example of a non-convex open set $A \subset \mathbb{R}^2$ such that $0 \notin A$ and there are no affine hyperplanes H such that $0 \in H$ and H isolates A.

2. Let $V = \mathbb{R}_\infty$ be the vector space of all infinite sequences $x = (\xi_1, \xi_2, \xi_3, \ldots)$ of real numbers such that all but finitely many terms ξ_i are zero. One can think of $\mathbb{R}_\infty$ as of the space of all univariate polynomials with real coefficients. Let $A \subset V \setminus \{0\}$ be the set of all such sequences x where the last non-zero term is strictly positive. Prove that $0 \notin A$, that A is convex, that A is not algebraically open and that there are no affine hyperplanes H such that $0 \in H$ and H isolates A.

3. Prove the following generalization of Theorem 1.6. Let V be a vector space, let $A \subset V$ be an algebraically open convex set and let $L \subset V$ be an affine subset such that $L \cap A = \emptyset$. Then there exists an affine hyperplane H containing L which strictly isolates A.

2. Convex Sets in Euclidean Space

In this section, we explore consequences of the Isolation Theorem for convex sets in Euclidean space. For finite-dimensional convex sets there is no difficulty in recognizing their interior and boundary.

(2.1) Definitions. Let $A \subset \mathbb{R}^d$ be a set. A point $u \in A$ is called an *interior* point of A if there exists an $\epsilon > 0$ such that the (open) ball $B(u, \epsilon) = \{x : \|x - u\| < \epsilon\}$ centered at u and of radius ϵ is contained in A: $B(u, \epsilon) \subset A$. The set of all interior points of A is called the *interior* of A and denoted $\operatorname{int}(A)$. The set of all non-interior points of A is called the *boundary* of A and denoted ∂A.

Now we prove that if, starting from any point of a convex set, we move towards an interior point of the set, we immediately get into the interior of the set.

(2.2) Lemma. *Let $A \subset \mathbb{R}^d$ be a convex set and let $u_0 \in \operatorname{int}(A)$ be an interior point of A. Then, for any point $u_1 \in A$ and any $0 \leq \alpha < 1$, the point $u_\alpha = (1-\alpha)u_0 + \alpha u_1$ is an interior point of A.*

Proof. Let $B(u_0, \epsilon) \subset A$ be a ball centered at u_0 and contained in A. Then elementary geometry shows that $B\big(u_\alpha, (1-\alpha)\epsilon\big) \subset A$; see Figure 12. □

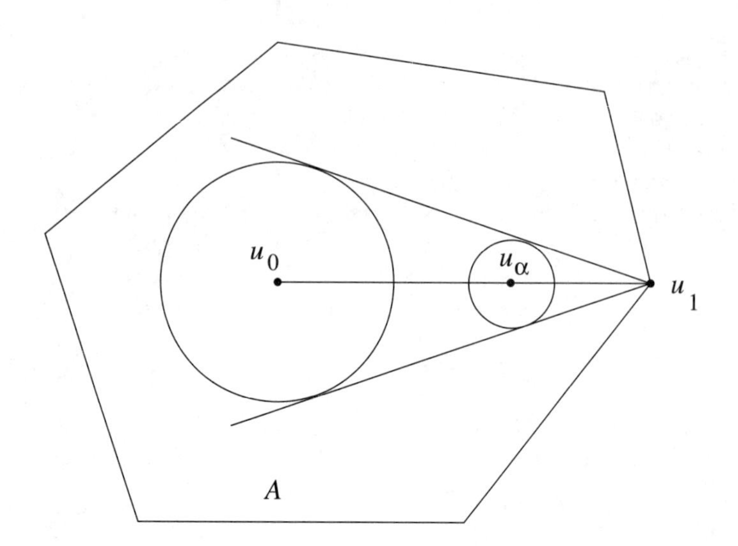

Figure 12

(2.3) Corollary. *Let $A \subset \mathbb{R}^d$ be a convex set. Then* $\text{int}(A)$ *is a convex set.*

Proof. Let $u, v \in \text{int}(A)$ be points and let $w = \alpha u + (1-\alpha)v$ for $0 \leq \alpha \leq 1$. If $\alpha < 1$, we apply Lemma 2.2 with $u_0 = v, u_1 = u$ and $w = u_\alpha$ to show that $w \in \text{int}(A)$. If $\alpha = 1$, then $w = u \in \text{int}(A)$. □

We note that Lemma 2.2 and Corollary 2.3 will be generalized to an infinite-dimensional situation in Section III.2.

PROBLEMS.

1. Let $v_1, \ldots, v_{d+1} \in \mathbb{R}^d$ be affinely independent points in $\mathbb{R}^d$. The polytope $\Delta = \text{conv}\big(v_1, \ldots, v_{d+1}\big)$ is called a d-dimensional simplex. Prove that Δ has a non-empty interior.

Hint: Let $u = (v_1 + \ldots + v_{d+1})/(d+1) \in \Delta$. We claim that for a sufficiently small $\epsilon > 0$ we have $B(u, \epsilon) \subset \Delta$. Indeed, the matrix of the system of $d+1$ linear equations in $d+1$ variables

$$\gamma_1 v_1 + \ldots + \gamma_{d+1} v_{d+1} = w \quad \text{and} \quad \gamma_1 + \ldots + \gamma_{d+1} = 1$$

is non-degenerate. Therefore, for each $w \in \mathbb{R}^d$, there is a unique solution $\gamma_1, \ldots, \gamma_{d+1}$ and the solution depends on w continuously. If $w = u$, then

$$\gamma_1 = \ldots = \gamma_{d+1} = 1/(d+1) > 0.$$

Therefore, if w is sufficiently close to u, all the γ's are non-negative and $w \in \Delta$.

2. Let $A \subset \mathbb{R}^d$ be a convex set such that $\text{int}(A) \neq \emptyset$ and let $H \subset \mathbb{R}^d$ be a hyperplane isolating $\text{int}(A)$. Prove that H isolates A.

We want to show that if a non-empty convex set in Euclidean space has empty interior, then we can pass to a smaller ambient space, where the set acquires an interior. This property makes the finite-dimensional situation radically different from the infinite-dimensional case.

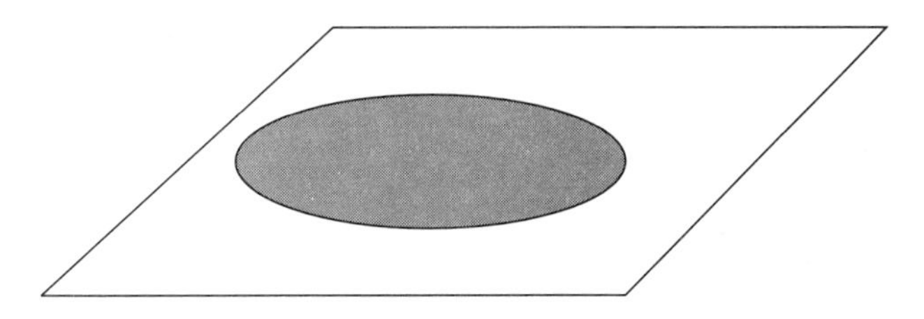

Figure 13. Example: a flat convex set in $\mathbb{R}^3$ acquires interior in the plane.

(2.4) Theorem. *Let $A \subset \mathbb{R}^d$ be a convex set. If* $\operatorname{int} A = \emptyset$, *then there exists an affine subspace $L \subset \mathbb{R}^d$ such that $A \subset L$ and* $\dim L < d$.

Proof. First, we claim there are no $d+1$ affinely independent points $v_1, \ldots, v_{d+1}$ in A. For if there were such points, then $\Delta = \operatorname{conv}(v_1, \ldots, v_{d+1}) \subset A$ and Problem 1, Section 2.3 would imply that Δ contains an interior point. Let $k < d+1$ be the maximum number of affinely independent points in A and let $v_1, \ldots, v_k$ be such points. Then, for each point $v \in A$ there is a solution to the system

$$\gamma_1 v_1 + \ldots + \gamma_k v_k + \gamma v = 0,$$
$$\gamma_1 + \ldots + \gamma_k + \gamma = 0$$

such that $\gamma \neq 0$. Then $v \in A$ can be expressed as an affine combination of $v_1, \ldots, v_k$,

$$v = \sum_{i=1}^{k} (-\gamma_i/\gamma) v_k.$$

Therefore, A is contained in the affine subspace L that is the affine hull of $v_1, \ldots, v_k$. So, $\dim L = k - 1 < d$. $\square$

(2.5) Definition. The *dimension* of a convex set $A \subset \mathbb{R}^d$ is the dimension of the smallest affine subspace that contains A. By convention, the dimension of the empty set is -1.

PROBLEM.

1. Let $A \subset \mathbb{R}_\infty$ be the set of Problem 2, Section 1.6. Prove that A does not contain any non-empty algebraically open subset and that A is not contained in any affine hyperplane of $\mathbb{R}_\infty$.

Let us take a closer look at the boundary of a convex set.

(2.6) Definitions. Let $K \subset \mathbb{R}^d$ be a closed convex set. A (possibly empty) set $F \subset K$ is called a *face* of K if there exists an affine hyperplane H which isolates K and such that $F = K \cap H$. If F is a point, then F is called an *exposed point* of K. A non-empty face $F \neq K$ is called a *proper* face of K.

PROBLEMS.

1°. Prove that a face is a closed convex set and that a face of a compact convex set is a compact convex set.

2. Find the faces of the unit ball $B = \{x \in \mathbb{R}^d : \ \|x\| \leq 1\}$.

✗ 3. Describe the faces of the d-dimensional unit cube

$$I = \Big\{x = (\xi_1, \ldots, \xi_d) : \quad 0 \leq \xi_k \leq 1 \quad \text{for} \quad k = 1, \ldots, d\Big\}.$$

4. Describe the faces of a d-dimensional simplex $\Delta = \operatorname{conv}(v_1, \ldots, v_{d+1})$, where $v_1, \ldots, v_{d+1}$ are affinely independent points in $\mathbb{R}^d$.

5. Let $K \subset \mathbb{R}^d$ be a closed convex set. Prove that the intersection of any two faces of K is a face of K.

6. Construct an example of a compact convex set $K \subset \mathbb{R}^2$, a face F of K and a face G of F such that G is not a face of K.

✗ 7. Prove that every non-empty compact convex set in $\mathbb{R}^d$ has an exposed point.

✗ 8. Construct a compact convex set $A \subset \mathbb{R}^2$ whose set of exposed points is not compact.

9* (Straszewicz' Theorem). Prove that every compact convex set $A \subset \mathbb{R}^d$ is the closure of the convex hull of the set of its exposed points.

Next, we prove that a boundary point lies in some face of a closed convex set.

(2.7) Theorem. *Let $K \subset \mathbb{R}^d$ be a convex set with a non-empty interior and let $u \in \partial K$ be a point. Then there exists an affine hyperplane H, called a support hyperplane at u, such that $u \in H$ and H isolates K.*

Proof. By Corollary 2.3, $\operatorname{int}(K)$ is a non-empty convex open set. Therefore, $\operatorname{int}(K)$ is a convex, algebraically open set such that $u \notin \operatorname{int}(K)$. Therefore, by Theorem 1.6, there is an affine hyperplane H containing u and isolating $\operatorname{int}(K)$. Then by Problem 2, Section 2.3, H isolates K, so H is a support hyperplane at u. □

PROBLEM.

1°. Construct an example of a closed convex set $K \subset \mathbb{R}^2$ with a non-empty interior and a point $u \in \partial K$ such that a support hyperplane of K at u is not unique.

2°. Let $B = \{x \in \mathbb{R}^d : \ \|x\| \leq 1\}$ be the unit ball and let $u \in \partial B$ be a point. Find the support hyperplane to B at u.

(2.8) Corollary. *Let $K \subset \mathbb{R}^d$ be a closed convex set with a non-empty interior and let $u \in \partial K$ be a point. Then there is a proper face F of K such that $u \in F$.*

Proof. Let H be a support hyperplane of K at u. Let $F = H \cap K$. $\square$

Now we prove a version of the Isolation Theorem for convex sets in $\mathbb{R}^d$.

(2.9) Theorem. *Let $A \subset \mathbb{R}^d$ be a non-empty convex set and let $u \notin A$ be a point. Then there is an affine hyperplane $H \subset \mathbb{R}^d$ such that $u \in H$ and H isolates A.*

Proof. Let us choose the minimal affine subspace $L \subset \mathbb{R}^d$ such that $A \subset L$. Theorem 2.4 implies that A has a non-empty interior as a subset of L. If $u \notin L$, we can choose H disjoint from L. Hence we may assume that $u \in L$. Thus, restricting ourselves to L, we see that $\operatorname{int}(A) \neq \emptyset$ (in L) and that $u \in L$. Then, by Theorem 2.7, there is an affine hyperplane $\widehat{H}$ in L, such that $u \in \widehat{H}$ and $\widehat{H}$ isolates A. Then we choose any hyperplane H such that $H \cap L = \widehat{H}$. $\square$

PROBLEM.

1. Let $A \subset \mathbb{R}^d$ be a convex set and let $L \subset \mathbb{R}^d$ be an affine subspace such that $L \cap A = \emptyset$. Prove that there exists an affine hyperplane H such that $L \subset H$ and H isolates A.

3. Extreme Points. The Krein-Milman Theorem for Euclidean Space

Certain points on the boundary of a convex set capture a lot of information about the set both in finite and infinite dimensions. Here is the central definition of this chapter.

(3.1) Extreme points. Let V be a vector space and let $A \subset V$ be a set. A point $a \in A$ is called an *extreme* point of A provided for any two points $b, c \in A$ such that $(b+c)/2 = a$ one must have $b = c = a$. The set of all extreme points of A is denoted $\operatorname{ex}(A)$.

Here is a simple and important theorem.

(3.2) Theorem. *Let V be a vector space, let $A \subset V$ be a non-empty set and let $f : V \longrightarrow \mathbb{R}$ be a linear functional.*

1. *Suppose that f attains its maximum (resp. minimum) on A at a unique point $u \in A$, that is, $f(u) > f(v)$ for all $v \neq u, v \in A$ (resp. $f(u) < f(v)$ for all $v \neq u, v \in A$). Then u is an extreme point of A.*
2. *Suppose that f attains its maximum (minimum) α on A and suppose that $B = \{x \in A : f(x) = \alpha\}$ is the set where the maximum (minimum) is attained. Let u be an extreme point of B. Then u is an extreme point of A.*

Proof. We will discuss the maximum case. The minimum is treated in a similar way. Let us prove the first part. If $u = (a+b)/2$, then $f(u) = (f(a)+f(b))/2$, where $f(a) \leq f(u)$ and $f(b) \leq f(u)$. Therefore, $f(a) = f(b) = f(u)$ and we must have $a = b = u$, because the maximum point is unique. For the second part, suppose that $u = (a+b)/2$ for $a, b \in A$. Then $\alpha = f(u) = (f(a)+f(b))/2$ and

$f(a), f(b) \leq \alpha$. Thus we must have $f(a) = f(b) = \alpha$, so $a, b \in B$. Then $a = b = u$ since u is an extreme point of B. $\square$

PROBLEMS.

1°. Let $K \subset \mathbb{R}^d$ be a closed convex set and let $F \subset K$ be a face. Prove that if $u \in F$ is an extreme point of F, then u is an extreme point of K.

2. Let $K \subset \mathbb{R}^d$ be a compact convex set and let $u \in K$ be a point such that $\|u\| \geq \|v\|$ for each $v \in K$. Prove that u is an extreme point of K.

We now prove a finite-dimensional version of a quite general and powerful result obtained by M.G. Krein and D.P. Milman in 1940.

(3.3) Theorem. *Let $K \subset \mathbb{R}^d$ be a compact convex set. Then K is the convex hull of the set of its extreme points:* $K = \operatorname{conv}\bigl(\operatorname{ex}(K)\bigr)$.

Proof. We proceed by induction on the dimension d. If $d = 0$, then K is a point and the result follows. Suppose that $d > 0$. Without loss of generality we may suppose that $\operatorname{int}(K) \neq \emptyset$. Otherwise, K lies in an affine subspace of a smaller dimension (cf. Theorem 2.4) and the result follows by the induction hypothesis. We must show that every point $u \in K$ can be represented as a convex combination of extreme points of K. If $u \in \partial K$, then, by Corollary 2.8, there exists a face F of K such that $u \in F$; see Figure 14 a). Then F lies in an affine subspace of a smaller dimension, and by the induction hypothesis $u \in \operatorname{conv}\bigl(\operatorname{ex}(F)\bigr)$, so the result follows since $\operatorname{ex}(F) \subset \operatorname{ex}(K)$ (see Problem 1, Section 3.2).

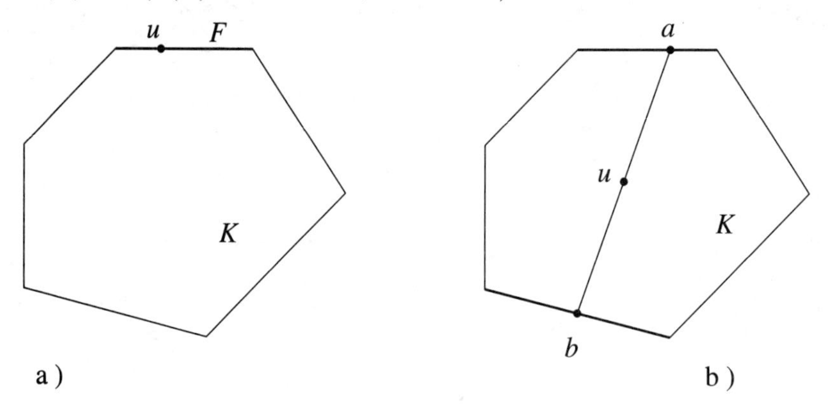

Figure 14

Suppose that $u \in \operatorname{int}(K)$. Let us draw a straight line L through u. The intersection $L \cap K$ is an interval $[a, b]$, where $a, b \in \partial K$ and u is an interior point of $[a, b]$; see Figure 14 b). As we already proved, $a, b \in \operatorname{conv}\bigl(\operatorname{ex}(K)\bigr)$. Since u is a convex combination of a and b, the result follows. $\square$

Theorem 3.3 is also known as Minkowski's Theorem.

PROBLEMS.

1. Prove that the set of extreme points of a closed convex set in $\mathbb{R}^2$ is closed.

2. Construct an example of a compact convex set $K \subset \mathbb{R}^3$ such that $\text{ex}(K)$ is not closed.

3. Let $A \subset \mathbb{R}^d$ be a set. Prove that u is an extreme point of $\text{conv}(A)$ if and only if $u \in A$ and $u \notin \text{conv}\big(A \setminus \{u\}\big)$.

4. Construct an example of a compact convex set $K \subset \mathbb{R}^2$ and a point $u \in K$ such that u is an extreme point of K, but not an exposed point of K.

5°. Prove that an exposed point is an extreme point.

6*. Let $A \subset \mathbb{R}^d$ be a closed convex set. Prove that each extreme point of A is a limit of exposed points of A.

The following corollary underscores the importance of extreme points for optimization.

(3.4) Corollary. *Let $K \subset \mathbb{R}^d$ be a compact convex set and let $f : \mathbb{R}^d \longrightarrow \mathbb{R}$ be a linear functional. Then there exists an extreme point u of K such that $f(u) \geq f(x)$ for all $x \in K$.*

Proof. Clearly, f attains its maximum value, say, α on K. Let $F = \big\{x \in K : f(x) = \alpha\big\}$ be the corresponding face of K. Then $\text{ex}(F) \neq \emptyset$ and any $u \in \text{ex}(F)$ is an extreme point of K; cf. Problem 1 of Section 3.2. □

Finally, a useful result whose proof resembles that of Theorem 3.3.

(3.5) Lemma. *Let $A \subset \mathbb{R}^d$ be a non-empty closed convex set which does not contain straight lines. Then A has an extreme point.*

Proof. We proceed by induction on d. If $d = 0$, the result obviously holds. Suppose that $d > 0$. Without loss of generality, we may assume that A has a non-empty interior. Otherwise, using Theorem 2.4, we reduce the dimension d. Let us choose a point $a \in A$ and let L be any straight line passing through a. The intersection $L \cap A$ is a non-empty, closed interval (bounded or unbounded) that cannot be the whole line L. Let b be a boundary point of that interval. Clearly $b \in \partial K$ and by Corollary 2.8 there is a proper face F of K containing b. We observe that F is a closed convex set which does not contain straight lines and that $\dim F < d$. Applying the induction hypothesis, we conclude that F has an extreme point u. Problem 1 of Section 3.2 implies then that u is an extreme point of A. □

4. Extreme Points of Polyhedra

For most of the rest of the chapter, we will be looking at the extreme points of various closed convex sets in Euclidean space. We start with a polyhedron (see Definition I.2.2), the set of solutions to finitely many linear inequalities in $\mathbb{R}^d$.

(4.1) Definition. An extreme point of a polyhedron is called a *vertex*.

Let us describe the vertices of a polyhedron.

(4.2) Theorem. *Let $P \subset \mathbb{R}^d$ be a polyhedron*

$$P = \Big\{ x \in \mathbb{R}^d : \ \langle c_i, x \rangle \leq \beta_i \quad for \quad i = 1, \ldots, m \Big\},$$

where $c_i \in \mathbb{R}^d$ and $\beta_i \in \mathbb{R}$ for $i = 1, \ldots, m$.

For $u \in P$ let

$$I(u) = \big\{ i : \ \langle c_i, u \rangle = \beta_i \big\}$$

be the set of the inequalities that are active on u. Then u is a vertex of P if and only if the set of vectors $\{c_i : i \in I(u)\}$ linearly spans the vector space $\mathbb{R}^d$. In particular, if u is a vertex of P, the set $I(u)$ contains at least d indices: $|I(u)| \geq d$.

Proof. Suppose that the vectors c_i with $i \in I(u)$ do not span $\mathbb{R}^d$. Then there is a non-zero $y \in \mathbb{R}^d$ such that $\langle y, c_i \rangle = 0$ for all $i \in I(u)$. We note that $\langle c_i, u \rangle < \beta_i$ for $i \notin I(u)$. For $\epsilon > 0$ let $u_+ = u + \epsilon y$ and let $u_- = u - \epsilon y$. Then $u = (u_+ + u_-)/2$, $u_+ \neq u_-$ and for sufficiently small $\epsilon > 0$ the points u_- and u_+ belong to the polyhedron P. Hence u is not an extreme point of P.

Suppose now that $u \in P$ and the vectors c_i with $i \in I(u)$ span $\mathbb{R}^d$. Suppose that $u = (v + w)/2$ for $v, w \in P$. Then $\langle c_i, v \rangle \leq \beta_i$ and $\langle c_i, w \rangle \leq \beta_i$. Since $\langle c_i, u \rangle = \beta_i$ for $i \in I(u)$, we must have $\langle c_i, v \rangle = \langle c_i, w \rangle = \beta_i$ for $i \in I(u)$. Since vectors c_i with $i \in I(u)$ span $\mathbb{R}^d$, the system $\langle c_i, x \rangle = \beta_i$, $i \in I(u)$, of linear equations must have a unique solution. Therefore, $v = w = u$ and u is an extreme point. □

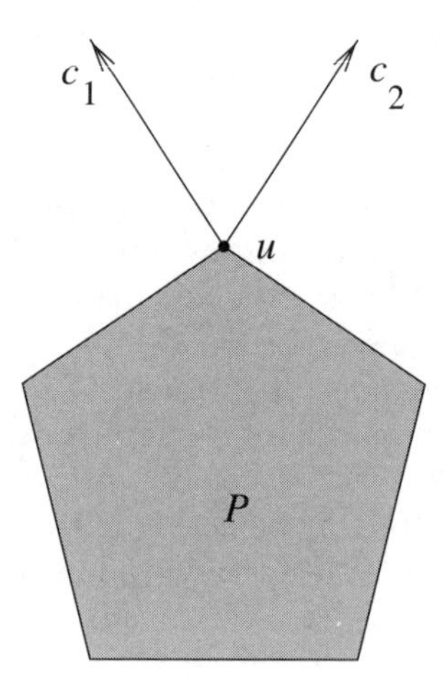

Figure 15. A polyhedron P, its vertex u and the vectors c_1 and c_2 of active constraints

PROBLEM.

1°. Prove that every vertex of a polyhedron is an exposed point.

The following corollary constitutes the first part of the Weyl-Minkowski Theorem.

(4.3) Corollary. *A bounded polyhedron is a polytope, that is, the convex hull of finitely many points.*

Proof. By Theorem 4.2, every vertex v of a polyhedron is a solution to a system $\langle c_i, x\rangle = \beta_i$, $i \in I(v)$, of linear equations where the vectors $c_i : \ i \in I(v)$ span $\mathbb{R}^d$. Every such system has at most one solution. Therefore, the number of vertices of a polyhedron in $\mathbb{R}^d$, defined by a set of m inequalities, does not exceed $\binom{m}{d}$ and hence is finite. By Theorem 3.3, P is the convex hull of the set of its extreme points and the result follows. □

PROBLEMS.

X1. Prove that a polyhedron has finitely many faces.

2°. Prove that a face of a polyhedron is a polyhedron.

3. Prove that polytopes have finitely many faces.

X 4*. Let $A \subset \mathbb{R}^d$ be a closed convex set. Prove that A has finitely many faces if and only if A is a polyhedron.

The effect of "unrealistic solutions" in linear programming problems

Let $P \subset \mathbb{R}^d$ be a polyhedron defined by a system of m linear inequalities. Suppose we want to solve a *linear programming problem*:

$$\begin{aligned} \text{Find} \quad & \gamma = \min\langle c, x\rangle \\ \text{Subject to} \quad & x \in P, \end{aligned}$$

where $c \in \mathbb{R}^d$ is the given vector of the objective function and $x \in P$ is a vector of variables. If the point $u \in P$ where the minimum is attained is unique, then by Part 1, Theorem 3.2, u must be a vertex of P. Theorem 4.2 then implies that at least d of the m inequality constraints are satisfied with equalities at u. This sometimes is not at all desired.

(4.4) Example. The Diet Problem. Suppose we have n different food ingredients, the unit price of the j-th ingredient being γ_j, $j = 1, \ldots, n$. We want the diet to be balanced with respect to m given nutrients. Suppose that α_{ij} is the content of the i-th nutrient in the j-th ingredient. Let $\xi_j : i = 1, \ldots, n$ be the quantity of the j-th ingredient in the diet and let $\beta_i : i = 1, \ldots, m$ be the target quantity of the i-th nutrient.

Trying to find the least expensive balanced diet, we come to a *linear programming* problem:

$$\text{Find} \quad \gamma = \min \sum_{j=1}^{n} \gamma_j \xi_j$$

$$\text{Subject to} \quad \sum_{j=1}^{n} \alpha_{ij}\xi_j = \beta_i \quad \text{for} \quad i = 1, \ldots, m \qquad \text{and}$$

$$\xi_j \geq 0 \quad \text{for} \quad j = 1, \ldots, n$$

in variables $(\xi_1, \ldots, \xi_n)$. Let $P \subset \mathbb{R}^n$ be the polyhedron of all feasible diets $x = (\xi_1, \ldots, \xi_n)$. Clearly, P lies in the affine subspace defined by the m balance constraints $\sum_{j=1}^{n} \alpha_{ij}\xi_j = \beta_i$. The dimension of the subspace, if it is non-empty, is at least $n - m$. If the optimal diet $x = (\xi_1, \ldots, \xi_n)$ is unique, it has to be a vertex of P, so at least $n - m$ of the coordinates $\xi_1, \ldots, \xi_n$ are zero. This, in turn, means that the optimal diet would consist of at most m ingredients. For example, if we are balancing the diet by the content of 5 nutrients, we should expect to get a menu consisting of 5 or fewer ingredients. Such a menu can hardly be considered realistic.

PROBLEM.

1°. One textbook on linear algebra describes the Cambridge diet. In particular, the book says: "In fact, the manufacturer of the Cambridge diet was able to supply 31 nutrients in precise amounts using only 33 ingredients". Prove that the manufacturer could have supplied the same 31 nutrients in precise amounts using only 31 or fewer ingredients.

5. The Birkhoff Polytope

In this section, we describe the vertices of an interesting polyhedron.

(5.1) Definitions. Let σ be a permutation of the set $\{1, \ldots, n\}$. The *permutation matrix* X^σ is the $n \times n$ matrix $X^\sigma = (\xi_{ij}^\sigma) : i, j = 1, \ldots, n$, defined as follows:

$$\xi_{ij}^\sigma = \begin{cases} 1 & \text{if } \sigma(j) = i, \\ 0 & \text{otherwise.} \end{cases}$$

For example,

$$\text{if} \quad \sigma = (123), \quad \text{that is,} \quad \sigma(1) = 2,\ \sigma(2) = 3 \text{ and } \sigma(3) = 1,$$

$$\text{then} \quad X^\sigma = \begin{pmatrix} 0 & 0 & 1 \\ 1 & 0 & 0 \\ 0 & 1 & 0 \end{pmatrix}.$$

An $n \times n$ matrix $X = (\xi_{ij}) : i, j = 1, \ldots, n$ is called *doubly stochastic* provided it is non-negative and the sum of entries in every row and every column is 1:

$$\sum_{i=1}^{n} \xi_{ij} = 1 \quad \text{for } j = 1, \ldots, n, \qquad \sum_{j=1}^{n} \xi_{ij} = 1 \quad \text{for } i = 1, \ldots, n \qquad \text{and}$$

$$\xi_{ij} \geq 0 \quad \text{for} \quad i, j = 1, \ldots, n.$$

The polyhedron B_n of all $n \times n$ doubly stochastic matrices is called the *Birkhoff Polytope.*

PROBLEMS.

1°. Prove that the set of integer doubly stochastic matrices is the set of permutation matrices.

2°. Prove that permutation matrices X^σ are extreme points of the Birkhoff Polytope.

3°. Check that B_n is bounded.

The following remarkable result was established independently by G. Birkhoff in 1946 and by J. von Neumann in 1953.

(5.2) Birkhoff - von Neumann Theorem. *The vertices of the Birkhoff Polytope B_n are exactly the $n \times n$ permutation matrices.*

Proof. Because of Problem 2, Section 5.1, it suffices to prove that if X is an extreme point of B_n, then $X = X^\sigma$ for some permutation σ. We prove this by induction on n. The case $n = 1$ is obvious. Suppose that $n > 1$. Let us consider the affine subspace $L \subset \mathbb{R}^{n^2}$ consisting of the $n \times n$ matrices $X = (\xi_{ij})$ such that

$$\sum_{i=1}^{n} \xi_{ij} = 1 \quad \text{for} \quad j = 1, \ldots, n \qquad \text{and} \qquad \sum_{j=1}^{n} \xi_{ij} = 1 \quad \text{for} \quad i = 1, \ldots, n.$$

We claim that $\dim L = (n-1)^2$. Indeed, a point (an $n \times n$ matrix X) from L is uniquely determined by an arbitrary choice of the $(n-1)^2$ entries ξ_{ij} for $i, j = 1, \ldots, n-1$, since the remaining entries of X are found as

$$\begin{aligned} \xi_{in} &= 1 - \sum_{j=1}^{n-1} \xi_{ij} \quad \text{for} \quad i = 1, \ldots, n-1, \\ \xi_{nj} &= 1 - \sum_{i=1}^{n-1} \xi_{ij} \quad \text{for} \quad j = 1, \ldots, n-1 \quad \text{and} \\ \xi_{nn} &= (2-n) + \sum_{i,j=1}^{n-1} \xi_{ij}. \end{aligned}$$

In the space L, the polytope B_n is defined by n^2 linear inequalities $\xi_{ij} \geq 0$. If X is an extreme point of B_n, by Theorem 4.2 some $(n-1)^2$ of these inequalities must be active on X. In other words, $\xi_{ij} = 0$ for some $(n-1)^2$ entries of X. Clearly, there cannot be a row containing zeros alone, and if every row contained at least two non-zero entries, the total number of zero entries would have been at most $n(n-2) < (n-1)^2$. Therefore, there must be a row, say, i_0 with $\xi_{i_0 j} = 0$ for all but one $j = j_0$. Now it is clear that $\xi_{i_0 j_0} = 1$ and that all other entries in the i_0-th row and in the j_0-th column must be zero. Crossing out the i_0-th row and the j_0-th column, we get an $(n-1) \times (n-1)$ doubly stochastic matrix, which must be an extreme point of B_{n-1}, so we may apply the induction hypothesis. □

PROBLEMS.

× 1. Prove that $\dim B_n = (n-1)^2$, so B_n has an interior point in the subspace L, constructed in the proof.

2. Find the radius of the ball in L centered at $\xi_{ij} = 1/n$ and touching the boundary of the polytope B_n.

3. Prove that the set $F = \big\{X \in B_n : \xi_{11} = 0\big\}$ is a face of B_n of dimension $(n-1)^2 - 1$ and that $G = \big\{X \in B_n : \xi_{11} = 1\big\}$ is a face of B_n of dimension $(n-2)^2$.

4°. Draw a picture of B_2.

5°. Let $U = (\zeta_{ij})$ be an $n \times n$ real orthogonal (that is, $UU^t = I$) or complex unitary (that is, $U\overline{U^t} = I$) matrix. Let $\beta_{ij} = |\zeta_{ij}|^2$. Prove that $B = (\beta_{ij})$ is a doubly stochastic matrix.

The problem of optimizing a linear function on the polytope B_n has an interesting combinatorial interpretation.

(5.3) The Assignment Problem. The *Assignment Problem* is formulated as follows: given an $n \times n$ matrix $C = (\gamma_{ij})$, find a permutation σ of the set $\{1, \dots, n\}$ such that $\sum_{i=1}^{n} \gamma_{i\sigma(i)}$ is maximum (or minimum). A typical interpretation of this problem is as follows: there are n candidates to fill n positions. Let γ_{ij} be the "benefit" (or the "damage") brought by the assignment of the i-th candidate to the j-th position. We are seeking to maximize (or minimize) the total benefit (or damage). Because of Theorem 5.2, the Assignment Problem can be posed as a problem of finding the maximum (minimum) value of a linear function on the polytope B_n, in short, as a *linear programming problem*:

$$\text{Find} \quad \gamma = \max \sum_{ij=1}^{n} \gamma_{ij}\xi_{ij}$$

$$\begin{aligned} \text{Subject to} \quad & \sum_{i=1}^{n} \xi_{ij} = 1 \quad \text{for} \quad j = 1, \dots, n, \\ & \sum_{j=1}^{n} \xi_{ij} = 1 \quad \text{for} \quad i = 1, \dots, n \qquad \text{and} \\ & \xi_{ij} \geq 0 \quad \text{for} \quad i, j = 1, \dots, n. \end{aligned}$$

Indeed, by Corollary 3.4 there is an optimal point (ξ_{ij}) which is an extreme point of the Birkhoff Polytope. By Theorem 5.2, such a point gives rise to a permutation (assignment) σ.

6. The Permutation Polytope and the Schur-Horn Theorem

A certain projection of the Birkhoff Polytope is of particular interest.

(6.1) Definition. Let us fix a point $x = (\xi_1, \dots, \xi_n)$ in $\mathbb{R}^n$. For a permutation σ of the set $\{1, \dots, n\}$, let $\sigma(x)$ be the vector $y = (\eta_1, \dots, \eta_n)$, where $\eta_i = \xi_{\sigma^{-1}(i)}$.

Let S_n be the symmetric group of all permutations of the set $\{1, \dots, n\}$. Let us define the *permutation polytope* $P(x)$ by

$$P(x) = \operatorname{conv}\big(\sigma(x) : \quad \sigma \in S_n\big).$$

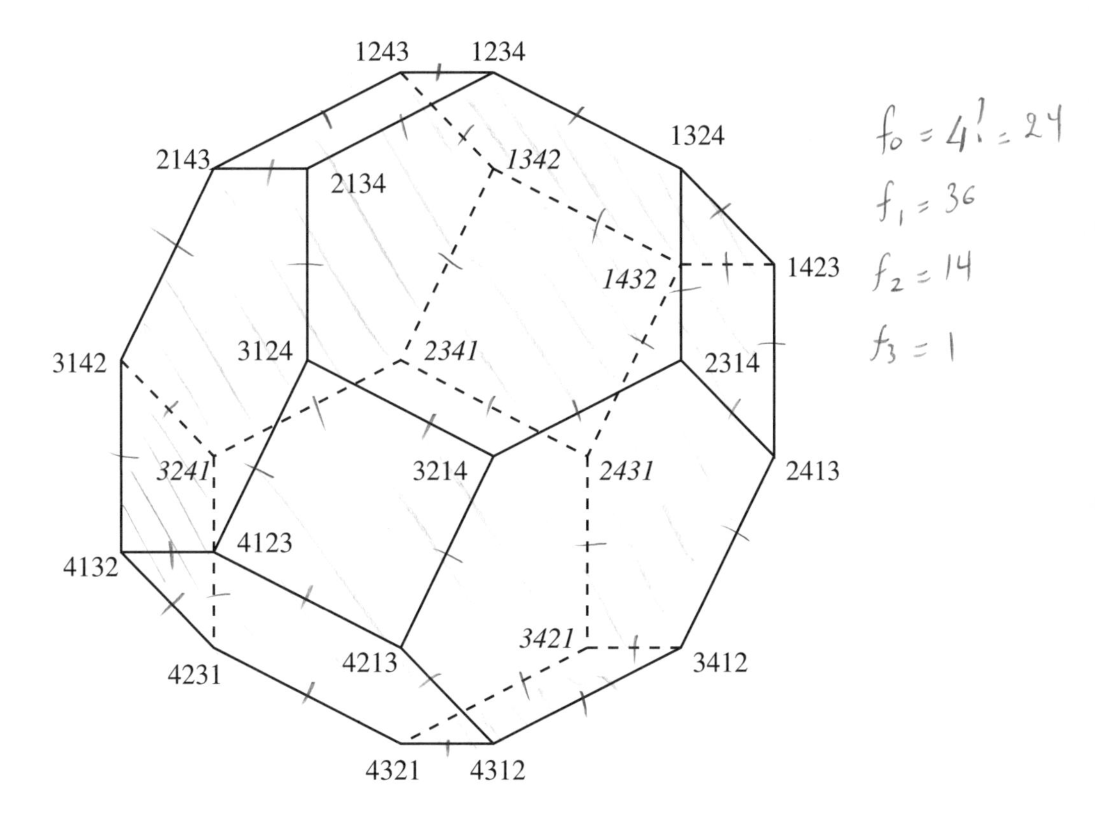

Figure 16. The permutation polytope $P(x)$ for $x = (1, 2, 3, 4)$

In words: we permute the coordinates of a given vector x in all possible ways and take the convex hull of the resulting vectors.

PROBLEMS.

1°. Prove that $\sigma(x) = X^\sigma x$, where X^σ is the permutation matrix corresponding to σ, and that $(\sigma\tau)(x) = \sigma\big(\tau(x)\big)$ for every two permutations σ and τ.

2°. Let us interpret $\mathbb{R}^{n^2}$ as the space of $n \times n$ matrices X. Let us fix a vector $a \in \mathbb{R}^n$. Consider the linear transformation $T : \mathbb{R}^{n^2} \longrightarrow \mathbb{R}^n$ defined by $T(X) = Xa$. Prove that $T(B_n) = P(a)$, where $B_n \subset \mathbb{R}^{n^2}$ is the Birkhoff Polytope.

3°. Prove that the permutation polytope $P(a)$, $a = (\alpha_1, \dots, \alpha_n)$, lies in the affine hyperplane $\Big\{(\xi_1, \dots, \xi_n) : \ \xi_1 + \ldots + \xi_n = \alpha_1 + \ldots + \alpha_n\Big\}$.

4. Suppose that not all the coordinates of a are equal. Prove that $\dim P(a) = n - 1$.

5°. Prove that $P(a)$ has $n!$ vertices if and only if the coordinates $\alpha_1, \ldots, \alpha_n$ of a are distinct.

6. Draw a picture of the permutation polytope $P(x)$ for $x = (1, 2, 3)$.

Permutation polytopes sometimes appear in quite unexpected situations. The first part of the following result was obtained by I. Schur in 1923, the second part by A. Horn in 1954.

(6.2) Schur-Horn Theorem. *Let us fix a positive integer n and real numbers $\lambda_1, \ldots, \lambda_n$. Let $l = (\lambda_1, \ldots, \lambda_n) \in \mathbb{R}^n$ be a vector.*

1. *Let $A = (\alpha_{ij})$ be an $n \times n$ real symmetric (or complex Hermitian) matrix with the eigenvalues $\lambda_1, \ldots, \lambda_n$. Then the diagonal $a = (\alpha_{11}, \ldots, \alpha_{nn})$ lies in the permutation polytope $P(l)$: $a \in P(l)$ (Schur's Theorem).*
2. *Let $a \in P(l)$ be a point from the permutation polytope. Then there exists an $n \times n$ real symmetric matrix $A = (\alpha_{ij})$ with the eigenvalues $\lambda_1, \ldots, \lambda_n$ and the diagonal $a = (\alpha_{11}, \ldots, \alpha_{nn})$ (Horn's Theorem).*

We will prove Schur's Theorem only (Part 1) using Schur's original approach. For a proof of Part 2, see, for example, [**MO79**].

Proof of Part 1. Let $D = \operatorname{diag}(\lambda_1, \ldots, \lambda_n)$ be diagonal matrix. Suppose that $A = (\alpha_{ij})$ is a real symmetric $n \times n$ matrix with the eigenvalues $\lambda_1, \ldots, \lambda_n$ (the proof for complex Hermitian matrices is completely analogous). Then $A = UDU^t$ for some orthogonal matrix $U = (\zeta_{ij})$. Hence the diagonal entries of A can be written as

$$\alpha_{kk} = \sum_{i=1}^{n} \zeta_{ki}^2 \lambda_i.$$

Let $B = (\beta_{ij})$ be the $n \times n$ matrix such that $\beta_{ij} = \zeta_{ij}^2$. Hence we may write $a = Bl$, where a and l are interpreted as n-columns of real numbers. Since U is an orthogonal matrix, the matrix B is doubly stochastic (cf. Problem 5, Section 5.2), that is, B is a non-negative matrix with all row and column sums equal to 1. By the Birkhoff - von Neumann Theorem (Theorem 5.2), B can be written as a convex combination of permutation matrices X^σ, $\sigma \in S_n$. Therefore we conclude that a is a convex combination of $\sigma(l) = X^\sigma l$, that is, $a \in P(l)$ by Problem 1 of Section 6.1. □

7. The Transportation Polyhedron

In this section, we describe a family of combinatorially defined polyhedra which includes, in particular, the Birkhoff Polytope.

Let $G = (V, E)$ be a directed (finite) graph with the set of vertices $V = \{1, \ldots, n\}$ and a set of edges $E \subset V \times V$. Any two vertices $i, j \in V$ can be either connected by an edge $i \to j$ or two edges $i \to j$ and $j \to i$ going in the opposite directions or not connected at all. We assume that the graph has no loops $i \to i$.

Suppose further that to each vertex i a real number β_i is assigned, which can be positive ("demand") or negative ("supply") or zero ("transit"). Suppose that to every edge $i \to j$ a number ξ_{ij} is assigned so that the following conditions are satisfied:

The balance requirement:

For every vertex $i \in V$

$$\sum_{j:\ (j \to i) \in E} \xi_{ji} - \sum_{j:\ (i \to j) \in E} \xi_{ij} = \beta_i.$$

No wrong way shipment:

For every edge $(i \to j) \in E$,

$$\xi_{ij} \geq 0.$$

An assignment of numbers ξ_{ij} satisfying the above requirements is called *a feasible flow* in G.

(7.1) Definition. Let us fix a graph $G = (V, E)$ with $|V| = n$ vertices and $|E| = m$ edges and a vector $b = (\beta_1, \ldots, \beta_n)$. Let us think of a feasible flow (ξ_{ij}) in G as a point in $\mathbb{R}^m$. The set of all feasible flows $x \in \mathbb{R}^m$ is called the *transportation polyhedron* and denoted $T(G, b)$.

PROBLEMS.

1. Prove that if the transportation polyhedron $T(G, b)$ is non-empty, then $\sum_{i=1}^n \beta_i = 0$.

2°. Construct an example where $\sum_{i=1}^n \beta_i = 0$ but the transportation polyhedron $T(G, b)$ is empty.

3°. Construct an example of an unbounded transportation polyhedron $T(G, b)$.

4. Suppose that G does not contain any directed cycle $i_1 \to i_2 \to \ldots \to i_l \to i_1$ for $l \geq 2$. Prove that $T(G, b)$ is a bounded polyhedron (polytope). It is called the *transportation polytope*.

There is a simple combinatorial description of the vertices of $T(G, b)$.

(7.2) Proposition. *Let $x = \big(\xi_{ij} : (i \to j) \in E\big)$ be an extreme point of the transportation polyhedron $T(G, b)$. Let $S \subset E$ be the set of all edges $i \to j$ where $\xi_{ij} > 0$. Then S does not contain any cycle $v_1 - v_2 - \ldots - v_l - v_1 : l \geq 2$, where the vertices $v_k, v_{k+1} : k = 1, \ldots, l-1$ and v_l, v_1 are connected by an edge (in either direction: $v_k \to v_{k+1}$ or $v_{k+1} \to v_k$).*

Proof. Suppose that there is such a cycle C consisting of edges with a strictly positive flow. Let us choose an $\epsilon > 0$ and let us construct two flows $y = (\eta_{ij})$ and $z = (\zeta_{ij})$ as follows:

$$\eta_{ij} = \begin{cases} \xi_{ij} & \text{if } (i \to j) \notin C \quad \text{and} \quad (j \to i) \notin C, \\ \xi_{ij} + \epsilon & \text{if } i = v_k, j = v_{k+1} \text{ or } i = v_l, j = v_1, \\ \xi_{ij} - \epsilon & \text{if } i = v_{k+1}, j = v_k \text{ or } i = v_1, j = v_l \end{cases}$$

and

$$\zeta_{ij} = \begin{cases} \xi_{ij} & \text{if } (i \to j) \notin C \quad \text{and} \quad (j \to i) \notin C, \\ \xi_{ij} - \epsilon & \text{if } i = v_k, j = v_{k+1} \text{ or } i = v_l, j = v_1, \\ \xi_{ij} + \epsilon & \text{if } i = v_{k+1}, j = v_k \text{ or } i = v_1, j = v_l. \end{cases}$$

In other words, we choose an orientation of the cycle C (say, clockwise). To construct y, we increase the flow on the edges of the cycle that go clockwise by ϵ and decrease the flow on the edges of the cycle that go counterclockwise by ϵ. To construct z, we decrease the flow on the clockwise edges by ϵ and increase the flow on the counterclockwise edges by ϵ; see Figure 17.

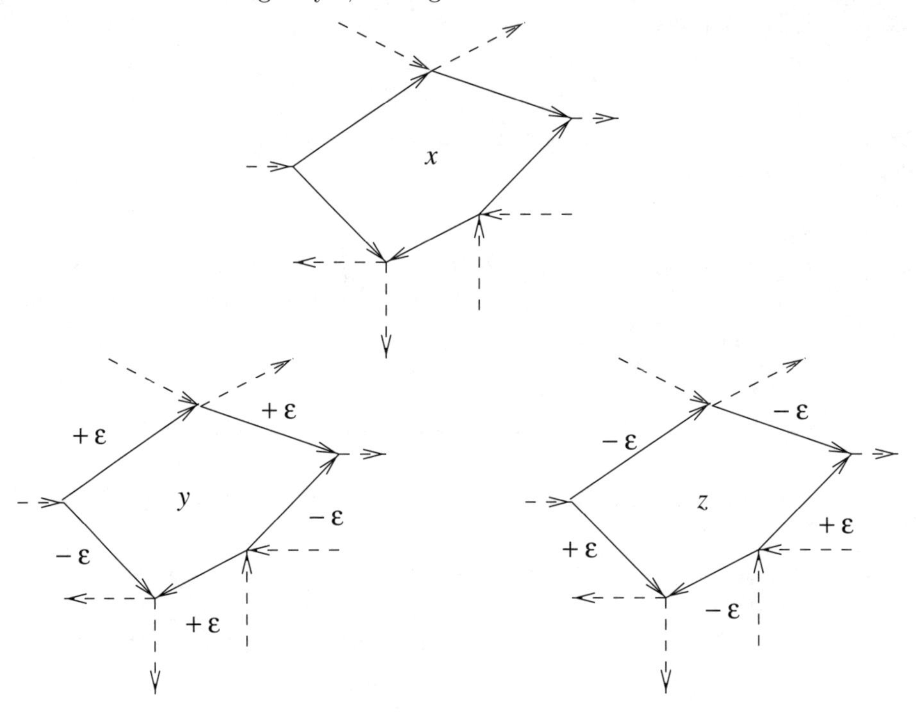

Figure 17. A decomposition of a circular flow $x = (y + z)/2$

Then the flows y and z satisfy the balance condition. Furthermore, if $\epsilon > 0$ is small enough, y and z are non-negative flows and hence feasible flows. Finally, $x = (y + z)/2$, which proves that x cannot be an extreme point of $T(G, b)$. □

(7.3) Definition. A graph without cycles is called a *forest.* A connected graph without cycles is called a *tree.*

PROBLEMS.

1°. Prove that a forest is a union of non-intersecting trees.

2. Prove that every (finite) forest has a vertex which is incident to at most one vertex of the forest.

(7.4) Corollary. *Suppose that all demands/supplies β_i, $i = 1, \ldots, n$, are integer numbers. Then every extreme point of the transportation polyhedron $T(G, b)$ is an integer flow $x = (\xi_{ij})$.*

Proof. Let $x = (\xi_{ij})$ be an extreme point. Proposition 7.2 asserts that the set $S \subset E$ with non-zero flows ξ_{ij} is a forest. We claim that once we know S, we can compute the flows ξ_{ij} from $\beta_1, \ldots, \beta_n$ by using addition and subtraction only. This, of course, would imply that all ξ_{ij} are integers.

If S is a forest with at least one edge, by Problem 2, Section 7.3, there is a vertex i such that there is only one edge with a non-zero flow incident to i. Let j be the other end of that edge. Clearly, if this edge is $i \to j$, we must have $\beta_i < 0$ and $\xi_{ij} = -\beta_i$. If this edge is $j \to i$, we must have $\beta_i > 0$ and $\xi_{ji} = \beta_i > 0$. Now we delete the vertex i with all edges of G incident to it, modify forest S accordingly, and adjust the demand/supply vector: if we had $\xi_{ij} > 0$, we let $\beta_j' := \beta_j + \beta_i$ and if we had $\xi_{ji} > 0$, we let $\beta_j' := \beta_j - \beta_i$. Hence we get a new graph G' with $n-1$ vertices and integer demands/supplies β_i', a new forest S' of G' and a new feasible flow ξ_{ij}' in G' such that S' is the set of edges where ξ_{ij}' are strictly positive. We proceed as above, until there are no edges in the forest. At that moment the flow $x = (\xi_{ij})$ is determined completely. □

PROBLEMS.

1. Deduce the Birkhoff - von Neumann Theorem (Theorem 5.2) from Corollary 7.4 as follows: consider the graph G with $2n$ vertices $1, \ldots, n$ and $1', \ldots, n'$ and the edges $i' \to j$, where $i' = 1', \ldots, n'$ and $j = 1, \ldots, n$. Let $\beta_i = 1$ be the demand for $i = 1, \ldots, n$ and let $\beta_{i'} = -1$ be the supply for $i' = 1', \ldots, n'$. Prove that the feasible flows $\xi_{i'j}$ are the doubly stochastic matrices and that the integral feasible flows are the permutation matrices.

2. Let us fix positive integers m and n and let us interpret a real $m \times n$ matrix as a point in $\mathbb{R}^{mn}$. Let $a = (\alpha_1, \ldots, \alpha_m)$ and $b = (\beta_1, \ldots, \beta_n)$ be two vectors of positive integers and let $P(a, b)$ be the set of all non-negative $m \times n$ matrices with the row sums $\alpha_1, \ldots, \alpha_m$ and the column sums $\beta_1, \ldots, \beta_n$. Prove that $P(a, b)$ is a bounded polyhedron (polytope) and that every vertex of $P(a, b)$ is an integer matrix.

3* (M.B. Gromova). Let us interpret the space $\mathbb{R}^d$ with $d = n^3$ as the space of all 3-dimensional matrices (ξ_{ijk}) where $1 \leq i, j, k \leq n$. Let us consider the polytope

$P_n \subset \mathbb{R}^d$ defined by the equations

$$\sum_{i,j=1}^{n} \xi_{ijk} = 1 \quad \text{for all} \quad k = 1, \dots, n,$$

$$\sum_{i,k=1}^{n} \xi_{ijk} = 1 \quad \text{for all} \quad j = 1, \dots, n \quad \text{and}$$

$$\sum_{j,k=1}^{n} \xi_{ijk} = 1 \quad \text{for all} \quad i = 1, \dots, n$$

and inequalities

$$\xi_{ijk} \geq 0 \quad \text{for all} \quad 1 \leq i, j, k \leq n.$$

Check that P_n is a polytope (it is called the polytope of 3-dimensional *polystochastic matrices*) and that $\dim P = n^3 - 3n + 1$.

Prove that for any sequence $1 > \sigma_1 > \sigma_2 \ldots > \sigma_p > 0$ of rational numbers there exists a positive integer b such that the numbers $(b-1)/b > \sigma_1 > \ldots > \sigma_p > 1/b$ compose the set of values of the non-zero coordinates (not counting multiplicities) of some vertex of P_n for some n.

Remark: This result, as well as its generalizations and extensions, is found in [**G92**].

The following problem can be considered as a generalization of the Assignment Problem; see Section 5.3.

(7.5) The Transportation Problem.

Suppose that γ_{ij} are (usually non-negative) costs on the edges of the graph $G = (V, E)$. The problem of finding a feasible flow $x \in T(G, b)$ minimizing the total cost $\sum_{i \to j \in E} \gamma_{ij} \xi_{ij}$ is called the *Transportation Problem.*

From Theorem 3.2 (Part 1) and Proposition 7.2 we deduce that if the optimal flow is unique, the set of edges $i \to j$ where the flow is positive must form a forest in G. Furthermore, by Corollary 3.4 we conclude that if $T(G, b)$ is bounded and non-empty, then there is an optimal flow with that property.

PROBLEMS.

1. Suppose that the transportation polyhedron $T(G, b)$ is non-empty and that the cost γ_{ij} is strictly positive for every edge $i \to j$ in E. Prove that there exists an optimal flow in the Transportation Problem and that the set of all optimal flows is a compact polyhedron (polytope), which is a face of the transportation polyhedron $T(G, b)$. Deduce that there exists an optimal solution such that the set of all edges where the flow is positive forms a forest in G.

2. Suppose that there is an optimal solution in Problem 7.5. Prove that there is an optimal solution such that the edges $i \to j$ with $\xi_{ij} > 0$ constitute a forest.

8. Convex Cones

We extend results of Section 3 on the structure of convex sets to convex cones. The theory that we develop here is parallel to that of Section 3.

(8.1) Cones, conic hulls and extreme rays. Let V be a vector space. A set $K \subset V$ is called a *cone* if $0 \in K$ and $\lambda x \in K$ for every $\lambda \geq 0$ and every $x \in K$. The cones we will be dealing with are convex. Alternatively, we can say that $K \subset V$ is a convex cone if $0 \in K$ and if for any two points $x, y \in K$ and any two numbers $\alpha, \beta \geq 0$, the point $z = \alpha x + \beta y$ is also in K.

Given points $x_1, \ldots, x_m \in V$ and non-negative numbers $\alpha_1, \ldots, \alpha_m$, the point

$$x = \sum_{i=1}^{m} \alpha_i x_i$$

is called a *conic* combination of the points $x_1, \ldots, x_m$. The set $\operatorname{co}(S)$ of all conic combinations of points from a set $S \subset V$ is called the *conic hull* of the set S. The conic hull $\operatorname{co}(x)$ of a non-zero point $x \in V$ is called the *ray* spanned by x.

Let $K \subset V$ be a cone and let $K_1 \subset K$ be a ray. We say that K_1 is an *extreme ray* of K if for any $u \in K_1$ and any $x, y \in K$, whenever $u = (x+y)/2$, we must have $x, y \in K_1$.

Let K be a cone and let $x \in K$ be a point. If $K_1 = \operatorname{co}(x)$ is an extreme ray of K, we say that x *spans* an extreme ray (of K).

PROBLEMS.

1°. Prove that $\operatorname{co}(S)$ is the smallest convex cone containing the set $S \subset V$, that is, the intersection of all convex cones in V that contain S.

2°. Let $K_1, K_2 \subset V$ be convex cones. Prove that the intersection $K_1 \cap K_2$ and the Minkowski sum $K_1 + K_2$ are convex cones.

3. Construct an example of two closed convex cones $K_1, K_2 \subset \mathbb{R}^3$ such that $K_1 + K_2$ is not closed.

4. Prove that the closure of a convex cone in $\mathbb{R}^d$ is a convex cone.

5. Prove that 0 is an extreme point of a convex cone K if K does not contain a straight line and construct an example of a convex cone $K \subset \mathbb{R}^2$, such that K contains a straight line but 0 is an extreme point of K.

6. Let $S \subset \mathbb{R}^d$ be a set. Prove that every point $x \in \operatorname{co}(S)$ is a conic combination of some d points from S.

We need a few technical lemmas, adjusting our results from previous sections for convex cones.

(8.2) Lemma. *Let $K \subset V$ be a cone and let $H \subset V$ be an affine hyperplane isolating K and such that $K \cap H \neq \emptyset$. Then $0 \in H$.*

Proof. Assume that for some (non-zero) linear functional $f: V \longrightarrow \mathbb{R}$ and some number α we have $H = \bigl\{x : f(x) = \alpha\bigr\}$ and that $K \subset \overline{H_-}$. Since $0 \in K$, we get

$\alpha \geq 0$. Suppose that $\alpha > 0$ and let $x \in K$ be a point such that $f(x) = \alpha$. Then, for $\lambda > 1$ we have $f(\lambda x) = \lambda\alpha > \alpha$ and $\lambda x \in K$, which contradicts the assumption that $K \subset \overline{H_-}$. Hence we must have $\alpha = 0$ and $0 \in H$. □

PROBLEM.

1°. Prove that a non-empty face of a cone is a cone.

(8.3) Definition. Let $K \subset V$ be a cone. A set $B \subset K$ is called a *base* of K if $0 \notin B$ and for every point $u \in K, u \neq 0$, there is a unique representation $u = \lambda v$ with $v \in B$ and $\lambda > 0$.

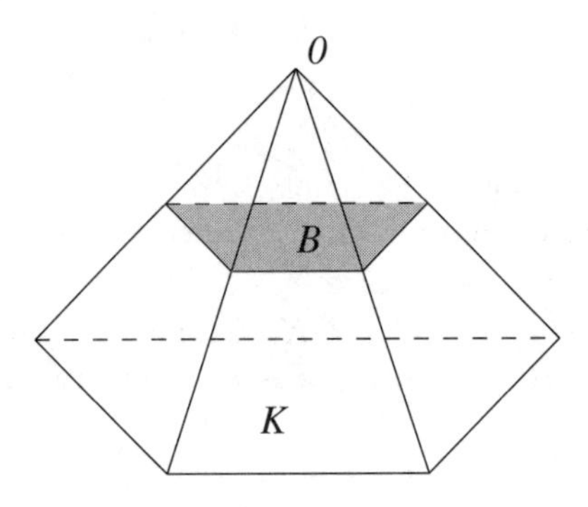

Figure 18. A base B of the cone K

(8.4) Lemma. *Let K be a cone with a convex base B and let $u \in K$, $u \neq 0$, be a non-zero point from K. Then u spans an extreme ray of K if and only if $u = \lambda v$, where $\lambda > 0$ and v is an extreme point of B.*

Proof. Suppose that u spans an extreme ray of K. Let $v \in B$ be a point such that $u = \lambda v$ for $\lambda > 0$. Suppose that $v = (v_1 + v_2)/2$. Then $u = (u_1 + u_2)/2$, where $u_1 = \lambda v_1$ and $u_2 = \lambda v_2$. Since u spans an extreme ray, we must have $u_1 = \mu_1 u$ and $u_2 = \mu_2 u$ for some $\mu_1, \mu_2 \geq 0$. Then $u_1 = (\mu_1\lambda)v$ and $u_2 = (\mu_2\lambda)v$. Since B is a base, we must have $v_1 = v_2 = v$, so v is an extreme point of B.

Suppose that $u = \lambda v$, where $\lambda > 0$ and v is an extreme point of B. Let us show that u spans an extreme ray of K. Suppose that $u = (u_1 + u_2)/2$. Then $u_1 = \lambda_1 v_1$ and $u_2 = \lambda_2 v_2$ for some $v_1, v_2 \in B$ and some non-negative λ_1, λ_2. Without loss of generality, we can assume that $\lambda_1, \lambda_2 > 0$. Then we can write $u = (\lambda_1 v_1 + \lambda_2 v_2)/2 = \beta(\alpha_1 v_1 + \alpha_2 v_2)$, where $\beta = (\lambda_1 + \lambda_2)/2$, $\alpha_1 = \lambda_1/(\lambda_1 + \lambda_2)$, and $\alpha_2 = \lambda_2/(\lambda_1 + \lambda_2)$. Since B is a base, we must have $\alpha_1 v_1 + \alpha_2 v_2 = v$. We note that $\alpha_1, \alpha_2 > 0$ and that $\alpha_1 + \alpha_2 = 1$. If $v_1 \neq v_2$, it follows that v lies inside the interval $[v_1, v_2]$, which contradicts the assumption that v is an extreme point of B. Therefore, we must have $v_1 = v_2$, so u spans an extreme ray of K. □

We obtain a conic version of Theorem 3.3.

(8.5) Corollary. *Let $K \subset \mathbb{R}^d$ be a cone with a compact base. Then every point $u \in K$ can be written as a conic combination*

$$u = \sum_{i=1}^{m} \lambda_i u_i, \quad \lambda_i \geq 0 : \ i = 1, \ldots , m,$$

where points u_i span extreme rays of K.

Proof. Let B be a base of K. Let us write $u = \lambda v$, where $v \in B$. By the Krein-Milman Theorem (Theorem 3.3), we can express v as a convex combination of extreme points $v_1, \ldots , v_m$ of B. Then u is a conic combination of $u_1 = \lambda_1 v_1, \ldots , u_m = \lambda_m v_m$. By Lemma 8.4, the points u_i span extreme rays of K. □

We will also need a topological fact.

(8.6) Lemma. *Let $K \subset \mathbb{R}^d$ be a cone with a compact base. Then K is closed.*

Proof. Let B be the compact base of K and let $u \notin K$ be a point. Our goal is to show that there is a neighborhood U of u such that $U \cap K = \emptyset$.

Let $\delta = \min\{\|x\| : \ x \in B\} > 0$ be the minimum distance from a point $x \in B$ to the origin. Let us choose $\lambda_0 = (\|u\| + 1)/\delta$ and let U_1 be the open ball of radius 1 centered at u. Then, for any $\lambda > \lambda_0$ we have $\lambda B \cap U_1 = \emptyset$.

Let $X = [0, \lambda_0] \times B$ and let us consider the map $\phi : X \longrightarrow \mathbb{R}^d$, $\phi(\lambda, x) = \lambda x$. Since B is compact, so is X. The image $\phi(X)$ is compact and hence closed in $\mathbb{R}^d$. Since $u \notin K$, we conclude that $u \notin \phi(X)$. Therefore, there is a neighborhood U_2 of u such that $U \cap \phi(X) = \emptyset$. Let $U = U_1 \cap U_2$. Then for any $\lambda \geq 0$, we have $U \cap \lambda B = \emptyset$ and the proof follows. □

We remark that the above result can be adjusted for infinite-dimensional spaces; see Lemma III.2.10.

PROBLEMS.

1. Let $K \subset \mathbb{R}^d$ be a cone with a compact base. Prove that 0 is a face of K.
2. Construct an example of a compact set $A \subset \mathbb{R}^2$ such that $\operatorname{co}(A)$ is not closed.
3. Prove that a closed cone in $\mathbb{R}^d$ without straight lines has a compact base.

9. The Moment Curve and the Moment Cone

We turn our attention to non-polyhedral convex sets. In this section, we discuss the boundary structure of an interesting non-polyhedral cone. Applications for problems of numerical integration are discussed in Section 10 and for probability problems in Sections III.9.3 and IV.2.

(9.1) The moment curve. Let us consider the space $\mathbb{R}^{d+1}$ with the coordinates $x = (\xi_0, \xi_1, \ldots , \xi_d)$ (we start with the zeroth coordinate). Given real numbers $\alpha < \beta$, the curve

$$g(\tau) = (1, \tau, \tau^2, \ldots , \tau^d) \in \mathbb{R}^{d+1} \quad \text{for } \alpha \leq \tau \leq \beta$$

is called *the moment curve.* Hence $g(\tau)$ lies in the affine hyperplane $\xi_0 = 1$ in $\mathbb{R}^{d+1}$.

Let $f(x) = \langle c, x \rangle$ be a linear function, where $c = (\gamma_0, \gamma_1, \ldots, \gamma_d)$. Then the value of f on the curve $g(\tau)$,

$$f\big(g(\tau)\big) = \gamma_d \tau^d + \gamma_{d-1}\tau^{d-1} + \ldots + \gamma_1 \tau + \gamma_0,$$

is a polynomial in τ of degree d.

PROBLEMS.

The problems below address some interesting properties of the moment curve and its relatives.

1. Prove that each hyperplane $H \subset \mathbb{R}^{d+1}$ such that $0 \in H$ intersects the moment curve $g(\tau)$ in at most d points.

2. Let $\delta_1 < \delta_2 < \ldots < \delta_d$ be real numbers. Consider the curve in $\mathbb{R}^d$

$$h(\tau) = \big(\exp\{\delta_1 \tau\}, \exp\{\delta_2 \tau\}, \ldots, \exp\{\delta_d \tau\}\big) \quad \text{for } \alpha \leq \tau \leq \beta.$$

Prove that each affine hyperplane $H \subset \mathbb{R}^d$ intersects $h(\tau)$ in at most d points.

3. Let $S^1 = \Big\{(\cos\tau, \sin\tau) : \ 0 \leq \tau \leq 2\pi\Big\}$ be the circle. Suppose that $d = 2k$ is even and let $h : S^1 \longrightarrow \mathbb{R}^d$ be the closed curve

$$h(\tau) = \big(\cos\tau, \sin\tau, \cos 2\tau, \sin 2\tau, \ldots, \cos k\tau, \sin k\tau\big), \quad 0 \leq \tau \leq 2\pi.$$

Prove that each affine hyperplane $H \subset \mathbb{R}^d$ intersects the curve $h(\tau)$ in at most d points.

4. Suppose that d is odd. Prove that one cannot embed the circle S^1 into $\mathbb{R}^d$ in such a way that every affine hyperplane intersects the circle in not more than d points.

5. Consider the set "Y" in the plane (three intervals having one common point). Prove that for any d one cannot embed Y into $\mathbb{R}^d$ in such a way that every affine hyperplane intersects Y in not more than d points.

Remark: A theorem of J.C. Mairhuber [**M56**] states that if a topological space X can be embedded into $\mathbb{R}^d$ as described above, then X must be a subset of a circle.

Now we define the main object of this section.

(9.2) Definition. The *moment cone*

$$M_{d+1} = \operatorname{co}\big(g(\tau) : \quad \alpha \leq \tau \leq \beta\big) \subset \mathbb{R}^{d+1}$$

is the conic hull of the curve $g(\tau)$. Sometimes we write $M_{d+1}[\alpha, \beta]$ instead of M_{d+1}.

PROBLEMS.

1°. Prove that $\operatorname{conv}\big(g(\tau) : \alpha \leq \tau \leq \beta\big)$ is a compact convex base of $M_{d+1}[\alpha, \beta]$.

2°. Prove that $\dim M_{d+1} = d + 1$.

(9.3) Lemma. *The moment cone M_{d+1} is closed.*

Proof. Follows from Lemma 8.6 and Problem 1 of Section 9.2. □

One interesting feature of the moment cone is that every point of M_{d+1} can be written as a conic combination of relatively few points of the moment curve. Moreover, if the point lies on the boundary, it *cannot* be written as a conic combination with positive coefficients of too many points of the moment curve.

(9.4) Proposition. *Let $u \in \partial M_{d+1}$ be a point on the boundary of M_{d+1}. Let us write u as a conic combination of points on the curve $g(\tau)$:*

$$u = \sum_{j=1}^{m} \lambda_j g(\tau_j), \quad \text{where } \lambda_j > 0 \text{ for } j = 1, \ldots, m \quad \text{and} \quad \alpha \leq \tau_1 < \ldots < \tau_m \leq \beta.$$

Then $m \leq (d+2)/2$. Furthermore, if $m = (d+2)/2$, then d is even and $\tau_1 = \alpha$, $\tau_m = \beta$.

Proof. By Problem 2 of Section 9.2, $\operatorname{int} M_{d+1} \neq \emptyset$, so by Theorem 2.7 there exists an affine hyperplane H that contains u and isolates M_{d+1}. Hence by Lemma 8.2 H contains the origin, so $H = \big\{x : \langle c, x\rangle = 0\big\}$ for some $c = (\gamma_0, \gamma_1, \ldots, \gamma_d) \neq 0$. We have $\langle c, x\rangle \geq 0$ for $x \in M_{d+1}$ and $\langle c, u\rangle = 0$. In particular, $\langle c, g(\tau)\rangle \geq 0$ for all $\alpha \leq \tau \leq \beta$. Let

$$p(\tau) = \langle c, g(\tau)\rangle = \gamma_0 + \gamma_1\tau + \ldots + \gamma_d\tau^d.$$

Hence $p(\tau)$ is a polynomial of degree d which is non-negative on the interval $[\alpha, \beta]$. Furthermore, since $\langle c, u\rangle = 0$ and $\lambda_j > 0$ we must have $p(\tau_j) = 0$ for $j = 1, \ldots, m$. Suppose that $\tau^* \in (\alpha, \beta)$ is a root of p, which lies strictly inside the interval (α, β). Then the multiplicity of the root must be an even number, since otherwise $p(\tau)$ would change sign in a neighborhood of τ^*. The only roots of multiplicity 1 can be $\tau^* = \alpha$ and $\tau^* = \beta$. The total number of roots of p, counting multiplicities, is at most d. If all the roots of p are strictly inside (α, β), then $2m \leq d$ and $m \leq d/2$. If only one endpoint of the interval $[\alpha, \beta]$ is a root of p, then there are $m - 1$ roots inside (α, β) and $2(m-1) + 1 \leq d$, so $m \leq (d+1)/2$. If both endpoints α and β are roots of p, then there are $(m-2)$ roots inside (α, β), so $2 + 2(m-2) \leq d$ and $m \leq (d+2)/2$. Thus, in any case, $m \leq (d+2)/2$. □

It follows then that every point $u \in M_{d+1}$ can be written as a convex combination of a small number of points on the moment curve, roughly a half of the number one would expect for the conic hull of a general set in $\mathbb{R}^{d+1}$; cf. Problem 6 of Section 8.1.

(9.5) Corollary. *Let*

$$m = \begin{cases} (d+1)/2 & \text{if } d \text{ is odd,} \\ (d+2)/2 & \text{if } d \text{ is even.} \end{cases}$$

Let $u \in M_{d+1}[\alpha, \beta]$ be a point. Then u can be represented as a conic combination of m (or fewer) points of the curve $g(\tau)$:

$$u = \sum_{i=1}^{m} \lambda_i g(\tau_i),$$

where $\lambda_i \geq 0$ for $i = 1, \ldots, m$ and $\alpha \leq \tau_1 < \ldots < \tau_m \leq \beta$.

Proof. If $u \in \partial M_{d+1}$, the result follows by Proposition 9.4. Suppose that $u \in \operatorname{int} M_{d+1}[\alpha, \beta]$. Let $t \geq 0$ be a parameter, and let us "shrink" the interval $[\alpha, \beta] \longrightarrow [\alpha + t, \beta - t]$ as t grows. Let us consider $M_{d+1}[\alpha + t, \beta - t]$. When $t = (\beta - \alpha)/2$, the curve g consists of a single point, so the cone $M_{d+1}[\alpha + t, \beta - t]$ consists of a single ray. If the point u is still in the cone, the result follows. Otherwise, there is a value t^* such that $u \in \partial M_{d+1}[\alpha + t^*, \beta - t^*]$. Now we use Proposition 9.4. □

10. An Application: "Double Precision" Formulas for Numerical Integration

Corollary 9.5 has a somewhat unexpected application. It implies the existence of some efficient formulas for numerical integration.

(10.1) Proposition. *Let us fix an interval $[\alpha, \beta]$ and a non-negative continuous density $\rho(\tau)$ on $[\alpha, \beta]$. Then, for any positive integer m, there exist m points $\tau_1^*, \ldots, \tau_m^*$ in the interval $[\alpha, \beta]$ and m non-negative numbers $\lambda_1, \ldots, \lambda_m$ such that*

$$\int_\alpha^\beta f(\tau)\rho(\tau)\, d\tau = \sum_{i=1}^{m} \lambda_i f(\tau_i^*)$$

for any polynomial f of degree at most $2m - 1$.

Proof. Let $d = 2m - 1$. Let $u = (\xi_0, \ldots, \xi_d) \in \mathbb{R}^{d+1}$, where $\xi_i = \displaystyle\int_\alpha^\beta \tau^i \rho(\tau)\, d\tau$ for $i = 0, \ldots, d$. Let us prove that $u \in M_{d+1}[\alpha, \beta]$. Indeed, since ρ is continuous, u can be written as a limit of Riemann sums:

$$u = \lim_{N \longrightarrow +\infty} \frac{\beta - \alpha}{N} \sum_{i=1}^{N} \rho(\tau_i) g(\tau_i),$$

where $\tau_1, \ldots, \tau_N$ are equally spaced points on $[\alpha, \beta]$. Since every Riemann sum is in the cone $M_{d+1}[\alpha, \beta]$, and by Lemma 9.3 the moment cone is closed, we get that

$u \in M_{d+1}$. Therefore, by Corollary 9.5, we can write u as a conic combination of m points on $g(\tau)$:

$$u = \sum_{i=1}^{m} \lambda_j g(\tau_i^*),$$

where $\lambda_i \geq 0$ and $\alpha \leq \tau_1^* < \ldots < \tau_m^* \leq \beta$. Now, let $f(\tau) = \gamma_d \tau^d + \ldots + \gamma_0$ be a polynomial of degree at most d. Let $c = (\gamma_0, \ldots, \gamma_d) \in \mathbb{R}^{d+1}$. Then

$$f(\tau_i^*) = \langle c, g(\tau_i^*) \rangle \quad \text{for} \quad i = 1, \ldots, m \quad \text{and} \quad \int_\alpha^\beta f(\tau)\rho(\tau)\, d\tau = \langle c, u \rangle,$$

which completes the proof. □

Formulas for numerical integration

$$\int_\alpha^\beta f(\tau)\rho(\tau)\, d\tau = \sum_{i=1}^{m} \lambda_i f(\tau_i^*)$$

that are exact for polynomials f of degree up to $d = 2m - 1$ are often called the *double precision* integration formulas (for obvious reasons). The proof of Proposition 10.1 explains the term "moment cone". The points of $M_{d+1}[\alpha, \beta]$ correspond to the *moments* of non-negative densities $\rho(\tau)$ on $[\alpha, \beta]$:

$$x = \left(\int_\alpha^\beta \rho(\tau)\, d\tau, \int_\alpha^\beta \tau\rho(\tau)\, d\tau, \ldots, \int_\alpha^\beta \tau^d \rho(\tau)\, d\tau \right).$$

As ρ varies, x ranges over the points in $M_{d+1}[\alpha, \beta]$.

PROBLEMS.

1. Prove that one cannot find m points $\tau_1^*, \ldots, \tau_m^*$ in the interval $[0, 1]$ and m real numbers $\lambda_1, \ldots, \lambda_m$ such that

$$\int_0^1 f(\tau)\, d\tau = \sum_{i=1}^{m} \lambda_i f(\tau_i^*)$$

for all polynomials f of degree $2m$.

2. Prove that for an interval $[\alpha, \beta]$, for every strictly positive continuous density $\rho(\tau)$ on $[\alpha, \beta]$ and for every positive integer m there is only one set of points $\tau_1^*, \ldots, \tau_m^*$ in the interval $[\alpha, \beta]$ and only one set of non-negative numbers $\lambda_1, \ldots, \lambda_m$ such that

$$\int_\alpha^\beta f(\tau)\rho(\tau)\, d\tau = \sum_{i=1}^{m} \lambda_i f(\tau_i^*)$$

for every polynomial f of degree at most $2m - 1$.

3*. A function

$$f(\tau) = \gamma_0 + \sum_{k=1}^{d} (\alpha_k \sin k\tau + \beta_k \cos k\tau) \quad \text{for} \quad 0 \leq \tau \leq 2\pi$$

is called a *trigonometric polynomial* of degree at most d. Let ρ be a non-negative continuous function on $[0, 2\pi]$ such that $\rho(0) = \rho(2\pi)$. Prove that there exist $d+1$ points $0 \leq \tau_0^* < \ldots < \tau_d^* < 2\pi$ and $d+1$ non-negative numbers $\lambda_0, \ldots, \lambda_d$ such that the formula

$$\int_0^{2\pi} f(\tau)\rho(\tau)\, d\tau = \sum_{i=0}^{d} \lambda_i f(\tau_i^*)$$

is exact for any trigonometric polynomial of degree at most d.

Hint: Use Problem 3 of Section 9.1.

4*. Let us fix $d = 2m$ distinct real numbers $\delta_1, \ldots, \delta_d$. A function

$$f(\tau) = \sum_{i=1}^{d} \alpha_i \exp\{\delta_i \tau\}$$

is called an *exponential polynomial* with exponents $\delta_1, \ldots, \delta_d$. Let ρ be a non-negative continuous density in the interval $[\alpha, \beta]$. Prove that there exist m points $\tau_1^*, \ldots, \tau_m^*$ in the interval $[\alpha, \beta]$ and m non-negative numbers $\lambda_1, \ldots, \lambda_m$, such that

$$\int_\alpha^\beta f(\tau)\rho(\tau)\, d\tau = \sum_{i=1}^{m} \lambda_i f(\tau_i^*)$$

for any exponential polynomial with exponents $\delta_1, \ldots, \delta_d$.

Hint: Use Problem 2 of Section 9.1.

Let us consider some small examples of double precision formulas for evaluating $\int_0^1 f(\tau)\, d\tau$.

(10.2) Example. The formula that uses one node and is exact for polynomials of degree at most 1 is

$$\int_0^1 f(\tau)\, d\tau = f\Big(\frac{1}{2}\Big).$$

PROBLEM.

1°. Prove that the formula of Example 10.2 is indeed exact on polynomials of degree at most 1 and that this is the only such formula.

(10.3) Example. Let us find formulas that use two nodes and are exact on polynomials of degree at most 2. In the xy plane, let us consider the parabola arc $\{g(\tau) = (\tau, \tau^2) : 0 \leq \tau \leq 1\}$. Let $u = (1/2, 1/3)$. There are infinitely many formulas

$$\int_0^1 f(\tau)\, d\tau = \lambda_1 f(\tau_1^*) + \lambda_2 f(\tau_2^*)$$

that are exact on polynomials of degree 2. The necessary and sufficient condition for $0 \leq \tau_1^* < \tau_2^* \leq 1$ is that the interval $[g(\tau_1^*), g(\tau_2^*)]$ in the xy plane contains u; see

Figure 19. Then λ_1 and λ_2 are found from $u = \lambda_1 g(\tau_1^*) + \lambda_2 g(\tau_2^*)$. For example, if we choose $\tau_1^* = 0$, then $\tau_2^* = 2/3$ and we get a formula

$$\int_0^1 f(\tau)\, d\tau = \frac{1}{4} f(0) + \frac{3}{4} f\Big(\frac{2}{3}\Big).$$

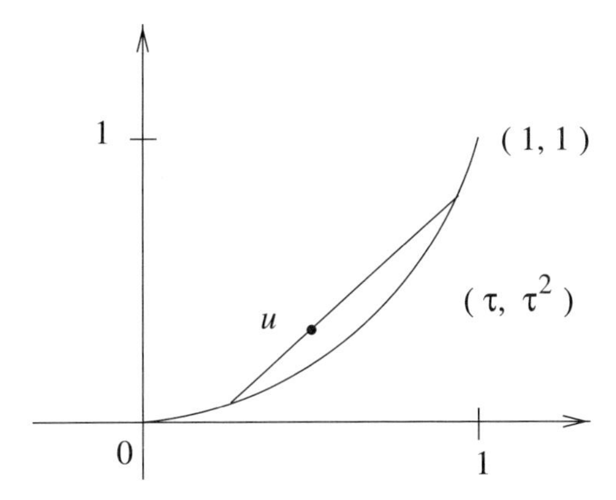

Figure 19

PROBLEM.

1. Find the one-parametric family of formulas with two nodes which are exact on polynomials of degree 2.

(10.4) Example. Here is a formula with two nodes which is exact for polynomials of degree at most 3:

$$\int_0^1 f(\tau)\, d\tau = \frac{1}{2} f\Big(\frac{1}{2} - \frac{1}{2\sqrt{3}}\Big) + \frac{1}{2} f\Big(\frac{1}{2} + \frac{1}{2\sqrt{3}}\Big).$$

PROBLEM.

1. Prove that the above formula is indeed exact on polynomials of degree at most 3 and that this is the only such formula which uses two nodes.

11. The Cone of Non-negative Polynomials

The cone we consider in this section is dual to the moment cone (in the sense rigorously described later; see Section IV.2). We recall that we have considered the set of positive multivariate polynomials in Section I.3.

(11.1) Definition. Let us interpret the space $\mathbb{R}^{d+1}$ as the space of all polynomials in one variable τ of degree at most d: a polynomial $\gamma_0 + \gamma_1\tau + \ldots + \gamma_d\tau^d$ is represented by the point $(\gamma_0, \ldots, \gamma_d)$. Let us fix numbers $\alpha < \beta$ and let $K_+[\alpha, \beta] \subset \mathbb{R}^{d+1}$ be the set of all polynomials p that are non-negative on the interval $[\alpha, \beta]$: $p(\tau) \geq 0$ for all $\tau \in [\alpha, \beta]$. Sometimes we write K_+ instead of $K_+[\alpha, \beta]$.

PROBLEMS.

1°. Prove that $K_+[\alpha, \beta]$ is a closed convex cone in $\mathbb{R}^{d+1}$.

2°. Prove that $K_+[\alpha, \beta]$ has a non-empty interior.

3. Prove that $p \in \partial K_+$ if and only if $p(\tau) \geq 0$ for all $\tau \in [\alpha, \beta]$ and $p(\tau_0) = 0$ for some $\tau_0 \in [\alpha, \beta]$.

Let us describe the extreme rays of $K_+[\alpha, \beta]$.

(11.2) Proposition. *The cone $K_+[\alpha, \beta]$ has a compact base. A polynomial $p \in K_+[\alpha, \beta]$ spans an extreme ray of $K_+[\alpha, \beta]$ if and only if the polynomial p is one of the following types:*

$$\begin{aligned} p(\tau) =& \delta \prod_{i=1}^{k} (\tau - \tau_i)^2, \quad 2k = d, \\ p(\tau) =& \delta (\tau - \alpha)(\beta - \tau) \prod_{i=1}^{k} (\tau - \tau_i)^2, \quad 2k + 2 = d \end{aligned}$$

for d even,

$$\begin{aligned} p(\tau) =& \delta (\tau - \alpha) \prod_{i=1}^{k} (\tau - \tau_i)^2, \quad 2k + 1 = d, \\ p(\tau) =& \delta (\beta - \tau) \prod_{i=1}^{k} (\tau - \tau_i)^2, \quad 2k + 1 = d \end{aligned}$$

for d odd, where $\delta > 0$ and $\tau_1, \ldots, \tau_k$ are (not necessarily distinct) points from the interval $[\alpha, \beta]$.

Proof. Let

$$B = \Big\{ q \in K_+ : \quad \int_\alpha^\beta q(\tau) \, d\tau = 1 \Big\}.$$

Obviously, B is a non-empty closed convex set. It is also clear that for any $p \in K_+$, $p \neq 0$, there is a unique representation $p = \lambda q$ for $q \in B$ and $\lambda > 0$; we take

$$\lambda = \int_\alpha^\beta p(\tau) \, d\tau.$$

We wish to show that B is compact. Let us define a norm $N : \mathbb{R}^{d+1} \longrightarrow \mathbb{R}$ by

$$N(p) = \int_\alpha^\beta |p(\tau)| \, d\tau.$$

Let $\epsilon > 0$ be the minimum value of the continuous function N on the unit sphere $\mathbb{S}^d = \{p : \ \|p\| = 1\}$, where $\| \cdot \|$ is the usual Euclidean norm. Then $\|p\| \leq 1/\epsilon$ for all $p \in B$, so B is compact. Hence B is a compact base of K_+.

Let $p \in K_+$ and suppose that co(p) is an extreme ray of K_+. If $\deg p < d$, we can write $p = (p_+ + p_-)/2$, where $p_+ = p + \epsilon(\tau - \alpha)p$ and $p_- = p - \epsilon(\tau - \alpha)p$. If $\epsilon > 0$ is sufficiently small, then $p_+, p_- \in K_+[\alpha, \beta]$ and p_- and p_+ are not proportional to p, which is a contradiction. Hence $\deg p = d$. We can factor $p = qr$, where $q, r \in K_+$ and r is a polynomial without any root in the interval $[\alpha, \beta]$. By continuity, we may choose a sufficiently small $\epsilon > 0$ such that $r_- = r - \epsilon$ and $r_+ = r + \epsilon$ are both non-negative on $[\alpha, \beta]$. We can write $p = (p_+ + p_-)/2$, where $p_+ = qr_+$ and $p_- = qr_-$. Since co(p) is an extreme ray, r_+ and r_- must be proportional to r and, therefore, r must be a constant. Summarizing, p has d roots in the interval $[\alpha, \beta]$. Finally, the multiplicity of every root of p which lies inside (α, β) must be even, since otherwise p changes its sign in a neighborhood of the root. So the polynomials that span extreme rays of K_+ must have the required structure.

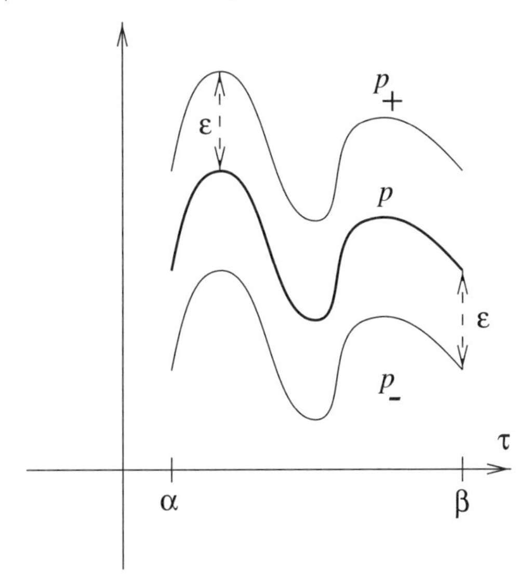

Figure 20. Decomposition $p = (p_+ + p_-)/2$ of a positive polynomial

It remains to prove that the polynomials having the required type indeed span extreme rays of K_+. Suppose that $p = (p_1 + p_2)/2$ for some $p_1, p_2 \in K_+$. We claim that every root τ^* of p of multiplicity m must be a root of both p_1 and p_2 of multiplicity at least m. Otherwise, for one of the polynomials p_i, $i = 1, 2$, we will have $p_i(\tau^*) = \ldots = p_i^{(k-1)}(\tau^*) = 0$, $p_i^{(k)}(\tau^*) < 0$ and for the other polynomial we will have $p_i(\tau^*) = \ldots = p_i^{(k-1)}(\tau^*) = 0$, $p_i^{(k)}(\tau^*) > 0$ for some $0 \leq k \leq m$. Therefore, one of the polynomials p_1 or p_2 would turn negative for some $\tau \in [\alpha, \beta]$ in a small neighborhood of the root τ^*. Since p has d roots, both p_1 and p_2 must be proportional to p.

□

Proposition 11.2 allows us to describe the structure of polynomials which are non-negative on a given interval $[\alpha, \beta]$.

(11.3) Corollary. *Let $p(\tau)$ be a polynomial such that $p(\tau) \geq 0$ for all $\tau \in [\alpha, \beta]$.*

If $d = 2k$ is even, then

$$p(\tau) = (\tau - \alpha)(\beta - \tau)\sum_{i \in I} q_i^2(\tau) + \sum_{j \in J} q_j^2(\tau)$$

for some polynomials q_i, q_j. Furthermore, we can choose q_i, q_j is such a way that $\deg q_i = k - 1$ for $i \in I$, $\deg q_j = k$ for $j \in J$ and all roots of q_i, q_j are real and belong to the interval $[\alpha, \beta]$.

If $d = 2k + 1$ is odd, then

$$p(\tau) = (\tau - \alpha)\sum_{i \in I} q_i^2(\tau) + (\beta - \tau)\sum_{j \in J} q_j^2(\tau)$$

for some polynomials q_i, q_j. Furthermore, we can choose q_i, q_j in such a way that $\deg q_i = \deg q_j = k$ for $i \in I$, $j \in J$ and all roots of q_i, q_j are real and belong to the interval $[\alpha, \beta]$.

Proof. Follows from Corollary 8.5 and Proposition 11.2. □

PROBLEMS.

1*. Prove that a polynomial $p(\tau)$ of degree d, which is non-negative on $[\alpha, \beta]$, has a unique representation

$$p(\tau) = \begin{cases} \delta \prod_{i=1}^{k}(\tau - \tau_{2i-1})^2 + \gamma(\tau - \alpha)(\beta - \tau)\prod_{i=1}^{k-1}(\tau - \tau_{2i})^2 & \text{if } d = 2k, \\ \delta(\tau - \alpha)\prod_{i=1}^{k}(\tau - \tau_{2i})^2 + \gamma(\beta - \tau)\prod_{i=1}^{k}(\tau - \tau_{2i-1})^2 & \text{if } d = 2k+1, \end{cases}$$

where $\gamma, \delta > 0$ and $\alpha \leq \tau_1 \leq \tau_2 \leq \ldots \leq \tau_{d-1} \leq \beta$.

Remark: See Chapter II, Section 10 of [**KS66**].

2. Let $K_+[0, +\infty) \subset \mathbb{R}^{d+1}$ be the set of all polynomials $p(\tau)$ of degree at most d such that $p(\tau) \geq 0$ for all $\tau \geq 0$. Prove that $K_+[0, +\infty)$ is a closed convex cone with a compact base and that the polynomials that span the extreme rays of $K_+[0, +\infty)$ are

$$p(\tau) = \delta \prod_{i=1}^{k}(\tau - \tau_i)^2 \quad (2k \leq d) \quad \text{and} \quad p(\tau) = \delta\tau \prod_{i=1}^{k}(\tau - \tau_i)^2 \quad (2k + 1 \leq d),$$

where $\delta > 0$ and $\tau_i \geq 0$ for $i = 1, \ldots, k$.

Deduce that every polynomial p which is non-negative on $[0, +\infty)$ can be represented in the form

$$p(\tau) = \tau \sum_{i \in I}^{m} q_i^2(\tau) + \sum_{j \in J} q_j^2(\tau),$$

where q_i and q_j are polynomials with all roots real and non-negative.

3. Let $K_+(-\infty, +\infty) \subset \mathbb{R}^{d+1}$ be the set of all polynomials $p(\tau)$ of degree at most d such that $p(\tau) \geq 0$ for all $\tau \in \mathbb{R}$. Prove that $K_+(-\infty, +\infty)$ is a closed

convex cone with a compact base and that the polynomials that span the extreme rays of $K_+(-\infty, +\infty)$ are

$$p(\tau) = \delta \prod_{i=1}^{k} (\tau - \tau_i)^2, \quad 2k \leq d,$$

where $\delta > 0$.

Deduce that every polynomial p which is non-negative on $(-\infty, +\infty)$ can be represented in the form

$$p(\tau) = \sum_{i \in I} q_i^2(\tau),$$

where q_i are polynomials with all real roots.

4. Let $d = 2$. Draw a picture of the cone of the quadratic polynomials $a\tau^2 + b\tau + c$ in $\mathbb{R}^3 = \{(a, b, c) : \ a, b, c \in \mathbb{R}\}$ that are non-negative on $(-\infty, +\infty)$ in the axes a, b and c, describe the sections of the cone by the planes $a = 0$, $b = 0$ and $c = 0$ and find a compact base of the cone.

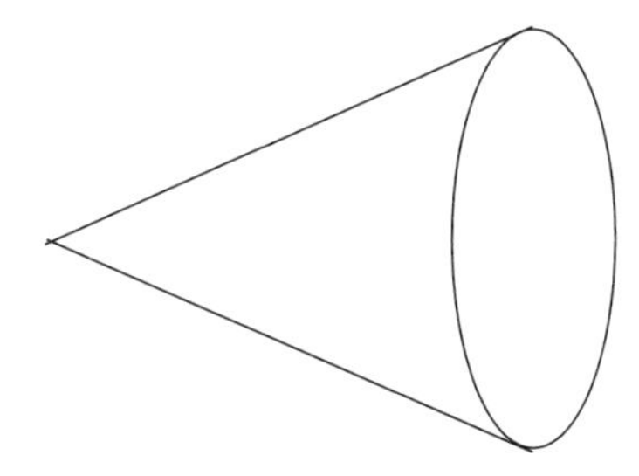

Figure 21. The cone of non-negative polynomials $a\tau^2 + b\tau + c$ (a general view)

5*. Prove that every polynomial $p(\tau)$ of degree $2k$ such that $p(\tau) \geq 0$ for all $\tau \in \mathbb{R}$ admits a unique representation of the form

$$p(\tau) = \delta \prod_{i=1}^{k} (\tau - \tau_i)^2 + \gamma \prod_{i=1}^{k-1} (\tau - \tau_i')^2,$$

where $\delta, \gamma > 0$ and $\tau_1 < \tau_1' < \tau_2 < \tau_2' < \ldots < \tau_{k-1} < \tau_{k-1}' < \tau_k$.

In particular, every non-negative polynomial is the sum of only two squares of polynomials.

Remark: See Chapter VI, Section 8 of [**KS66**].

For Problems 6–8 see Section 6.3 of [**BC98**] and [**R00**].

6* (Hilbert's Theorem). Let $p(x, y, x)$ be a homogeneous polynomial of degree 4 in three real variables such that $p(x, y, z) \geq 0$ for all real x, y and z. Prove that p can be written as a sum of squares of quadratic forms in x, y and z.

7* (T. Motzkin). Let

$$p(x,y,z) = x^4y^2 + x^2y^4 + z^6 - 3x^2y^2z^2.$$

Prove that $p(x,y,z) \geq 0$ for all real x, y and z and yet p cannot be written as a sum of squares of polynomials.

8* (M.-D. Choi and T.-Y. Lam). Let

$$p(x,y,z,w) = w^4 + x^2y^2 + y^2z^2 + z^2x^2 - 4xyzw.$$

Prove that $p(x,y,z,w) \geq 0$ for all real x, y, z and w and yet p cannot be written as a sum of squares of polynomials.

12. The Cone of Positive Semidefinite Matrices

The cone of positive semidefinite matrices studied in this section is arguably the most important of all non-polyhedral cones whose facial structure we completely understand. It is the central object in semidefinite programming (see Section IV.10) and various questions regarding the moment cone and the cone of non-negative polynomials can be reduced to positive semidefiniteness of certain matrices (see Section IV.2).

As usual, we review some linear algebra first.

(12.1) The space of symmetric matrices. An $n \times n$ matrix $A = (a_{ij})$ is called *symmetric*, provided $a_{ij} = a_{ji}$ for $i, j = 1, \ldots, n$. We identify the vector space Sym_n of all $n \times n$ symmetric matrices A with the Euclidean space $\mathbb{R}^d$, where $d = n(n+1)/2$. Let A be an $n \times n$ symmetric matrix and let U be an $n \times n$ orthogonal matrix (that is, $U^t = U^{-1}$). Then $U^{-1}AU$ is a symmetric matrix. For every symmetric $n \times n$ matrix A there is an orthogonal matrix U such that $U^{-1}AU$ is a diagonal matrix, having the eigenvalues of A on the diagonal. The number of non-zero eigenvalues is equal to the rank of A.

The scalar product of two symmetric matrices $A = (a_{ij})$ and $B = (b_{ij})$ is defined as

$$\langle A, B \rangle = \sum_{i,j=1}^{n} a_{ij} b_{ij}.$$

An important formula for the scalar product is

$$\langle A, B \rangle = \mathrm{tr}(AB) = \mathrm{tr}(BA),$$

where tr is the trace, that is, the sum of all diagonal entries. In particular, it follows that if U is an $n \times n$ orthogonal matrix, then

$$\langle U^{-1}AU,\ U^{-1}BU \rangle = \langle A, B \rangle,$$

since

$$\langle U^{-1}AU,\ U^{-1}BU \rangle = \mathrm{tr}(U^{-1}AUU^{-1}BU) = \mathrm{tr}(U^{-1}ABU) = \mathrm{tr}(AB) = \langle A, B \rangle$$

(we use that $\operatorname{tr}(C) = \operatorname{tr}(U^{-1}CU)$ for any invertible U). With a symmetric matrix $A = (a_{ij})$ we associate the quadratic form $q_A : \mathbb{R}^n \longrightarrow \mathbb{R}$:

$$q_A(x) = \sum_{i,j=1}^{n} a_{ij}\xi_i\xi_j,$$

where $x = (\xi_1, \ldots, \xi_n) \in \mathbb{R}^n$. If $x = (\xi_1, \ldots, \xi_n)$, let us denote by $x \otimes x$ the $n \times n$ symmetric matrix $X = (x_{ij})$ where $x_{ij} = \xi_i\xi_j$. Then one can write

$$q_A(x) = \langle A, x \otimes x\rangle.$$

PROBLEMS.

1°. Prove that the dimension of the space of symmetric $n \times n$ matrices is indeed $n(n+1)/2$.

2°. Prove that $\sum_{i,j=1}^{n} a_{ij}b_{ij} = \operatorname{tr}(AB)$, where $A = (a_{ij})$ and $B = (b_{ij})$ are symmetric $n \times n$ matrices.

3°. Check that $U^t AU$ is a symmetric matrix, provided U is any $n \times n$ matrix and A is an $n \times n$ symmetric matrix.

Now we define the cone we are interested in.

(12.2) Positive semidefinite matrices. An $n \times n$ symmetric matrix A is called *positive semidefinite* provided $q_A(x) \geq 0$ for all $x \in \mathbb{R}^n$. An $n \times n$ symmetric matrix A is called *positive definite* provided A is positive semidefinite and $q_A(x) = 0$ only if $x = 0$. We denote the set of all positive semidefinite $n \times n$ symmetric matrices by $\mathcal{S}_+$.

A symmetric matrix is positive semidefinite if and only if all eigenvalues of A are non-negative and positive definite if and only if all eigenvalues are positive. In particular, if A is a positive (semi)definite matrix and U is an orthogonal matrix, then $U^{-1}AU$ is a positive (semi)definite matrix. In other words, for any orthogonal matrix U, the linear transformation $X \longmapsto U^{-1}XU$ of Sym_n maps the set $\mathcal{S}_+$ onto itself. If A is positive semidefinite, then all diagonal entries are non-negative and $a_{ij}^2 \leq a_{ii}a_{jj}$ for every pair $1 \leq i \neq j \leq n$.

We recall that X is positive semidefinite of $\operatorname{rank} X \leq 1$ if and only if X can be written as $X = x \otimes x$ for some $x \in \mathbb{R}^n$.

PROBLEMS.

1°. Prove that $\mathcal{S}_+ \subset \operatorname{Sym}_n$ is a closed convex cone which does not contain straight lines.

2°. Prove that A is an interior point of $\mathcal{S}_+$ if and only if A is positive definite.

3. Prove that for every two points $x, y \in \operatorname{int} \mathcal{S}_+$ there exists a non-degenerate linear transformation T of Sym_n, such that $T(\mathcal{S}_+) = \mathcal{S}_+$ and $T(x) = y$. In other words, the cone $\mathcal{S}_+$ is *homogeneous*. Prove that the cone $\mathbb{R}^d_+ = \Big\{(\xi_1, \ldots, \xi_d) : \xi_i \geq 0$ for $i = 1, \ldots, d\Big\}$ is also homogeneous.

4. Prove that $B = \{A \in \mathcal{S}_+ : \operatorname{tr}(A) = 1\}$ is a compact base of $\mathcal{S}_+$.

5. The space Sym_2 is identified with $\mathbb{R}^3$. Draw a picture of the cone of positive semidefinite 2×2 matrices.

6°. Check that if A is a positive semidefinite 2×2 matrix, then $a_{11}, a_{22} \geq 0$ and $a_{12}^2 \leq a_{11}a_{22}$.

We arrive at the main result of this section.

(12.3) Proposition. *Let A be an $n \times n$ positive semidefinite matrix. Suppose that $\operatorname{rank} A = r$. If $r = n$, then A is an interior point of $\mathcal{S}_+$. If $r < n$, then A is an interior point of a face $\mathcal{F}$ of $\mathcal{S}_+$, where $\dim \mathcal{F} = r(r+1)/2$. There is a rank-preserving isometry identifying the face $\mathcal{F}$ with the cone of positive semidefinite $r \times r$ matrices.*

Proof. If $\operatorname{rank} A = n$, then A is positive definite, so the result follows by Problem 2, Section 12.2.

Suppose that $\operatorname{rank} A = r < n$. We will construct a hyperplane $H \subset \operatorname{Sym}_n$ which contains A and isolates $\mathcal{S}_+$. Let $\lambda_1, \ldots, \lambda_r > 0$ be the non-zero eigenvalues of A. Let us find an orthogonal matrix U, such that $U^{-1}AU = D$, where $D = \operatorname{diag}(\lambda_1, \ldots, \lambda_r, 0, \ldots, 0)$. Let $C = \operatorname{diag}(0, \ldots, 0, 1, \ldots, 1)$ be the matrix whose first r diagonal entries are 0 and the last $n - r$ diagonal entries are 1's, and let $Q = UCU^{-1}$.

Figure 22. The structure of matrices D and C

Then Q is a non-zero positive semidefinite matrix and

$$\langle Q, A \rangle = \langle UCU^{-1}, UDU^{-1} \rangle = \langle C, D \rangle = 0.$$

Furthermore, for any positive semidefinite $n \times n$ matrix X, the matrix $Y = U^{-1}XU$ is positive semidefinite and

$$\langle Q, X \rangle = \langle UCU^{-1}, UYU^{-1} \rangle = \langle C, Y \rangle \geq 0,$$

since the diagonal entries of Y must be non-negative. Therefore, the hyperplane $H = \{X \in \mathrm{Sym}_n : \langle Q, X\rangle = 0\}$ isolates $\mathcal{S}_+$ and contains A. Let us describe the corresponding face

$$\mathcal{F} = \{X \in \mathcal{S}_+ : \quad \langle Q, X\rangle = 0\}.$$

The map $X \longmapsto Y = U^{-1}XU$ is a non-degenerate linear transformation which maps $\mathcal{S}_+$ onto itself, maps Q onto C and A onto D. Then the face $\mathcal{F}$ is mapped onto a face $\mathcal{F}'$, containing D and consisting of all positive semidefinite matrices Y such that $\langle Y, C\rangle = 0$:

$$\mathcal{F}' = \{Y \in \mathcal{S}_+ : \quad \langle Y, C\rangle = 0\}.$$

Clearly, Y must have the last $n - r$ diagonal entries equal to zero. Since Y is positive semidefinite, all entries in the last $n - r$ rows and last $n - r$ columns must be 0 (see Section 12.2). The upper left $r \times r$ submatrix of Y can be an arbitrary positive semidefinite matrix.

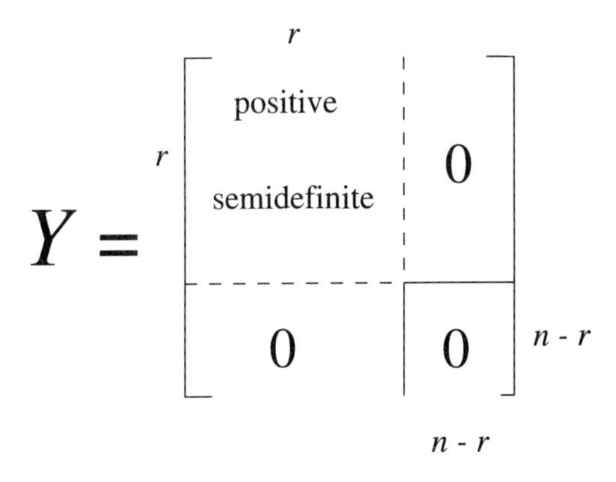

Figure 23. The structure of matrix Y

Thus the face $\mathcal{F}'$ may be identified with the cone of all $r \times r$ positive semidefinite matrices (in particular, $\dim \mathcal{F}' = (r + 1)r/2$) and it is seen that $\mathcal{F}'$ contains D in its interior. Since $Y \longmapsto X = UYU^{-1}$ is a non-degenerate linear transformation, which maps D onto A and $\mathcal{F}'$ onto $\mathcal{F}$, we conclude that $\dim \mathcal{F} = r(r + 1)/2$ and $\mathcal{F}$ contains A in its interior. □

PROBLEMS.

1. Prove that the dimensions of faces $\mathcal{F}$ of $\mathcal{S}_+$ are $0, 1, 3, \ldots, r(r + 1)/2, \ldots$. Prove that if $\mathcal{F}$ is a face of $\mathcal{S}$ and $\dim \mathcal{F} = r(r + 1)/2$, then there is a matrix $A \in \operatorname{int} \mathcal{F}$ such that $\operatorname{rank} A = r$.

2. Let $\mathcal{S}_+$ be the cone of $n \times n$ positive semidefinite matrices, let $\mathcal{F} \subset \mathcal{S}_+$ be a face and let r be a positive integer such that $\dim \mathcal{F} < r(r+1)/2 \leq n(n+1)/2$. Prove that there is a face $\mathcal{F}'$ of $\mathcal{S}_+$ such that $\mathcal{F}$ is a face of $\mathcal{F}'$ and $\dim \mathcal{F}' = r(r + 1)/2$.

3. Let us choose positive integers $0 < r < n$, let $\mathcal{S}_1$ be the cone of positive semidefinite $n \times n$ matrices and let $\mathcal{S}_2$ be the cone of positive semidefinite $r \times r$

matrices. Let $\mathcal{F} \subset \mathcal{S}_1$ be a face such that $\dim \mathcal{F} = r(r+1)/2$. Construct an isometry (that is, a distance-preserving map) $\mathcal{S}_2 \longrightarrow \mathcal{F}$.

4. Prove that $A \in \mathcal{S}_+$ spans an extreme ray of $\mathcal{S}_+$ if and only if $\operatorname{rank} A = 1$.

Using Proposition 12.3, we get the following nice description of the facial structure of the cone of positive semidefinite matrices $\mathcal{S}_+$.

(12.4) Corollary. *The faces of $\mathcal{S}_+ \subset \operatorname{Sym}_n$ are parameterized by the subspaces of $\mathbb{R}^n$. For a subspace $L \subset \mathbb{R}^n$, let*

$$\mathcal{F}_L = \Big\{ Y \in \mathcal{S}_+ : \ L \subset \ker Y \Big\}.$$

Then $\mathcal{F}_L$ is a face of $\mathcal{S}_+$ and $\dim \mathcal{F}_L = r(r+1)/2$, where $r = \operatorname{codim} L$. As L ranges over all subspaces of codimension r, $\mathcal{F}_L$ ranges over all faces of dimension $r(r+1)/2$.

Proof. Given a subspace L of codimension r, let us choose the coordinates so that $L = \big\{(0, \ldots, 0, \xi_{r+1}, \ldots, \xi_n)\big\}$. Then $\mathcal{F}_L$ consists of the matrices Y depicted in Figure 23. The supporting hyperplane for $\mathcal{F}_L$ is $H = \big\{X : \langle C, X \rangle = 0\big\}$, where C is depicted in Figure 22. If $\mathcal{F}$ is a face of $\mathcal{S}_+$, then $\mathcal{F} = \mathcal{F}_L$, where $L = \ker A$ and A is a matrix in the interior of $\mathcal{F}$. □

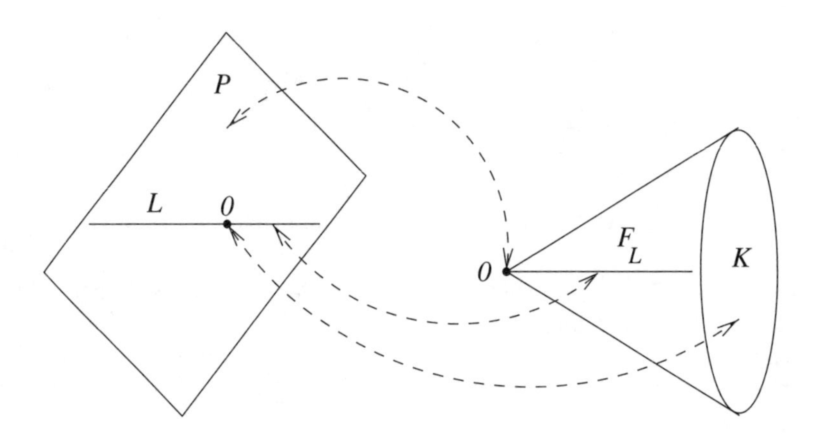

Figure 24. The correspondence between subspaces of $\mathbb{R}^n$ and faces of $\mathcal{S}_+$ for $n = 2$

PROBLEM.

1. Let L_1 and L_2 be subspaces of $\mathbb{R}^n$. Prove that $\mathcal{F}_{L_1}$ is a face of $\mathcal{F}_{L_2}$ if and only if $L_2 \subset L_1$.

Problem 1 asserts that the *face lattice* of the cone of $n \times n$ positive semidefinite matrices is (anti)isomorphic to the lattice of all subspaces of $\mathbb{R}^n$.

13. Linear Equations in Positive Semidefinite Matrices

In the next three sections we discuss various applications of the results of Section 12. We will be dealing with systems of linear equations in matrices and it is convenient to adopt some notation. To express that X is positive semidefinite, instead of writing $X \in \mathcal{S}_+$, we write $X \succeq 0$. To express that X is positive definite, we write $X \succ 0$. Proposition 12.3 has an interesting implication: if a system of linear equations in positive semidefinite matrices has a solution, it has a solution of a small rank.

(13.1) Proposition. *Let $\mathcal{A} \subset \mathrm{Sym}_n$ be an affine subspace such that the intersection $\mathcal{S}_+ \cap \mathcal{A}$ is non-empty and $\operatorname{codim} \mathcal{A} < (r+2)(r+1)/2$ for some non-negative integer r. Then there is a matrix $X \in \mathcal{S}_+ \cap \mathcal{A}$ such that $\operatorname{rank} X \leq r$.*

Equivalently, let us fix k symmetric $n \times n$ matrices $A_1, \dots, A_k$ and k real numbers $\alpha_1, \dots, \alpha_k$. If there is a matrix $X \succeq 0$ such that

$$\langle A_i, X \rangle = \alpha_i \quad \textit{for} \quad i = 1, \dots, k,$$

then there is a matrix $X_0 \succeq 0$ such that

$$\langle A_i, X_0 \rangle = \alpha_i \quad \textit{for} \quad i = 1, \dots, k$$

and, additionally

$$\operatorname{rank} X_0 \leq \left\lfloor \frac{\sqrt{8k+1}-1}{2} \right\rfloor.$$

Proof. To see that the second statement is indeed equivalent to the first one, let

$$\mathcal{A} = \Big\{ X \in \mathrm{Sym}_n : \quad \langle A_i, X \rangle = \alpha_i \quad \text{for} \quad i = 1, \dots, k \Big\}$$

be the affine subspace of symmetric $n \times n$ matrices which satisfy the given k matrix equations. Then $\operatorname{codim} \mathcal{A} \leq k$ and $k < (r+2)(r+1)/2$ if and only if

$$r \leq \left\lfloor \frac{\sqrt{8k+1}-1}{2} \right\rfloor.$$

We prove the first statement. Let $\mathcal{K} = \mathcal{S}_+ \cap \mathcal{A}$. The set $\mathcal{K}$ is non-empty, closed and does not contain straight lines (cf. Problem 1 of Section 12.2). Therefore, by Lemma 3.5, $\mathcal{K}$ contains an extreme point X_0.

Suppose that $\operatorname{rank} X_0 = m$. Then, by Proposition 12.3, X_0 must be an interior point of a face $\mathcal{F}$ of $\mathcal{S}_+$ of dimension $m(m+1)/2$. We observe that X_0 is an interior point of the intersection $\mathcal{F} \cap \mathcal{A}$. Since X_0 is an extreme point, we must have $\dim(\mathcal{F} \cap \mathcal{A}) = 0$, which implies that $\operatorname{codim} \mathcal{A} > \dim \mathcal{F}$, so $\operatorname{codim} \mathcal{A} > m(m+1)/2$. Hence $m \leq r$ and the proof follows. $\square$

Here is an immediate consequence for systems of two matrix equations:

(13.2) Corollary. *Let us fix two symmetric $n \times n$ matrices $A = (a_{ij})$ and $B = (b_{ij})$ and two real numbers α and β. The system of two quadratic equations*

$$\sum_{i,j=1}^{n} a_{ij}\xi_i\xi_j = \alpha \quad \text{and} \quad \sum_{i,j=1}^{n} b_{ij}\xi_i\xi_j = \beta$$

has a real solution $x = (\xi_1, \ldots, \xi_n)$ if and only if the system of two linear matrix equations

$$\langle A, X\rangle = \alpha \quad \text{and} \quad \langle B, X\rangle = \beta$$

has a positive semidefinite solution $X \succeq 0$.

Proof. Let $x = (\xi_1, \ldots, \xi_n)$ be a solution to the system of quadratic equations. Let us define $X = (x_{ij})$ by $x_{ij} = \xi_i\xi_j$. Then X is a positive semidefinite matrix and $\langle A, X\rangle = \alpha$ and $\langle B, X\rangle = \beta$.

On the other hand, suppose there is a solution $X \succeq 0$ to the system of equations $\langle A, X\rangle = \alpha$ and $\langle B, X\rangle = \beta$. Proposition 13.1 implies that there is a solution such that $\operatorname{rank} X \leq 1$ (substitute $k = 2$ in the formula). Then $X = (x_{ij})$ can be written as $x_{ij} = \xi_i\xi_j$ for some set of numbers $\xi_1, \ldots, \xi_n$. Then $x = (\xi_1, \ldots, \xi_n)$ is a solution to the system of quadratic equations. $\square$

The following corollary is an example of a *hidden convexity* result: the image of a (possibly non-convex) set under a (possibly non-linear) map turns out to be convex with "no obvious reason". One (hidden) reason why this might happen is that the image in question coincides with the image of some convex set under some linear transformation.

(13.3) Corollary. *Let $q_1, q_2 : \mathbb{R}^n \longrightarrow \mathbb{R}$ be quadratic forms. Consider the map $\phi : \mathbb{R}^n \longrightarrow \mathbb{R}^2$, $\phi(x) = \big(q_1(x), q_2(x)\big)$. Then the image $\phi(\mathbb{R}^n)$ is a convex cone in $\mathbb{R}^2$.*

Proof. Let $A = (a_{ij})$ be the matrix of q_1 and let $B = (b_{ij})$ be the matrix of q_2, so

$$q_1(x) = \sum_{i,j=1}^{n} a_{ij}\xi_i\xi_j \quad \text{and} \quad q_2(x) = \sum_{i,j=1}^{n} b_{ij}\xi_i\xi_j$$

for $x = (\xi_1, \ldots, \xi_n)$. By Corollary 13.2, the image $\phi(\mathbb{R}^n)$ can be viewed as the image of the convex cone $\mathcal{S}_+$ of positive semidefinite matrices under the linear transformation $X \longmapsto \big(\langle A, X\rangle, \langle B, X\rangle\big)$ and hence is a convex cone. $\square$

This result is due to L.L. Dines (1941).

PROBLEMS.

1°. Construct an example of a system of three quadratic equations

$$\sum_{i,j=1}^{n} a_{ij}\xi_i\xi_j = \alpha, \quad \sum_{i,j=1}^{n} b_{ij}\xi_i\xi_j = \beta, \quad \sum_{i,j=1}^{n} c_{ij}\xi_i\xi_j = \gamma$$

which does not have a solution $(\xi_1, \ldots, \xi_n)$, but such that the corresponding system of linear matrix equations

$$\langle A, X \rangle = \alpha, \quad \langle B, X \rangle = \beta, \quad \langle C, X \rangle = \gamma$$

has a positive semidefinite solution $X \succeq 0$.

2°. Check that $X = (x_{ij})$ has the form $x_{ij} = \xi_i \xi_j$ if and only if $\operatorname{rank} X \leq 1$ and X is positive semidefinite.

3 (C-K. Li and B-S. Tam). Let $A_1, \ldots, A_k$ be $n \times n$ symmetric matrices and let $\alpha_1, \ldots, \alpha_k$ be real numbers. Let

$$\mathcal{K} = \Big\{ X \succeq 0 : \quad \langle A_i, X \rangle = \alpha_i, \ i = 1, \ldots, k \Big\}.$$

Suppose that $X \in \mathcal{K}$ and that $\operatorname{rank} X = r$. Let us decompose $X = QQ^t$, where Q is an $n \times r$ matrix of rank r. Prove that the dimension of the smallest face of $\mathcal{K}$ containing X is equal to the codimension of $\operatorname{span}\big(Q^t A_1 Q, \ldots, Q^t A_k Q\big)$ in the space of $r \times r$ symmetric matrices.

Remark: See Theorem 31.5.3 of [**DL97**].

4. Prove the following strengthening of Corollary 13.2. Let us fix two symmetric $n \times n$ matrices $A = (a_{ij})$ and $B = (b_{ij})$ and three real numbers α, β and γ. The system of two quadratic equations

$$\sum_{i,j=1}^{n} a_{ij} \xi_i \xi_j = \alpha \quad \text{and} \quad \sum_{i,j=1}^{n} b_{ij} \xi_i \xi_j = \beta$$

has a real solution $x = (\xi_1, \ldots, \xi_n)$ such that $\sum_{i=1}^n \xi_i^2 \leq \gamma$ if and only if the system of two linear matrix equations

$$\langle A, X \rangle = \alpha \quad \text{and} \quad \langle B, X \rangle = \beta$$

has a positive semidefinite solution $X \succeq 0$ such that $\operatorname{tr}(X) \leq \gamma$.

5. Prove the following strengthening of Corollary 13.3. Let $q_1, q_2 : \mathbb{R}^n \longrightarrow \mathbb{R}$ be quadratic forms. Consider the map $\phi : \mathbb{R}^n \longrightarrow \mathbb{R}^2$, $\phi(x) = \big(q_1(x), q_2(x)\big)$. Then the image $\phi(B)$ of the unit ball $B = \big\{x : \|x\| \leq 1\big\}$ is a compact convex set in $\mathbb{R}^2$.

In general, as we will see shortly, the bound of Proposition 13.1 is the best possible. However, there is one special case where it can be sharpened.

(13.4) Proposition. *Let $\mathcal{A} \subset \operatorname{Sym}_n$ be an affine subspace such that the intersection $\mathcal{S}_+ \cap \mathcal{A}$ is non-empty and bounded. Suppose that* $\operatorname{codim} \mathcal{A} = (r+2)(r+1)/2$ *for some positive $r > 0$ and that $n \geq r+2$. Then there is a matrix $X \in \mathcal{S}_+ \cap \mathcal{A}$ such that* $\operatorname{rank} X \leq r$.

Equivalently, for some $r > 0$, let us fix $k = (r+2)(r+1)/2$ symmetric $n \times n$ matrices $A_1, \ldots, A_k$ where $n \geq r+2$ and k real numbers $\alpha_1, \ldots, \alpha_k$. If there is a solution $X \succeq 0$ to the system

$$\langle A_i, X \rangle = \alpha_i \quad \textit{for} \quad i = 1, \ldots, k$$

and the set of all such solutions is bounded, then there is a matrix $X_0 \succeq 0$ *such that*

$$\langle A_i, X_0 \rangle = \alpha_i \quad \text{for} \quad i = 1, \ldots, k$$

and, additionally

$$\operatorname{rank} X_0 \leq r.$$

The proof requires some algebraic topology. Below, we explicitly state what we need.

(13.5) Topological Fact. *Let us consider the set* $\mathbb{RP}^{n-1}$ *of all straight lines* l *in* $\mathbb{R}^n$ *passing through the origin. We make* $\mathbb{RP}^{n-1}$ *a metric space by letting the distance* $d(l_1, l_2)$ *between two lines be the angle between* l_1 *and* l_2. *Let* $\mathbb{S}^{n-1} = \{x : \|x\| = 1\}$ *be the unit sphere in* $\mathbb{R}^n$. *Then, for* $n > 2$ *there is no continuous map* $\phi : \mathbb{S}^{n-1} \longrightarrow \mathbb{RP}^{n-1}$ *such that* $\phi(x) \neq \phi(y)$ *for every pair of distinct points* $x, y \in \mathbb{S}^{n-1}$.

The space $\mathbb{RP}^{n-1}$ is called the *projective space*. It is an $(n-1)$-dimensional compact connected manifold without boundary. The fact follows from the observation that $\mathbb{RP}^{n-1}$ and $\mathbb{S}^{n-1}$ are not homeomorphic for $n > 2$ and from the Invariance of Domain Theorem which implies that such an embedding $\phi : \mathbb{S}^{n-1} \longrightarrow \mathbb{RP}^{n-1}$ would have been a homeomorphism (see, for example, Chapter III, Section 6 of [**Ma80**]).

PROBLEM.

1. Construct a continuous map $\phi : \mathbb{S}^1 \longrightarrow \mathbb{RP}^1$ such that $\phi(x) \neq \phi(y)$ for every pair of distinct points $x, y \in \mathbb{S}^1$. That is, $\mathbb{S}^1$ and $\mathbb{RP}^1$ *are* homeomorphic.

Proposition 13.4 will be deduced from the following special case.

(13.6) Lemma. *Let* $r > 0$ *and let* $\mathcal{A} \subset \operatorname{Sym}_{r+2}$ *be an affine subspace such that* $\dim \mathcal{A} = r + 2$, *i.e.,* $\operatorname{codim} \mathcal{A} = (r+2)(r+1)/2$. *Suppose that the intersection* $\mathcal{A} \cap \mathcal{S}_+$ *is non-empty and bounded. Then there is a matrix* $X \in \mathcal{S}_+ \cap \mathcal{A}$ *such that* $\operatorname{rank} X \leq r$.

Proof. Suppose that $\mathcal{A} \cap \operatorname{int} \mathcal{S}_+ = \emptyset$. Problem 1 of Section 2.9 implies that $\mathcal{A}$ lies in the support hyperplane of a proper face $\mathcal{F}$ of $\mathcal{S}_+$. By Proposition 12.3, the face $\mathcal{F}$ can be identified with the cone of positive semidefinite matrices of a smaller rank $s < r + 2$. Now the result follows from Proposition 13.1

Hence we may assume that $\mathcal{A} \cap \operatorname{int} \mathcal{S}_+ \neq \emptyset$. Let $\mathcal{B} = \mathcal{A} \cap \mathcal{S}_+$. Then $\mathcal{B}$ is an $(r+2)$-dimensional convex compact set. We are going to prove that for some matrix $X_0 \in \partial \mathcal{B}$ one has $\operatorname{rank} X_0 \leq r$. Let us suppose this is not so and obtain a contradiction. For every $X \in \partial \mathcal{B}$ we must have $\operatorname{rank} X < r + 2$ (cf. Problem 2 of Section 12.2). Assuming that $\operatorname{rank} X > r$, we conclude that we must have $\operatorname{rank} X = r + 1$ for every $X \in \partial \mathcal{B}$. Therefore, for every $X \in \partial \mathcal{B}$, the set $\ker X$ is a straight line passing through the origin in $\mathbb{R}^{r+2}$. Let us construct a map $\phi : \mathbb{S}^{r+1} \longrightarrow \mathbb{RP}^{r+1}$ as follows. We consider $\mathbb{S}^{r+1}$ to be centered at a point $o \in \operatorname{int} \mathcal{B}$.

For every $y \in \mathbb{S}^{r+1}$, the ray $[o, y)$ intersects $\partial\mathcal{B}$ at a single matrix $X(y)$ (we use the fact that $\mathcal{B}$ is compact and Lemma 2.2). Let $\phi(y) = \ker X(y)$.

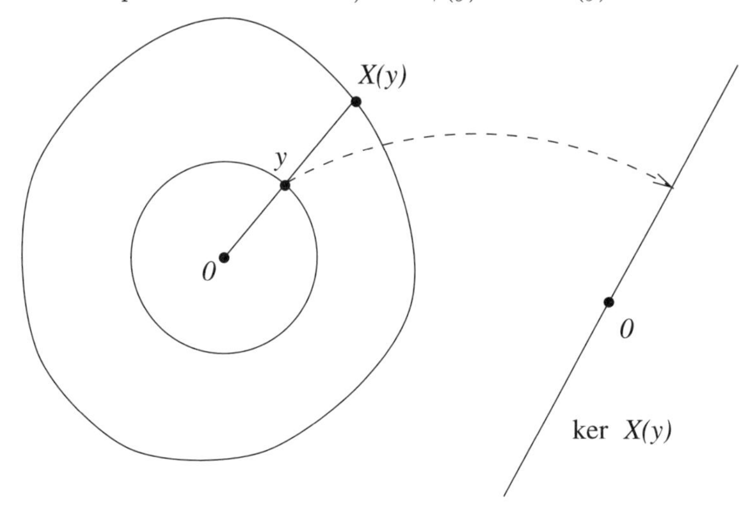

Figure 25. The map ϕ

Note that ϕ is a continuous map. Since $r > 0$, by using (13.5), we conclude that there must be two distinct points $y, z \in \mathbb{S}^{r+1}$ such that $\phi(x) = \phi(y)$. In other words, there will be two distinct matrices Y and Z in $\partial\mathcal{B}$ such that $\ker Y = \ker Z$. Using Corollary 12.4, we conclude that Y and Z lie in the same face $\mathcal{F}_L$ of $\mathcal{S}_+$, where $L \subset \mathbb{R}^{r+1}$ is a straight line. Hence, by Proposition 12.3 and Corollary 12.4, the interior of $\mathcal{F}_L$ consists of matrices of rank $r+1$. We draw the straight line (YZ) through Y and Z; see Figure 26.

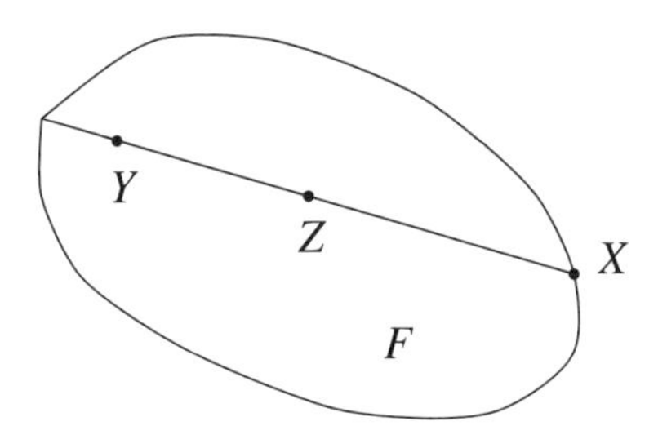

Figure 26

Since $\mathcal{A}$ is an affine subspace, $(YZ) \subset \mathcal{A}$. Since $\mathcal{B}$ is bounded, the line (YZ) intersects $\partial\mathcal{F}$ at some point X. We must have $\operatorname{rank} X \leq r$. Clearly, $X \in \mathcal{A} \cap \mathcal{S}_+$. $\square$

Now we are ready to prove Proposition 13.4.

Proof of Proposition 13.4. By Proposition 13.1, there is a matrix $Y \in \mathcal{A} \cap \mathcal{S}_+$ such that $\operatorname{rank} Y \leq r+1$. Let us choose a linear subspace $L \subset \mathbb{R}^n$ such that $L \subset \ker Y$ and $\operatorname{codim} L = r+2$ and let $\mathcal{F}_L$ be the corresponding face of $\mathcal{S}_+$ (see Corollary 12.4). Hence $Y \in \mathcal{F}_L$ and, therefore, $\mathcal{F}_L \cap \mathcal{A} \neq \emptyset$. Since there is a rank-preserving isometry between $\mathcal{F}_L$ and the cone of positive semidefinite $(r+2) \times (r+2)$ matrices, the proof follows by Lemma 13.6. □

PROBLEMS.

1. Show by examples that in Proposition 13.4 none of the conditions: $\mathcal{S}_+ \cap \mathcal{A}$ is bounded, $r > 0$ and $n \geq r+2$ can be dropped.

2°. Let $q_k : \mathbb{R}^n \longrightarrow \mathbb{R}$ be quadratic forms with matrices $A_k = (a_{ij,k})$, $k = 1, \ldots, m$, so

$$q_k(x) = \sum_{i,j=1}^{n} a_{ij,k} \xi_i \xi_j \quad \text{for} \quad x = (\xi_1, \ldots, \xi_n).$$

Let α_k: $k = 1, \ldots, m$ be real numbers. Suppose that the system of quadratic equations $q_k(x) = \alpha_k$ for $k = 1, \ldots, m$ has a solution $x \in \mathbb{R}^n$. Prove that there exists a positive semidefinite $n \times n$ matrix X such that $\langle A_k, X \rangle = \alpha_k$ for $k = 1, \ldots, m$ and such that $\operatorname{rank} X \leq 1$.

3. Let $A = (a_{ij}), B = (b_{ij})$ and $C = (c_{ij})$ be $n \times n$ symmetric matrices, where $n \geq 3$, and let α, β and γ be real numbers. Suppose that the system of linear matrix equations

$$\langle A, X \rangle = \alpha, \quad \langle B, X \rangle = \beta \quad \text{and} \quad \langle C, X \rangle = \gamma$$

has a positive semidefinite solution $X \succeq 0$. Suppose further that for some numbers τ_1, τ_2 and τ_3 the linear combination $\tau_1 A + \tau_2 B + \tau_3 C$ is (strictly) positive definite. Prove that there exists a solution $x \in \mathbb{R}^n$ to the system of quadratic equations

$$q_1(x) = \alpha, \quad q_2(x) = \beta \quad \text{and} \quad q_3(x) = \gamma,$$

where

$$q_1(x) = \sum_{i,j=1}^{n} a_{ij} \xi_i \xi_j, \quad q_2(x) = \sum_{i,j=1}^{n} b_{ij} \xi_i \xi_j \quad \text{and} \quad q_3(x) = \sum_{i,j=1}^{n} c_{ij} \xi_i \xi_j$$

$$\text{for} \quad x = (\xi_1, \ldots, \xi_n)$$

are the corresponding quadratic forms.

The last problem requires some probability theory.

4. Let $q_k : \mathbb{R}^n \longrightarrow \mathbb{R}$ be quadratic forms with matrices A_k and let α_k be real numbers as in Problem 2 above. Suppose that there exists a positive semidefinite matrix X such that $\langle A_k, X \rangle = \alpha_k$ for $k = 1, \ldots, m$. Let T be a matrix such that $TT^* = X$ and let us consider a probability distribution of a vector y in $\mathbb{R}^n$ such that $\mathbf{E}\,(y) = 0$ and $\mathbf{E}\,(y \otimes y) = I$, where I is the identity matrix. Let $x = Ty$, so x is a random variable. Prove that $\mathbf{E}\,\big(q_k(x)\big) = \alpha_k$ for $k = 1, \ldots, m$.

Remark: Thus the existence of a positive semidefinite solution X to a system of linear matrix equations $\langle A_k, X\rangle = \alpha_k$ can be interpreted as the existence of a probability measure for the vector x of variables in the corresponding system $q_k(x) = \alpha_k$ of quadratic equations, such that the expected value of every quadratic form $q_k(x)$ is equal to the right-hand side α_k. This observation gives rise to a method of finding an approximate solution to the system of quadratic equations: an appropriate probability measure in $\mathbb{R}^n$ is constructed and a vector x is sampled at random. This is the idea of *randomized rounding*; see [**MR95**]. We develop this method in Sections V.5-6.

14. Applications: Quadratic Convexity Theorems

In this section, we continue our study of hidden quadratic convexity results initiated by Corollary 13.3. The following result was proved by L. Brickman [**Br61**]. Brickman's original proof is sketched in Problem 6 below. We use Proposition 13.4 as our main tool.

(14.1) Theorem. *Let $n > 2$ and let $\mathbb{S}^{n-1} = \{x \in \mathbb{R}^n : \|x\| = 1\}$ be the unit sphere. Let $q_1, q_2 : \mathbb{R}^n \longrightarrow \mathbb{R}$ be quadratic forms and let $\phi : \mathbb{R}^n \longrightarrow \mathbb{R}^2$ be the corresponding quadratic map, $\phi(x) = (q_1(x), q_2(x))$. Then the image $\phi(\mathbb{S}^{n-1}) \subset \mathbb{R}^2$ is a convex set.*

Proof. Let $A = (a_{ij})$ be the matrix of q_1 and let $B = (b_{ij})$ be the matrix of q_2, so

$$q_1(x) = \sum_{i,j=1}^{n} a_{ij}\xi_i\xi_j \quad \text{and} \quad q_2(x) = \sum_{i,j=1}^{n} b_{ij}\xi_i\xi_j \quad \text{for} \quad x = (\xi_1, \ldots, \xi_n).$$

Let $\mathcal{B} = \{X \succeq 0 : \operatorname{tr}(X) = 1\}$. Clearly, $\mathcal{B} \subset \operatorname{Sym}_n$ is a convex set. Let us define a linear transformation $\psi : \operatorname{Sym}_n \longrightarrow \mathbb{R}^2$ by $\psi(X) = (\langle A, X\rangle, \langle B, X\rangle)$. Clearly, $\psi(\mathcal{B}) \subset \mathbb{R}^2$ is a convex set.

We claim that $\phi(\mathbb{S}^{n-1}) = \psi(\mathcal{B})$. Indeed, if $(\alpha, \beta) \in \phi(\mathbb{S}^{n-1})$, then there is a vector $x \in \mathbb{R}^n$ such that $q_1(x) = \alpha$, $q_2(x) = \beta$ and $\|x\| = 1$. Let us define $X = (x_{ij})$ by $x_{ij} = \xi_i\xi_j$ for $i, j = 1, \ldots, n$. Then $X \in \mathcal{B}$ and $\psi(X) = (\alpha, \beta)$. Conversely, let $(\alpha, \beta) \in \psi(\mathcal{B})$ be a point. Then there exists an $X \succeq 0$ such that

$$\langle A, X\rangle = \alpha, \quad \langle B, X\rangle = \beta \quad \text{and} \quad \operatorname{tr}(X) = 1. \tag{14.1.1}$$

We observe that the set of solutions $X \succeq 0$ to the system (14.1.1) of three linear equations is non-empty and bounded (cf. Problem 4, Section 12.2). Applying Proposition 13.4 with $r = 1$, we conclude that there is a matrix $X_0 \succeq 0$ satisfying (14.1.1) and such that $\operatorname{rank} X_0 \leq 1$. Such a matrix can be written in the form $X_0 = (x_{ij})$, $x_{ij} = \xi_i\xi_j$ for some vector $x = (\xi_1, \ldots, \xi_n)$. We have $\|x\| = \operatorname{tr}(X_0) = 1$, $q_1(x) = \alpha$ and $q_2(x) = \beta$. Hence $(\alpha, \beta) \in \phi(\mathbb{S}^{n-1})$. □

PROBLEMS.

1°. Show that Theorem 14.1 does not hold for $n = 2$.

2. Show that for any $n \geq 1$ one can find three quadratic forms $q_1, q_2, q_3 : \mathbb{R}^n \longrightarrow \mathbb{R}$ such that if $\phi : \mathbb{R}^n \longrightarrow \mathbb{R}^3$ is the corresponding quadratic map $x \longmapsto \big(q_1(x), q_2(x), q_3(x)\big)$, then the image $\phi(\mathbb{S}^{n-1}) \subset \mathbb{R}^3$ is not convex.

3. Deduce Corollary 13.3 from Theorem 14.1.

4. Deduce the result of Problem 5, Section 13.3 from Theorem 14.1.

5. Let $q_1, q_2, q_3 : \mathbb{R}^n \longrightarrow \mathbb{R}$, $n \geq 3$, be quadratic forms and let $\phi : \mathbb{R}^n \longrightarrow \mathbb{R}^3$ be the corresponding quadratic map, $\phi(x) = \big(q_1(x), q_2(x), q_3(x)\big)$. Suppose that for some numbers α_1, α_2 and α_3, the form $q = \alpha_1 q_1 + \alpha_2 q_2 + \alpha_3 q_3$ is (strictly) positive definite. Prove that the image $\phi(\mathbb{R}^n)$ is a convex cone in $\mathbb{R}^3$.

6. Find a different proof of Theorem 14.1 along the following lines. First, show that it would suffice to prove Theorem 14.1 for $n = 3$. Next, observe that to prove that $\phi(\mathbb{S}^2)$ is convex, it suffices to prove that the intersection of $\phi(\mathbb{S}^2)$ with every straight line $l \subset \mathbb{R}^2$, $l = \big\{(\xi, \eta) : \alpha\xi + \beta\eta = \gamma\big\}$, is connected. Now, let $q = \alpha q_1 + \beta q_2$ and prove that $\phi(\mathbb{S}^2) \cap l$ is the image of the set $\big\{x \in \mathbb{S}^2 : q(x) = \gamma\big\}$, which consists of at most two connected components (circles) symmetric about the origin.

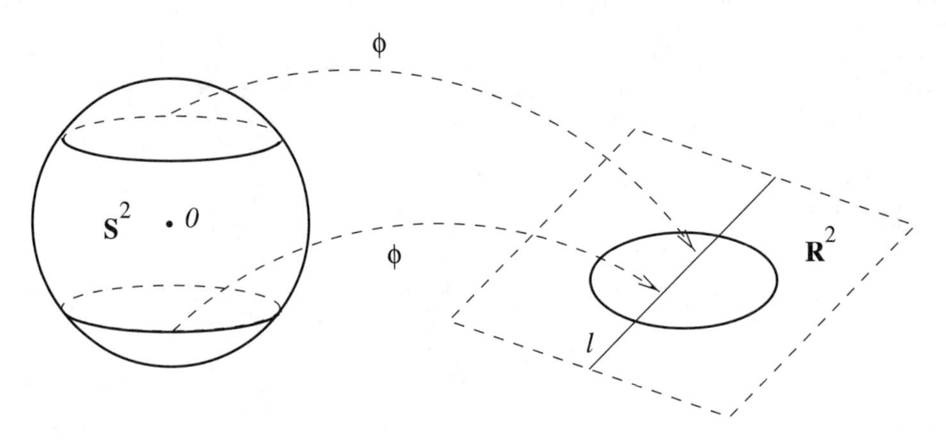

Figure 27

Given an $n \times n$ complex matrix $A = (a_{ij})$, the set $R(A) \subset \mathbb{C}$ in the complex plane

$$R(A) = \Big\{ \sum_{i,j=1}^{n} a_{ij}\zeta_i\overline{\zeta}_j : \quad |\zeta_1|^2 + \ldots + |\zeta_n|^2 = 1 \Big\}$$

is called the *numerical range* of A. Results of O. Toeplitz (1918) and F. Hausdorff (1919) establish convexity of the numerical range.

(14.2) Corollary (Toeplitz-Hausdorff Theorem). *The numerical range of a matrix is a convex set in the complex plane.*

Proof. For $n = 1$ the numerical range is just a point $a_{11} \in \mathbb{C}$. For $n > 1$, the set of n-tuples $(\zeta_1, \ldots, \zeta_n)$ of complex numbers $\zeta_k = \xi_k + i\eta_k$, $k = 1, \ldots, n$, such that $|\zeta_1|^2 + \ldots + |\zeta_n|^2 = 1$ can be identified with the $(2n-1)$-dimensional sphere

$$\mathbb{S}^{2n-1} = \Big\{(\xi_1, \eta_1, \xi_2, \eta_2, \ldots, \xi_n, \eta_n) : \quad \sum_{k=1}^{n} \xi_k^2 + \sum_{k=1}^{n} \eta_k^2 = 1\Big\}.$$

The proof now follows from Theorem 14.1 since the numerical range can be viewed as the image of $\mathbb{S}^{2n-1}$ under a quadratic map $\phi : \mathbb{S}^{2n-1} \longrightarrow \mathbb{R}^2 = \mathbb{C}$. □

A theory parallel to that of Sections 12–13 for real symmetric matrices can be developed for complex Hermitian matrices and even for quaternionic Hermitian matrices. The corresponding results in the complex case are sketched in the problems below.

PROBLEMS.

1°. An $n \times n$ complex matrix $A = (a_{ij})$ is called *Hermitian* provided $a_{ij} = \overline{a_{ji}}$ for all $1 \leq i, j \leq n$. Prove that all $n \times n$ Hermitian matrices form a real vector space Her_n of dimension n^2 with the scalar product $\langle A, B \rangle = \sum_{i,j=1}^{n} a_{ij}\overline{b_{ij}}$. Prove that $\langle U^*AU, U^*BU \rangle = \langle A, B \rangle$ for any two Hermitian matrices A and B and any unitary matrix U, where * denotes the conjugate matrix.

2. A Hermitian matrix $A = (a_{ij})$ is called *positive semidefinite* provided $\sum_{i,j=1}^{n} a_{ij}\zeta_i\overline{\zeta_j} \geq 0$ for all n-tuples $z = (\zeta_1, \ldots, \zeta_n)$ of complex numbers. Let $\mathcal{H}_+ \subset \text{Her}_n$ be the set of all positive semidefinite Hermitian $n \times n$ matrices. Prove that $\mathcal{H}_+$ is a closed convex n^2-dimensional cone with a compact base consisting of positive semidefinite matrices of trace 1. Draw a picture of the base of the cone for $n = 2$ (it is a 3-dimensional object).

3. Let A be an $n \times n$ positive semidefinite Hermitian matrix. Suppose that $\text{rank}\, A = r$. Prove that A is an interior point of a face $\mathcal{F}$ of $\mathcal{H}_+$, where $\dim \mathcal{F} = r^2$. Prove that there is a rank-preserving isometry identifying face $\mathcal{F}$ with the cone of positive semidefinite $r \times r$ Hermitian matrices. Prove that the faces of $\mathcal{H}_+$ are parameterized by complex linear subspaces $L \subset \mathbb{C}^n$:

$$\mathcal{F}_L = \big\{A \in \mathcal{H}_+ : \ L \subset \ker A\big\}.$$

4. Let us fix a number $r \geq 0$, a number $k \leq r^2 + 2r$ and a number $n \geq r$. Let $A_1, \ldots, A_k$ be $n \times n$ Hermitian matrices and let $\alpha_1, \ldots, \alpha_k$ be real numbers. Suppose that there is a positive semidefinite solution $X \in \text{Her}_n$ to the system of equations $\langle A_i, X \rangle = \alpha_i$ for $i = 1, \ldots, k$. Prove that there is a positive semidefinite solution X_0 to the above system such that $\text{rank}\, X_0 \leq r$.

5. Let us fix a number $r \geq 1$, let $k = (r+1)^2$ and let $n \geq r+2$. Let $A_1, \ldots, A_k$ be $n \times n$ Hermitian matrices and let $\alpha_1, \ldots, \alpha_k$ be real numbers. Suppose that there is a positive semidefinite solution $X \in \text{Her}_n$ to the system of equations $\langle A_i, X \rangle = \alpha_i$ for $i = 1, \ldots, k$ and that the set of all such solutions is bounded. Prove that there is a positive semidefinite solution X_0 to the above system such that $\text{rank}\, X_0 \leq r$.

6. Let $A=(a_{ij})$, $B=(b_{ij})$ and $C=(c_{ij})$ be $n \times n$ complex Hermitian matrices, where $n \geq 3$ and let $q_1, q_2, q_3 : \mathbb{C}^n \longrightarrow \mathbb{R}$ be the corresponding Hermitian forms:

$$q_1(z) = \sum_{i,j=1}^{n} a_{ij}\zeta_i\overline{\zeta_j}, \quad q_2(z) = \sum_{i,j=1}^{n} b_{ij}\zeta_i\overline{\zeta_j} \quad \text{and} \quad q_3(z) = \sum_{i,j=1}^{n} c_{ij}\zeta_i\overline{\zeta_j}$$

$$\text{for} \quad z = (\zeta_1, \ldots, \zeta_n).$$

Let $\phi : \mathbb{C}^n \longrightarrow \mathbb{R}^3$ be the map $\phi(z) = \big(q_1(z), q_2(z), q_3(z)\big)$ and let

$$\mathbb{S}^{2n-1} = \Big\{(\zeta_1, \ldots, \zeta_n) : \quad \sum_{i=1}^{n} |\zeta_i|^2 = 1\Big\}$$

be the unit sphere. Prove that the image $\phi(\mathbb{S}^{2n-1})$ is a convex set in $\mathbb{R}^3$.

Finally, let us describe the convex hull of a general quadratic image of the sphere. The following result was obtained by Y.H. Au-Yeung and Y.T. Poon [**AP79**].

(14.3) Theorem. *Let us fix a number $r \geq 1$, a number $k < (r+2)(r+1)/2$ and a number $n \geq r+2$. Let $q_1, \ldots, q_k : \mathbb{R}^n \longrightarrow \mathbb{R}$ be quadratic forms and let $\phi : \mathbb{R}^n \longrightarrow \mathbb{R}^k$ be the corresponding quadratic map, $\phi(x) = \big(q_1(x), \ldots, q_k(x)\big)$. Let $\mathbb{S}^{n-1} = \big\{x \in \mathbb{R}^n : \|x\| = 1\big\}$ be the unit sphere. Then every point of* $\operatorname{conv}\big(\phi(\mathbb{S}^{n-1})\big)$ *can be represented as a convex combination of r (not necessarily distinct) points from $\phi(\mathbb{S}^{n-1})$.*

Proof. The proof is parallel to that of Theorem 14.1. Let $y = (\eta_1, \ldots, \eta_k)$ be a point from the convex hull of $\phi(\mathbb{S}^{n-1})$. Hence we can write $y = \alpha_1\phi(x_1) + \ldots + \alpha_m\phi(x_m)$ for some points $x_1, \ldots, x_m \in \mathbb{S}^{n-1}$ and some non-negative α_i such that $\sum_{i=1}^m \alpha_i = 1$. Let A_i be the matrix of q_i for $i = 1, \ldots, k$ and let

$$X = \sum_{i=1}^{m} \alpha_i (x_i \otimes x_i),$$

where $x \otimes x$ is the matrix $(\xi_i\xi_j)$ for $x = (\xi_1, \ldots, \xi_n)$. Then $X \succeq 0$ and

$$\text{(14.3.1)} \qquad \langle A_i, X\rangle = \eta_i \quad \text{for} \quad i = 1, \ldots, k \quad \text{and} \quad \operatorname{tr}(X) = 1.$$

Hence the set of all positive semidefinite matrices X satisfying (14.3.1) is non-empty and bounded. Applying Proposition 13.4, we conclude that there exists a solution $X_0 \succeq 0$ of (14.3.1) such that $\operatorname{rank} X_0 \leq r$. Such a matrix X_0 can be decomposed as $X_0 = \sum_{i=1}^r \beta_i(u_i \otimes u_i)$, where $u_i \in \mathbb{S}^{n-1}$, β_i are non-negative numbers for $i = 1, \ldots, r$ and $\sum_{i=1}^r \beta_i = 1$. It follows from (14.3.1) that

$$y = \sum_{i=1}^{r} \beta_i\phi(u_i)$$

and the proof follows. $\square$

PROBLEMS.

1°. Deduce Theorem 14.1 from Theorem 14.3.

2. Let $A = (a_{ij})$, $B = (b_{ij})$, $C = (c_{ij})$, $D = (d_{ij})$ and $E = (e_{ij})$ be $n \times n$ real symmetric matrices. Consider the quadratic map $\phi : \mathbb{C}^n \longrightarrow \mathbb{R}^5$, where

$$(\zeta_1, \ldots, \zeta_n) \longmapsto \Big(\sum_{i,j=1}^{n} a_{ij}\zeta_i\overline{\zeta_j}, \sum_{i,j=1}^{n} b_{ij}\zeta_i\overline{\zeta_j}, \sum_{i,j=1}^{n} c_{ij}\zeta_i\overline{\zeta_j}, \sum_{i,j=1}^{n} d_{ij}\zeta_i\overline{\zeta_j}, \sum_{i,j=1}^{n} e_{ij}\zeta_i\overline{\zeta_j} \Big).$$

Let

$$\mathbb{S}^{2n-1} = \Big\{ (\zeta_1, \ldots, \zeta_n) \in \mathbb{C}^n : \quad \sum_{k=1}^{n} |\zeta_k|^2 = 1 \Big\}$$

be the sphere. Prove that the image $\phi(\mathbb{S}^{2n-1})$ is a convex set in $\mathbb{R}^5$.

3. Let us fix a number $r \geq 1$. Let $\mathbb{S}^{n-1} \subset \mathbb{R}^n$ be the unit sphere, $n \geq r+2$. Let us fix a Borel measure μ in $\mathbb{S}^{n-1}$ such that $\mu(\mathbb{S}^{n-1}) < \infty$ and a subspace L in the space of quadratic forms $q : \mathbb{R}^n \longrightarrow \mathbb{R}$ such that $\dim L \leq (r+1)(r+2)/2 - 1$. Prove that there exist r points $x_1, \ldots, x_r \in \mathbb{S}^{n-1}$ and r non-negative numbers $\lambda_1, \ldots, \lambda_r$ such that

$$\int_{\mathbb{S}^{n-1}} f \, d\mu = \sum_{i=1}^{r} \lambda_i f(x_i) \quad \text{for any} \quad f \in L.$$

4°. Let $q_1, \ldots, q_k : \mathbb{R}^n \longrightarrow \mathbb{R}$ be quadratic forms whose matrices are diagonal. Let $\phi : \mathbb{R}^n \longrightarrow \mathbb{R}^k$ be the corresponding quadratic map. Prove that $\phi(\mathbb{S}^{n-1})$ is a convex set in $\mathbb{R}^k$.

5*. Let us call a symmetric matrix $A = (a_{ij})$ *r-diagonal* if $a_{ij} = 0$ unless $|i-j| < r$. Let $q_1, \ldots, q_k : \mathbb{R}^n \longrightarrow \mathbb{R}$ be quadratic forms whose matrices are r-diagonal matrices and let $\phi : \mathbb{R}^n \longrightarrow \mathbb{R}^k$ be the corresponding quadratic map, $\phi(x) = \big(q_1(x), \ldots, q_k(x)\big)$. Let $B = \big\{x \in \mathbb{R}^n : \|x\| \leq 1\big\}$ be the unit ball in $\mathbb{R}^n$. Prove that every point from $\operatorname{conv}\big(\phi(B)\big)$ is a convex combination of some r points from $\phi(B)$.

Hint: Cf. Problem 3 of Section IV.10.3.

6*. In Problem 5 above, is it true that every point from $\operatorname{conv}\big(\phi(\mathbb{S}^{n-1})\big)$ is a convex combination of some r points from $\phi(\mathbb{S}^{n-1})$ if n is sufficiently large?

7. Let $q_1, \ldots, q_k : \mathbb{C}^n \longrightarrow \mathbb{R}$ be Hermitian forms whose matrices are real and 2-diagonal. Let $\phi : \mathbb{C}^n \longrightarrow \mathbb{R}^k$ be the corresponding quadratic map and let $B = \Big\{(\zeta_1, \ldots, \zeta_n) : \sum_{i=1}^{n} |\zeta_i|^2 \leq 1\Big\}$ be the unit ball in $\mathbb{C}^n$. Deduce from Problem 4 that the image $\phi(B)$ is a convex set in $\mathbb{R}^k$.

8*. Prove the following result of S. Friedland and R. Loewy [**FL76**], which is essentially equivalent to Proposition 13.4: suppose that $2 \leq r \leq n-1$. Let $L \subset \operatorname{Sym}_n$ be a subspace in the space of symmetric matrices such that $\dim L \geq (r-1)(2n-r+2)/2$. Then L contains a non-zero matrix whose largest eigenvalue is at least of multiplicity r.

9*. Prove the following result of F. Bohnenblust (see [**FL76**]), which is essentially equivalent to Proposition 13.4. Suppose that $r > 0$ and that $n \geq r+2$.

Let L be a subspace in the space of quadratic forms $q : \mathbb{R}^n \longrightarrow \mathbb{R}$ such that $\dim L < (r+2)(r+1)/2 - 1$. Suppose that the following condition is satisfied: whenever for some vectors $x_1, \ldots, x_r \in \mathbb{R}^n$ one has

$$\sum_{i=1}^{r} q(x_i) = 0 \quad \text{for all} \quad q \in L,$$

one must have $x_i = 0$ for $i = 1, \ldots, r$. Then L contains a positive definite form.

15. Applications: Problems of Graph Realizability

We discuss further applications of our results regarding linear equations in positive semidefinite matrices. They allow us to visualize the rank restrictions of Propositions 13.1 and 13.4.

Once again, we review some linear algebra first.

(15.1) Some linear algebra: Gram matrices. Let $v_1, \ldots, v_n$ be vectors in $\mathbb{R}^d$. Let us define an $n \times n$ matrix $X = (x_{ij})$ by $x_{ij} = \langle v_i, v_j \rangle$ (we consider the usual scalar product in $\mathbb{R}^d$). Matrix X is called the *Gram matrix* of vectors $v_1, \ldots, v_n$. It is known that X is positive semidefinite and that $\operatorname{rank} X \leq d$; in fact, $\operatorname{rank} X$ is the dimension of $\operatorname{span}(v_1, \ldots, v_n)$. Conversely, if X is a positive definite matrix such that $\operatorname{rank} X \leq d$, then X is the Gram matrix of some n vectors $v_1, \ldots, v_n$ in $\mathbb{R}^d$: $x_{ij} = \langle v_i, v_j \rangle$.

Now we state the problem.

(15.2) The graph realization problem. Suppose we are given an (undirected) weighted graph $G = (V, E; \rho)$, where $V = \{v_1, \ldots, v_n\}$ is the set of vertices, E is the set of edges and $\rho : E \longrightarrow \mathbb{R}_+$ is a function, which assigns to every edge $(i,j) \in E$ a non-negative number ("length") ρ_{ij}. We say that G is *d-realizable* if one can place the vertices $v_1, \ldots, v_n$ in $\mathbb{R}^d$ in such a way that $\|v_i - v_j\| = \rho_{ij}$ for every edge $(i,j) \in E$. We say that G is *realizable* if it is d-realizable for some d. The most intuitive case is 3-realizability: some problems of robotics ("linkages") and computational chemistry ("molecules") lead to problems of 3-realizability of certain graphs; see [**CH88**]. Questions of d-realizability for the smallest possible d turn out to be relevant to problems in statistics, archaeology, genetics and geography. We will consider this problem again in Section V.6.

PROBLEMS.

Some classical problems of the field:

1* (The "cycloheptane problem"). Prove that one can place seven points $v_1, \ldots, v_7$ in $\mathbb{R}^3$ in such a way that $\|v_1 - v_2\| = \|v_2 - v_3\| = \ldots = \|v_6 - v_7\| = \|v_7 - v_1\| = 1$ and $\|v_1 - v_3\| = \|v_2 - v_4\| = \ldots = \|v_5 - v_7\| = \|v_1 - v_6\| = \|v_2 - v_7\| = \sqrt{8/3}$.

Remark: The constants 1 and $\sqrt{8/3}$ are chosen in such a way that the angles between consecutive intervals v_{i-1}, v_i and v_i, v_{i+1} are equal to $\arccos(-1/3)$, that is, to the angle between two intervals connecting vertices of a regular tetrahedron with its center; see Figure 28.

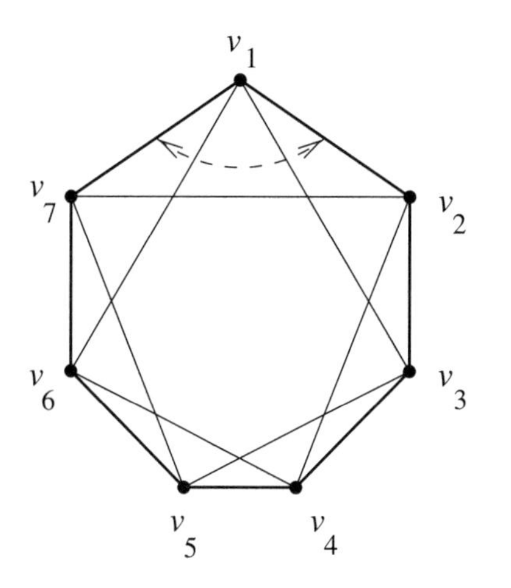

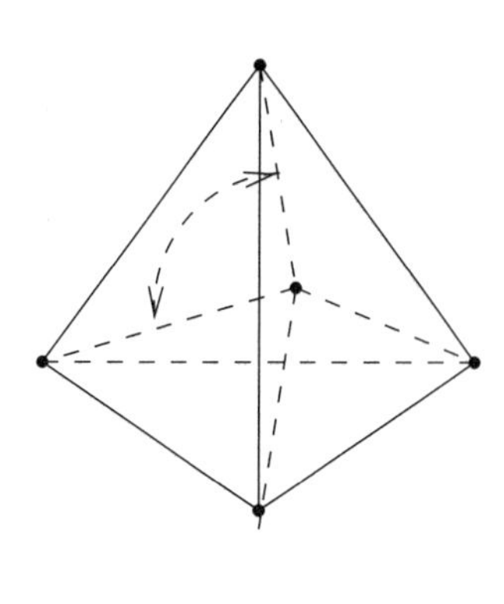

Figure 28

2*. Prove that every such configuration of seven points in $\mathbb{R}^3$ has one degree of freedom (modulo rigid motions).

Remark: Apparently, it is very difficult to prove that, modulo rigid motions, there are exactly two connected components in the configuration space.

3* (The "cyclohexane problem"). Prove that one can place six points $v_1, \ldots, v_6$ in $\mathbb{R}^3$ in such a way that $\|v_1 - v_2\| = \|v_2 - v_3\| = \ldots = \|v_5 - v_6\| = \|v_6 - v_1\| = 1$ and $\|v_1 - v_3\| = \|v_2 - v_4\| = \ldots = \|v_4 - v_6\| = \|v_1 - v_5\| = \|v_2 - v_6\| = \sqrt{8/3}$.

4*. Prove that there are two such configurations of six points, one of which has one degree of freedom (modulo rigid motions) and the other is rigid.

5*. Consider the X set of all configurations of five points $v_1, \ldots, v_5$ in $\mathbb{R}^2$ such that $\|v_1 - v_2\| = \|v_2 - v_3\| = \|v_3 - v_4\| = \|v_4 - v_5\| = \|v_5 - v_1\| = 1$. Thus X can be viewed as a subset of $\left(\mathbb{R}^2\right)^5 = \mathbb{R}^{10}$. We observe that if $x \in X$ and g is an orientation-preserving isometry of $\mathbb{R}^2$, then $g(x) \in X$. Let Y be the factor space of X modulo all orientation-preserving isometries of $\mathbb{R}^2$. Prove that Y is a 2-dimensional manifold homeomorphic to the sphere with four handles.

An easy problem.

6°. Prove that if a weighted graph with n vertices is realizable, it is $(n-1)$-realizable.

Let us relate the problem of graph realizability to linear equations in positive semidefinite matrices.

(15.3) A straightforward reformulation. Let $v_1, \ldots, v_n$ be a realization of the graph in $\mathbb{R}^d$, and let $X = (x_{ij})$, $x_{ij} = \langle v_i, v_j \rangle$ be the Gram matrix of $v_1, \ldots, v_n$. Then $X \succeq 0$ and for any edge $e = (i, j)$ of G, we have

$$\rho_{ij}^2 = \|v_i - v_j\|^2 = \langle v_i, v_i \rangle - 2\langle v_i, v_j \rangle + \langle v_j, v_j \rangle = x_{ii} - 2x_{ij} + x_{jj}.$$

Hence we conclude that the problem of realizability of G is equivalent to the following problem:

Is there an $n \times n$ matrix $X = (x_{ij})$ such that $X \succeq 0$ and

$$x_{ii} - 2x_{ij} + x_{jj} = \rho_{ij}^2$$

for every edge $(i, j) \in E$?

The problem of d-realizability is equivalent to the above problem with one additional constraint:

$$\operatorname{rank} X \leq d.$$

It turns out that if d is large enough, realizability is equivalent to d-realizability.

(15.4) Proposition. *Suppose that the number k of edges of G satisfies the inequality $k < (d+2)(d+1)/2$. Then G is d-realizable if and only if it is realizable. In particular, if $k \leq 9$, the graph is realizable if and only if it is 3-realizable.*

Proof. By Proposition 13.1, if the system of linear matrix equations

$$x_{ii} - 2x_{ij} + x_{jj} = \rho_{ij}^2 : \quad (i, j) \in E$$

has a positive semidefinite solution $X = (x_{ij})$, it has a positive semidefinite solution X_0, such that additionally, $\operatorname{rank} X_0 \leq d$. □

PROBLEMS.

1. Let G be the complete graph with $d+2$ vertices (and $(d+2)(d+1)/2$ edges) such that the length of every edge (v_i, v_j) is 1. Prove that G is realizable but not d-realizable.

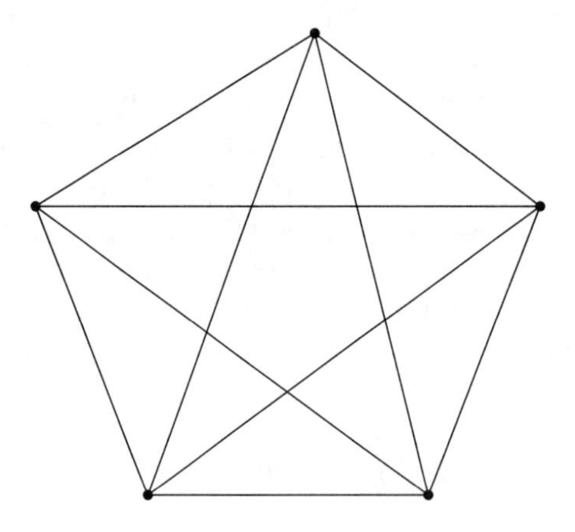

Figure 29. The complete graph with five vertices. If every edge is to have the unit length, the graph is 4-realizable but not 3-realizable.

2. Suppose that G is a cycle $v_1 - v_2 - \ldots - v_n - v_1$ with some weights on the edges. Prove that G is realizable if and only if it is 2-realizable.

3*. Suppose that G has n vertices $v_1, \ldots, v_n$ and $2n$ edges: $(v_1, v_2), (v_2, v_3), \ldots, (v_{n-1}, v_n), (v_n, v_1)$ and $(v_1, v_3), (v_2, v_4), \ldots, (v_{n-2}, v_n), (v_{n-1}, v_1), (v_n, v_2)$ with some weights on the edges. Is it true that if G is realizable, it is 4-realizable?

4* (M. Bakonyi and C.R. Johnson). A graph $G = (V, E)$ is called *chordal* provided for any cycle $v_1 - v_2 - \ldots - v_k - v_1$, where $k \geq 4$, there is a chord $v_i - v_j$, where $1 \leq i < j \leq k$. A subset $K \subset V$ is called a *clique* if every two vertices from K are connected by an edge. Prove that G is chordal if and only if it has the following property:

for any choice of weights on the edges, the graph is realizable provided every clique is realizable.

Remark: See Section 31.4 of [**DL97**].

5. Suppose that we want to place six points $v_1, \ldots, v_6$ in $\mathbb{R}^d$ with prescribed distances $\|v_1 - v_2\|$, $\|v_2 - v_3\|$, $\|v_3 - v_4\|$, $\|v_4 - v_5\|$, $\|v_5 - v_6\|$ and $\|v_6 - v_1\|$ and prescribed angles between the pairs of opposite edges: (v_1, v_2) and (v_4, v_5), (v_2, v_3) and (v_5, v_6) and (v_3, v_4) and (v_6, v_1); see Figure 30.

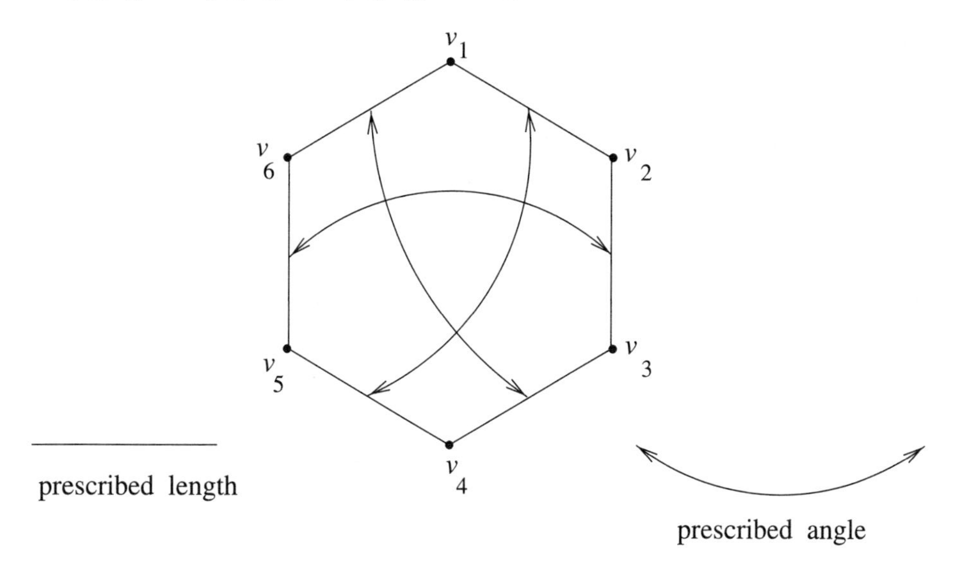

Figure 30

Prove that if such a placement exists for some d, it exists for $d = 3$.

Problem 1 of Section 15.4 shows that the bounds of Proposition 15.4 (and hence the bounds of Proposition 13.1) are the best possible. It turns out, however, that the graph of Problem 1 is the only graph with $(d + 2)(d + 1)/2$ edges which is realizable, but not d-realizable. To see that, we need to refine our reduction of the realizability question to systems of linear equations in positive semidefinite matrices.

(15.5) The economical reformulation. Let $v_1, \ldots, v_n$ be a realization of the graph $G = (V, E)$ with n vertices in $\mathbb{R}^d$. We can always assume that $v_n = 0$. Let $X = (x_{ij})$ be the $(n-1) \times (n-1)$ Gram matrix of the vectors $v_1, \ldots, v_{n-1}$. Then $X \succeq 0$ and we have the following affine constraints:

$$x_{ii} = \rho_{in}^2 \quad \text{if } (i,n) \text{ is an edge} \quad \text{and}$$

(15.5.1)

$$x_{ii} - 2x_{ij} + x_{jj} = \rho_{ij}^2 \quad \text{if } (i,j) \text{ is an edge and } 1 \leq i,j \leq n-1.$$

Hence we conclude that the problem of realizability of G is equivalent to the following problem:

is there an $(n-1) \times (n-1)$ matrix $X = (x_{ij})$ such that $X \succeq 0$ and X satisfies (15.5.1)?

The problem of d-realizability is equivalent to the above problem with one additional constraint:

$$\operatorname{rank} X \leq d.$$

(15.6) Proposition. *Suppose that G has $k = (d+2)(d+1)/2$ edges and that G is not a union of a complete graph with $d+2$ vertices and 0 or more isolated vertices. Then G is d-realizable if and only if it is realizable.*

Proof. Since G is d-realizable if and only if its connected components are realizable, without loss of generality we may assume that G is connected. Since G is not a complete graph, we must have $n \geq d+3$, and so $n-1 \geq d+2$. Now we use Proposition 13.4. We claim that the set of positive semidefinite solutions X to the system 15.5.1 is bounded. Indeed, since G is connected, each vertex v_i can be connected to v_n by a path. Since v_n is fixed at 0, the length of $\|v_i\| = \sqrt{x_{ii}}$ is bounded by the length of the path. In particular, $\sqrt{x_{ii}}$ is bounded by the sum of all ρ_{ij}. Therefore, the set of feasible matrices $X \succeq 0$ is bounded (cf. Section 12.2). Hence the proof follows by Proposition 13.4. □

PROBLEMS.

1. Let G be a graph with $n \geq d+2$ vertices consisting of a complete graph with $d+2$ vertices and $n-d-2$ isolated vertices. Let us assign length 1 to every edge. Prove that G is realizable, but not d-realizable. Letting $n = d+2$, deduce that the condition $n \geq d+2$ in Proposition 13.4 cannot be dropped. Letting $n > d+2$, deduce that the condition of boundedness in Proposition 13.4 cannot be dropped either.

2. Given a graph $G = (V, E)$ with $|V| = n$ vertices and $k = |E|$ edges and a number d, let us consider the following *rigidity map* $\phi_G : \mathbb{R}^{nd} \longrightarrow \mathbb{R}^k$. A point x from $\mathbb{R}^{nd} = \mathbb{R}^d \times \ldots \times \mathbb{R}^d$ is interpreted as an n-tuple of d vectors $(v_1, \ldots, v_n)$

and $\phi(x)$ is the k-tuple of squared distances $\|v_i - v_j\|^2$, where (i,j) runs over all edges of G. Let $\mathbb{S}^{nd-1}$ be the unit sphere in $\mathbb{R}^{nd}$. Prove that if $n \geq d+3$ and $k < (d+2)(d+1)/2$, then the image $\phi(\mathbb{S}^{nd-1})$ is a convex set in $\mathbb{R}^k$.

3. In Problem 5 of Section 15.4, suppose we want to place $v_1, \ldots, v_6$ so that additionally we have $\sum_{i=1}^{6} \|v_i\|^2 = 1$. Prove that such a placement exists if and only if it exists for $d = 3$.

16. Closed Convex Sets

We conclude the chapter by discussing some general structural properties of closed convex sets in $\mathbb{R}^d$.

Let us define a *ray* as a set R of points of the type

$$R = \{v + \tau u : \quad \tau \geq 0\},$$

where v and u are given points in $\mathbb{R}^d$ and $u \neq 0$. We say that R *emanates from* v *in the direction of* u. In Section 8, we define rays as emanating from the origin only. Our current extended definition should not lead to confusion since the starting point of any ray under consideration will be clear from the context.

We show that if a closed convex set contains a ray, the ray "replicates" itself all over the set. Our proof works equally well for infinite-dimensional spaces.

(16.1) Lemma. *Let $A \subset \mathbb{R}^d$ be a closed convex set which contains a ray. Then there exists a closed convex cone $K \subset \mathbb{R}^d$, called the recession cone of A, such that for every point $a \in A$, the union of all rays that emanate from a and are contained in A is the translation $a + K$.*

Proof. Without loss of generality, we assume that $0 \in A$. First, we prove that the union K of all rays that emanate from 0 and are contained in A, if non-empty, is a closed convex cone in $\mathbb{R}^d$. Indeed, suppose that R_1 and R_2 are two rays emanating from 0 and contained in A. Let us choose a point $x \in R_1$ and a point $y \in R_2$. Hence $\tau x \in R_1 \subset A$ and $\tau y \in R_2 \subset A$ for all $\tau \geq 0$. Let $z = \alpha x + \beta y$ for some $\alpha, \beta \geq 0$. If $z = 0$, then obviously $z \in K$. If $z \neq 0$, then let $\gamma = \alpha + \beta > 0$ and for any $\tau \geq 0$ we have that $\tau z = \alpha(\tau x) + \beta(\tau y) = (\alpha\gamma^{-1})(\gamma\tau x) + (\beta\gamma^{-1})(\gamma\tau y)$ is a convex combination of points from A. Hence $\tau z \in A$ and the ray emanating from the origin in the direction of z is contained in A. Hence K is a convex cone (we did not use yet that A is closed).

Let us prove that K is closed, or, equivalently, that the complement $\mathbb{R}^d \setminus K$ is open. Let $u \in \mathbb{R}^d \setminus K$ be a point. The ray in the direction of u is not contained in A and hence there is a point $w = \tau u$ for some $\tau > 0$ such that $w \notin A$. Since A is closed, there is a neighborhood W of w such that $W \cap A = \emptyset$. Then no point x in the neighborhood $U = \tau^{-1} W$ of u belongs to the cone K. Thus we have proven that K is a closed convex cone.

Now we show that if for some $a \in A$ the ray emanating from a in some direction u is contained in A, then for any point $b \in A$, the ray emanating from b in the same direction u is also contained in A. As before, without loss of generality, we may assume that $a = 0$. Hence $\tau u \in A$ for all $\tau \geq 0$. Let us choose a $\tau \geq 0$ and an $0 < \epsilon < 1$. Then $(\epsilon^{-1}\tau)u \in A$ and since A is convex, the point

$$(1-\epsilon)b + \tau u = (1-\epsilon)b + \epsilon\big(\epsilon^{-1}\tau u\big)$$

is contained in A. Since A is closed, we conclude that $b + \tau u \in A$, which completes the proof. □

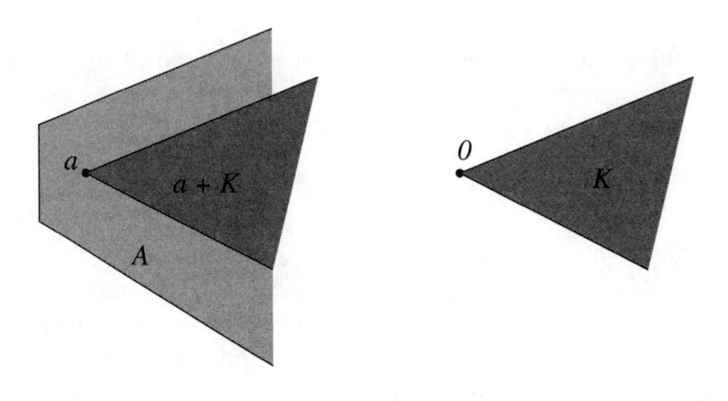

Figure 31. A set A and its recession cone K

If the set $A \subset \mathbb{R}^d$ does not contain rays, we say that its recession cone is $\{0\}$.

PROBLEMS.

1°. Let $A \subset \mathbb{R}^d$ be a closed convex set which contains a straight line. Prove that there exists a subspace $L \subset \mathbb{R}^d$ such that for every point $a \in A$, the union of all straight lines passing through a and contained in A is the affine subspace $a + L$.

2°. Let $A \subset \mathbb{R}^d$ be a closed convex set and let $a \in A$ be a point. Let us define the set K of *feasible directions* from a by $K = \big\{u \in \mathbb{R}^d : a + \epsilon u \in A \text{ for some } \epsilon > 0\big\}$. Prove that K is a convex cone. Does K have to be closed?

3°. Let

$$P = \big\{x \in \mathbb{R}^d : \quad \langle c_i, x \rangle \leq \beta_i \quad \text{for} \quad i = 1, \ldots, m\big\}$$

be a polyhedron.

Prove that the recession cone K of P is defined by

$$K = \big\{x \in \mathbb{R}^d : \quad \langle c_i, x \rangle \leq 0, \quad i = 1, \ldots, m\big\}.$$

Assuming that $0 \in A$, prove the union of all straight lines that pass through 0 and are contained in A, if non-empty, is the subspace

$$L = \big\{x \in \mathbb{R}^d : \quad \langle c_i, x \rangle = 0, \quad i = 1, \ldots, m\big\}.$$

4. Construct an example of a convex (but not closed) set $A \subset \mathbb{R}^2$ and two points $a, b \in A$ such that A contains a ray emanating from a in some direction u but does not contain the ray emanating from b in the same direction u.

5. Let $P \subset \mathbb{R}^d$ be a polyhedron, $P = \{x : \langle c_i, x \rangle \leq \beta_i, i = 1, \ldots, m\}$ for some vectors c_i and numbers β_i. Let $v \in P$ be a point and let $I = \{i : \langle a_i, v \rangle = \beta_i\}$ be the set of inequalities active on v. Prove that the cone K of feasible directions from v (see Problem 2) is defined by $K = \{u : \langle a_i, u \rangle \leq 0 \text{ for } i \in I\}$.

6. Let $A \subset \mathbb{R}^d$ be a closed convex set which does not contain rays. Prove that A is compact.

Now we prove that straight lines can be "factored out" from a closed convex set in Euclidean space.

(16.2) Lemma. *Let $A \subset \mathbb{R}^d$ be a closed convex set containing straight lines. Then there exists a subspace $L \subset \mathbb{R}^d$ such that for the orthogonal projection A' of A onto the orthogonal complement $L^\perp$ of L we have:*

1. *A' is a closed convex set which does not contain straight lines;*
2. *$A = A' + L$.*

Proof. Let us define L as the subspace of $\mathbb{R}^d$ such that for every point $a \in A$, the union of all straight lines passing through a and contained in A is $a + L$; see Problem 1 of Section 16.1. Let $pr : \mathbb{R}^d \longrightarrow L^\perp$ be the orthogonal projection onto $L^\perp$ so that $A' = pr(A)$. Thus for every $x \in A'$ we have $pr^{-1}(x) = x + L \subset A$.

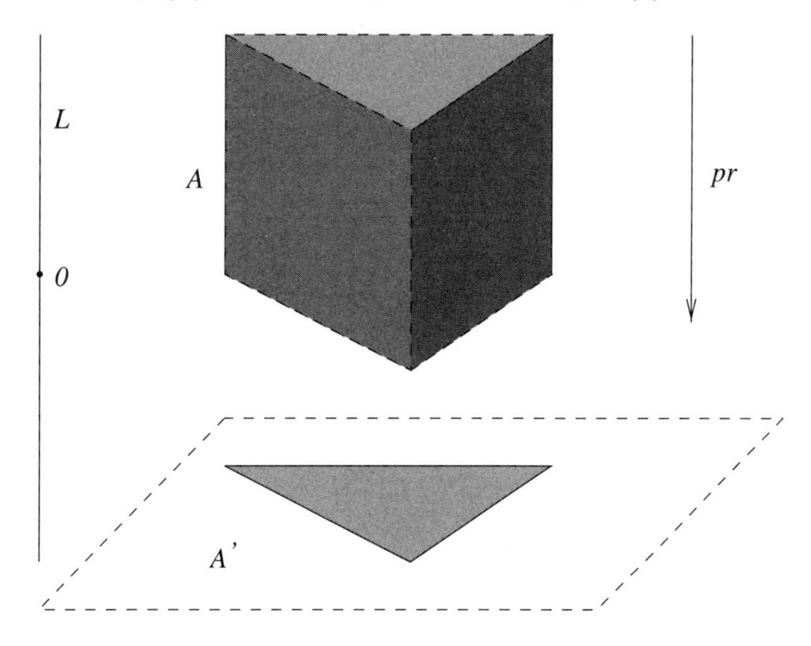

Figure 32

Clearly, A' is a convex set and $A = A' + L$. Moreover, A' does not contain straight lines for if $l \subset A'$ is a straight line, then $l + L \subset A$ is an affine subspace whose dimension is greater than that of L, which contradicts the definition of L. Finally, A' is closed, for if $\{x_n\}$ is a sequence in A' converging to a point x, then for any $u \in L$, $y_n = x_n + u$ is a sequence of points in A converging to $x + u$. Since A is closed, we have $x + u \in A$ and hence $x \in A'$. □

Last, a useful lemma.

(16.3) Lemma. *Let $A \subset \mathbb{R}^d$ be a closed convex set which does not contain straight lines. Then every point $x \in A$ can be written in the form $x = y + z$, where y is a convex combination of extreme points of A and z is a point from the recession cone K of A.*

Proof. We proceed by induction on d. The result is clear for $d = 1$. Suppose that $d > 1$. Without loss of generality we may assume that A has a non-empty interior (otherwise, we reduce the dimension by considering A in its affine hull). Let us choose a straight line L passing through x. The intersection $L \cap A$ is either a closed ray $a + \tau u, \tau \geq 0$, with the endpoint $a \in \partial A$ or a closed interval $[a, b]$ with $a, b \in \partial A$ (possibly with $a = b$).

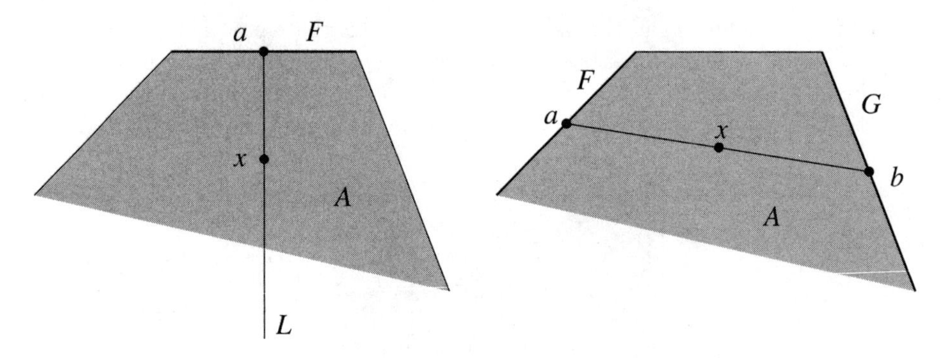

Figure 33

In the first case, let us choose a proper face F of A containing a (see Corollary 2.8). Hence $\dim F < d$ and by the induction hypothesis we can write $a = y + z'$, where y is a convex combination of extreme points of F and z' is in the recession cone of F. Since $x = a + \tau u$, we get $x = y + (z' + \tau u)$. Now we observe that y is a convex combination of extreme points of A (Problem 1 of Section 3.2) and $z = z' + \tau u$ is in the recession cone of A.

In the second case, we choose proper faces F containing a and G containing b. As above, we can write $a = y' + z'$, where y' is a convex combination of extreme points of F and z' is in the recession cone of A and $b = y'' + z''$, where y'' is a convex combination of extreme points of G and z'' is in the recession cone of A. Since x is a convex combination of a and b, the proof follows. □

17. Remarks

Our general reference for convexity in Euclidean space is [**W94**]. The Isolation Theorem (Theorem 1.6) is an algebraic form of the Hahn-Banach Theorem; see [**Bou87**] and [**Ru91**]. The author learned the proof of Theorem 1.6 from A.M. Vershik. Although topology seems to be completely "evicted" from our proof, it is present under cover. Indeed, if we declare a subset A of a vector space V open if it is a union of convex algebraically open sets, we make V a topological vector space with the convenient property that every linear functional $f : V \longrightarrow \mathbb{R}$ is continuous. Algebraically open and closed sets were used earlier by V. Klee [**Kl63**] to define the Euler characteristic in the abstract setting of a real vector space. The Birkhoff Polytope, the permutation polytopes, transportation polyhedra and many related polytopes (polyhedra) are discussed in detail in [**YKK84**] and [**BS96**]. The polytope of polystochastic matrices, also called the multiindex transportation polytope (see Problem 3 of Section 7.4) is one possible generalization of the Birkhoff Polytope. A.M. Vershik proposed a different generalization: one can think of the Birkhoff polytope as the convex hull of the matrices representing the action of the symmetric group S_n in $\mathbb{R}^n$ by permutations of the coordinates. Similarly, for any representation of the symmetric group (and any finite group for that matter), one can define a polytope that is the convex hull of the operators of the representation; see [**Barv92**]. For the Diet Problem, the Assignment Problem and the Transportation Problem (also called Min-Cost Problem), see [**PS98**]. For the Schur-Horn Theorem and many related topics consult [**MO79**]. The moment cone and the cone of univariate non-negative polynomials are thoroughly treated in [**KS66**]. The facial structure of the cone of positive semidefinite matrices is described, for example, in [**DL97**]. Our presentation of quadratic convexity results is based on some original papers: [**Br61**], [**AP79**], [**Da71**], [**Ve84**], [**FL76**], [**Barv95**], [**Barv01**] and [**DL97**]. The approach to "hidden convexity" based on supplanting the image of a (non-convex) set under a (non-linear) map by the image of a convex set under a linear map was demonstrated first in [**Li66**] and in the context of quadratic convexity in [**Da71**]. The problem of graph realizability is also known as the Euclidean matrix completion problem; see [**DL97**] for references and results. For problems of distance geometry related to graph realizability and their applications see [**CH88**] and [**H95**].

Chapter III

Convex Sets in Topological Vector Spaces

We extend our methods to study convex sets in topological vector spaces. We prove the Krein-Milman Theorem for locally convex topological vector spaces and explore the extreme points of some convex sets which can be considered as infinite-dimensional extensions of familiar Euclidean objects. In particular, we consider an L^∞-analogue of a polyhedron and a "simplex" of probability measures. Applications include problems of optimal control and probability and some "hidden convexity" results based on Lyapunov's Theorem. Our approach is geometric and, whenever possible, we stress similarities between finite- and infinite-dimensional situations. Exercises address some of the peculiar features of the infinite dimension: existence of dense hyperplanes, discontinuous linear functionals and disjoint convex sets that cannot be separated by a hyperplane.

1. Separation Theorems in Euclidean Space and Beyond

In this section, while being mostly in the Euclidean setting, we develop some general techniques that also work in infinite-dimensional situations. We prove separation theorems in $\mathbb{R}^d$ and discuss how they can be extended to general vector spaces. Later in this chapter, separation theorems will become our main tool to handle infinite-dimensional convex sets.

Let V be a vector space. Recall (see Definition II.1.4) that sets $A, B \subset V$ are separated by a hyperplane H if A and B belong to different closed halfspaces $\overline{H_+}$ and $\overline{H_-}$. Equivalently, the sets A and B are separated by a hyperplane if there exists a non-zero linear functional $f : V \longrightarrow \mathbb{R}$ and a number α such that $f(x) \leq \alpha$ for each $x \in A$ and $f(x) \geq \alpha$ for each $x \in B$. Sets $A, B \subset V$ are strictly separated by a hyperplane H if A and B belong to different open halfspaces H_+ and H_-. Equivalently, the sets A and B are strictly separated by a hyperplane if there exists a non-zero linear functional $f : V \longrightarrow \mathbb{R}$ and a number α such that $f(x) < \alpha$ for all $x \in A$ and $f(x) > \alpha$ for all $x \in B$.

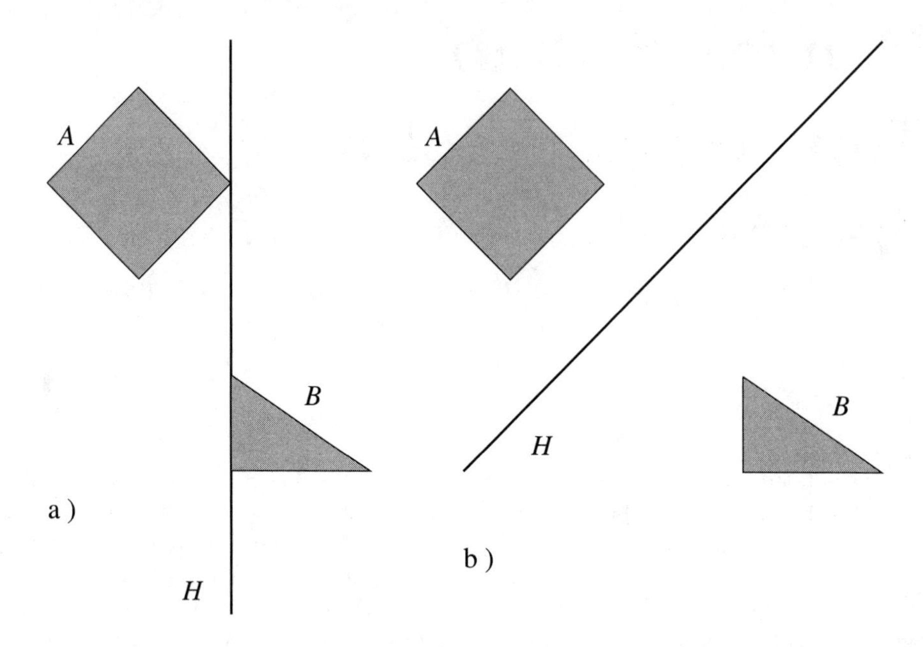

Figure 34. Example: a) A and B separated by H, b) A and B strictly separated by H

(1.1) Definition. Let V be a vector space and let $A, B \subset V$ be sets. We define the set $A - B \subset V$ by

$$A - B = \{x - y : \ x \in A, \ y \in B\}.$$

PROBLEM.

1°. Prove that if sets A and B are convex, then $A - B$ is convex as well.

Our first result is a corollary of Theorem II.2.9.

(1.2) Theorem. *Let A and B be non-empty convex sets in $\mathbb{R}^d$ such that $A \cap B = \emptyset$. Then there exists an affine hyperplane H which separates A and B.*

Proof. Let $C = A - B$. By Problem 1 of Section 1.1 the set C is convex. Since $A \cap B = \emptyset$, we have $0 \notin C$. Therefore, by Theorem II.2.9, there exists a hyperplane H such that $0 \in H$ and H isolates C. In other words, there exists a non-zero linear functional $f : \mathbb{R}^d \longrightarrow \mathbb{R}$, such that $f(x - y) \leq 0$ for all $x \in A$, $y \in B$. We have $f(x) \leq f(y)$ for each $x \in A$ and each $y \in B$. Therefore, there exists an $\alpha \in \mathbb{R}$, such that $f(x) \leq \alpha$ for all $x \in A$ and $f(y) \geq \alpha$ for all $y \in B$ (we can choose $\alpha = \sup\{f(x) : x \in A\}$, for example). The hyperplane $H = \{x \in \mathbb{R}^d : f(x) = \alpha\}$ separates A and B. □

PROBLEMS.

1°. Suppose that one of the sets $A, B \subset \mathbb{R}^d$ is open. Prove that $C = A - B$ is open.

2. Construct an example of two closed sets $A, B \subset \mathbb{R}^2$ such that $C = A - B$ is not closed.

3. Let $A, B \subset \mathbb{R}^d$ be sets. Suppose that A is closed and B is compact. Prove that the set $C = A - B$ is closed.

4. Prove that if $A, B \subset \mathbb{R}^d$ are compact sets, then $C = A - B$ is a compact set.

5°. Let $A, B \subset \mathbb{R}^d$ be open sets. Suppose that a hyperplane H separates A and B. Prove that H strictly separates A and B.

The following result as well as its infinite-dimensional version (Theorem 3.4) will be used extensively.

(1.3) Theorem. *Let $A \subset \mathbb{R}^d$ be a closed convex set and let $u \notin A$ be a point. Then there exists an affine hyperplane H which strictly separates A and u.*

Proof. Since A is a closed set and $u \notin A$, there exists a $\rho > 0$ such that the open ball $B(u, \rho) = \{x : \|x - u\| < \rho\}$ does not intersect A. Let

$$B = B(0, \rho/2) = \{x : \|x\| < \rho/2\}.$$

We claim that $A + B$ and $u + B$ are open non-intersecting convex sets. Indeed,

$$A + B = \bigcup_{x \in A} B(x, \rho/2)$$

is a union of open sets, so it is open. Similarly, $u + B = B(u, \rho/2)$ is an open ball centered at u. It follows by Problem 4, Section I.1.5, that $A + B$ and $u + B$ are convex. Suppose that $(A + B) \cap (B + u) \neq \emptyset$. For a point $x \in (A + B) \cap (B + u)$, we have $\|y - x\| < \rho/2$ for some $y \in A$ and $\|u - x\| < \rho/2$. Therefore, $\|u - y\| < \rho$, which is a contradiction. Therefore, the sets $A + B$ and $B + u$ do not intersect.

By Theorem 1.2, there is a hyperplane H that separates $A + B$ and $u + B$. Since the sets $A + B$ and $u + B$ are open, H must strictly separate the sets (see Problem 5, Section 1.2). □

PROBLEMS.

1. Construct an example of two disjoint non-empty closed convex sets A and B in $\mathbb{R}^2$ that cannot be strictly separated by a hyperplane.

2. Let $A \subset \mathbb{R}^d$ be a non-empty closed convex set and let $u \notin A$ be a point. Prove that there exists a unique point $v \in A$ such that $\|u - v\| \leq \|u - x\|$ for all $x \in A$. Furthermore, prove that the hyperplane H orthogonal to $u - v$ and passing through the point $(u+v)/2$ strictly separates A and u.

3. Using Problem 3, Section 1.2, prove that if $A, B \subset \mathbb{R}^d$ are disjoint non-empty convex sets, where A is closed and B is compact, then A and B can be strictly separated by a hyperplane H.

4. Let $A, B \subset \mathbb{R}^d$ be disjoint non-empty convex sets. Suppose that B is compact and A is closed. Prove that there is a pair of points $u \in A$ and $v \in B$ such that $\|u - v\| \leq \|x - y\|$ for all $x \in A$ and all $y \in B$. Furthermore, prove that the hyperplane H orthogonal to $u - v$ and passing through $(u+v)/2$ strictly separates A and B.

(1.4) What can go wrong in infinite dimension?

While Theorem 1.3 can be generalized to a wide class of inifinite-dimensional spaces, Theorem 1.2 apparently lacks such a generalization.

PROBLEM.

1. Let $V = \mathbb{R}_\infty$ be the vector space of all infinite sequences $x = (\xi_1, \xi_2, \xi_3, \dots)$ of real numbers such that all but finitely many terms ξ_i are zero (see Problem 2, Section II.1.6). Let $A \subset V \setminus \{0\}$ be the set of all such sequences x whose last non-zero term is strictly positive and let $B = -A$ be the set of sequences whose last non-zero term is strictly negative. Prove that A and B are convex, that $A \cap B = \emptyset$ and that A and B cannot be separated by a hyperplane.

The ultimate reason why a straightforward extension of Theorem 1.2 fails in infinite dimension is that infinite-dimensional convex sets can be amazingly "shallow": they can have an empty interior and yet not be contained in any hyperplane (see Problem 1 of Section II.2.5).

If we require at least one set to be algebraically open (see Definition II.1.5), we can extend the separation theorem to an arbitrary vector space.

(1.5) Theorem. *Let V be a vector space and let $A, B \subset V$ be non-empty convex sets such that $A \cap B = \emptyset$. Suppose that A is algebraically open. Then A and B can be separated by an affine hyperplane.*

Proof. Let $C = A - B$. We can write C as a union of algebraically open sets:

$$C = \bigcup_{b \in B} (A - b).$$

Thus C is an algebraically open convex set. Furthermore, $0 \notin C$, so by Theorem II.1.6 there exists a hyperplane H such that $0 \in H$ and H isolates C. The proof is completed as in Theorem 1.2. $\square$

(1.6) Definition. Let V be a vector space and let $A \subset V$ be a convex set. We say that $v \in A$ *lies in the algebraic interior* of A if for any straight line l passing through v the point v lies in the interior of the intersection $A \cap l$. The set of all points v that lie in the algebraic interior of A is called the *algebraic interior* of A.

PROBLEM.

1. Let V be a vector space and let $A \subset V$ be a convex set. Let u_0 be a point in the algebraic interior of A. Prove that for any point $u_1 \in A$ and any $0 \leq \alpha < 1$, the point $u_\alpha = (1 - \alpha)u_0 + \alpha u_1$ lies in the algebraic interior of A.

Hint: Cf. Lemma II.2.2.

2. Prove that the algebraic interior of a convex set is an algebraically open convex set.

Hint: Use Problem 1.

3. Let V be a vector space, let $A, B \subset V$ be convex sets and let $H \subset V$ be a hyperplane. Suppose that H separates B and the algebraic interior of A. Prove that H separates A and B.

Hint: Use Problem 1.

4. Let $A \subset \mathbb{R}_\infty$ be the set of Problem 1 of Section 1.4. Prove that the algebraic interior of A is empty.

Lastly, we conclude that two non-intersecting convex sets can be separated by a hyperplane if one of them is sufficiently "solid".

(1.7) Corollary. *Let V be a vector space and let $A, B \subset V$ be non-empty convex sets such that $A \cap B = \emptyset$. Suppose that A has a non-empty algebraic interior. Then A and B can be separated by an affine hyperplane.*

Proof. Let A_1 be the algebraic interior of A. Then, by Problem 2, Section 1.6, A_1 is a non-empty algebraically open convex set. By Theorem 1.5, there is a hyperplane H which separates A_1 and B. Then, by Problem 3, Section 1.6, H separates A and B. □

2. Topological Vector Spaces, Convex Sets and Hyperplanes

We intend to study convex sets in a richer setting of topological vector spaces. In this section, we introduce topological vector spaces and review with or without proofs some basic facts concerning them (see [**Ru91**] and [**Co90**]). We review some topology first.

(2.1) Topological spaces. We recall that a *topological space* is a set X together with a family $F \subset 2^X$ of its subsets, called *open sets*, such that

- $\emptyset$ and X are open sets;

• the intersection of any two (equivalently, of finitely many) open sets is an open set;

• the union of open sets is an open set.

The family F is called a *topology* on X. An open subset containing a point $x \in X$ is called a *neighborhood* of that point. A set $C \subset X$ is called *closed* if $X \setminus C$ is open. A map $\phi : X \longrightarrow Y$, where X and Y are topological spaces, is called *continuous* if for any open set $U \subset Y$ the preimage $\phi^{-1}(U) = \{x \in X : \phi(x) \in U\}$ is an open set in X. Equivalently, ϕ is continuous if for every $x \in X$ and for every neighborhood U of $\phi(x)$ in Y there exists a neighborhood W of x such that for each $x' \in W$ we have $\phi(x') \in U$. If $F_1, F_2 \subset 2^X$ are two topologies on X, we say that F_2 is *stronger* (F_1 is *weaker*) if $F_1 \subset F_2$. We recall that a set $C \subset X$ is called *compact* if for any family of open subsets $\{U_i \subset X,\ i \in I\}$ such that $C \subset \bigcup_{i \in I} U_i$ there is a finite subfamily $U_{i_1}, \ldots, U_{i_n}$ such that $C \subset U_{i_1} \cup \ldots \cup U_{i_n}$.

We recall the construction of the direct product. Let X and Y be topological spaces. The *direct product* $Z = X \times Y$ is made a topological space by declaring a set $U \subset Z$ open if it can be represented as a union of direct products $U_1 \times U_2$, where U_1 is an open subset of X and U_2 is an open subset of Y. Equivalently, this topology is the weakest among all topologies that make the projections $Z \longrightarrow X$, $(x, y) \longmapsto x$ and $Z \longrightarrow Y$, $(x, y) \longmapsto y$ continuous. Similarly, if $\{X_i : i \in I\}$ is a (possibly infinite) family of topological spaces, the *direct product* $Z = \prod_{i \in I} X_i$ is identified with the space of all functions f on the set of indices I such that $f(i) \in X_i$ for all $i \in I$. The topology of the direct product is the weakest topology on Z for which all projections $Z \longrightarrow X_i$ are continuous. Equivalently, we declare a set open if and only if it is a union of basic open sets U of the type $U = \{f \in Z : f(i_1) \in U_{i_1}, \ldots, f(i_n) \in U_{i_n}\}$, where $i_1, \ldots, i_n \in I$ and $U_{i_1} \subset X_{i_1}, \ldots, U_{i_n} \subset X_{i_n}$ are open sets. Tikhonov's Theorem asserts that if each X_i is compact, then Z is compact; cf. [**Ru91**] and [**Co90**].

Next, we introduce the central notion of this chapter.

(2.2) Topological vector spaces. Let V be a vector space. Suppose that V is also a topological space so that the following properties hold:

• For every vector $v \in V$ the set $\{v\}$ is closed.

• The map $V \times V \longrightarrow V$, $(x, y) \longmapsto x + y$ is continuous. Equivalently, for every $w_1, w_2 \in V$ and every neighborhood U of $u = w_1 + w_2$, there is a neighborhood W_1 of w_1 and a neighborhood W_2 of w_2 such that $W_1 + W_2 \subset U$.

• The map $\mathbb{R} \times V \longrightarrow V$, $(\alpha, x) \longmapsto \alpha x$ is continuous. Equivalently, for every w in V and every $\alpha \in \mathbb{R}$, for every neighborhood U of $u = \alpha w$, there is a neighborhood W of w and a number $\epsilon > 0$ such that $\beta x \in U$ provided $x \in W$ and $|\alpha - \beta| < \epsilon$.

Then V is called a *topological vector space.*

In particular, for any given $u \in V$, the *translation* $x \longmapsto u + x$ is a continuous transformation, and for any given $\alpha \in \mathbb{R}$, the *scaling* $x \longmapsto \alpha x$ is a continuous transformation.

PROBLEMS.

1. Let V be a topological vector space and let $x \in V$ be a vector. Prove that if $U \subset V$ is an open (closed) set, then the translation $U + x$ is an open (closed) set.

2. Let V be a topological vector space and let $\alpha \neq 0$ be a number. Prove that if $U \subset V$ is an open (closed) set, then $\alpha U = \{\alpha x : x \in U\}$ is an open (closed) set in V.

3. Let V be a topological vector space and let $w_1 \neq w_2$ be two distinct points in V. Prove that there exist neighborhoods W_1 of w_1 and W_2 of w_2 such that $W_1 \cap W_2 = \emptyset$. Deduce that compact sets in V are closed.

4*. Let V be a topological vector space and let $L \subset V$ be a finite-dimensional affine subspace. Prove that L is a closed subset of V.

5. A set $A \subset V$ in a vector space V is called *balanced* provided $\alpha A \subset A$ for all α such that $|\alpha| \leq 1$. Prove that every neighborhood of the origin in a topological vector space contains a balanced neighborhood of the origin.

(2.3) Definitions. Let V be a topological vector space, $A \subset V$ be a set and $u \in V$ be a point. We say that u lies in the *interior* of A provided there is a neighborhood $U \subset A$ of u. We say that u lies in the *closure* of A provided for every neighborhood U of u we have $U \cap A \neq \emptyset$. The set of all points in the interior of A is denoted $\operatorname{int}(A)$. The set of all points in the *closure* of A is denoted $\operatorname{cl}(A)$.

The following result is very similar to Lemma II.2.2.

(2.4) Lemma. *Let V be a topological vector space and let $A \subset V$ be a convex set. Let $u_0 \in \operatorname{int}(A)$ and $u_1 \in A$. Then, for every $0 \leq \alpha < 1$ and $u_\alpha = (1-\alpha)u_0 + \alpha u_1$, we have $u_\alpha \in \operatorname{int}(A)$.*

Proof. Let $U_0 \subset A$ be a neighborhood of u_0. Let us consider a map $T : V \longrightarrow V$,

$$x \longmapsto \frac{1}{1-\alpha}(x - u_1) + u_1.$$

Then T is continuous and $T(u_\alpha) = u_0$ (see Figure 12). Therefore, the preimage $U_\alpha = T^{-1}(U_0)$ is a neighborhood of u_α. Let us show that $U_\alpha \subset A$. Indeed, if $x \in U_\alpha$, then $T(x) = y \in A$. Solving for x, we get $x = (1-\alpha)y + \alpha u_1$. Since A is convex, the result follows. □

(2.5) Theorem. *Let V be a topological vector space and let $A \subset V$ be a convex set. Then $\operatorname{int}(A)$ and $\operatorname{cl}(A)$ are convex sets.*

Proof. The proof that $\operatorname{int}(A)$ is a convex set follows the proof of Corollary II.2.3. Let us prove that $\operatorname{cl}(A)$ is a convex set. Let $u_0, u_1 \in \operatorname{cl}(A)$ and let $u_\alpha = \alpha u_0 + (1-\alpha)u_1$ for $0 < \alpha < 1$. We know that every neighborhood of u_0 or u_1 intersects

A and we must prove that every neighborhood of u_α intersects A. Let U be a neighborhood of u_α; see Figure 35.

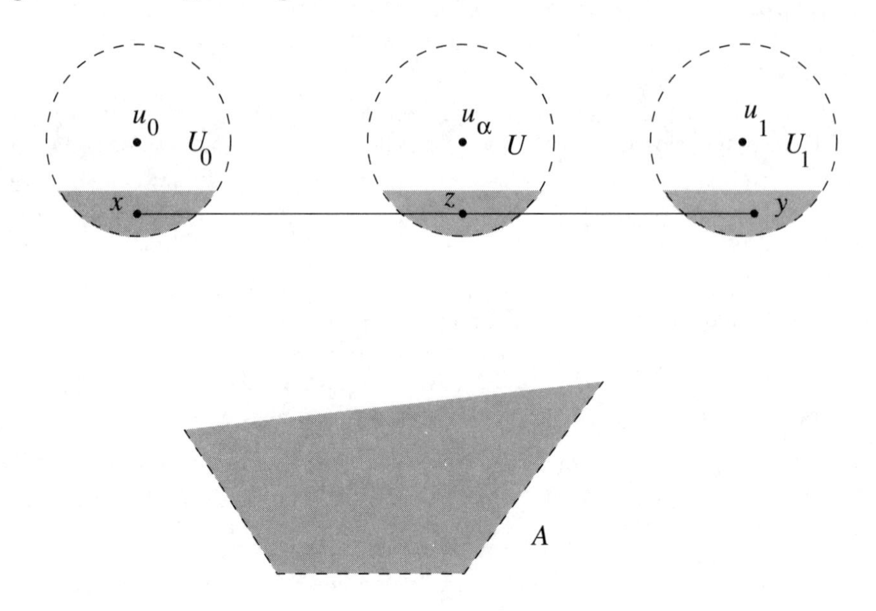

Figure 35

Since the operation $(x, y) \longmapsto \alpha x + (1 - \alpha)y$ is continuous, there must be neighborhoods U_0 of u_0 and U_1 of u_1 such that $\alpha x + (1-\alpha)y \in U$ for every $x \in U_0$ and every $y \in U_1$. Since $u_0 \in \operatorname{cl}(A)$, there is a point $x \in U_0 \cap A$ and since $u_1 \in \operatorname{cl}(A)$, there is a point $y \in U_1 \cap A$. Since A is convex, for $z = \alpha x + (1-\alpha)y$ we have $z \in A$. Therefore, $U \cap A \neq \emptyset$. □

PROBLEMS.

1°. Prove that $\operatorname{int}(A + x) = \operatorname{int}(A) + x$ and $\operatorname{cl}(A + x) = \operatorname{cl}(A) + x$ for each $A \subset V$ and each $x \in V$.

2°. Prove that $\operatorname{int}(\alpha A) = \alpha \operatorname{int}(A)$ and $\operatorname{cl}(\alpha A) = \alpha \operatorname{cl}(A)$ for each $A \subset V$ and each $\alpha \neq 0$.

Among other counter-intuitive things that can happen in infinite dimension, hyperplanes may be everywhere dense.

(2.6) Theorem. *Let V be a topological vector space and let $H \subset V$ be an affine hyperplane. Then either $\operatorname{cl}(H) = H$ (that is, H is closed) or $\operatorname{cl}(H) = V$ (that is, H is dense in V).*

Proof. It suffices to prove the result assuming that $0 \in H$. Let $f : V \longrightarrow \mathbb{R}$ be a linear functional such that

$$H = \{x \in V : \quad f(x) = 0\}.$$

Suppose that H is not closed. Then there is a point $u \in \operatorname{cl}(H) \setminus H$. Let $\alpha = f(u) \neq 0$. Let us choose any $w \in V$. Suppose that $\beta = f(w)$. Let $\gamma = \beta/\alpha$ and let $h = w - \gamma u$. Then $f(h) = f(w) - \gamma f(u) = 0$, so $h \in H$. In other words, $w = h + \gamma u$. Since multiplication and addition are continuous operations, for every neighborhood W of w, there is a neighborhood U of u, such that $h + \gamma U \subset W$. Since $u \in \operatorname{cl}(H)$, there must be a point $h_1 \in U \cap H$. Then the point $h + \gamma h_1$ lies in W. Hence $W \cap H \neq \emptyset$ and $w \in \operatorname{cl}(H)$. It follows that H is dense in V. □

PROBLEMS.

1. Consider the space $C[0,1]$ of all continuous functions on the interval $[0,1]$. Prove that we can make $C[0,1]$ a topological vector space by declaring a set $U \subset C[0,1]$ open if for every $f \in U$ there is an $\epsilon > 0$ such that the set

$$U(f, \epsilon) = \Big\{g \in C[0,1] : \quad |f(\tau) - g(\tau)| < \epsilon \quad \text{for all} \quad \tau \in [0,1]\Big\}$$

is contained in U.

Let $L \subset C[0,1]$ be a subspace consisting of all smooth functions (a function is called smooth if it is differentiable at every point and the derivative is continuous). Prove that L has infinite codimension and that $\operatorname{cl}(L) = C[0,1]$.

2*. Using Zorn's Lemma and Problem 1, show that there exists a dense hyperplane in $C[0,1]$.

3. Prove that every hyperplane in $\mathbb{R}^d$ is closed.

4. Consider the space V of all continuous functions on the interval $[0,1]$. Prove that we can make V a topological vector space by declaring a set $U \subset V$ open if for every $f \in U$ there is an $\epsilon > 0$ such that the set

$$U(f, \epsilon) = \Big\{g \in V : \quad \int_0^1 \sqrt{|f(\tau) - g(\tau)|} \, d\tau < \epsilon\Big\}$$

is contained in U. Prove that all hyperplanes $H \subset V$ are dense in V.

5. Let V be the space of all smooth functions (functions with continuous derivative) on the interval $[0,1]$. Prove that we can make V a topological vector space by declaring a set $U \subset V$ open if for every $f \in U$ there is an $\epsilon > 0$ such that the set

$$U(f, \epsilon) = \Big\{g \in V : \quad |f(\tau) - g(\tau)| < \epsilon \quad \text{for all} \quad \tau \in [0,1]\Big\}$$

is contained in U (cf. Problem 1). Let $H = \{f : f'(1/2) = 0\}$. Prove that H is a dense hyperplane in V.

6. Let V be a topological vector space and let $L \subset V$ be an (affine) subspace. Prove that $\operatorname{cl}(L)$ is an (affine) subspace.

Closed hyperplanes correspond to continuous linear functionals, as the following result shows.

(2.7) Theorem. *Let V be a topological vector space, let $f : V \longrightarrow \mathbb{R}$ be a non-zero linear functional and let $\alpha \in \mathbb{R}$ be a number. Then the affine hyperplane $H(\alpha) = \{x \in V : \ f(x) = \alpha\}$ is closed if and only if f is continuous.*

Proof. If f is continuous, then $H(\alpha) = f^{-1}(\alpha)$ is a closed set as the preimage of a closed set $\{\alpha\}$.

Let us prove that if $H(\alpha)$ is closed, then f is continuous. Since all the hyperplanes

$$H(\gamma) = \{x : \ f(x) = \gamma\}$$

are translations of each other, it follows that all $H(\gamma)$ are closed. The core of the argument is to prove that both halfspaces

$$H_+(\gamma) = \{x : \ f(x) > \gamma\} \quad \text{and} \quad H_-(\gamma) = \{x : \ f(x) < \gamma\}$$

are open. Since all the halfspaces $H_-(\gamma)$ are translations of each other and all the halfspaces $H_+(\gamma)$ are translations of each other, it suffices to prove that for some γ both $H_+(\gamma)$ and $H_-(\gamma)$ are open.

Suppose, for example, that $H_+(\gamma)$ is not open. Then there is a point $u \in H_+(\gamma)$ such that every neighborhood of u intersects

$$\overline{H_-}(\gamma) = \{x : \ f(x) \leq \gamma\}.$$

Thus $f(u) > \gamma$, but each neighborhood U of u contains a point x such that $f(x) \leq \gamma$.

Applying a translation, if necessary, we may assume that $u = \mathbf{0}$ is the origin (of course, the translation may change γ). Let us choose any neighborhood U of $u = \mathbf{0}$. Since $0 \cdot \mathbf{0} = \mathbf{0}$ and multiplication by a scalar is continuous, there is a neighborhood W of the origin and a number $0 < \delta < 1$ such that $\tau x \in U$ for any $|\tau| < \delta$ and any $x \in W$. Let $U_1 = \bigcup_{0<\tau<\delta} \tau W$. Then $U_1 \subset U$ is a neighborhood of $u = \mathbf{0}$ and $[u, x] \subset U_1$ for any $x \in U_1$ (cf. Problem 5 of Section 2.2 and Figure 36). Thus U_1 is a neighborhood of u and hence U_1 must intersect $\overline{H_-}(\gamma)$. We claim that, in fact, U_1 must intersect the hyperplane $H(\gamma)$. Indeed, let us choose an $x \in U_1 \cap \overline{H_-}(\gamma)$. If $x \in H(\gamma)$, then U_1 intersects $H(\gamma)$ as claimed. If $x \in H_-(\gamma)$, then $f(x) < \gamma$. Since $f(u) > \gamma$ and f is linear, for some $y \in [u, x]$ we have $f(y) = \gamma$ and hence the interval $[u, x]$ intersects H. Since $[u, x] \subset U_1$, we conclude that U_1 intersects H. Since $U_1 \subset U$, we conclude that every neighborhood U of u intersects $H(\gamma)$, which contradicts the assumption that $H(\gamma)$ is closed. Therefore, both $H_+(\gamma)$ and $H_-(\gamma)$ are open.

Let us choose an $x \in V$ and let $\gamma = f(x)$. Given an $\epsilon > 0$, let

$$U = f^{-1}(\gamma - \epsilon, \gamma + \epsilon) = H_+(\gamma - \epsilon) \cap H_-(\gamma + \epsilon).$$

Then U is a neighborhood of x and for every $y \in U$ we have $|f(y) - f(x)| < \epsilon$. Hence f is continuous. $\square$

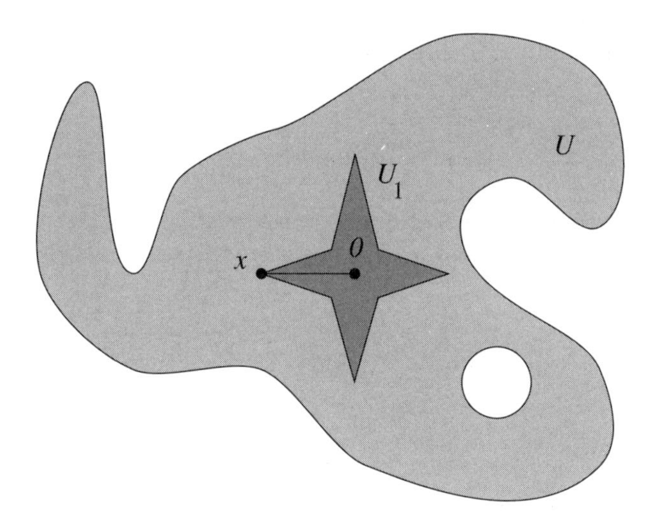

Figure 36. Any neighborhood U of the origin contains a neighborhood U_1 with the property that $[0, x] \subset U_1$ for all $x \in U_1$.

The set of all continuous linear functionals on a given topological vector space can itself be made a topological vector space.

(2.8) The Dual Space. Let V be a topological vector space. The dual space V^*, as a vector space, consists of all continuous linear functionals $f : V \longrightarrow \mathbb{R}$ with addition: $g = f_1 + f_2$ provided $g(x) = f_1(x) + f_2(x)$ for all $x \in V$ and scalar multiplication: $g = \alpha f$ provided $g(x) = \alpha f(x)$ for all $x \in V$. There is a remarkable topology, called weak* (pronounced "weak star") topology, on V^*. Open sets in V^* are unions of elementary open sets of the type

$$\begin{aligned} &U(x_1, \ldots, x_n; \alpha_1, \ldots, \alpha_n; \beta_1, \ldots, \beta_n) \\ &\quad = \Big\{ f \in V^* : \quad \alpha_i < f(x_i) < \beta_i \quad \text{for} \quad i = 1, \ldots, n \Big\}, \end{aligned}$$

where $x_1, \ldots, x_n \in V$ and $\alpha_1, \beta_1, \ldots, \alpha_n, \beta_n \in \mathbb{R}$.

PROBLEMS.

1. Prove that for any $x \in V$ the function $\phi_x : V^* \longrightarrow \mathbb{R}$, $\phi(f) = f(x)$ is a continuous linear functional on V^*.

2*. Prove that every continuous linear functional $\phi : V^* \longrightarrow \mathbb{R}$ has the form $\phi(f) = f(x)$ for some $x \in V$.

Hint: See Section IV.4 and Theorem IV.4.2.

3. Prove that for any two distinct points $f, g \in V^*$, there is a continuous linear functional $\phi : V^* \longrightarrow \mathbb{R}$, such that $\phi(f) \neq \phi(g)$.

4. Suppose that V^* is infinite-dimensional. Prove that every non-empty open set in V^* contains an infinite-dimensional affine subspace.

We are going to use the following important fact.

(2.9) Alaoglu's Theorem. *Let V be a topological vector space, let $U \subset V$ be a neighborhood of the origin and let V^* be the dual space endowed with the weak* topology. The set*

$$U^\circ = \Big\{ f \in V^* : \quad |f(x)| \leq 1 \quad \textit{for all} \quad x \in U \Big\}$$

is compact in V^*.

Sketch of Proof. For every $x \in U$, let $I_x = [-1, 1]$ be a copy of the interval $[-1, 1]$ indexed by x. Let

$$C = \prod_{x \in U} I_x$$

be the direct product identified with the set of *all* functions $\phi : U \longrightarrow [-1, 1]$. We introduce the topology of the direct product on C in the standard way; see Section 2.1. Then, by the Tikhonov Theorem, C is compact. Now we identify U° with a subset of continuous linear functionals of C and prove that U° is a closed subset of C. Hence U° is compact. □

We conclude this section with a useful lemma, which is a straightforward generalization of Lemma II.8.6.

(2.10) Lemma. *Let V be a topological vector space and let $C \subset V$ be a compact convex set such that $0 \notin C$. Then $K = \operatorname{co}(C)$ is a closed convex cone.*

Proof. The proof is completely analogous to that of Lemma II.8.6. Clearly, K is a convex cone such that every point $x \in K$ can be represented in the form $x = \lambda u$ for some $u \in C$ and some $\lambda \geq 0$. Let us prove that K is closed.

Let us choose a point $u \notin K$. Our goal is to find a neighborhood U of u such that $U \cap K = \emptyset$. Since C is closed and $0 \notin C$, there is a neighborhood W of the origin such that $W \cap C = \emptyset$. Let us choose a neighborhood U_1 of u and a number $\delta > 0$ such that $\alpha U_1 \subset W$ for all $|\alpha| < \delta$ (such U_1 and δ exist because scalar multiplication is a continuous operation). In particular, $\alpha U_1 \cap C = \emptyset$ for all $|\alpha| < \delta$. Therefore, for all $\lambda > \delta^{-1}$ we have $U_1 \cap \lambda C = \emptyset$ (see Figure 37).

Let $X = [0, \delta^{-1}] \times C$ and let $\phi : X \longrightarrow V$ be the map $\phi(\tau, x) = \tau x$. Since C and X are compact and ϕ is continuous, the image $\phi(X)$ is compact in V. Therefore, $\phi(X)$ is a closed subset of V (cf. Problem 3 of Section 2.2). Since $u \notin K$, we have $u \notin \phi(X)$ and there is a neighborhood U_2 of u with the property that $U_2 \cap \phi(X) = \emptyset$. Then $U = U_1 \cap U_2$ is the desired neighborhood of u. □

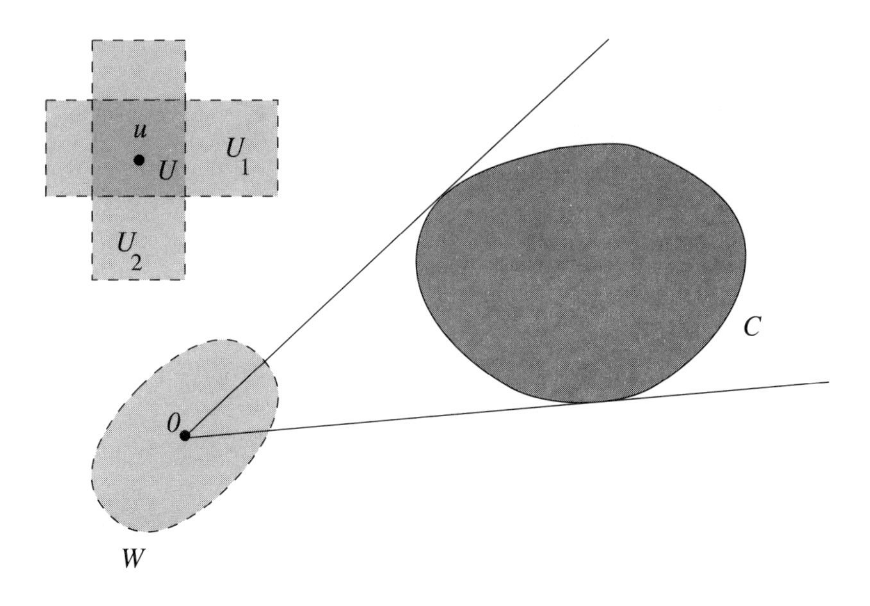

Figure 37. For large λ we have $U_1 \cap \lambda C = \emptyset$ and for small λ we have $U_2 \cap \lambda C = \emptyset$.

3. Separation Theorems in Topological Vector Spaces

In this section, we adapt separation theorems in Euclidean space (Section 1) to the infinite-dimensional situation.

(3.1) Lemma. *Let V be a topological vector space and let $A \subset V$ be an open set. Then A is an algebraically open set.*

Proof. Let $L = \{u + \tau v : \tau \in \mathbb{R}\}$ be a straight line in V $(v \neq 0)$. We must prove that the intersection $A \cap L$ is an open set in L, possibly empty. If $A \cap L$ is not empty, let $w \in A \cap L$, $w = u + \tau_0 v$. Since A is open, there is a neighborhood W of w, such that $W \subset A$. Since addition and scalar multiplication are continuous operations, there exists an $\epsilon > 0$, such that if $|\tau - \tau_0| < \epsilon$, then $u + \tau v \in W$. This implies that w is an interior point of the intersection $A \cap L$. Therefore, A is an algebraically open set. □

PROBLEMS.

1. Let $V = \mathbb{R}_\infty$ be a vector space of all infinite sequences of real numbers $x = (\xi_1, \xi_2, \xi_3, \dots)$, such that only finitely many terms ξ_i are non-zero (see Problem 1 of Section 1.4 and Problem 2, Section II.1.6). Prove that we can make V a topological vector space by declaring a set $U \subset V$ open if for every $x \in U$ there is

an $\epsilon > 0$ such that the set

$$U(x, \epsilon) = \left\{ y = (\eta_1, \eta_2, \dots) : \quad \sum_{i=1}^{\infty} |\xi_i - \eta_i|^2 < \epsilon \right\}$$

is contained in U. Let $A = \{x \in V : \ |\xi_k| < 1/k \text{ for } k = 1, \dots\}$. Prove that A is convex and algebraically open, but not open.

2. Let V be a vector space. Let us declare a set $A \subset V$ open if and only if it is a union of algebraically open convex sets. Prove that this converts V into a topological vector space (cf. Problem 6 of Section II.1.5).

3. Let V be a topological vector space of Problem 2. Prove that every hyperplane in V is closed.

4. Let V be a topological vector space and let $A, B \subset V$ be non-empty open sets. Suppose that a hyperplane $H \subset V$ separates A and B. Prove that H strictly separates A and B and that H is closed.

(3.2) Theorem. *Let V be a topological vector space and let $A, B \subset V$ be convex sets. Suppose that $A \cap B = \emptyset$ and that $\operatorname{int}(A) \neq \emptyset$. Then there is a closed affine hyperplane $H \subset V$ which separates A and B. Equivalently, there is a continuous, not identically zero, linear functional $f : V \longrightarrow \mathbb{R}$ such that $f(x) \leq f(y)$ for all $x \in A$ and $y \in B$.*

Proof. By Lemma 3.1, the algebraic interior of A is not empty. Therefore, by Corollary 1.7, there is a hyperplane $H \subset V$ which separates A and B. By Theorem 2.6, either H is closed or H is dense in V. In the latter case, H must have a non-empty intersection with any open subset, in particular, with the interior of A. Therefore, H has a non-empty intersection with the algebraic interior of A.

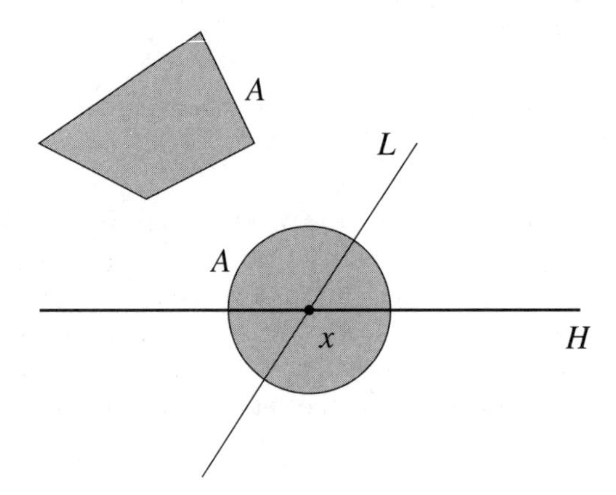

Figure 38

If $x \in H \cap A$ is a point in the algebraic interior of A, then we can choose a straight line L passing through x and not contained in H. The intersection $L \cap A$

contains an open interval around x and hence H cannot isolate A and thus cannot separate A and B. The contradiction shows that H must be a closed hyperplane. By Theorem 2.7, the corresponding linear functional f is continuous. □

We define an important class of topological vector spaces with the abundance of continuous linear functionals (there are topological vector spaces with no non-zero continuous linear functionals; cf. Problem 4 of Section 2.6).

(3.3) Definition. A topological vector space V is called *locally convex* provided for every point $u \in V$ and every neighborhood U of u there is a convex neighborhood $W \subset U$ of u.

PROBLEMS.

1. Let V be a vector space. A function $p : V \longrightarrow \mathbb{R}$ is called a *norm* provided

- $p(u) \geq 0$ for each $u \in V$ and $p(u) = 0$ only if $u = 0$,
- $p(\alpha u) = |\alpha| \cdot p(u)$ for each $u \in V$ and each $\alpha \in \mathbb{R}$,
- $p(u + v) \leq p(u) + p(v)$ for each $u \in V$ and each $v \in V$.

Let us make V a topological vector space by declaring a set $U \subset V$ open if for every $u \in U$ there is an $\epsilon > 0$ such that the set

$$U(u, \epsilon) = \Big\{ v \in V : \quad p(u - v) < \epsilon \Big\}$$

is contained in U. Prove that V is a locally convex topological vector space. Such a space V is called a *normed space.*

2. Let V be a topological vector space and let $A \subset V$ be an open set. Prove that $\operatorname{conv}(A)$ is an open set.

3. Let V be a vector space. Prove that the strongest topology that makes V a locally convex topological vector space is the topology where a set $U \subset V$ is open if and only if it is a union of convex algebraically open sets. Prove that every linear functional $f : V \longrightarrow \mathbb{R}$ is continuous in this topology.

4. Let V be the topological vector space of Problem 4, Section 2.6. Prove that the only open convex sets in V are the empty set and the whole space V.

Now we can generalize Theorem 1.3.

(3.4) Theorem. *Let V be a locally convex topological vector space. Let $A \subset V$ be a closed convex set and let $u \notin A$ be a point. Then there exists a closed hyperplane H that strictly separates A and u. Equivalently, there exists a continuous linear functional $f : V \longrightarrow \mathbb{R}$ such that $f(x) < f(u)$ for all $x \in A$.*

Proof. Since A is closed, the complement $U = V \setminus A$ is a neighborhood of u. Let $U_0 = U - u$ be a translation of U, so U_0 is a neighborhood of the origin. Because the transformation $(x, y) \longmapsto x - y$ is continuous, there are neighborhoods W_1 and W_2 of the origin, such that $x - y \in U_0$ for each $x \in W_1$ and each $y \in W_2$. Since V is locally convex, we can choose W_1 and W_2 to be convex.

Let us consider $A + W_2$ and $u + W_1$. The sets are convex (see Problem 4, Section I.1.5) and open, since $A + W_2 = \bigcup_{x \in A} (W_2 + x)$ is a union of open sets. Furthermore, $(A + W_2) \cap (u + W_1) = \emptyset$.

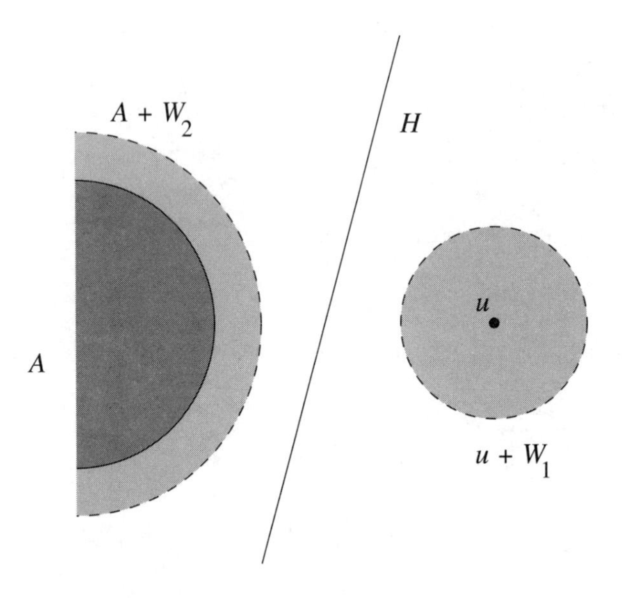

Figure 39

Indeed, if $(A + W_2) \cap (u + W_1) \neq \emptyset$, then for points $a \in A$, $y \in W_2$ and $x \in W_1$, we will have $a + y = u + x$, that is $a = u + (x - y)$. Since $x - y \in U_0$, this would imply that A and U intersect, which is a contradiction. Theorem 3.2 implies that there is a closed hyperplane that separates $A + W_2$ and $u + W_1$. Since $A + W_1$ and $u + W_1$ are open, the hyperplane H must strictly separate $A + W_1$ and $u + W_1$ (see Problem 4 of Section 3.1). $\square$

PROBLEMS.

1°. Prove that in a locally convex topological vector space V, any two points $x \neq y$ can be strictly separated by a closed hyperplane.

2. Let V be a locally convex topological vector space and let $A, B \subset V$ be convex sets such that A is closed, B is compact and $A \cap B = \emptyset$. Prove that A and B can be separated by a closed hyperplane.

Hint: It suffices to construct convex neighborhoods W_1 and W_2 of the origin, such that $(A + W_1) \cap (B + W_2) = \emptyset$ and then use Theorem 3.2. For every point $x \in B$, construct convex neighborhoods $W_1(x)$ and $W_2(x)$ of the origin, such that $(A + W_1(x)) \cap (x + W_2(x)) = \emptyset$. Since B is compact, there is a finite set of points $x_1, \ldots, x_n \in B$, such that the sets $x_i + W_2(x_i)$, $i = 1, \ldots, n$, cover B. Let $W_1 = \bigcap_{i=1}^{n} W_1(x)$ and $W_2 = \bigcap_{i=1}^{n} W_2(x)$.

4. The Krein-Milman Theorem for Topological Vector Spaces

Our next goal is to extend Theorem II.3.3 to the infinite-dimensional situation. The Krein-Milman Theorem that we prove below allows us to relate topological (compactness) and geometric (convexity, extreme points) properties. For the rest of the chapter, we will be studying the extreme points of some particular infinite-dimensional compact convex sets.

(4.1) The Krein-Milman Theorem. *Let V be a locally convex topological vector space and let $K \subset V$ be a compact convex set. Then K is the closure of the convex hull of the set of its extreme points, $K = \operatorname{cl}\Big(\operatorname{conv}\big(\operatorname{ex}(K)\big)\Big)$.*

Proof. First, we establish that every non-empty compact set K in V has an extreme point. Let us call a non-empty compact convex subset $A \subset K$ *extreme* provided for any two points $x, y \in K$ and $z = (x + y)/2$, whenever $z \in A$, we must have $x, y \in A$. Clearly, K is an extreme set. Let $X \subset K$ be the smallest extreme subset (that is, not containing any extreme subset of K other than itself). The existence of X is established via Zorn's Lemma or the axiom of choice. Let us prove that X is a point. Suppose that X contains two different points, say x_1 and x_2. Let us choose a closed hyperplane H that strictly separates x_1 and x_2; see Theorem 3.4. In other words, there is a continuous linear functional $f : V \longrightarrow \mathbb{R}$ such that $f(x_1) < f(x_2)$. Let $\alpha = \min\big\{f(y) : y \in X\big\}$. Since X is compact (a closed subset of a compact set) the minimum is attained. Let $Y = \big\{x \in X : f(x) = \alpha\big\}$ be a face of X. Clearly, Y is a compact convex subset of K and Y is an extreme set (cf. Theorem II.3.2). On the other hand, Y does not contain x_2, so Y is strictly smaller than X. The contradiction shows that X must be a point, that is, an extreme point.

Now, we prove that K is the closure of the convex hull of the set of its extreme points. Let $A = \operatorname{cl}\Big(\operatorname{conv}\big(\operatorname{ex}(K)\big)\Big)$. Then A is a closed convex subset of K.

Suppose that there is a point $u \in K \setminus A$. Let us choose a closed hyperplane H that strictly separates u from A (Theorem 3.4). In other words, there is a continuous linear functional $f : V \longrightarrow \mathbb{R}$ such that $f(x) > f(u)$ for any $x \in A$. Let $\alpha = \min\big\{f(x) : x \in K\big\}$ (the minimum is attained since f is continuous and K is compact) and let $F = \big\{x \in K : f(x) = \alpha\big\}$ be the corresponding face of K. So, F is a compact convex set and, as we proved, must contain an extreme point v, which will be an extreme point of K (cf. Theorem II.3.2). On the other hand, $v \notin A$, which is a contradiction; see Figure 40. □

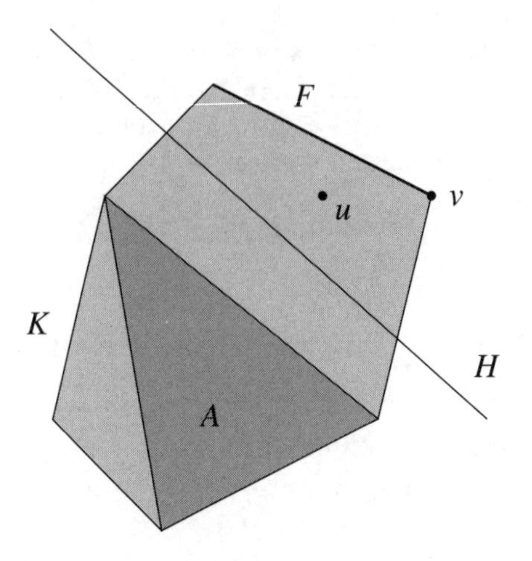

Figure 40

PROBLEMS.

1. Let $C[0,1]$ be the vector space of all continuous functions on the interval $[0,1]$. Let $B = \{f : |f(\tau)| \leq 1 \text{ for all } \tau \in [0,1]\}$. Find the extreme points of B.

2. Let A be the following subset of $C[0,1]$:

$$A = \left\{ f \in C[0,1] : \int_0^1 f(\tau)\, d\tau = 0 \quad \text{and} \quad |f(\tau)| \leq 1 \quad \text{for all} \quad \tau \in [0,1] \right\}.$$

Let us make $C[0,1]$ a topological vector space as in Problem 1, Section 2.6. Check that $C[0,1]$ is locally convex. Prove that A is a closed convex set which does not contain straight lines and that A has no extreme points. Thus there is no straightforward infinite-dimensional generalization of Lemma II.3.5.

3. Let V be a topological vector space and let A be a convex set such that $\operatorname{int}(A) \neq \emptyset$. Prove that for each point $u \in \partial A$, there is a closed support hyperplane at u, that is, a closed hyperplane H that contains u and isolates A.

4. Let $A \subset V$ be a convex set in a vector space V and let F be a face of A. Prove that F is an extreme set of A, that is, for any two $x, y \in A$, whenever $z = (x+y)/2 \in F$, we must have $x, y \in F$.

5. Let V be the topological vector space of Problem 5, Section 2.6. Let

$$A = \{f : \quad f'(1/2) > 0\} \quad \text{and} \quad B = \{f : \quad f'(1/2) < 0\}$$

be subsets of V.

Prove that A and B are disjoint convex, algebraically open subsets of V, so there is an affine hyperplane $H \subset V$ strictly separating them. Prove that A and B are dense in V. Deduce that there is no closed hyperplane $H \subset L$ strictly separating A and B.

We obtain an infinite-dimensional counterpart of Corollary II.3.4.

(4.2) Corollary. *Let V be a locally convex topological vector space and let $K \subset V$ be a compact convex set. Let $f : V \longrightarrow \mathbb{R}$ be a continuous linear functional. Then there exists an extreme point u of K such that $f(u) \geq f(x)$ for all $x \in K$.*

Proof. Since f is continuous and K is compact, f attains its maximum value α on K. Let $F = \{x \in K : f(x) = \alpha\}$ be the corresponding face of K. Hence F is a compact convex set and by the Krein-Milman Theorem (Theorem 4.1), F has an extreme point u. Then by Theorem II.3.2 (Part 2), u is an extreme point of K. We have $\alpha = f(u) \geq f(x)$ for all $x \in K$. □

5. Polyhedra in L^∞

In this section, we study some infinite-dimensional convex sets which may be viewed as analogues of polyhedra. First, we describe the ambient space.

(5.1) Spaces L^1 and L^∞. Let $L^1[0,1]$ be the vector space of all integrable functions on the interval $[0,1]$, that is, Lebesgue measurable functions f such that

$$\int_0^1 |f(\tau)| \, d\tau < +\infty.$$

As usual, we do not distinguish between functions that differ on a set of measure 0. We make $L^1[0,1]$ a topological vector space by declaring a set $U \subset L^1[0,1]$ open if for every $f \in U$ there is an $\epsilon > 0$ such that the set

$$U(f,\epsilon) = \left\{ g \in L^1[0,1] : \quad \int_0^1 |f(\tau) - g(\tau)| d\tau < \epsilon \right\}$$

is contained in U.

Let $L^\infty[0,1]$ be the vector space of all Lebesgue measurable functions f on the interval $[0,1]$ such that $|f(\tau)| \leq C$ for some constant C and for almost all (that is, for all except a set of zero measure) $\tau \in [0,1]$. As usual, we do not distinguish between functions that differ on a set of measure 0.

It is known that every continuous linear functional $\phi : L^1[0,1] \longrightarrow \mathbb{R}$ has the form

$$\phi(f) = \int_0^1 f(\tau) g(\tau) \, d\tau$$

for some $g \in L^\infty[0,1]$; see, for example, Appendix B of [**Co90**]. This allows us to view $L^\infty[0,1]$ as the dual space to $L^1[0,1]$ and introduce the weak* topology on L^∞; see Section 2.8. Thus a set $U \subset L^\infty[0,1]$ is open if and only if it is a union of some basic open sets

$$\begin{aligned} &U(g_1, \ldots, g_n; \alpha_1, \ldots, \alpha_n; \beta_1, \ldots, \beta_n) \\ &= \left\{ f \in L^\infty[0,1] : \quad \alpha_i < \int_0^1 g_i(\tau) f(\tau) \, d\tau < \beta_i \quad \text{for} \quad i = 1, \ldots, n \right\}, \end{aligned}$$

where $g_1, \ldots, g_n \in L^1[0,1]$ are functions and $\alpha_1, \ldots, \alpha_n$ and $\beta_1, \ldots, \beta_n$ are numbers.

PROBLEM.

1. Check that $L^1[0,1]$ is a locally convex topological vector space.

(5.2) Proposition. *Let $B \subset L^\infty[0,1]$ be the set*

$$B = \Big\{u \in L^\infty[0,1] : \quad 0 \leq u(\tau) \leq 1 \quad \textit{for almost all} \quad \tau \in [0,1]\Big\}.$$

The set B is compact in the weak topology.*

PROBLEMS.

1. Prove that if u is an extreme point of B, then $u(\tau) \in \{0,1\}$ for almost all $u \in [0,1]$.
2. Deduce Proposition 5.2 from Theorem 2.9.

Hint: Consider a neighborhood U of the origin in $L^1[0,1]$:

$$U = \Big\{g \in L^1[0,1] : \quad \int_0^1 |g(\tau)| \, d\tau < 1\Big\}.$$

Let

$$K = \Big\{f \in L^\infty[0,1] : \quad \int_0^1 (fg) \, d\tau \leq 1 \quad \text{for all} \quad g \in U\Big\}.$$

Prove that K consists of the functions $f \in L^\infty[0,1]$ such that $|f(\tau)| \leq 1$ for almost all $\tau \in [0,1]$. Use Theorem 2.9 to show that K is compact. Show that $B \subset K$ is a closed subset.

3. Prove that $\operatorname{conv}\big(\operatorname{ex}(B)\big)$ is not closed.
4. Deduce from Proposition 5.2 that the set

$$K = \Big\{f \in L^\infty[0,1] : \quad |f(\tau)| \leq 1 \quad \text{for almost all} \quad \tau \in [0,1]\Big\}$$

is weak* compact.

Next, we introduce sets which may be considered as an L^∞ version of polyhedra (Problem 1 of Section 5.3 explains the relationship of our sets to polyhedra in Euclidean space). The sets are defined by finitely many linear equations and infinitely many inequalities in $L^\infty[0,1]$.

(5.3) Proposition. *Let us fix m functions $f_1(\tau), \ldots, f_m(\tau) \in L^1[0,1]$ and m numbers $\beta_1, \ldots, \beta_m \in \mathbb{R}$. Let $B \subset L^\infty[0,1]$ be the set*

$$B = \Big\{u \in L^\infty[0,1] : \quad 0 \leq u(\tau) \leq 1 \quad \textit{for almost all} \quad \tau \in [0,1]\Big\},$$

and let

$$A = \Big\{u \in B : \int_0^1 f_i(\tau)u(\tau) \, d\tau = \beta_i \quad \textit{for } i = 1, \ldots, m\Big\}.$$

Then A is a convex weak compact subset of $L^\infty[0,1]$. If u is an extreme point of A, then $u(\tau) \in \{0,1\}$ for almost all τ.*

Proof. It is obvious that A is convex. Furthermore, A is a closed subset of B and so A is compact as follows from Proposition 5.2.

Let u be an extreme point of A. Suppose that the set $\{\tau : u(\tau) \notin \{0,1\}\}$ has a positive measure. Then for some $\delta > 0$ and $X = \{\tau : \delta \leq |u(\tau)| \leq 1-\delta\}$, the measure of X is positive. Let us find $m+1$ pairwise disjoint subsets $X_1, \ldots, X_{m+1}$ of X of positive measure and let $[X_i]$ be the indicator function of X_i:

$$[X_i](\tau) = \begin{cases} 1 & \text{if } \tau \in X_i, \\ 0 & \text{if } \tau \notin X_i. \end{cases}$$

Let $\epsilon_1, \ldots, \epsilon_{m+1}$ be real numbers (to be specified later) and let

$$v = \epsilon_1[X_1] + \ldots + \epsilon_{m+1}[X_{m+1}].$$

Furthermore, let $u_+ = u+v$ and let $u_- = u-v$. Then $u = (u_+ + u_-)/2$. Obviously, if $|\epsilon_1| < \delta, \ldots, |\epsilon_m| < \delta$, we will have $0 \leq u_+(\tau), u_-(\tau) \leq 1$ for almost all τ.

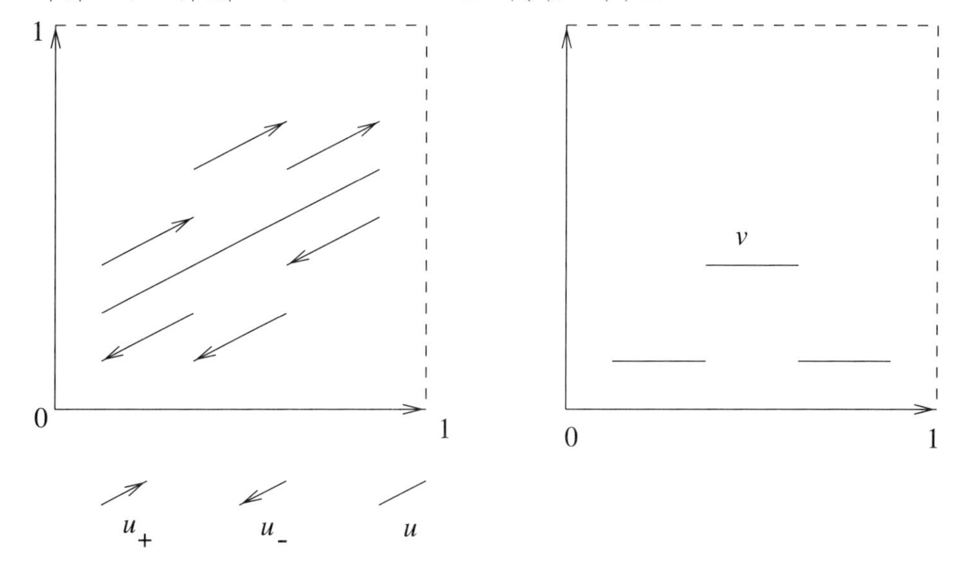

Figure 41. Decomposing $u = (u_+ + u_-)/2$

Now we show that we can choose a non-zero v so that $u_+, u_- \in A$. It suffices to choose $\epsilon_1, \ldots, \epsilon_{m+1}$ so that the system of m homogeneous equations

$$\sum_{j=1}^{m+1} \epsilon_j \int_0^1 f_i(\tau)[X_j](\tau)\, d\tau = 0, \quad i = 1, \ldots, m,$$

in $m+1$ variables $\epsilon_1, \ldots, \epsilon_{m+1}$ is satisfied. Since the number of variables exceeds the number of equations, there must be a solution $\epsilon_1, \ldots, \epsilon_{m+1}$, where not all ϵ's are zero. Scaling, if necessary, we make $|\epsilon_1|, \ldots, |\epsilon_{m+1}| < \delta$. Then $u_-, u_+ \in A$. That contradicts the assumption that u is an extreme point of A. □

PROBLEMS.

1. Consider the following "discretization" of Proposition 5.3. Let us choose N points $0 \leq \tau_1 < \ldots < \tau_N \leq 1$ in the interval $[0,1]$ and let $f_i(\tau_j)$, $i = 1, \ldots, m$, $j = 1, \ldots, N$, be real numbers. Consider the set

$$A = \Big\{ \big(u(\tau_1), \ldots, u(\tau_N)\big) : \quad 0 \leq u(\tau_j) \leq 1 \quad \text{for} \quad j = 1, \ldots, N \qquad \text{and} \qquad \sum_{j=1}^{N} f_i(\tau_j) u(\tau_j) = \beta_i \quad \text{for} \quad i = 1, \ldots, m \Big\}$$

as a polyhedron in $\mathbb{R}^N$. Suppose that $u = \big(u(\tau_1), \ldots, u(\tau_N)\big)$ is an extreme point of A. Prove that at least $N - m$ of the numbers $u(\tau_1), \ldots, u(\tau_N)$ are either 0 or 1.

2. This is an extension of Proposition 5.3.

Let us consider the space $L^\infty([0,1], \mathbb{R}^d)$ of all Lebesgue measurable bounded vector-valued functions $f : [0,1] \longrightarrow \mathbb{R}^d$ and the space $L^1([0,1], \mathbb{R}^d)$ of all Lebesgue integrable vector-valued functions $f : [0,1] \longrightarrow \mathbb{R}^d$. Let us fix a bounded polyhedron (polytope) $P \subset \mathbb{R}^d$, m functions $f_1, \ldots, f_m \in L^1([0,1], \mathbb{R}^d)$ and m numbers $\beta_1, \ldots, \beta_m$. Let

$$A = \Big\{ u \in L^\infty([0,1], \mathbb{R}^d) : \quad u(\tau) \in P \quad \text{for almost all} \quad \tau \quad \text{and} \quad \int_0^1 \langle f_i(\tau), u(\tau) \rangle \, d\tau = \beta_i \quad \text{for } i = 1, \ldots, m \Big\}.$$

Prove that A is a convex set and that if u is an extreme point of A, then $u(\tau)$ is a vertex of P for almost all τ.

6. An Application: Problems of Linear Optimal Control

In this section, we show that a problem of optimal control can be considered as a problem of optimizing a linear functional over an L^∞-polyhedron. This is an example of an infinite-dimensional linear program; cf. II.4.4. We will go back to this problem again in Section IV.12.

We review some differential equations first.

(6.1) Solving linear systems of differential equations with control. Let $A(\tau)$ and $B(\tau)$ be $n \times n$ matrices, where $\tau \in [0,1]$ is a real parameter. We assume that $A(\tau)$ and $B(\tau)$ are smooth functions of τ and consider a system of linear differential equations

$$\frac{d}{d\tau} \mathbf{x}(\tau) = A(\tau)\mathbf{x}(\tau) + B(\tau)\mathbf{u}(\tau), \tag{6.1.1}$$

where $\mathbf{x}(\tau) = \big(x_1(\tau), \ldots, x_n(\tau)\big)$ and $\mathbf{u}(\tau) = \big(u_1(\tau), \ldots, u_n(\tau)\big)$ are vectors from $\mathbb{R}^n$. The function $\mathbf{x}(\tau)$ is a *solution* of the system (6.1.1), whereas $\mathbf{u}(\tau)$ is a *control*.

If the control is chosen, under mild assumptions the solution $\mathbf{x}$ is determined by the *initial condition*:

$$\mathbf{x}(0) = \mathbf{x}_0,$$

where $\mathbf{x}_0 \in \mathbb{R}^n$ is a given vector. The solution $\mathbf{x}$ can be found as follows. Let $X(\tau)$ be an $n \times n$ matrix, which is a solution to the matrix system

$$\frac{d}{d\tau}X(\tau) = A(\tau)X(\tau) \quad \text{with the intial condition} \quad X(0) = I,$$

where I is the $n \times n$ identity matrix. Then

$$\mathbf{x}(\tau) = X(\tau)\Big(\mathbf{x}_0 + \int_0^\tau X^{-1}(t)B(t)\mathbf{u}(t)\, dt\Big). \tag{6.1.2}$$

In general, we want to choose the control $\mathbf{u}$ in the space $L^\infty\big([0,1], \mathbb{R}^n\big)$ of vector-valued functions $[0,1] \longrightarrow \mathbb{R}^n$. Then, if $X(\tau)$ and $B(\tau)$ are continuous, the integral (6.1.2) is well defined. Let us consider $\mathbf{x}(\tau)$ defined by (6.1.2) as a solution to the original system of differential equations, even though it may be, say, non-differentiable at some points. In particular, if we are to choose the control $\mathbf{u}(\tau)$ in such a way that some *terminal* condition $\mathbf{x}(1) = \mathbf{x}_1$ is satisfied, we get the following integral constraint:

$$\int_0^1 X^{-1}(\tau)B(\tau)\mathbf{u}(\tau)\, d\tau = X^{-1}(1)\mathbf{x}_1 - \mathbf{x}_0. \tag{6.1.3}$$

PROBLEM.

1°. Check that formula (6.1.2) indeed holds.

(6.2) Example. A problem of linear optimal control. Let us fix smooth real functions $a_0(\tau), a_1(\tau), b(\tau), c_0(\tau), c_1(\tau)$ and $d(\tau)$: $\tau \in [0,1]$. Consider a differential equation for a function $x(\tau)$

$$x''(\tau) = a_0(\tau)x(\tau) + a_1(\tau)x'(\tau) + b(\tau)u(\tau),$$

where $u(\tau)$ is a control, which we would like to choose in such a way that the initial conditions

$$x(0) = x_0, \quad x'(0) = v_0$$

and the terminal conditions

$$x(1) = x_1, \quad x'(1) = v_1$$

are satisfied and the functional

$$\int_0^1 \Big(c_0(\tau)x(\tau) + c_1(\tau)x'(\tau) + d(\tau)u(\tau)\Big)\, d\tau$$

is minimized. In addition, the control must satisfy the condition

$$0 \le u(\tau) \le 1 \quad \text{for all} \quad \tau \in [0,1].$$

This is a simple example of a linear optimal control problem. The differential equations describe the movement of an "object", thus relating the coordinate $x(\tau)$, the velocity $x'(\tau)$ and the acceleration $x''(\tau)$. The control u is a "force" we may apply. We want to transfer the "object" from the initial position (coordinate, velocity) to the final position (coordinate, velocity), so that the total cost we pay for the coordinate ("gravity"), for the velocity ("friction") and for the control ("fuel") is minimized.

Our aim is to show that this optimization problem can be considered as a *linear programming problem* of the type:

$$\begin{aligned} \text{Find} \quad & \gamma = \inf \int_0^1 g(\tau)u(\tau)\, d\tau \\ \text{Subject to} \quad & \int_0^1 f_1(\tau)u(\tau)\, d\tau = \beta_1, \\ & \int_0^1 f_2(\tau)u(\tau)\, d\tau = \beta_2 \quad \text{and} \\ & 0 \leq u(\tau) \leq 1 \quad \text{for almost all} \quad \tau \in [0,1], \end{aligned}$$

where $u \in L^\infty[0,1]$ is a variable and the functions $f_1, f_2, g \in L^1[0,1]$ and the numbers β_1, β_2 can be explicitly computed.

To see this, let us write the differential equation in the form (6.1.1) by introducing a new variable $v(\tau) = x'(\tau)$. We get

$$\frac{d}{d\tau}x(\tau) = v(\tau) \quad \text{and} \quad \frac{d}{d\tau}v(\tau) = a_0(\tau)x(\tau) + a_1(\tau)v(\tau) + b(\tau)u(\tau)$$

with the conditions

$$\begin{pmatrix} x(0) \\ v(0) \end{pmatrix} = \begin{pmatrix} x_0 \\ v_0 \end{pmatrix} \quad \text{and} \quad \begin{pmatrix} x(1) \\ v(1) \end{pmatrix} = \begin{pmatrix} x_1 \\ v_1 \end{pmatrix}.$$

Thus, for vectors

$$\mathbf{x}(\tau) = \begin{pmatrix} x(\tau) \\ v(\tau) \end{pmatrix} \quad \text{and} \quad \mathbf{u}(\tau) = \begin{pmatrix} 0 \\ u(\tau) \end{pmatrix}$$

we have

$$A(\tau) = \begin{pmatrix} 0 & 1 \\ a_0(\tau) & a_1(\tau) \end{pmatrix} \quad \text{and} \quad B(\tau) = \begin{pmatrix} 0 & 0 \\ 0 & b(\tau) \end{pmatrix}$$

in (6.1.1). Now the formulas for f_1, f_2 and β_1, β_2 are obtained from (6.1.3). The formula for g can be obtained from (6.1.2).

(6.3) Corollary. *If the problem of linear optimal control is feasible, there exists an optimal solution $u \in L^\infty[0,1]$ such that $u(\tau) \in \{0,1\}$ for almost all $\tau \in [0,1]$. If the optimal solution $u \in L^\infty[0,1]$ is unique, then $u(\tau) \in \{0,1\}$ for almost all $\tau \in [0,1]$.*

Proof. By Proposition 5.3, the set of all feasible controls $u(\tau)$ is weak* compact. By Corollary 4.2, there exists an optimal control u that is an extreme point of the set of all feasible solutions. If u is unique, by Theorem II.3.2 (Part 1), u necessarily is an extreme point. By Proposition 5.3, we must have $u(\tau) \in \{0, 1\}$ for almost all $\tau \in [0, 1]$. □

The conclusion of Corollary 6.3 is something akin to "unrealistic solutions" in the Diet Problem; see Example II.4.4. Indeed, it turns out that the optimal control u at all times is either "hit the brakes" ($u = 0$) or "press the gas pedal to the floor" ($u = 1$), which is not always acceptable in practice.

PROBLEMS.

1°. Consider the equation

$$x''(\tau) = a_0(\tau)x(\tau) + a_1(\tau)x'(\tau) + b(\tau)u(\tau)$$

with the initial conditions

$$x(0) = 0 \quad \text{and} \quad x'(0) = 0.$$

Show that the solution $x(\tau)$ depends linearly on the control $u(\tau)$: if $x_1(\tau)$ is the solution for $u_1(\tau)$, $x_2(\tau)$ is the solution for $u_2(\tau)$, then $x(\tau) = \alpha_1 x_1(\tau) + \alpha_2 x_2(\tau)$ is the solution for $u(\tau) = \alpha_1 u_1(\tau) + \alpha_2 u_2(\tau)$, where α_1 and α_2 are real numbers.

2°. Let $y(\tau)$ be a solution to the equation

$$y''(\tau) = a_0(\tau)y(\tau) + a_1(\tau)y'(\tau)$$

with some initial conditions

$$y(0) = x_0 \quad \text{and} \quad y'(0) = v_0$$

and let $z(\tau)$ be a solution of the equation

$$z''(\tau) = a_0(\tau)z(\tau) + a_1(\tau)z'(\tau) + b(\tau)u(\tau)$$

with the initial conditions $z'(0) = z(0) = 0$. Prove that $x(\tau) = y(\tau) + z(\tau)$ is a solution to the equation

$$x''(\tau) = a_0(\tau)x(\tau) + a_1(\tau)x'(\tau) + b(\tau)u(\tau)$$

with the initial conditions $x(0) = x_0$ and $x'(0) = v_0$.

Problems 1 and 2 provide some intuition for why problems of linear optimal control can be written as linear programs of optimizing $\int_0^1 g(\tau)u(\tau)\, d\tau$ subject to integral constraints $\int_0^1 f_i(\tau)u(\tau)\, d\tau = \beta_i$ and the "domain" constraint $0 \leq u(\tau) \leq 1$. First, Problem 2 allows us to reduce (in "nice" cases) the general case to the case of zero initial conditions. Next, if the initial conditions $x(0)$, $x'(0)$ are zero, the terminal values $x(1)$ and $x'(1)$ are linear functions of the control u. A "reasonable" linear function (say, weak* continuous) has the form $\int_0^1 f(\tau)u(\tau)\, d\tau$ for some $f \in L^1[0, 1]$. Thus fixing the terminal values of the solution x amounts to fixing some integral constraints on u.

7. An Application: The Lyapunov Convexity Theorem

We apply Proposition 5.3 to obtain a theorem by A.A. Lyapunov (1940) which, in full generality, asserts that the range of a non-atomic countably additive vector-valued measure is convex. The proof below belongs to J. Lindenstrauss [**Li66**]. We adapt it to the special case of a vector-valued measure on $[0,1]$.

(7.1) Theorem. *Let us fix m functions $f_1, \ldots, f_m \in L^1[0,1]$ and let $\mathcal{A}$ be the family of all Lebesgue measurable subsets of $[0,1]$. For $A \in \mathcal{A}$, let $\phi(A) \in \mathbb{R}^m$ be the point*

$$\phi(A) = (\xi_1, \ldots, \xi_m), \quad \textit{where} \quad \xi_i = \int_A f_i(\tau) \, d\tau \quad \textit{for} \quad i = 1, \ldots, m.$$

Then the set

$$X = \Big\{ \phi(A) : \quad A \in \mathcal{A} \Big\}$$

is a compact convex set in $\mathbb{R}^m$.

Proof. The idea is to prove that X can be represented as the image of a compact convex set under a continuous linear map.

Let us extend ϕ to a map $\phi : L^\infty[0,1] \longrightarrow \mathbb{R}^m$,

$$\phi(u) = (\xi_1, \ldots, \xi_m), \quad \text{where} \quad \xi_i = \int_0^1 u(\tau) f_i(\tau) \, d\tau \quad \text{for} \quad i = 1, \ldots, m.$$

Let

$$B = \Big\{ u \in L^\infty[0,1] : \quad 0 \leq u(\tau) \leq 1 \quad \text{for almost all} \quad \tau \in [0,1] \Big\}.$$

Then B is convex and weak* compact (see Proposition 5.2). Since ϕ is linear and continuous, the image $\phi(B) \subset \mathbb{R}^m$ is a compact convex set.

Next, we claim that $\phi(B) = X$. Clearly, $X \subset \phi(B)$.

Conversely, let us choose $a = (\alpha_1, \ldots, \alpha_m) \in \phi(B)$. Then the set

$$B_a = \Big\{ u \in B : \quad \int_0^1 u(\tau) f_i(\tau) \, d\tau = \alpha_i, \quad \text{for} \quad i = 1, \ldots, m \Big\}$$

is non-empty and weak* compact (Proposition 5.3). Therefore, by the Krein-Milman Theorem (Theorem 4.1), there is an extreme point $u_a \in B_a$; cf. Figure 42. By Proposition 5.3, we have $u_a(\tau) \in \{0, 1\}$ for almost all τ.

Let $A = \big\{ \tau : u_a(\tau) = 1 \big\}$. Then $\phi(A) = a$. Therefore, $\phi(B) \subset X$ and the result follows. □

Recall that we used a similar "convexification" argument in Sections II.13–14 on quadratic convexity.

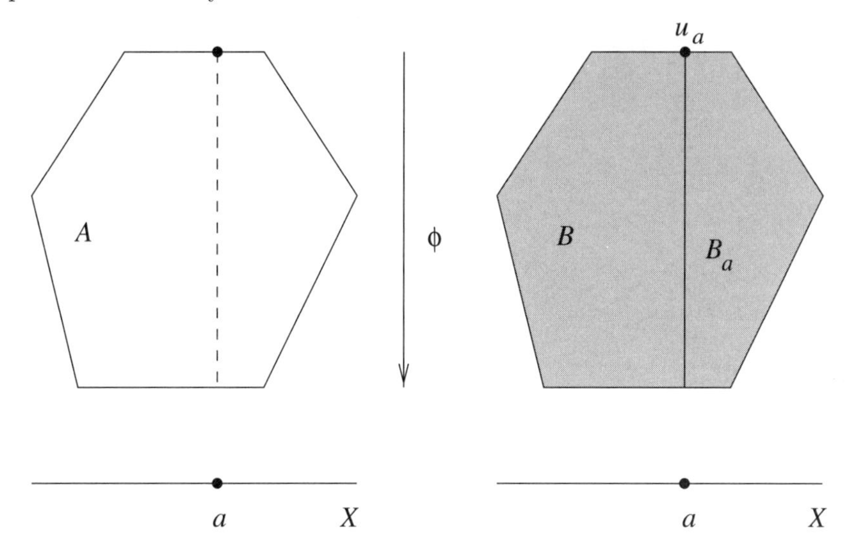

Figure 42. "Convexification": given a set A and a map ϕ, we find a convex set B such that $\phi(B) = \phi(A)$ and ϕ is linear on B.

Here is an interesting corollary.

(7.2) Corollary. *Let $S \subset \mathbb{R}$ be a Lebesgue measurable set and suppose that $F_i(x, \tau) : S \times [0,1] \longrightarrow \mathbb{R}$, $i = 1, \ldots, m$, are Lebesgue integrable functions. Let Γ denote the set of all Lebesgue measurable functions $x : [0,1] \longrightarrow S$. For a function $x \in \Gamma$, let us define $\psi(x) \in \mathbb{R}^m$ by*

$$\psi(x) = (\xi_1, \ldots, \xi_m), \quad \textit{where} \quad \xi_i = \int_0^1 F_i\big(x(\tau), \tau\big)\, d\tau \quad \textit{for} \quad i = 1, \ldots, m.$$

Then the set $X \subset \mathbb{R}^m$,

$$X = \big\{\psi(x) : \ x \in \Gamma\big\},$$

is convex.

Proof. Let us choose two points $a, b \in X$. Hence we have $a = \psi(x)$ and $b = \psi(y)$ for some Lebesgue measurable functions x and y on the interval $[0,1]$, such that $x(\tau), y(\tau) \in S$ for all τ. We will construct a convex set $Y \subset \mathbb{R}^m$, such that $Y \subset X$ and $a, b \in Y$. This will prove that X is convex. To do that, for a Lebesgue measurable subset $A \subset [0,1]$, let us define a function $z_A(\tau)$:

$$z_A(\tau) = \begin{cases} x(\tau) & \text{if } \tau \in A, \\ y(\tau) & \text{if } \tau \notin A; \end{cases}$$

cf. Figure 43.

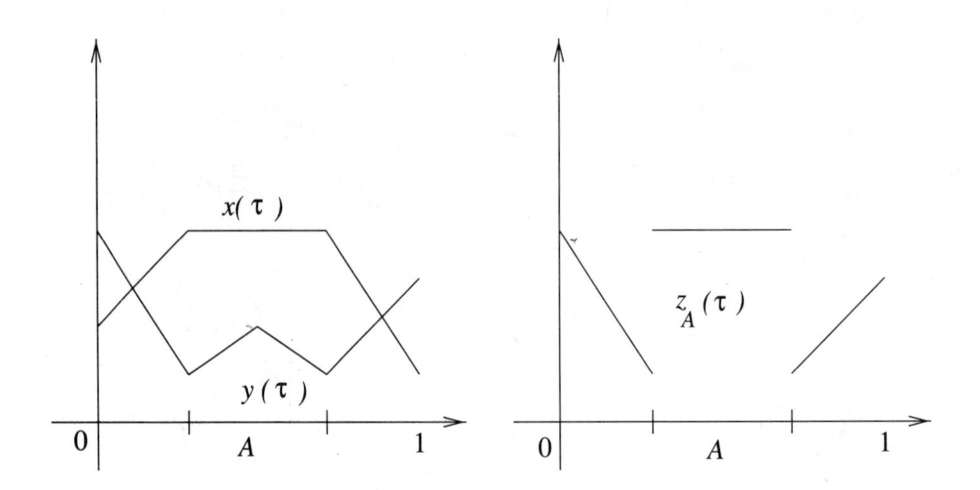

Figure 43

Obviously, $z_A \in \Gamma$. Let

$$Y = \Big\{\psi(z_A) \quad \text{for some Lebesgue measurable} \quad A \subset [0,1]\Big\}.$$

Since $z_{[0,1]}(\tau) = x(\tau)$ and $z_{\emptyset}(\tau) = y(\tau)$, we have $a, b \in Y$. Since $z_A \in \Gamma$, we conclude that $Y \subset X$. To see that Y is convex, let us define functions $f_i \in L^1[0,1]$, $i = 1, \ldots, m$, by the formulas:

$$f_i(\tau) = F_i\big(x(\tau), \tau\big) - F_i\big(y(\tau), \tau\big) \quad \text{for} \quad i = 1, \ldots, m.$$

For a Lebesgue measurable set $A \subset [0,1]$, let

$$\phi(A) = (\xi_1, \ldots, \xi_m), \quad \text{where} \quad \xi_i = \int_A f_i(\tau)\, d\tau \quad \text{for} \quad i = 1, \ldots, m.$$

Let

$$c = (\gamma_1, \ldots, \gamma_m), \quad \text{where} \quad \gamma_i = \int_0^1 F_i\big(y(\tau), \tau\big)\, d\tau \quad \text{for} \quad i = 1, \ldots m.$$

Then $\psi(z_A) = \phi(A) + c$. Therefore, the set Y is a translation (by c) of the set

$$\Big\{\phi(A) : \quad A \subset [0,1] \quad \text{is Lebesgue measurable}\Big\}.$$

The latter set is convex by Theorem 7.1. □

PROBLEMS.

1°. Let us consider the set $X \subset \mathbb{R}^3$ consisting of the points $a = (\alpha_1, \alpha_2, \alpha_3)$, such that

$$\alpha_1 = \int_0^1 \cos\big(\tau x(\tau)\big)\, d\tau, \quad \alpha_2 = \int_0^1 \tau^3 x^2(\tau)\, d\tau, \quad \alpha_3 = \int_0^1 \tau^2 2^{x(\tau)}\, d\tau,$$

where $x(\tau)$ ranges over the set of all Lebesgue measurable functions $x : [0,1] \longrightarrow (0, e) \cup (\pi, 4)$. Prove that X is a convex set.

2*. Draw a picture of the set X from Problem 1.

8. The "Simplex" of Probability Measures

In this section, we present a certain infinite-dimensional analogue of a standard simplex; cf. Section I.2.2. First, we construct the ambient space.

(8.1) The space of continuous functions and its dual. Let $C[0,1]$ be the vector space of all real-valued continuous functions on the interval $[0,1]$. We make $C[0,1]$ a topological vector space by declaring a set $U \subset C[0,1]$ open if for every $f \in U$ there is an $\epsilon > 0$ such that the set

$$U(f,\epsilon) = \Big\{ g \in C[0,1] : \quad |f(\tau) - g(\tau)| < \epsilon \quad \text{for all} \quad 0 \leq \tau \leq 1 \Big\}$$

is contained in U; cf. Problem 1, Section 2.6. Let $V[0,1]$ be the space of all continuous linear functionals $\phi : C[0,1] \longrightarrow \mathbb{R}$. The space $V[0,1]$ is often called the *space of signed Borel measures* on [0,1]. The reason for such a name is that every continuous linear functional $\phi : C[0,1] \longrightarrow \mathbb{R}$ can be represented in the form

$$\phi(f) = \int_0^1 f \, d\mu,$$

where μ is a signed Borel measure; see, for example, Appendix C of [**Co90**]. Thus μ may be a regular measure, like $\mu = \tau^2 \, d\tau$,

$$\phi(f) = \int_0^1 f(\tau)\tau^2 \, d\tau$$

or a δ-measure, like $\mu = \delta_{1/2}$,

$$\phi(f) = f(1/2).$$

We make $V[0,1]$ a topological vector space by introducing the weak* topology (see Section 2.8).

PROBLEMS.

1°. Check that $C[0,1]$ is a locally convex topological vector space.

2. Let us define the *norm* $p(\phi)$ of a linear functional $\phi : C[0,1] \longrightarrow \mathbb{R}$ by

$$p(\phi) = \sup\Big\{ |\phi(f)| : \quad f \in C[0,1] \quad \text{and} \quad |f(\tau)| \leq 1 \quad \text{for all} \quad \tau \in [0,1] \Big\}.$$

Find the norms of the linear functionals

$$\phi(f) = \int_0^1 f(\tau) \, d\tau, \quad \phi(f) = \int_0^1 \tau f(\tau) \, d\tau \quad \text{and} \quad \phi_\epsilon(f) = \frac{f(1/2+\epsilon) - f(1/2-\epsilon)}{2\epsilon},$$

where $\epsilon > 0$ is a parameter.

We introduce the central object of this section.

(8.2) Definitions. A linear functional $\phi : C[0,1] \longrightarrow \mathbb{R}$ is called *positive* if $\phi(f) \geq 0$ for all f such that $f(\tau) \geq 0$ for all $0 \leq \tau \leq 1$. Let $\Delta \subset V[0,1]$ be the set of all positive linear functionals $\phi : C[0,1] \longrightarrow \mathbb{R}$ such that $\phi(\mathbf{1}) = 1$, where $\mathbf{1}$ is the function that is identically 1 on $[0,1]$. The set Δ is called the set of all *Borel probability measures on* $[0,1]$. Let us fix a $\tau \in [0,1]$. The linear functional $\delta_\tau : C[0,1] \longrightarrow \mathbb{R}$, where $\delta_\tau(f) = f(\tau)$, is called the *delta-measure.*

The set Δ of Borel probability measures is our infinite-dimensional analogue of the standard simplex; see Problem 1 of Section 8.4 for some justification.

PROBLEMS.

1. Prove that every positive linear functional is continuous.

2. Prove that every continuous linear functional $\phi : C[0,1] \longrightarrow \mathbb{R}$ can be represented in the form $\phi = \phi_1 - \phi_2$, where ϕ_1 and ϕ_2 are positive linear functionals.

Hint: Let $f \in C[0,1]$ be a non-negative function, $f(\tau) \geq 0$ for all $\tau \in [0,1]$. Let $\phi_1(f) = \sup\{\phi(g) : \ 0 \leq g(\tau) \leq f(\tau) \text{ for all } \tau \in [0,1]\}$.

3. Let ϕ be a positive linear functional such that $\phi(\mathbf{1}) = 0$. Prove that $\phi = 0$.

4°. Prove that Δ is a convex set and that $\delta_\tau \in \Delta$ for any $\tau \in [0,1]$.

5°. Prove that $\delta_\tau(fg) = \delta_\tau(f)\delta_\tau(g)$ for any $\tau \in [0,1]$ and any two functions $f, g \in C[0,1]$.

6. Let $\phi : C[0,1] \longrightarrow \mathbb{R}$ be a linear functional such that $\phi(fg) = \phi(f)\phi(g)$ for any two functions $f, g \in C[0,1]$. Prove that either $\phi = 0$ or $\phi = \delta_\tau$ for some $\tau \in [0,1]$.

7. Prove that δ_τ is an extreme point of Δ for any $\tau \in [0,1]$.

(8.3) Proposition. *The set Δ is compact in the weak* topology of $V[0,1]$.*

PROBLEM.

1. Deduce Proposition 8.3 from Theorem 2.9.

(8.4) Proposition. *The extreme points of the set Δ of Borel probability measures on the interval $[0,1]$ are the delta-measures δ_τ for $\tau \in [0,1]$.*

Proof. Let us choose a $\tau^* \in [0,1]$. First, we prove that $\delta_* = \delta_{\tau^*}$ is indeed an extreme point of Δ. Clearly, $\delta_* \in \Delta$ (see Problem 4, Section 8.2). Suppose that $\delta_* = (\phi_1 + \phi_2)/2$ for some $\phi_1, \phi_2 \in \Delta$. Let f be any continuous function such that $f(\tau^*) = 1$ and $f(\tau) \leq 1$ for all $\tau \in [0,1]$. Since ϕ_1 and ϕ_2 are positive linear functionals, we have

$$\phi_i(f) = \phi_i\big(\mathbf{1} - (\mathbf{1} - f)\big) = \phi_i(\mathbf{1}) - \phi_i(\mathbf{1} - f) \leq 1.$$

On the other hand, $\delta_*(f) = 1$. Therefore, we must have $\phi_1(f) = \phi_2(f) = 1$. By linearity, it then follows that $\phi_1(f) = \phi_2(f) = f(\tau^*)$ for any function $f \in C[0,1]$ such that $f(\tau^*) \geq f(\tau)$ for all $\tau \in [0,1]$. That is, ϕ_1 and ϕ_2 agree with δ_{τ^*} on every function f attaining its maximum at τ^*. Now, for any $f \in C[0,1]$, we have

$f = g_1 - g_2$, where $g_1(\tau) = \min\{f(\tau^*), f(\tau)\}$ and $g_2(\tau) = \min\{0, f(\tau^*) - f(\tau)\}$.

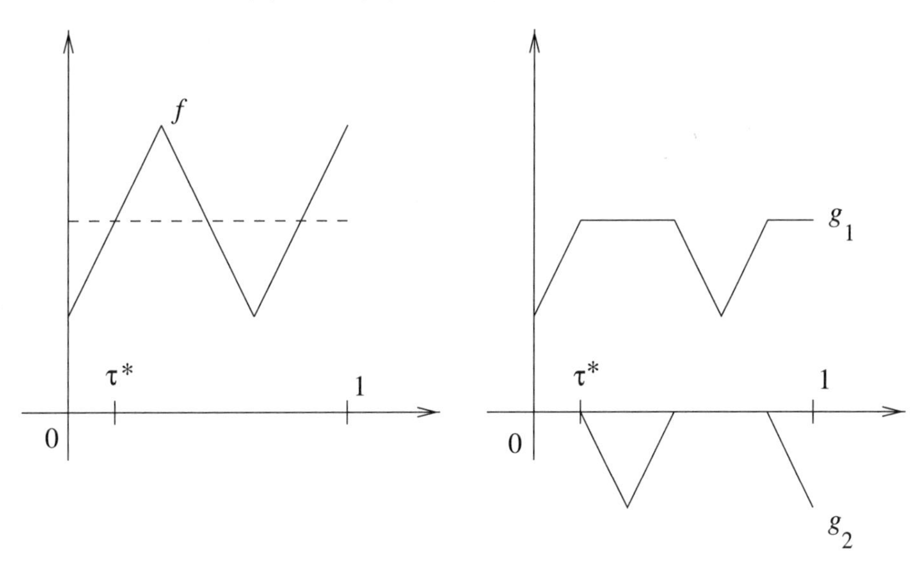

Figure 44. Representing a given function f as a difference of two functions g_1 and g_2 that attain their maximum at a given point τ^*

Hence g_1 and g_2 are continuous functions on the interval $[0, 1]$ that attain their maximum value at τ^*. Therefore, $\phi_i(g_j) = g_j(\tau^*)$ for $i, j = 1, 2$. By linearity, we conclude that $\phi_i(f) = \phi_i(g_1) - \phi_i(g_2) = f(\tau^*)$. Therefore, $\phi_1 = \phi_2 = \delta_*$, so δ_* is an extreme point.

Suppose that ϕ is an extreme point of Δ. First, we establish that $\phi(fg) = \phi(f)\phi(g)$ for any two functions $f, g \in C[0, 1]$ and then we deduce that $\phi = \delta_\tau$ for some $\tau \in [0, 1]$. Let us fix a function $h \in C[0, 1]$. Then $\psi : C[0, 1] \longrightarrow \mathbb{R}$ defined by $\psi(f) = \phi(hf)$ is a continuous linear functional on $C[0, 1]$. Let us choose h such that $0 < h(\tau) < 1$ for all $\tau \in [0, 1]$. Since ϕ is positive, we have $0 < \phi(h) < 1$. We define $\psi_1, \psi_2 : C[0, 1] \longrightarrow \mathbb{R}$ by

$$\psi_1(f) = \frac{\phi(hf)}{\phi(h)} \quad \text{and} \quad \psi_2(f) = \frac{\phi((\mathbf{1} - h)f)}{\phi(\mathbf{1} - h)}.$$

It is easy to see that $\psi_i \in \Delta$. We can write ϕ as a convex combination

$$\phi = \phi(h)\psi_1 + \phi(\mathbf{1} - h)\psi_2.$$

Since ϕ is an extreme point of Δ, we must have $\psi_1 = \psi_2 = \phi$. In particular, $\phi(fh) = \phi(h)\phi(f)$ for any $f \in C[0, 1]$ and any h such that $0 < h(\tau) < 1$ for each $\tau \in [0, 1]$. By linearity, it follows that $\phi(fg) = \phi(f)\phi(g)$ for any two $f, g \in C[0, 1]$.

Let $H = \big\{f \in C[0, 1] : \phi(f) = 0\big\}$ be the kernel of ϕ. Hence H is a (closed) hyperplane in $C[0, 1]$. We observe that for every $f \in H$, there is a $\tau \in [0, 1]$

such that $f(\tau) = 0$. Indeed, if $f \in H$ is a function which is nowhere 0, then $1/f$ is a continuous function and we would have had $1 = \phi(\mathbf{1}) = \phi(f \cdot (1/f)) = \phi(f)\phi(1/f) = 0$, which is a contradiction. Next, we observe that any set of finitely many functions $f_1, \ldots, f_m \in H$ has a common zero τ^*: $f_1(\tau^*) = \ldots = f_m(\tau^*) = 0$. Otherwise, the function $f = f_1^2 + \ldots + f_m^2$ is everywhere positive and $f \in H$ since $\phi(f) = \phi(f_1^2) + \ldots + \phi(f_m^2) = \phi^2(f_1) + \ldots + \phi^2(f_m) = 0$. But we already proved that every function $f \in H$ must have a zero. Finally, we conclude that there is a point $\tau^* \in [0,1]$ such that $f(\tau^*) = 0$ for any $f \in H$. Indeed, for any $f \in C[0,1]$, the set $X_f = \big\{\tau \in [0,1] : f(\tau) = 0\big\}$ is a closed set and any finite intersection $X_{f_1} \cap \ldots \cap X_{f_m}$ for $f_1, \ldots, f_m \in H$ is non-empty. Since $[0,1]$ is a compact interval, the intersection $\bigcap_{f \in H} X_f$ is non-empty. Now we see that H is a subset of the kernel of δ_{τ^*}, which implies that $\phi = \alpha\delta_{\tau^*}$ for some $\alpha \in \mathbb{R}$. Hence $\phi = \delta_{\tau^*}$, which completes the proof. □

PROBLEMS.

1. Consider the "discretization" of spaces $C[0,1]$ and $V[0,1]$. Namely, fix a set T of d points $0 \leq \tau_1 < \ldots < \tau_d \leq 1$ in the interval $[0,1]$. Interpret the space $C(T)$ of functions continuous on T as $\mathbb{R}^d$. Identify the space $V(T)$ of all continuous linear functionals on $C(T)$ with $\mathbb{R}^d$. Identify the set Δ of all non-negative functionals $\phi \in V(T)$ such that $\phi(\mathbf{1}) = 1$ with the simplex

$$\Delta_d = \Big\{(\gamma_1, \ldots, \gamma_d) : \quad \sum_{i=1}^{d} \gamma_i = 1 \quad \text{and} \quad \gamma_i \geq 0 \quad \text{for} \quad i = 1, \ldots, d\Big\};$$

cf. Problem 1 of I.2.2. Hence $\Delta \subset V[0,1]$ may be considered as an infinite-dimensional version of the simplex.

2. Let $O \subset V[0,1]$ be the set of functionals ϕ such that $|\phi(f)| \leq 1$ for any $f \in C[0,1]$ with the property that $|f(\tau)| \leq 1$ for each $\tau \in [0,1]$. Prove that the extreme points of O are δ_τ and $-\delta_\tau$ for $\tau \in [0,1]$. Hence O may be considered as an infinite-dimensional analogue of the (hyper)octahedron; see Section I.2.2.

Notation. For $\mu \in V[0,1]$ and $f \in C[0,1]$ we often write $\int_0^1 f \, d\mu$ instead of $\mu(f)$.

9. Extreme Points of the Intersection. Applications

We need a simple and useful result which describes the extreme points of the intersection of a convex set with an affine subspace. First, we describe the intersection with a hyperplane.

(9.1) Lemma. *Let V be a vector space and let $K \subset V$ be a convex set, such that for every straight line $L \subset V$ the intersection $K \cap L$ is a closed bounded interval, possibly empty or a point. If $H \subset V$ is an affine hyperplane, then every extreme point of $K \cap H$ can be expressed as a convex combination of at most two extreme points of K.*

Proof. Let u be an extreme point of $K \cap H$. If u is an extreme point of K, the result follows. Otherwise, there are two distinct points $u_+, u_- \in K$, such that $u = (u_+ + u_-)/2$. Let L be the straight line passing through u_+ and u_-. The intersection $L \cap K = [u_1, u_2]$ is a closed interval, containing u in its interior. We claim that u_1 and u_2 are extreme points of K. Suppose, for example, that u_1 is not an extreme point of K. Then there are two distinct points $v_+, v_- \in K$, such that $u_1 = (v_+ + v_-)/2$. Clearly, $v_+ \notin L$ and $v_- \notin L$. Consider the 2-dimensional plane A passing through the points v_+, v_-, and u_2 and the triangle $\Delta = \text{conv}(v_+, v_-, u_2)$ in the plane A; see Figure 45.

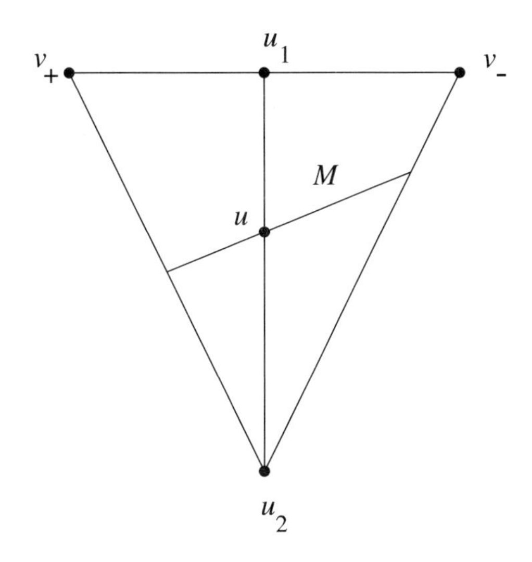

Figure 45

The point u is an interior point of the triangle. Since H is a hyperplane, the intersection $M = A \cap H$ is a straight line passing through u. Since u is an interior point of Δ, the intersection $\Delta \cap H$ is an interval, containing u as its interior point. Since K is convex, $\Delta \cap H \subset K \cap H$, so the intersection $K \cap H$ contains an interval, containing u as its interior point, which contradicts the assumption that u is an extreme point of $K \cap H$.

The contradiction shows that u_1 and u_2 are extreme points of $K \cap H$. Since u is a convex combination of u_1 and u_2, the result follows. □

Carathéodory's Theorem (Theorem I.2.3) determines the number of points of a set A which are needed to represent a point of the set B as a convex combination, provided B is the *convex hull* of A. The following result determines the number of points which are needed if B is a *section* of A. We call it the *dual* Carathéodory Theorem (we discuss the general concept of duality in Chapter IV and the duality between intersections and convex hulls in Section IV.1).

(9.2) Theorem. *Let K be a convex subset of a vector space V such that for any straight line L the intersection $K \cap L$ is a closed bounded interval, possibly empty or a point. Then every extreme point of the intersection of K with m hyperplanes $H_1, \ldots, H_m$ can be expressed as a convex combination of at most $m+1$ extreme points of K. Equivalently, if $A \subset V$ is an affine subspace such that* $\operatorname{codim} A = m$, *then every extreme point of the intersection $K \cap A$ can be expressed as a convex combination of at most $m+1$ extreme points of K.*

Proof. First, we prove that every extreme point of the intersection $K \cap H_1 \cap \ldots \cap H_m$ is a convex combination of extreme points of K. Let us define $K_0 = K$ and $K_i = K_{i-1} \cap H_i$, $i = 1, \ldots, m$. Since the intersection of a straight line L with any affine subspace in V is either L itself or a point or empty, we can apply Lemma 9.1 to K_i. Hence we conclude that every extreme point of K_i is a convex combination of at most two extreme points of K_{i-1}. Therefore, every extreme point of $K_m = K \cap H_1 \cap \ldots \cap H_m$ is a convex combination of at most 2^m extreme points of $K_0 = K$.

Let u be an extreme point of $K \cap H_1 \cap \ldots \cap H_m$. Let us write

$$\begin{aligned} &u = \alpha_1 u_1 + \ldots + \alpha_n u_n, \quad \text{where} \\ &\alpha_i \geq 0 \quad \text{and} \quad u_i \in \operatorname{ex}(K) \quad \text{for} \quad i = 1, \ldots, n \qquad \text{and} \\ &\alpha_1 + \ldots + \alpha_n = 1, \end{aligned}$$

with the smallest possible n (we know that we can choose $n \leq 2^m$). Clearly, $\alpha_i > 0$ for all $i = 1, \ldots, n$. Furthermore, the points $u_1, \ldots, u_n$ are affinely independent, since otherwise we could have reduced n as in the proof of Carathéodory's Theorem (see Theorem I.2.3). Therefore, u is an interior point of the $(n-1)$-dimensional simplex $\Delta = \operatorname{conv}(u_1, \ldots, u_n) \subset K$ (cf. Problem 1, Section II.2.3). If $n > m+1$, then the intersection $\Delta \cap H_1 \ldots \cap H_m$ contains an interval containing u as its interior point, which contradicts the assumption that u is an extreme point of $K \cap H_1 \cap \ldots \cap H_m$. Hence $n \leq m+1$ and the proof follows. □

PROBLEMS.

1. Show by example that, in general, the constant $m+1$ in Theorem 9.2 cannot be reduced.

2°. Let $K = \{x \in \mathbb{R}^d : \|x\| \leq 1\}$ be a ball. Prove that every extreme point of $K \cap H_1 \cap \ldots \cap H_m$ is an extreme point of K.

3. Let $\mathbb{R}^d : d = n(n+1)/2$ be the space of $n \times n$ symmetric matrices and let $K = \{X \succeq 0 \text{ and } \operatorname{tr}(X) = 1\}$. Prove that every extreme point of $K \cap H_1 \cap \ldots \cap H_m$ is a convex combination of not more than $\lfloor (\sqrt{8m+1} - 1)/2 \rfloor + 1$ extreme points of K.

Hint: Cf. Proposition II.13.1.

4. Let $\mathbb{R}^d$, $d = n^2$, be the space of all $n \times n$ matrices and let $K \subset \mathbb{R}^d$ be the Birkhoff Polytope (the polytope of doubly stochastic matrices; see Section II.5). Let $S \subset \mathbb{R}^d$ be the subspace of all $n \times n$ symmetric matrices. Prove that every extreme point of $K \cap S$ can be represented as a convex combination of not more than two extreme points of K.

A generalization of Problem 4:

5. Let G be a finite group of linear transformations of $\mathbb{R}^d$. Let $K \subset \mathbb{R}^d$ be a compact convex set, which is G-invariant: $g(x) \in K$ for all $g \in G$ and all $x \in K$. Let $L = \{x \in \mathbb{R}^d : g(x) = x \text{ for all } g \in G\}$ be the subspace of G-invariant vectors (check that L is indeed a subspace). Prove that every extreme point of $K \cap L$ is a convex combination of not more than $|G|$ (the cardinality of G) extreme points of K.

Theorem 9.2 looks intuitively obvious in small dimensions; see Figure 46.

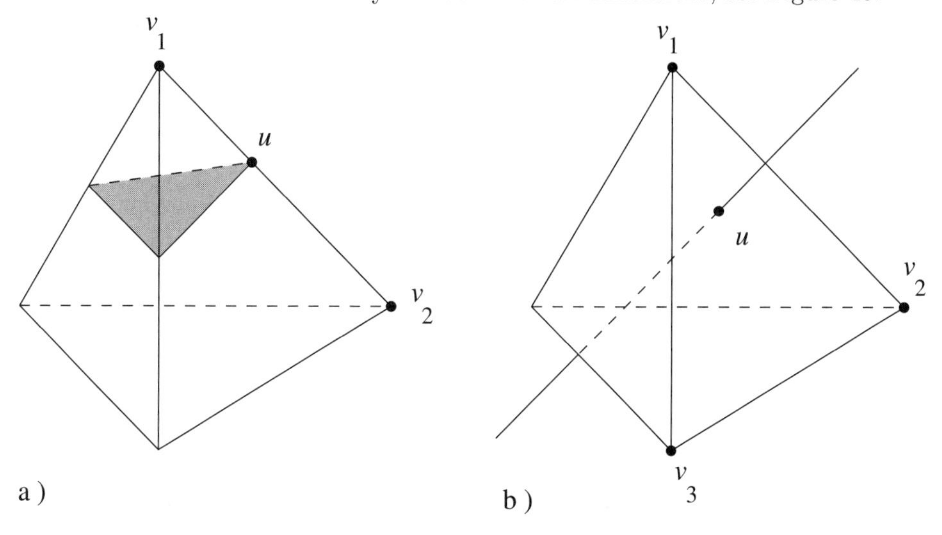

Figure 46. a) u is a convex combination of v_1 and v_2; b) u is a convex combination of v_1, v_2 and v_3.

It has some interesting infinite-dimensional applications.

(9.3) Application: extreme points of a set of probability measures. Let $f_1, \ldots, f_m \in C[0,1]$ be continuous functions on the interval $[0,1]$. Suppose that μ is an unknown Borel probability measure on the interval $[0,1]$ (see Section 8) but we know the expectations

$$\int_0^1 f_i \, d\mu = \alpha_i, \quad i = 1, \ldots, m.$$

We want to estimate the expectation

$$\alpha_0 = \int_0^1 g \, d\mu$$

of yet another known function $g \in C[0,1]$. For instance, if we have $f_i(\tau) = \tau^i$, $i = 1, \ldots, m$, then α_i are the *moments* of μ, and g is some other function of interest. Theorem 9.2 and Proposition 8.4 lead to the following useful *discretization principle.*

(9.4) Proposition. *Let us fix continuous functions $g, f_1, \dots, f_m$ on the interval $[0,1]$ and real numbers $\alpha_1, \dots, \alpha_m$. Suppose that the set B of Borel probability measures μ on $[0,1]$ satisfying the equations*

$$\int_0^1 f_i \, d\mu = \alpha_i \quad \textit{for} \quad i = 1, \dots, m$$

is non-empty. Then there exist measures $\mu^+, \mu^- \in B$ such that

1. *the measures μ^+ and μ^- are convex combinations of at most $m+1$ delta-measures:*

$$\mu^+ = \sum_{i=1}^{m+1} \lambda_i^+ \delta_{\tau_i^+}, \quad \mu^- = \sum_{i=1}^{m+1} \lambda_i^- \delta_{\tau_i^-},$$

where

$$\sum_{i=1}^{m+1} \lambda_i^+ = \sum_{i=1}^{m+1} \lambda_i^- = 1, \qquad \lambda_i^+, \lambda_i^- \geq 0 \quad \textit{for} \quad i = 1, \dots, m+1,$$

and $\tau_i^+, \tau_i^- \in [0,1]$ for $i = 1, \dots, m+1$;

2. *the set of values*

$$\alpha_0 = \int_0^1 g \, d\mu \quad \textit{for} \quad \mu \in B$$

is the interval $[\alpha^-, \alpha^+]$, where

$$\alpha^- = \int_0^1 g \, d\mu^- \quad \textit{and} \quad \alpha^+ = \int_0^1 g \, d\mu^+.$$

Proof. The set B can be represented as the intersection $B = \Delta \cap H_1 \cap \ldots \cap H_m$, where Δ is the simplex of Borel probability measures (see Section 8) and the affine hyperplanes H_i are defined by the equations:

$$H_i = \Big\{ \mu \in V[0,1] : \quad \int_0^1 f_i \, d\mu_i = \alpha_i \Big\}.$$

Clearly, B is convex. Since Δ is weak* compact (Proposition 8.3) and H_i are closed in the weak* topology, the set B is compact. The function

$$\alpha_0 : \mu \longmapsto \int_0^1 g \, d\mu$$

is weak* continuous and hence by Corollary 4.2 there is an extreme point μ^- of B where α_0 attains its minimum on B and there is an extreme point μ^+ of B where α_0 attains its maximum on B. By Theorem 9.2, μ^- and μ^+ can be represented as convex combinations of some $m+1$ extreme points of Δ. Proposition 8.4 implies that the extreme points of Δ are the delta-measures. $\square$

PROBLEMS.

1. Let us fix an $\alpha \in [0,1]$. Find a probability measure μ on $[0,1]$ such that $\int_0^1 \tau \, d\mu = \alpha$ and $\int_0^1 (\tau - \alpha)^2 \, d\mu$ is maximized.

2. Let μ be a probability measure on $[0,1]$. Let

$$D(\mu) = \int_0^1 \tau^2 \, d\mu - \left(\int_0^1 \tau \, d\mu\right)^2$$

be the *variance* of μ. Prove that $0 \leq D(\mu) \leq 1/4$.

Hint: Use Problem 1.

3. Let us consider the set S_{d+1} of all $(d+1)$-tuples $(\xi_0, \dots, \xi_d)$, where $\xi_i = \int_0^1 \tau^i \, d\mu$ for some $\mu \in \Delta$ and $i = 0, \dots, d$. Prove that S_{d+1} is the section of the moment cone M_{d+1} (see Section II.9) by the hyperplane $\xi_0 = 1$.

10. Remarks

For topological vector spaces, the Krein-Milman Theorem and spaces L^1, L^∞, $C[0,1]$ and $V[0,1]$ see [**Bou87**], [**Ru91**] and [**Co90**]. We note that many of the results of Sections 5, 7, 8 and 9 can be generalized in a straightforward way if the interval $[0,1]$ is replaced by a compact metric space X. The author learned Corollary 7.2 and its proof from A. Megretski; see also [**MT93**]. A general reference for optimal control theory is [**BH75**].

Chapter IV

Polarity, Duality and Linear Programming

Duality is a powerful technique in convexity theory which emerges as the most symmetric way to state separation theorems. Often, non-trivial facts are obtained from trivial ones by simple "translation" using the language of duality. We start with polarity in Euclidean space, prove that it extends to a valuation on the algebra of closed convex sets, complete the proof of the Weyl-Minkowski Theorem and prove a necessary and sufficient condition for a point to belong to the moment cone. We proceed to develop the duality theory for linear programming in topological vector spaces ordered by cones. We revisit many of the familiar problems such as the Diet Problem, the Transportation Problem, the problem of uniform (Chebyshev) approximation and the L^∞ linear programming problems related to optimal control and study some new problems, such as problems of semidefinite programming and the Mass-Transfer Problem. We obtain characterizations of optimal solutions in these problems.

1. Polarity in Euclidean Space

Let us define the central object of this section.

(1.1) Definition. Let $A \subset \mathbb{R}^d$ be a non-empty set. The set

$$A^\circ = \Big\{ c \in \mathbb{R}^d : \quad \langle c, x \rangle \leq 1 \quad \text{for all} \quad x \in A \Big\}$$

is called the *polar* of A.

PROBLEMS.

1°. Prove that A° is a closed convex set containing the origin.

2°. Prove that $\left(\mathbb{R}^d\right)^\circ = \{0\}$ and that $\{0\}^\circ = \mathbb{R}^d$.

3°. Prove that if $A \subset B$, then $B^\circ \subset A^\circ$.

4°. Prove that $\left(\bigcup_{i \in I} A_i\right)^\circ = \bigcap_{i \in I} A_i^\circ$.

5°. Let $A \subset \mathbb{R}^d$ be a set and let $\alpha > 0$ be a number. Prove that $(\alpha A)^\circ = \alpha^{-1} A^\circ$.

6. Let $L \subset \mathbb{R}^d$ be a linear subspace. Prove that L° is the orthogonal complement of L.

7°. Let $A = \operatorname{conv}\left(v_1, \ldots, v_m\right)$ be a polytope. Prove that

$$A^\circ = \left\{x \in \mathbb{R}^n : \quad \langle v_i, x \rangle \leq 1, \quad \text{for} \quad i = 1, \ldots, m\right\}.$$

8°. Prove that $A \subset (A^\circ)^\circ$.

Here is one of our main tools which we use to "translate" properties and statements about convex sets.

(1.2) The Bipolar Theorem. *Let $A \subset \mathbb{R}^d$ be a closed convex set containing the origin. Then $(A^\circ)^\circ = A$.*

Proof. From Problem 8, Section 1.1, we saw that $A \subset (A^\circ)^\circ$. It remains to show, therefore, that $(A^\circ)^\circ \subset A$. Suppose that there is a point u such that $u \in (A^\circ)^\circ$ and yet $u \notin A$. Since A is closed and convex, by Theorem III.1.3 there exists an affine hyperplane strictly separating u from A. In other words, there exists a vector $c \neq 0$ and a number α such that $\langle c, x \rangle < \alpha$ for all $x \in A$ and $\langle c, u \rangle > \alpha$. Since $0 \in A$, we conclude that $\alpha > 0$. Let us consider $b = \alpha^{-1} c$. We have $\langle b, x \rangle < 1$ for all $x \in A$ and hence $b \in (A^\circ)$. However, $\langle b, u \rangle > 1$, which contradicts the assumption that $u \in (A^\circ)^\circ$. Therefore $(A^\circ)^\circ \subset A$ and the result follows. □

PROBLEMS.

1°. Prove that $((A^\circ)^\circ)^\circ = A^\circ$ for every non-empty set $A \subset \mathbb{R}^d$.

2. Let $A \subset \mathbb{R}^d$ be a non-empty set. Prove that $(A^\circ)^\circ$ is the closure of $\operatorname{conv}\left(A \cup \{0\}\right)$.

3. Let $A = \left\{x \in \mathbb{R}^d : \langle c_i, x \rangle \leq 1 \text{ for } i = 1, \ldots, m\right\}$. Prove that $A^\circ = \operatorname{conv}(0, c_1, \ldots, c_m)$.

4. Let $A \subset \mathbb{R}^d$ be a non-empty set such that $A^\circ = A$. Prove that

$$A = \left\{x \in \mathbb{R}^d : \quad \|x\| \leq 1\right\},$$

the unit ball.

5. Let $A = \left\{(\xi_1, \ldots, \xi_d) : \quad -1 \leq \xi_i \leq 1 \text{ for } i = 1, \ldots, d\right\}$ and let $B = \left\{(\xi_1, \ldots, \xi_d) : \quad |\xi_1| + \ldots + |\xi_d| \leq 1\right\}$. Prove that $A^\circ = B$ and $B^\circ = A$.

6. Let us fix $p, q > 0$ such that $1/p + 1/q = 1$. Let

$$A = \Big\{(\xi_1, \dots, \xi_d) : \quad \sum_{i=1}^d |\xi_i|^p \le 1\Big\} \quad \text{and} \quad B = \Big\{(\xi_1, \dots, \xi_d) : \quad \sum_{i=1}^d |\xi_i|^q \le 1\Big\}.$$

Prove that $A^\circ = B$ and that $B^\circ = A$.

7°. Let $B(0, \lambda) = \big\{x \in \mathbb{R}^d : \|x\| \le \lambda\big\}$ be the ball of radius λ. Prove that $B(0, \lambda)^\circ = B(0, 1/\lambda)$.

Now we are ready to prove the second part of the Weyl-Minkowski Theorem (the first part is Corollary II.4.3), which will be the first time we obtain a result by translating a known result.

(1.3) Corollary. *A polytope is a polyhedron.*

Proof. Let $P = \operatorname{conv}\big(v_1, \dots, v_m\big)$ be a polytope in $\mathbb{R}^d$. Without loss of generality we may assume that $\operatorname{int} P \ne \emptyset$ (otherwise, we consider the smallest affine subspace containing P – see Theorem II.2.4) and that $0 \in \operatorname{int} P$ (otherwise, we shift P). In other words, $B(0, \epsilon) \subset P$ for some $\epsilon > 0$, where $B(0, \lambda)$ is the ball of radius λ centered at the origin. By Problem 3, Section 1.1 and Problem 7, Section 1.2, we have $P^\circ \subset B(0, \epsilon)^\circ = B(0, 1/\epsilon)$, so P° is bounded. Furthermore,

$$P^\circ = \Big\{c \in \mathbb{R}^d : \quad \langle c, v_i \rangle \le 1 \quad \text{for} \quad i = 1, \dots, m\Big\},$$

so P° is a polyhedron (cf. Problem 7, Section 1.1). Hence P° is a bounded polyhedron and therefore by Corollary II.4.3 it is a polytope:

$$P^\circ = \operatorname{conv}\big(u_1, \dots, u_n\big) \quad \text{for some} \quad u_i \in \mathbb{R}^d.$$

Applying the Bipolar Theorem (Theorem 1.2), we conclude that

$$P = (P^\circ)^\circ = \Big\{x \in \mathbb{R}^d : \quad \langle x, u_i \rangle \le 1 \quad \text{for} \quad i = 1, \dots, n\Big\}$$

is a polyhedron. □

PROBLEM.

1. Prove that the polar of a polyhedron is a polyhedron.

Here are some interesting dualities:

2. Let us define the *projective plane* $\mathbb{RP}^2$ as follows: the points of $\mathbb{RP}^2$ are the straight lines in $\mathbb{R}^3$ passing through the origin. The straight lines in $\mathbb{RP}^2$ are the planes in $\mathbb{R}^3$ passing through the origin. As usual, a line in $\mathbb{RP}^2$ (a plane in $\mathbb{R}^3$) consists of points of $\mathbb{RP}^2$ (lines in $\mathbb{R}^3$).

Prove that for every two distinct points A and B of $\mathbb{RP}^2$ there is a unique straight line $c \subset \mathbb{RP}^2$ that contains A and B.

Prove that every two distinct lines $a, b \in \mathbb{RP}^2$ intersect at a unique point C.

There is the *polarity correspondence* between points and lines of $\mathbb{RP}^2$: a point $A \in \mathbb{RP}^2$ corresponds to a line $a \subset \mathbb{RP}^2$ if and only if A, considered as a line in $\mathbb{R}^3$, is orthogonal to a, considered as a plane in $\mathbb{R}^3$. Prove that a point C is the intersection of two distinct lines a and b if and only if the distinct points A and B lie on the straight line c.

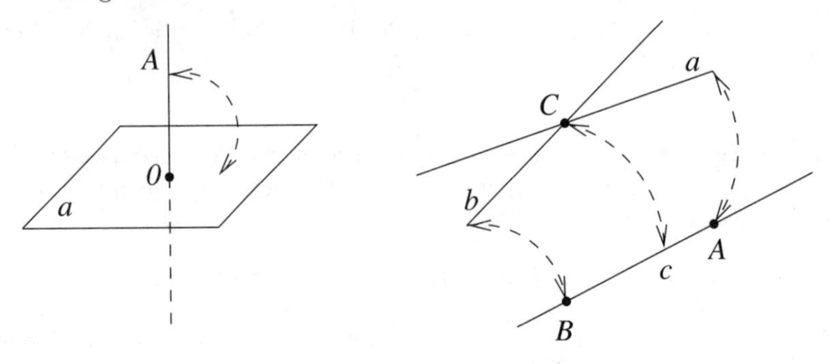

Figure 47. The polarity correspondence between points and straight lines in $\mathbb{RP}^2$

The Euclidean plane $\mathbb{R}^2$ can be embedded in $\mathbb{RP}^2$ in the following way. Let us identify $\mathbb{R}^2$ with an affine plane in $\mathbb{R}^3$, not passing through the origin. A point $x \in \mathbb{R}^2$ is identified with the straight line $X \subset \mathbb{R}^3$, passing through x and the origin. Hence, x is identified with a point of $\mathbb{RP}^2$; see Figure 48.

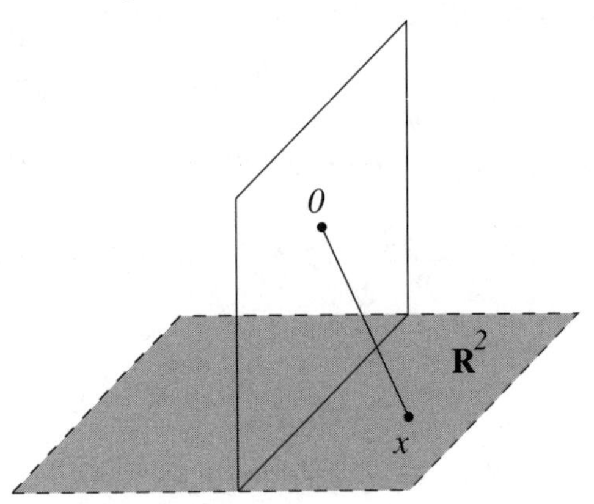

Figure 48. A point in $\mathbb{R}^2$ is identified with a point in $\mathbb{RP}^2$ and a line in $\mathbb{R}^2$ is identified with a line in $\mathbb{RP}^2$.

Prove that straight lines in $\mathbb{R}^2$ are identified with straight lines in $\mathbb{RP}^2$. Describe the points of $\mathbb{RP}^2$ that are not identified with any point of $\mathbb{R}^2$ and the straight lines in $\mathbb{RP}^2$ that are not identified with any straight line of $\mathbb{R}^2$.

3. A polytope $P \subset \mathbb{R}^d$ is called *self-dual* if the origin is an interior point of P and P° can be obtained from P by an invertible linear transformation. Suppose that $P = \operatorname{conv}(v_1, \dots, v_{d+1})$, where $v_1, \dots, v_{d+1}$ are affinely independent and $0 = v_1 + \dots + v_{d+1}$. Prove that P is self-dual.

4. Prove that the cube $I_d = \{(\xi_1, \dots, \xi_d) : \quad -1 \leq \xi_i \leq 1, \ i = 1, \dots, d\}$ is not self-dual for $d > 2$. Prove that I_1 and I_2 are self-dual.

5. Prove that a regular polygon in $\mathbb{R}^2$, containing the origin as its center, is self-dual.

6. Let e_1, e_2, e_3, e_4 be the standard orthonormal basis of $\mathbb{R}^4$. The polytope

$$P = \operatorname{conv}\big(e_i + e_j, -e_i - e_j, e_i - e_j : \quad 1 \leq i \neq j \leq 4\big)$$

is called the *24-cell*. Prove that P is self-dual.

The polytope P is a regular 4-dimensional polytope having twenty-four vertices and twenty-four facets; see also Problem 9 of Section VII.3.2.

Recall (Section II.8) that a convex set $K \subset \mathbb{R}^d$ is called a *convex cone* if $0 \in K$ and $\lambda x \in K$ for every $x \in K$ and $\lambda \geq 0$. Polars of cones look especially natural.

(1.4) Lemma. *Let $K \subset \mathbb{R}^d$ be a convex cone. Then*

$$K^\circ = \Big\{x \in \mathbb{R}^d : \quad \langle x, y \rangle \leq 0 \quad \textit{for every} \quad y \in K\Big\}.$$

Proof. By Definition 1.1, K° consists of all the points x in $\mathbb{R}^d$ such that $\langle x, y \rangle \leq 1$ for all $y \in K$. Suppose that for some $x \in K^\circ$ and some $y \in K$ we have $\langle x, y \rangle > 0$. Then, for a sufficiently large $\lambda > 0$, one has $\langle x, \lambda y \rangle > 1$. Since K is a cone, $\lambda y \in K$, which contradicts the definition of K°. □

PROBLEMS.

1°. Prove that the polar of a cone is a cone.

2°. Let $K_1, K_2 \subset \mathbb{R}^d$ be cones. Prove that $(K_1 + K_2)^\circ = K_1^\circ \cap K_2^\circ$.

3. Let $K_1, K_2 \subset \mathbb{R}^d$ be closed convex cones. Prove that $(K_1 \cap K_2)^\circ$ is the closure of $K_1^\circ + K_2^\circ$.

4. Prove that a convex cone $K \subset \mathbb{R}^d$ is the conic hull of a finite set if and only if K can be represented in the form $K = \{x \in \mathbb{R}^d : \langle c_i, x \rangle \leq 0, i \in I\}$ for a finite, possibly empty, set of vectors $\{c_i, i \in I\} \subset \mathbb{R}^d$. Such cones are called *polyhedral.*

It turns out that polarity can be extended to a valuation on the algebra $\mathcal{C}(\mathbb{R}^d)$ of closed convex sets; cf. Definition I.7.3. The following result is due to J. Lawrence; see [**L88**].

(1.5) Theorem. *There exists a linear transformation $\mathcal{D} : \mathcal{C}(\mathbb{R}^d) \longrightarrow \mathcal{C}(\mathbb{R}^d)$ such that $\mathcal{D}([A]) = [A^\circ]$ for any non-empty closed convex set $A \subset \mathbb{R}^d$.*

Proof. For $\epsilon > 0$, let us define the function $F_\epsilon : \mathbb{R}^d \times \mathbb{R}^d \longrightarrow \mathbb{R}$,

$$F_\epsilon(x, y) = \begin{cases} 1 & \text{if } \langle x, y \rangle \geq 1 + \epsilon, \\ 0 & \text{otherwise.} \end{cases}$$

For a function $g \in \mathcal{C}(\mathbb{R}^d)$ and a fixed $y \in \mathbb{R}^d$, let us consider the function $h_{y,\epsilon}(x) = g(x)F_\epsilon(x, y)$. We claim that $h_{y,\epsilon} \in \mathcal{C}(\mathbb{R}^d)$. By linearity, it suffices to check this in the case when $g = [A]$ is the indicator function of a closed convex set $A \subset \mathbb{R}^d$. In this case, $h_{y,\epsilon} = [A \cap H_{y,\epsilon}]$, where

$$H_{y,\epsilon} = \big\{x \in \mathbb{R}^d : \quad \langle x, y \rangle \geq 1 + \epsilon\big\}$$

is a closed halfspace. Since $[A \cap H_{y,\epsilon}]$ is a closed convex set, we have $h_{y,\epsilon} \in \mathcal{C}(\mathbb{R}^d)$ and we can apply the Euler characteristic χ (cf. Section I.7):

$$\chi(h_{y,\epsilon}) = \begin{cases} 1 & \text{if } \langle x, y \rangle \geq 1 + \epsilon \quad \text{for some} \quad x \in A, \\ 0 & \text{otherwise.} \end{cases}$$

For $g \in \mathcal{C}(\mathbb{R}^d)$ let us define $f_\epsilon = \mathcal{D}_\epsilon(g)$ by the formula:

$$f_\epsilon(y) = \chi(g) - \chi(h_{y,\epsilon}) = \chi(g) - \chi\big(g(x)F_\epsilon(x, y)\big) \quad \text{for all} \quad y \in \mathbb{R}^d.$$

By Theorem I.7.4, we conclude that

$$\mathcal{D}_\epsilon(\alpha_1 g_1 + \alpha_2 g_2) = \alpha_1 \mathcal{D}_\epsilon(g_1) + \alpha_2 \mathcal{D}_\epsilon(g_2) \tag{1.5.1}$$

for all $g_1, g_2 \in \mathcal{C}(\mathbb{R}^d)$ and all $\alpha_1, \alpha_2 \in \mathbb{R}$. Suppose that $g = [A]$ is the indicator function of a non-empty closed convex set A. Then for $f_\epsilon = \mathcal{D}_\epsilon(g)$ we have

$$f_\epsilon(y) = \begin{cases} 1 & \text{if } \langle x, y \rangle < 1 + \epsilon \quad \text{for all} \quad x \in A, \\ 0 & \text{otherwise.} \end{cases}$$

Therefore,

$$\lim_{\epsilon \longrightarrow +0} f_\epsilon(y) = \begin{cases} 1 & \text{if } \langle x, y \rangle \leq 1 \quad \text{for all} \quad x \in A, \\ 0 & \text{otherwise.} \end{cases} \tag{1.5.2}$$

Now, for $g \in \mathcal{C}(\mathbb{R}^d)$ we define $f = \mathcal{D}(g)$ by

$$f(y) = \lim_{\epsilon \longrightarrow +0} f_\epsilon(y) \quad \text{where} \quad f_\epsilon = \mathcal{D}_\epsilon(g).$$

By (1.5.1) and (1.5.2) it follows that $\mathcal{D}(g)$ is well defined, that $\mathcal{D}[A] = [A^\circ]$ for all non-empty closed convex sets $A \subset \mathbb{R}^d$ and that $\mathcal{D}(\alpha_1 g_1 + \alpha_2 g_2) = \alpha_1 \mathcal{D}(g_1) + \alpha_2 \mathcal{D}(g_2)$ for all $g_1, g_2 \in \mathcal{C}(\mathbb{R}^d)$ and all $\alpha_1, \alpha_2 \in \mathbb{R}$. $\square$

PROBLEMS.

1°. Check that the subspace $\mathcal{C}_0(\mathbb{R}^d) \subset \mathcal{C}(\mathbb{R}^d)$ spanned by the indicator functions $[A]$ of closed convex set $A \subset \mathbb{R}^d$ containing the origin is a subalgebra of $\mathcal{C}(\mathbb{R}^d)$ and that $\mathcal{D}$ maps $\mathcal{C}_0(\mathbb{R}^d)$ onto itself. Consider the restriction $\mathcal{D} : \mathcal{C}_0(\mathbb{R}^d) \longrightarrow \mathcal{C}_0(\mathbb{R}^d)$.

Prove that $\mathcal{D}^2 = \text{id}$, where id is the identity transformation of $\mathcal{C}_0(\mathbb{R}^d)$.

2. Find the eigenvalues of $\mathcal{D}$ as an operator $\mathcal{D} : \mathcal{C}_0(\mathbb{R}^d) \longrightarrow \mathcal{C}_0(\mathbb{R}^d)$.

3. Consider the subalgebra $\mathcal{C}op(\mathbb{R}^d)$ generated by the indicator functions $[K]$ of polyhedral cones K (see Problem 4 of Section 1.4). Prove that there exists a bilinear operation $\star$ such that $[K_1] \star [K_2] = [K_1 + K_2]$ for any two polyhedral cones K_1 and K_2 (cf. Problem 1 of Section I.8.2). Prove that the operator $\mathcal{D}$ maps the space $\mathcal{C}op(\mathbb{R}^d)$ onto itself. Furthermore, prove that $\mathcal{D}(fg) = \mathcal{D}(f) \star \mathcal{D}(g)$. Thus, polarity on polyhedral cones plays a role similar to that of the Fourier transform in analysis.

4. Let $I \cup J = \{1, \ldots , d\}$ be a partition and let us define two convex cones $K_1, K_2 \subset \mathbb{R}^d$ by

$$K_1 = \Big\{(\xi_1, \ldots , \xi_d) : \xi_i \geq 0 \quad \text{for} \quad i \in I \quad \text{and} \quad \xi_j > 0 \quad \text{for} \quad j \in J\Big\} \qquad \text{and}$$

$$K_2 = \Big\{(\xi_1, \ldots , \xi_d) : \xi_i \leq 0 \quad \text{for} \quad i \in I \quad \text{and} \quad \xi_j > 0 \quad \text{for} \quad j \in J\Big\}.$$

Prove that $\mathcal{D}\big([K_1]\big) = (-1)^{|J|}[K_2]$.

Theorem 1.5 allows us to associate with a valuation $\mu : \mathcal{C}(\mathbb{R}^d) \longrightarrow \mathbb{R}$ the dual valuation $\mu^* : \mathcal{C}(\mathbb{R}^d) \longrightarrow \mathbb{R}$ defined by the formula: $\mu^*(g) = \mu\big(\mathcal{D}(g)\big)$. Some interesting valuations defined on certain subspaces of $\mathcal{C}(\mathbb{R}^d)$ arise this way. For example, if $A \subset \mathbb{R}^d$ is a compact convex set containing the origin in its interior, then so is A° and we can define the *dual volume* of A by $\text{vol}^*(A) = \text{vol}(A^\circ)$. Hence vol^* extends to a valuation (linear functional) on the subspace of $\mathcal{C}(\mathbb{R}^d)$ spanned by the indicator functions of compact convex sets containing the origin in their interiors. Some interesting properties of volumes and dual volumes are described in Problems 3–6 of Section V.1.3.

If $K \subset \mathbb{R}^d$ is a closed convex cone, then so is K°. We can define the *spherical angle* $\gamma(K)$ of K as follows: let $\mathbb{S}^{d-1} \subset \mathbb{R}^d$ be the unit sphere endowed with the rotation invariant probability measure ν. We let $\gamma(K) = \nu(K \cap \mathbb{S}^{d-1})$. We can define the *exterior spherical angle* $\gamma^*(K)$ of K by $\gamma^*(K) = \gamma(K^\circ)$. In particular, if $d = 2$, then $\gamma(K)$ is the usual angle divided by 2π and $\gamma^*(K) = 0.5 - \gamma(K)$, provided $K \neq \{0\}, \mathbb{R}^2$. Hence γ^* extends to a valuation on the subspace of $\mathcal{C}(\mathbb{R}^d)$ spanned by the indicator functions of closed convex cones.

Theorem 1.5 implies that if the indicator functions of some non-empty closed convex sets satisfy a linear relation, then the indicator functions of their polars satisfy the same linear relation; see Figure 49.

(1.6) Corollary. *Let $A_i : i = 1, \ldots , m$ be non-empty closed convex sets in $\mathbb{R}^d$ and let $\alpha_i : i = 1, \ldots , m$ be real numbers such that*

$$\sum_{i=1}^{m} \alpha_i [A_i] = 0.$$

Then

$$\sum_{i=1}^{m} \alpha_i [A_i^\circ] = 0.$$

Proof. We apply the operator $\mathcal{D}$ of Theorem 1.5 to both sides of the identity $\sum_{i=1}^{m} \alpha_i [A_i] = 0$. □

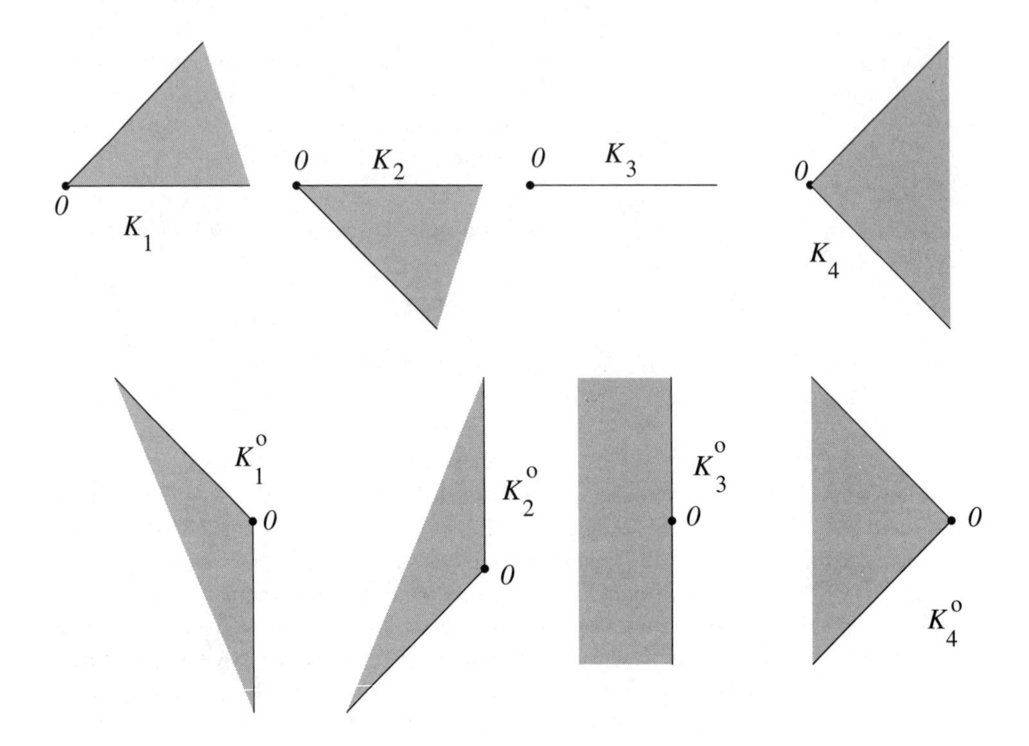

Figure 49. Example: $[K_1] + [K_2] - [K_3] = [K_4]$ and $[K_1^\circ] + [K_2^\circ] - [K_3^\circ] = [K_4^\circ]$

2. An Application: Recognizing Points in the Moment Cone

As an application of the Bipolar Theorem (Theorem 1.2), we show how to decide whether a given point belongs to the moment cone (see Section II.9). This is yet another demonstration of the "duality principle" which allows us to obtain some useful information without extra work.

Let us consider the space $\mathbb{R}^{d+1}$ of all $(d+1)$-tuples $x = (\xi_0, \ldots, \xi_d)$. Let us fix the interval $[0, 1] \subset \mathbb{R}$ (the case of a general interval $[\alpha, \beta] \subset \mathbb{R}$ is treated similarly). Let

$$g(\tau) = (1, \tau, \tau^2, \ldots, \tau^d) \quad \text{for} \quad 0 \leq \tau \leq 1$$

be the moment curve and let

$$M = \operatorname{co}\big(g(\tau) : \quad 0 \leq \tau \leq 1\big)$$

be the corresponding moment cone. Given a point $a = (\alpha_0, \ldots, \alpha_d)$, we want to decide whether $a \in M$. In other words (cf. Sections II.9, II.10 and III.9), we want to decide whether there exists a Borel measure μ on the interval $[0, 1]$ with the prescribed moments:

$$\int_0^1 \tau^i \, d\mu = \alpha_i, \quad i = 1, \ldots, d.$$

Let $K \subset \mathbb{R}^{d+1}$ be the cone of polynomials that are non-negative on $[0, 1]$ (see Section II.11):

$$K = \Big\{(\gamma_0, \ldots, \gamma_d) : \quad \gamma_0 + \gamma_1 \tau + \ldots + \gamma_d \tau^d \geq 0 \quad \text{for all} \quad \tau \in [0, 1]\Big\}.$$

(2.1) Lemma. *We have*

$$M = \Big\{x \in \mathbb{R}^{d+1} : \quad \langle x, c \rangle \geq 0 \quad \textit{for all} \quad c \in K\Big\}.$$

Proof. We can write

$$\begin{aligned} K &= \Big\{c \in \mathbb{R}^{d+1} : \quad \langle c, \ g(\tau) \rangle \geq 0 \quad \text{for all} \quad \tau \in [0, 1]\Big\} \\ &= \Big\{c \in \mathbb{R}^{d+1} : \quad \langle c, x \rangle \geq 0 \quad \text{for all} \quad x \in M\Big\}. \end{aligned}$$

Denoting $-K = \{-c : c \in K\}$ and using Lemma 1.4, we can write $-K = M^\circ$. Since both K and M are closed convex sets containing the origin (cf. Lemma II.9.3), applying the Bipolar Theorem (Theorem 1.2), we get

$$M = (-K)^\circ = \Big\{x \in \mathbb{R}^{d+1} : \quad \langle x, c \rangle \geq 0 \quad \text{for all} \quad c \in K\Big\},$$

which completes the proof. □

PROBLEM.

1. Let $H_{2k,n}$ be the real vector space of all homogeneous polynomials of degree $2k$ in n real variables $x = (\xi_1, \ldots, \xi_n)$. Let us introduce the scalar product $\langle f, g \rangle$ as in Problem 3 of Section I.3.5, thus making $H_{2k,n}$ a Euclidean space. For a vector $c \in \mathbb{R}^n$, let $p_c(x) = \langle c, x \rangle^{2k}$.

Let

$$K_1 = \Big\{p \in H_{2k,n} : \quad p(x) \geq 0 \quad \text{for all} \quad x \in \mathbb{R}^n\Big\}$$

be the cone of non-negative polynomials and let

$$K_2 = \operatorname{co}\Big(p_c \in H_{2k,n} : \quad c \in \mathbb{R}^n\Big)$$

be the conic hull of the powers of linear forms. Prove that $K_1^\circ = -K_2$ and that $K_2^\circ = -K_1$.

(2.2) Proposition. *Let $a = (\alpha_0, \ldots, \alpha_d)$ be a point.*

1. *Suppose that $d = 2m$ is even. Then $a \in M$ if and only if*

$$\sum_{i,j=0}^{m} \alpha_{i+j}\xi_i\xi_j \geq 0 \quad \text{for all} \quad x = (\xi_0, \ldots, \xi_m) \in \mathbb{R}^{m+1}$$

and

$$\sum_{i,j=0}^{m-1} \big(\alpha_{i+j+1} - \alpha_{i+j+2}\big)\xi_i\xi_j \geq 0 \quad \text{for all} \quad x = (\xi_0, \ldots, \xi_{m-1}) \in \mathbb{R}^{m}.$$

2. *Suppose that $d = 2m + 1$ is odd. Then $a \in M$ if and only if*

$$\sum_{i,j=0}^{m} \alpha_{i+j+1}\xi_i\xi_j \geq 0 \quad \text{for all} \quad x = (\xi_0, \ldots, \xi_m) \in \mathbb{R}^{m+1}$$

and

$$\sum_{i,j=0}^{m} \big(\alpha_{i+j} - \alpha_{i+j+1}\big)\xi_i\xi_j \geq 0 \quad \text{for all} \quad x = (\xi_0, \ldots, \xi_m) \in \mathbb{R}^{m+1}.$$

Proof. Suppose that $d = 2m$. Corollary II.11.3 asserts that the polynomial $p(\tau) = \gamma_0 + \gamma_1\tau + \ldots + \gamma_d\tau^d$ is non-negative on the interval $[0, 1]$ if and only if p can be written as a convex combination of polynomials q_i^2, where $q_i = \xi_0 + \xi_1\tau + \ldots + \xi_m\tau^m$ are polynomials of degree m, and polynomials $\tau(1-\tau)q_j^2$, where $q_j(\tau) = \xi_0 + \xi_1\tau + \ldots + \xi_{k-1}\tau^{m-1}$ are polynomials of degree $m - 1$.

Applying Lemma 2.1, we conclude that $a \in M$ if and only if

$$\begin{aligned} &\langle a, c\rangle \geq 0 \quad \text{for all points} \quad c = (\gamma_0, \ldots, \gamma_d) \quad \text{such that} \\ &\gamma_0 + \gamma_1\tau + \ldots + \gamma_d\tau^d = (\xi_0 + \xi_1\tau + \ldots + \xi_m\tau^m)^2 \end{aligned}$$

for some $x = (\xi_0, \ldots, \xi_m) \in \mathbb{R}^{m+1}$ and

$$\begin{aligned} &\langle a, c\rangle \geq 0 \quad \text{for all points} \quad c = (\gamma_0, \ldots, \gamma_d) \quad \text{such that} \\ &\gamma_0 + \gamma_1\tau + \ldots + \gamma_d\tau^d = \tau(1-\tau)(\xi_0 + \xi_1\tau + \ldots + \xi_{m-1}\tau^{m-1})^2 \end{aligned}$$

for some $x = (\xi_1, \ldots, \xi_{m-1}) \in \mathbb{R}^{m-1}$.

In other words, $a \in M$ if and only if

$$\sum_{k=0}^{d} \alpha_k \Big(\sum_{i+j=k} \xi_i\xi_j \Big) \geq 0 \quad \text{for all} \quad x = (\xi_0, \ldots, \xi_m) \quad \text{and}$$

$$\sum_{k=0}^{d} \alpha_k \Big(\sum_{i+j=k-1} \xi_i\xi_j - \sum_{i+j=k-2} \xi_i\xi_j \Big) \geq 0 \quad \text{for all} \quad x = (\xi_0, \ldots, \xi_{m-1}).$$

The first part now follows.

Suppose that $d = 2m+1$. Corollary II.11.3 asserts that the polynomial $p(\tau) = \gamma_0 + \gamma_1\tau + \ldots + \gamma_d\tau^d$ is non-negative on the interval $[0,1]$ if and only if p can be written as a convex combination of polynomials τq_i^2 and $(1-\tau)q_j^2$, where $q_i, q_j = \xi_0 + \xi_1\tau + \ldots + \xi_m\tau^m$ are polynomials of degree m.

Applying Lemma 2.1, we conclude that $a \in M$ if and only if

$$\langle c, a\rangle \geq 0 \quad \text{for all points} \quad c = (\gamma_0, \ldots, \gamma_d) \quad \text{such that}$$
$$\gamma_0 + \gamma_1\tau + \ldots + \gamma_d\tau^d = \tau(\xi_0 + \xi_1\tau + \ldots + \gamma_m\tau^m)^2$$

for some $x = (\xi_0, \ldots, \xi_m) \in \mathbb{R}^{m+1}$ and

$$\langle c, a\rangle \geq 0 \quad \text{for all points} \quad c = (\gamma_0, \ldots, \gamma_d) \quad \text{such that}$$
$$\gamma_0 + \gamma_1\tau + \ldots + \gamma_d\tau^d = (1-\tau)(\xi_0 + \xi_1\tau + \ldots + \gamma_m\tau^m)^2$$

for some $x = (\xi_0, \ldots, \xi_m) \in \mathbb{R}^{m+1}$. In other words, $a \in M$ if and only if

$$\sum_{k=0}^{d} \alpha_k \Big(\sum_{i+j=k-1} \xi_i\xi_j \Big) \geq 0 \quad \text{for all} \quad x = (\xi_0, \ldots, \xi_m) \quad \text{and}$$
$$\sum_{k=0}^{d} \alpha_k \Big(\sum_{i+j=k} \xi_i\xi_j - \sum_{i+j=k-1} \xi_i\xi_j \Big) \geq 0 \quad \text{for all} \quad x = (\xi_0, \ldots, \xi_m)$$

and the proof of Part 2 follows. □

The necessary and sufficient condition for a point to lie in the moment cone is often stated as follows.

(2.3) Corollary. *Let $a = (\alpha_0, \ldots, \alpha_d)$ be a point.*

1. *Suppose that $d = 2m$ is even.*

For $n = 0, \ldots, m$ let A_n be the $(n+1) \times (n+1)$ matrix whose (i,j)-th entry is α_{i+j-2}:

$$A_n = \begin{pmatrix} \alpha_0 & \alpha_1 & \ldots & \alpha_n \\ \alpha_1 & \alpha_2 & \ldots & \alpha_{n+1} \\ \ldots & \ldots & \ldots & \ldots \\ \alpha_n & \alpha_{n+1} & \ldots & \alpha_{2n} \end{pmatrix}.$$

For $n = 0, \ldots, m-1$ let A'_n be the $(n+1) \times (n+1)$ matrix whose (i,j)-th entry is $\alpha_{i+j-1} - \alpha_{i+j}$,

$$A'_n = \begin{pmatrix} \alpha_1 - \alpha_2 & \ldots & \alpha_{n+1} - \alpha_{n+2} \\ \alpha_2 - \alpha_3 & \ldots & \alpha_{n+2} - \alpha_{n+3} \\ \ldots & \ldots & \ldots \\ \alpha_{n+1} - \alpha_{n+2} & \ldots & \alpha_{2n+1} - \alpha_{2n+2} \end{pmatrix}.$$

Then $a \in M$ if and only if the matrices A_n for $n = 0, \ldots, m$ and A'_n for $n = 0, \ldots, m-1$ are positive semidefinite.

2. *Suppose that* $d = 2m + 1$ *is odd.*

For $n = 0, \ldots, m$ *let* A_n *be the* $(n+1) \times (n+1)$ *matrix whose* (i,j)*-th entry is* α_{i+j-1},

$$A_n = \begin{pmatrix} \alpha_1 & \alpha_2 & \ldots & \alpha_{n+1} \\ \alpha_2 & \alpha_3 & \ldots & \alpha_{n+2} \\ \ldots & \ldots & \ldots & \ldots \\ \alpha_{n+1} & \alpha_{n+2} & \ldots & \alpha_{2n+1} \end{pmatrix}.$$

For $n = 0, \ldots, m$ *let* A'_n *be the* $(n+1) \times (n+1)$ *matrix whose* (i,j)*-th entry is* $\alpha_{i+j-2} - \alpha_{i+j-1}$,

$$A'_n = \begin{pmatrix} \alpha_0 - \alpha_1 & \ldots & \alpha_n - \alpha_{n+1} \\ \alpha_1 - \alpha_2 & \ldots & \alpha_{n+1} - \alpha_{n+2} \\ \ldots & \ldots & \ldots \\ \alpha_n - \alpha_{n+1} & \ldots & \alpha_{2n} - \alpha_{2n+1} \end{pmatrix}.$$

Then $a \in M$ *if and only if the matrices* A_n *and* A'_n *for* $n = 0, \ldots, m$ *are positive semidefinite.*

Proof. Follows from Proposition 2.2. □

PROBLEM.

1. Draw a picture of M and check the conditions of Corollary 2.3 when $d = 2$.

3. Duality of Vector Spaces

In this section, we begin to introduce the general framework of duality. We define it first for vector spaces and extend to topological vector spaces in the next section. We start with a key definition.

(3.1) Definition. Let E and F be real vector spaces. A non-degenerate bilinear form $\langle\rangle : E \times F \longrightarrow \mathbb{R}$ is called a *duality* of E and F.

In other words, for each $e \in E$ and each $f \in F$, a real number $\langle e, f \rangle$ is defined such that

$$\langle \alpha_1 e_1 + \alpha_2 e_2,\ f \rangle = \alpha_1 \langle e_1,\ f \rangle + \alpha_2 \langle e_2,\ f \rangle \qquad \text{and}$$
$$\langle e,\ \alpha_1 f_1 + \alpha_2 f_2 \rangle = \alpha_1 \langle e,\ f_1 \rangle + \alpha_2 \langle e,\ f_2 \rangle$$

for all $e, e_1, e_2 \in E$, for all $f, f_1, f_2 \in F$ and for all $\alpha_1, \alpha_2 \in \mathbb{R}$ (that is, $\langle\rangle$ is a bilinear form). Moreover,

$$\text{if} \quad \langle e, f \rangle = 0 \quad \text{for all} \quad e \in E, \quad \text{then} \quad f = 0 \qquad \text{and}$$
$$\text{if} \quad \langle e, f \rangle = 0 \quad \text{for all} \quad f \in F, \quad \text{then} \quad e = 0$$

(that is, $\langle\rangle$ is non-degenerate).

Next, we list our main examples.

(3.2) Examples.

(3.2.1) Euclidean spaces. Let $E = \mathbb{R}^d$, let $F = \mathbb{R}^d$ and let

$$\langle x, y \rangle = \sum_{i=1}^{d} \xi_i \eta_i, \quad \text{where} \quad x = (\xi_1, \ldots, \xi_d) \quad \text{and} \quad y = (\eta_1, \ldots, \eta_d).$$

(3.2.2) Spaces of symmetric matrices. Let $E = F = \mathrm{Sym}_n$ be the space of $n \times n$ symmetric matrices (see Section II.12.1) and let

$$\langle A, B \rangle = \mathrm{tr}(AB) = \sum_{i,j=1}^{n} \alpha_{ij} \beta_{ij}, \quad \text{where} \quad A = (\alpha_{ij}) \quad \text{and} \quad B = (\beta_{ij}).$$

Of course, this can be considered as a particular case of (3.2.1).

(3.2.3) Spaces L^1 and L^∞. Let $E = L^1[0,1]$, let $F = L^\infty[0,1]$ (see Section III.5.1) and let

$$\langle f, g \rangle = \int_0^1 f(\tau) g(\tau) \, d\tau, \quad \text{where} \quad f \in L^1[0,1] \quad \text{and} \quad g \in L^\infty[0,1].$$

Similarly, one can define dualities of $L^p(X, \mu)$ and $L^q(X, \mu)$ with $1/p + 1/q = 1$ and a space X with a measure μ.

(3.2.4) Spaces of continuous functions and spaces of signed measures.

Let $E = C[0,1]$ be the space of all continuous functions on the interval $[0,1]$, let $F = V[0,1]$ be the space of signed Borel measures (see Section III.8.1) and let

$$\langle f, \mu \rangle = \int_0^1 f \, d\mu, \quad \text{where} \quad f \in C[0,1] \quad \text{and} \quad \mu \in V[0,1].$$

Similarly, one can consider a more general duality of $C(X)$ and $V(X)$ for a compact metric space X.

PROBLEMS.

1°. Check that (3.2.1)–(3.2.4) are indeed dualities.

2. Prove that there is no duality with $E = \mathbb{R}^n$ and $F = \mathbb{R}^m$, where $n \neq m$.

3°. Let $E = \mathbb{R}_\infty$ be the vector space of all sequences $x = (\xi_i : i \in \mathbb{N})$ of real numbers such that all but finitely many ξ_i's are 0 and let $F = \mathbb{R}^\infty$ be the space of all sequences $x = (\xi_i : i \in \mathbb{N})$ of real numbers. Let

$$\langle x, y \rangle = \sum_{i=1}^{\infty} \xi_i \eta_i, \quad \text{where} \quad x = (\xi_i) \in \mathbb{R}_\infty \quad \text{and} \quad y = (\eta_i) \in \mathbb{R}^\infty.$$

Note that the sum is well defined since it contains only finitely many non-zero terms. Prove that $\langle \rangle$ is a duality.

4°. Let $\langle \rangle_1 : E_1 \times F_1 \longrightarrow \mathbb{R}$ and $\langle \rangle_2 : E_2 \times F_2 \longrightarrow \mathbb{R}$ be dualities. Let $E = E_1 \oplus E_2$ and let $F = F_1 \oplus F_2$. Prove that $\langle \rangle : E \times F \longrightarrow \mathbb{R}$ defined by $\langle e_1 + e_2, \ f_1 + f_2 \rangle = \langle e_1, f_1 \rangle_1 + \langle e_2, f_2 \rangle_2$ is a duality.

We introduce another crucial definition.

(3.3) Definition. Let $\langle\rangle_1 : E_1 \times F_1 \longrightarrow \mathbb{R}$ and $\langle\rangle_2 : E_2 \times F_2 \longrightarrow \mathbb{R}$ be dualities of vector spaces. Let $A : E_1 \longrightarrow E_2$ and $A^* : F_2 \longrightarrow F_1$ be linear transformations. We say that A^* is *dual* (also called *adjoint*) to A provided

$$\langle A(e),\ f\rangle_2 = \langle e,\ A^*(f)\rangle_1$$

for all $e \in E_1$ and $f \in F_2$.

PROBLEMS.

1°. Prove that if a dual linear transformation A^* exists, it is necessarily unique.

2°. Let $\langle\rangle_1 : \mathbb{R}^n \times \mathbb{R}^n \longrightarrow \mathbb{R}$ and $\langle\rangle_2 : \mathbb{R}^m \times \mathbb{R}^m \longrightarrow \mathbb{R}$ be the dualities of Example 3.2.1. Let us choose the standard bases in $\mathbb{R}^n$ and $\mathbb{R}^m$. Prove that the matrix of the dual linear transformation $A^* : \mathbb{R}^m \longrightarrow \mathbb{R}^n$ is the transpose of the matrix of the transformation $A : \mathbb{R}^n \longrightarrow \mathbb{R}^m$.

3°. Let $\langle\rangle_1 : E_1 \times F_1 \longrightarrow \mathbb{R}$ be a duality of vector spaces and let $\langle\rangle_2 : \mathbb{R}^m \times \mathbb{R}^m \longrightarrow \mathbb{R}$ be the duality of Example 3.2.1. Let us fix some $f_1, \ldots, f_m \in F$ and let $A : E_1 \longrightarrow \mathbb{R}^m$ be a linear transformation defined by

$$A(e) = \big(\langle e, f_1\rangle_1, \ldots, \langle e, f_m\rangle_1\big) \quad \text{for all} \quad e \in E_1.$$

Prove that the transformation $A^* : \mathbb{R}^m \longrightarrow F_1$ defined by

$$A^*(x) = \xi_1 f_1 + \ldots + \xi_m f_m, \quad \text{where} \quad x = (\xi_1, \ldots, \xi_m)$$

is dual to A.

One can define polarity in this general framework.

(3.4) Polarity. Let $\langle\rangle : E \times F \longrightarrow \mathbb{R}$ be a duality of vector spaces. Let $A \subset E$ be a non-empty set. The set $A^\circ \subset F$,

$$A^\circ = \Big\{f \in F : \ \langle e, f\rangle \leq 1 \quad \text{for all} \quad e \in A\Big\},$$

is called the *polar* of A. Similarly, if $A \subset F$ is a non-empty set, the set $A^\circ \subset E$,

$$A^\circ = \Big\{e \in E : \ \langle e, f\rangle \leq 1 \quad \text{for all} \quad f \in A\Big\},$$

is called the polar of A.

Many properties of polars in Euclidean space are extended in a straightforward way to the general situation of spaces in duality.

PROBLEMS.

Let 0_E be the origin of E and let 0_F be the origin of F.

1°. Prove that the polar $A^\circ \subset F$ (resp. $A^\circ \subset E$) of a non-empty set $A \subset E$ (resp. $A \subset F$) is a convex set containing the origin 0_F (resp. 0_E).

2°. Prove that $E^\circ = \{0_F\}$, $\{0_E\}^\circ = F$, $F^\circ = \{0_E\}$ and that $\{0_F\}^\circ = E$.

3°. Prove that if $A \subset B$, then $B^\circ \subset A^\circ$.

4° Prove that $\left(\bigcup_{i\in I} A_i\right)^\circ = \bigcap_{i\in I} A_i^\circ$.

5°. Let A be a set and let $\alpha \neq 0$ be a number. Prove that $(\alpha A)^\circ = \alpha^{-1} A^\circ$.

6°. Let $L \subset E$ be a subspace. Prove that

$$L^\circ = \Big\{ f \in F : \ \langle e, f \rangle = 0 \quad \text{for all} \quad e \in L \Big\}$$

and similarly for subspaces $L \subset F$.

7°. Let $K \subset E$ be a convex cone. Prove that

$$K^\circ = \Big\{ f \in F : \ \langle e, f \rangle \leq 0 \quad \text{for all} \quad e \in K \Big\}$$

and similarly for cones $K \subset F$.

8°. Prove that $A \subset (A^\circ)^\circ$.

4. Duality of Topological Vector Spaces

There is a standard way to introduce topology from a duality.

(4.1) The topology of a duality. Let $\langle\rangle : E \times F \longrightarrow \mathbb{R}$ be a duality. We can make E and F topological vector spaces in the following way. A *basic open set* $U \subset E$ is a set of the type

$$U = \Big\{ e \in E : \quad \alpha_i < \langle e, f_i \rangle < \beta_i \quad \text{for} \quad i = 1, \ldots, m \Big\},$$

where $f_1, \ldots, f_m \in F$ are some vectors and $\alpha_1, \beta_1, \ldots, \alpha_m, \beta_m \in \mathbb{R}$ are some numbers. An open set in E is the union of some basic open sets. Similarly, a *basic open set* $W \subset F$ is a set of the type

$$W = \Big\{ f \in F : \quad \alpha_i < \langle f, e_i \rangle < \beta_i \quad \text{for} \quad i = 1, \ldots, m \Big\},$$

where $e_1, \ldots, e_m \in E$ are some vectors and $\alpha_1, \beta_1, \ldots, \alpha_m, \beta_m \in \mathbb{R}$ are some numbers. An open set in F is a union of some basic open sets.

We call this topology the *weak topology of the duality.*

PROBLEMS.

1°. Check that in the weak topology of the duality, E and F become locally convex topological vector spaces.

2. Prove that in Examples 3.2.1 and 3.2.2, the weak topology of the duality coincides with the standard topology in $\mathbb{R}^d$ and Sym_n, respectively.

3°. Prove that in Examples 3.2.3 and 3.2.4 the weak topology of the duality in the space F is the weak* topology.

4. Let V be a vector space and let $\phi, \phi_1, \ldots, \phi_m : V \longrightarrow \mathbb{R}$ be linear functionals such that $\phi(x) = 0$ whenever $\phi_1(x) = \ldots = \phi_m(x) = 0$. Prove that $\phi = \alpha_1\phi_1 + \ldots + \alpha_m\phi_m$ for some $\alpha_1, \ldots, \alpha_m \in \mathbb{R}$.

Hint: Consider the linear transformation

$$\Phi : V \longrightarrow \mathbb{R}^m, \quad \Phi(x) = \big(\phi_1(x), \ldots, \phi_m(x)\big).$$

Prove that there is a linear functional $\psi : \mathbb{R}^m \longrightarrow \mathbb{R}$, such that $\phi(x) = \psi\big(\Phi(x)\big)$ for all $x \in V$. Now use that $\psi(y)$ is a linear combination of the coordinates of y.

5. Let $\langle\rangle_1 : E_1 \times F_1 \longrightarrow \mathbb{R}$ and $\langle\rangle_2 : E_2 \times F_2 \longrightarrow \mathbb{R}$ be dualities. Let $E = E_1 \oplus E_2$ and $F = F_1 \oplus F_2$. Let us define a duality $\langle\rangle : E \times F \longrightarrow \mathbb{R}$, where $\langle e_1 + e_2,\ f_1 + f_2\rangle = \langle e_1, f_1\rangle_1 + \langle e_2, f_2\rangle_2$ as in Problem 4 of Section 3.2. Prove that the weak topology of the duality $\langle\rangle$ is the direct product of the topologies defined by the dualities $\langle\rangle_1$ and $\langle\rangle_2$; cf. Section III.2.1.

The next result underlines a perfect symmetry between spaces in duality.

(4.2) Theorem. *Let $\langle\rangle : E \times F \longrightarrow \mathbb{R}$ be a duality and let us make E and F topological vector spaces by introducing the weak topology of the duality $\langle\rangle$.*

Then for every $f \in F$ the function $\phi(e) = \langle e, f\rangle$ is a continuous linear functional $\phi : E \longrightarrow \mathbb{R}$ and for every $e \in E$ the function $\psi(f) = \langle e, f\rangle$ is a continuous linear functional $\psi : F \longrightarrow \mathbb{R}$. Moreover, every continuous linear functional $\phi : E \longrightarrow \mathbb{R}$ can be written as $\phi(e) = \langle e, f\rangle$ for some unique $f \in F$ and every continuous linear functional $\psi : F \longrightarrow \mathbb{R}$ can be written as $\psi(f) = \langle e, f\rangle$ for some unique $e \in E$.

Proof. Let us prove that $\phi(e) = \langle e, f\rangle$ is a continuous linear functional $\phi : E \longrightarrow \mathbb{R}$. Clearly, ϕ is linear. Let us choose $e_0 \in E$ and $\epsilon > 0$. Let $\alpha = \phi(e_0) = \langle e_0, f\rangle$ and let

$$U = \big\{e \in E : \quad \alpha - \epsilon < \langle e, f\rangle < \alpha + \epsilon\big\}.$$

Then U is a (basic) neighborhood of e_0 and for every $e \in U$ we observe that $|\phi(e) - \phi(e_0)| < \epsilon$. Hence ϕ is continuous at e_0. Since e_0 was arbitrary, ϕ is continuous. Similarly, we prove that the function $\psi : F \longrightarrow \mathbb{R}$ defined by $\psi(f) = \langle e, f\rangle$ for some fixed $e \in E$ is linear and continuous.

Let $\phi : E \longrightarrow \mathbb{R}$ be a continuous linear functional. In particular, ϕ is continuous at $e = 0$. This implies that there exists a neighborhood U of 0 such that $|\phi(x)| < 1$ for all $x \in U$. We can choose U to be a basic open set

$$U = \Big\{x \in E : \quad \alpha_i < \langle x, f_i\rangle < \beta_i \quad \text{for} \quad i = 1, \ldots, m\Big\},$$

where $\alpha_i < 0 < \beta_i$ and $f_i \in F$ for $i = 1, \ldots, m$. Let us choose an $\epsilon > 0$. Since ϕ is linear, for any

$$x \in \epsilon U = \Big\{x \in E : \quad \epsilon\alpha_i < \langle x, f_i\rangle < \epsilon\beta_i \quad \text{for} \quad i = 1, \ldots, m\Big\},$$

we have $|\phi(x)| < \epsilon$. Let

$$L = \Big\{x \in E : \quad \langle x, f_i \rangle = 0 \quad \text{for} \quad i = 1, \dots, m\Big\}$$

be a subspace in E. We have $L \subset \epsilon U$ for any $\epsilon > 0$. Therefore, $|\phi(x)| < \epsilon$ for all $x \in L$ and any $\epsilon > 0$. Hence $\phi(x) = 0$ for any $x \in L$. In other words, $\phi(x) = 0$ provided $\phi_i(x) = \langle x, f_i \rangle = 0$ for $i = 1, \dots, m$. Problem 4 of Section 4.1 implies that ϕ can be written as a linear combination $\phi = \alpha_1 \phi_1 + \ldots + \alpha_m \phi_m$. In other words, $\phi(e) = \langle e, f \rangle$, where $f = \alpha_1 f_1 + \ldots + \alpha_m f_m$. Similarly, we prove that every continuous linear functional $\psi : F \longrightarrow \mathbb{R}$ can be written as $\psi(f) = \langle e, f \rangle$ for some fixed $e \in E$.

It remains to show that representations $\phi = \langle \cdot, f \rangle$ and $\psi = \langle e, \cdot \rangle$ are unique. Indeed, suppose that there are two vectors $f_1, f_2 \in F$ such that $\phi(e) = \langle e, f_1 \rangle = \langle e, f_2 \rangle$ for all $e \in E$. Then $\langle e, f_1 - f_2 \rangle = 0$ for all $e \in E$ and since the bilinear form is non-degenerate, we must have $f_1 - f_2 = 0$ and $f_1 = f_2$. The uniqueness of representations for linear functionals $\psi : F \longrightarrow \mathbb{R}$ is proved similarly. □

Theorem 4.2 prompts the following definition.

(4.3) Definition. Let E and F be *topological* vector spaces. A non-degenerate bilinear form $\langle\rangle : E \times F \longrightarrow \mathbb{R}$ is called a *duality* of E and F if the following conditions are satisfied:

- for every $f \in F$ the linear functional $\phi : E \longrightarrow \mathbb{R}$ defined by $\phi = \langle \cdot, f \rangle$ is continuous and every continuous linear functional $\phi : E \longrightarrow \mathbb{R}$ can be written as $\phi = \langle \cdot, f \rangle$ for some unique $f \in F$

and

- for every $e \in E$ the linear functional $\phi : F \longrightarrow \mathbb{R}$ defined by $\phi = \langle e, \cdot \rangle$ is continuous and every continuous linear functional $\phi : E \longrightarrow \mathbb{R}$ can be written as $\phi = \langle e, \cdot \rangle$ for some unique $e \in E$.

PROBLEMS.

1. Let $\langle\rangle_1 : E_1 \times F_1 \longrightarrow \mathbb{R}$ and $\langle\rangle_2 : E_2 \times F_2 \longrightarrow \mathbb{R}$ be dualities of topological vector spaces. Let $A : E_1 \longrightarrow E_2$ be a continuous linear transformation. Prove that there exists a dual transformation $A^* : F_2 \longrightarrow F_1$.

Hint: Let us choose a vector $f \in F_2$. Then $\phi(e) = \langle A(e), f \rangle_2$ is a continuous linear functional $\phi : E_1 \longrightarrow \mathbb{R}$ and hence can be written in the form $\phi(e) = \langle e, f' \rangle$ for some $f' \in F_1$. Let $A^*(f) = f'$.

2°. Let $\langle\rangle : E \times F \longrightarrow \mathbb{R}$ be a duality of topological vector spaces. Prove that the polar of a set $A \subset E, F$ is a closed set.

3. Let $\langle\rangle : E \times F \longrightarrow \mathbb{R}$ be a duality of topological locally convex vector spaces. Let $A \subset E, F$ be a closed convex set containing the origin. Prove that $(A^\circ)^\circ = A$ (Bipolar Theorem).

4°. Let $\langle\rangle_1 : E_1 \times F_1 \longrightarrow \mathbb{R}$ and $\langle\rangle_2 : E_2 \times F_2 \longrightarrow \mathbb{R}$ be dualities of topological vector spaces. Let us introduce the topology of the direct product in $E = E_1 \oplus E_2$

and $F = F_1 \oplus F_2$. Prove that $\langle \rangle : E \times F \longrightarrow \mathbb{R}$,

$$\langle e_1 + e_2, f_1 + f_2 \rangle = \langle e_1, f_1 \rangle_1 + \langle e_2, f_2 \rangle_2,$$

is a duality of E and F.

5. Ordering a Vector Space by a Cone

We introduce a structure in a vector space which generalizes the concept of a system of linear inequalities in Euclidean space.

(5.1) Cones and orders. Let V be a (topological) vector space and let $K \subset V$ be a convex cone. The cone K defines an *order* on V as follows: we say that $x \leq y$ (sometimes we write $x \leq_K y$) provided $y - x \in K$. Similarly, we write $x \geq y$ provided $x - y \in K$. Thus, $K = \big\{x \in E : \ x \geq 0\big\}$.

PROBLEMS.

Let V be a vector space, let $K \subset V$ be a convex cone and let $\leq$ be the corresponding order on V.

1°. Prove that $x \geq x$ for any $x \in V$.

2°. Prove that if $x \leq y$ and $y \leq z$, then $x \leq z$.

3°. Prove that if $x \leq y$ and $\alpha \geq 0$, then $\alpha x \leq \alpha y$.

4°. Prove that if $x_1 \leq y_1$ and $x_2 \leq y_2$, then $x_1 + x_2 \leq y_1 + y_2$.

5°. Suppose that the cone K does not contain straight lines. Prove that if $x \leq y$ and $y \leq x$, then $x = y$.

6°. Suppose that $K = \{0\}$. Prove that $x \leq y$ if and only if $x = y$. Suppose that $K = V$. Prove that $x \leq y$ for any two $x, y \in V$.

7°. Suppose that V is a topological vector space and that $K \subset V$ is a closed cone. Prove that the order $\leq$ is *continuous*: if $\{x_n\}$ and $\{y_n\}$ are two sequences such that $x = \lim_{n \longrightarrow \infty} x_n$ and $y = \lim_{n \longrightarrow \infty} y_n$ and $x_n \leq y_n$ for all n, then $x \leq y$.

Here are our main examples of cones and associated orders.

(5.2) Examples.

(5.2.1) Euclidean space. Let $V = \mathbb{R}^d$ and let

$$\mathbb{R}^d{}_+ = \Big\{(\xi_1, \ldots, \xi_d) : \ \xi_i \geq 0 \quad \text{for} \quad i = 1, \ldots, d\Big\}.$$

Consequently, $x \leq y$ if $\xi_i \leq \eta_i$ for $i = 1, \ldots, d$, where $x = (\xi_1, \ldots, \xi_d)$ and $y = (\eta_1, \ldots, \eta_d)$.

(5.2.2) Symmetric matrices. Let $V = \operatorname{Sym}_n$ be the space of all $n \times n$ symmetric matrices and let

$$\mathcal{S}_+ = \Big\{X \in \operatorname{Sym}_n : \ X \quad \text{is positive semidefinite}\Big\}.$$

Consequently, $X \leq Y$ if $Y - X$ is a positive semidefinite matrix.

(5.2.3) Spaces L^1 and L^∞. For $V = L^1[0,1]$ (see Section III.5.1), we let

$$L^1_+ = \Big\{ f \in L^1[0,1] : \ f(\tau) \geq 0 \quad \text{for almost all} \quad \tau \in [0,1] \Big\}.$$

Similarly, for $V = L^\infty[0,1]$, we let

$$L^\infty_+ = \Big\{ f \in L^\infty[0,1] : \ f(\tau) \geq 0 \quad \text{for almost all} \quad \tau \in [0,1] \Big\}.$$

Hence $f \leq g$ if and only if $f(\tau) \leq g(\tau)$ for almost all $\tau \in [0,1]$.

(5.2.4) Spaces $C[0,1]$ and $V[0,1]$. Let $V = C[0,1]$ be the space of all continuous functions on the interval $[0,1]$. We define

$$C_+ = \Big\{ f \in C[0,1] : \ f(\tau) \geq 0 \quad \text{for every} \quad \tau \in [0,1] \Big\}.$$

Hence $f \leq g$ if and only if $f(\tau) \leq g(\tau)$ for all $\tau \in [0,1]$.

Let $V = V[0,1]$ be the space of all signed Borel measures on the interval $[0,1]$; see Section III.8.1. We let

$$V_+ = \Big\{ \mu \in V[0,1] : \ \int_0^1 f \, d\mu \geq 0 \quad \text{for every} \quad f \in C_+ \Big\}.$$

In other words, the cone V_+ consists of all positive linear functionals $\mu : C[0,1] \longrightarrow \mathbb{R}$; cf. Definition III.8.2. Thus $\mu \leq \nu$ if and only if $\int_0^1 f \, d\mu \leq \int_0^1 f \, d\nu$ for all continuous non-negative functions f.

PROBLEMS.

1°. Prove that the cones of Examples 5.2.1–5.2.4 are convex cones without straight lines.

2°. Prove that the cones of Examples 5.2.1–5.2.4 are closed in the weak topology defined by the corresponding duality; see Section 4.1.

For a cone $K \subset V$, let $K - K = \big\{ x - y : \quad x \in K, y \in K \big\}$; cf. Definition III.1.1.

3. Prove that $K - K = V$ in Examples 5.2.1–5.2.3.

4. Prove that $K - K = V$ in Example 5.2.4.

5. Let us consider the following *lexicographic order* $\succ$ in $\mathbb{R}^d$: we say that $x \succ y$, where $x = (\xi_1, \ldots, \xi_d)$ and $y = (\eta_1, \ldots, \eta_d)$, if there is a $1 \leq k \leq d$ such that $\xi_k > \eta_k$ and $\xi_i = \eta_i$ for all $i < k$. Prove that there is a convex cone $K \subset \mathbb{R}^d$ such that $x \succ y$ if and only if $x \geq_K y$. Prove that the cone K is not closed.

For the theory of linear programming that we are going to develop now, it is more convenient to deal with dual cones rather than with polars.

(5.3) Definition. Let $\langle\rangle : E \times F \longrightarrow \mathbb{R}$ be a duality of vector spaces. Let $K \subset E$ be a convex cone. The cone $K^* \subset F$,

$$K^* = \big\{ f \in F : \quad \langle e, f \rangle \geq 0 \quad \text{for all} \quad e \in K \big\},$$

is called *dual* to K. Similarly, if $K \subset F$ is a convex cone, the cone $K^* \subset E$,

$$K^* = \big\{ e \in E : \quad \langle e, f \rangle \geq 0 \quad \text{for all} \quad f \in K \big\},$$

is called *dual* to K.

PROBLEMS.

1°. Suppose that E and F are locally convex topological vector spaces and $K \subset E, F$ is a closed convex cone. Prove that $(K^*)^* = K$.

Hint: Use Problem 3 of Section 4.3.

2°. In Example 5.2 prove that

$$\begin{aligned}
&(\mathbb{R}^d_+)^* = \mathbb{R}^d_+, \\
&(\mathcal{S}_+)^* = \mathcal{S}_+, \\
&(L^\infty_+)^* = L^1_+ \quad \text{and} \quad (L^1_+)^* = L^\infty_+, \\
&(C_+)^* = V_+ \quad \text{and} \quad (V_+)^* = C_+.
\end{aligned}$$

6. Linear Programming Problems

From now on until the end of the chapter we are going to consider various *linear programming problems.* Linear programming grew out of a great variety of pure and applied problems; see [**Schr86**]. It may be considered as a theory of linear inequalities, thus extending linear algebra. We adopt the approach of L.V. Kantorovich who considered linear programming within the general framework of functional analysis. In this approach, linear inequalities are encoded by cones in the appropriate spaces; see [**AN87**]. Sometimes, this general theory is called "conic linear programming", "linear programming in spaces with cones" or "linear programming in ordered spaces" to distinguish it from the more traditional theory of linear inequalities in Euclidean space.

It is very useful to consider linear programming problems in pairs. To formulate a pair of linear programming problems, we need vector spaces in duality ordered by their respective cones and two linear transformations dual to each other.

(6.1) The problems. Let $\langle\rangle_1 : E_1 \times F_1 \longrightarrow \mathbb{R}$ and $\langle\rangle_2 : E_2 \times F_2 \longrightarrow \mathbb{R}$ be dualities of vector spaces. We fix a convex cone $K_1 \subset E_1$ and a convex cone $K_2 \subset E_2$. Let $K_1^* \subset F_1$ and $K_2^* \subset F_2$ be the dual cones:

$$\begin{aligned}
K_1^* &= \big\{ f \in F_1 : \quad \langle e, f \rangle \geq 0 \quad \text{for all} \quad e \in K_1 \big\} \quad \text{and} \\
K_2^* &= \big\{ f \in F_2 : \quad \langle e, f \rangle \geq 0 \quad \text{for all} \quad e \in K_2 \big\}.
\end{aligned}$$

Let $A : E_1 \longrightarrow E_2$ be a linear transformation. Suppose that $A^* : F_2 \longrightarrow F_1$ is the dual linear transformation, so that

$$\langle Ax, l\rangle_2 = \langle x, A^* l\rangle_1$$

for all $x \in E_1$ and $l \in F_2$. As is customary in linear programming, we write simply Ax and A^*l instead of $A(x)$ and $A^*(l)$.

Let us choose a $c \in F_1$ and a $b \in E_2$. We consider a pair of *linear programming problems*:

(6.1.1) Primal Problem.

$$\begin{aligned} \text{Find} \quad & \gamma = \inf\langle x, c\rangle_1 \\ \text{Subject to} \quad & Ax \geq_{K_2} b \qquad \text{and} \\ & x \geq_{K_1} 0 \end{aligned}$$

with a variable $x \in E_1$

and

(6.1.2) Dual Problem.

$$\begin{aligned} \text{Find} \quad & \beta = \sup\langle b, l\rangle_2 \\ \text{Subject to} \quad & A^*l \leq_{K_1^*} c \qquad \text{and} \\ & l \geq_{K_2^*} 0 \end{aligned}$$

with a variable $l \in F_2$.

A point $x \in E_1$ which satisfies the conditions $Ax \geq_{K_2} b$ and $x \geq_{K_1} 0$ is called a *feasible plan* in Problem 6.1.1 or a *primal feasible plan.* A point $l \in F_2$ which satisfies the conditions $A^*l \leq_{K_1^*} c$ and $l \geq_{K_2^*} 0$ is called a *feasible plan* in Problem 6.1.2 or a *dual feasible plan.* If Problem 6.1.1 does not have a feasible plan, we say that $\gamma = +\infty$. If Problem 6.1.2 does not have a feasible plan, we say that $\beta = -\infty$. If the infimum γ in Problem 6.1.1 is attained, a feasible plan x such that $\gamma = \langle x, c\rangle_1$ is called an *optimal plan* or a *primal optimal plan* (solution). Similarly, if the supremum β in Problem 6.1.2 is attained, a feasible plan l such that $\beta = \langle b, l\rangle_2$ is called an *optimal plan* or a *dual optimal plan* (solution).

PROBLEM.

1. If $(K_1^*)^* = K_1$ and $(K_2^*)^* = K_2$, then the primal and dual problems are interchangeable. To show that, make the substitution $l = -q$ in Problem 6.1.2:

$$\begin{aligned} \text{Find} \quad & -\inf\langle b, q\rangle_2 \\ \text{Subject to} \quad & A^*q \geq_{K_1^*} -c \qquad \text{and} \\ & q \geq_{-K_2^*} 0. \end{aligned}$$

Show that the dual problem may be interpreted as the primal Problem 6.1.1.

The following general result is often known under a unifying name of the "Weak Duality Theorem".

(6.2) Theorem.

1. *For any primal feasible plan x and any dual feasible plan l, we have*

$$\langle x, c\rangle_1 \geq \langle b, l\rangle_2.$$

In particular, $\gamma \geq \beta$ ("weak duality"). If $\langle x, c\rangle_1 = \langle b, l\rangle_2$, then x is a primal optimal plan, l is a dual optimal plan and $\gamma = \beta$.

2. *Suppose that x is a primal feasible plan and that l is a dual feasible plan such that*

$$\langle x,\ c - A^* l\rangle_1 = 0 \quad \text{and} \quad \langle Ax - b,\ l\rangle_2 = 0.$$

Then x is a primal optimal plan, l is a dual optimal plan and $\gamma = \beta$ ("optimality criterion").

3. *Suppose that x is a primal optimal plan, l is a dual optimal plan so that $\langle x, c\rangle_1 = \gamma$ and $\langle b, l\rangle_2 = \beta$. Suppose that $\gamma = \beta$. Then*

$$\langle x,\ c - A^* l\rangle_1 = 0 \quad \text{and} \quad \langle Ax - b,\ l\rangle_2 = 0$$

("complementary slackness").

Proof. Let us prove Part 1. Since x is a feasible plan in the primal problem, we have $Ax \geq_{K_2} b$, that is, $Ax - b \in K_2$. Since l is a feasible plan in the dual problem, we have $l \geq_{K_2^*} 0$, that is, $l \in K_2^*$. Therefore, $\langle Ax - b, l\rangle_2 \geq 0$ and hence $\langle Ax, l\rangle_2 \geq \langle b, l\rangle_2$. On the other hand, $\langle Ax, l\rangle_2 = \langle x, A^* l\rangle_1$ and hence

$$\langle x, A^* l\rangle_1 \geq \langle b, l\rangle_2.$$

Since l is a feasible plan in the dual problem, we have $A^* l \leq_{K_1^*} c$, hence $c - A^* l \in K_1^*$. Since x is a feasible plan in the primal problem, we have $x \geq_{K_1} 0$, that is, $x \in K_1$. Therefore, $\langle x, c - A^* l\rangle_1 \geq 0$ and hence

$$\langle x, c\rangle_1 \geq \langle x, A^* l\rangle_1.$$

Finally, we conclude that

$$\langle x, c\rangle_1 \geq \langle b, l\rangle_2.$$

If $\langle x, c\rangle_1 = \langle b, l\rangle_2$, then x has to be a primal optimal plan since for every primal feasible plan x' we must have $\langle x', c\rangle_1 \geq \langle b, l\rangle_2 = \langle x, c\rangle_1$. Similarly, l has to be a dual optimal plan since for every dual feasible plan l' we must have $\langle b, l'\rangle_2 \leq \langle x, c\rangle_1 = \langle b, l\rangle_2$.

To prove Part 2, we note that

$$\langle x, c\rangle_1 = \langle x, A^* l\rangle_1 = \langle Ax, l\rangle_2 = \langle b, l\rangle_2,$$

which by Part 1 implies that x and l are optimal plans in the primal and dual problems, respectively.

To prove Part 3, we observe that in the course of the proof of Part 1, we established a chain of inequalities

$$\langle x, c\rangle_1 \geq \langle x, A^*l\rangle_1 = \langle Ax, l\rangle_2 \geq \langle b, l\rangle_2$$

for any primal feasible plan x and any dual feasible plan l. Hence if $\langle x, c\rangle_1 = \langle b, l\rangle_2$, we must have $\langle x, c\rangle_1 = \langle x, A^*l\rangle_1$, which proves that $\langle x,\ c - A^*l\rangle_1 = 0$ and $\langle Ax, l\rangle_2 = \langle b, l\rangle_2$, which proves that $\langle Ax - b,\ l\rangle_2 = 0$. □

While Theorem 6.2 is very simple and may even seem tautological, it is quite powerful. By demonstrating a primal feasible plan x, we establish an upper bound on the optimal value of γ. By demonstrating a dual feasible plan l, we establish a lower bound for β and hence, by Theorem 6.2, for γ. The difference $\gamma - \beta$ is called the *duality gap*. The most interesting (and not infrequent) situation is when the duality gap is zero, that is, when $\gamma = \beta$. In this case, we can estimate the common optimal value with an arbitrary precision just by demonstrating appropriate primal and dual feasible plans. If the duality gap is zero and both primal and dual problems have solutions, the complementary slackness conditions often allow us to extract some useful information about optimal solutions and, in many cases, provides a way to find them. We address the situation of the zero duality gap in the next section.

PROBLEMS.

1°. Suppose that $\gamma = -\infty$ in Problem 6.1.1. Prove that Problem 6.1.2 does not have a feasible plan. Similarly, suppose that $\beta = +\infty$ in Problem 6.1.2. Prove that Problem 6.1.1 does not have a feasible plan.

2. Let us consider a problem of linear programming in $C[0,1]$:

$$\begin{aligned} \text{Find} \quad & \gamma = \inf \int_0^1 \tau x(\tau)\ d\tau \\ \text{Subject to} \quad & \int_0^1 x(\tau)\ d\tau = 1 \qquad \text{and} \\ & x(\tau) \geq 0 \quad \text{for all} \quad 0 \leq \tau \leq 1 \end{aligned}$$

in the primal variable $x \in C[0,1]$. Prove that $\gamma = 0$ but that there are no optimal solutions $x \in C[0,1]$. Using the duality (3.2.4) between continuous functions and measures, show that the dual problem is

$$\begin{aligned} \text{Find} \quad & \beta = \sup \lambda \\ \text{Subject to} \quad & \tau\ d\tau - \lambda\ d\tau \quad \text{is a non-negative measure on the interval} \quad [0,1] \end{aligned}$$

in the dual variable $\lambda \in \mathbb{R}$. Prove that $\beta = 0$ and that $\lambda = 0$ is the optimal solution.

3 (D. Gale). Let $E_1 = \mathbb{R}_\infty$, $F_1 = \mathbb{R}^\infty$ (see Problem 3, Section 3.2), $E_2 = \mathbb{R}^2$ and $F_2 = \mathbb{R}^2$. Let us consider the following linear programming problem for $x =$

$(\xi_i : i \in \mathbb{N}) \in \mathbb{R}_\infty$.

$$\begin{aligned} \text{Find} \quad & \gamma = \inf \xi_1 \\ \text{Subject to} \quad & \xi_1 + \sum_{k=2}^{\infty} k\xi_k = 1, \\ & \sum_{k=2}^{\infty} \xi_k = 0 \qquad \text{and} \\ & \xi_k \geq 0 \quad \text{for} \quad k \in \mathbb{N}. \end{aligned}$$

Interpret the problem:

$$\begin{aligned} \text{Find} \quad & \beta = \sup \lambda_1 \\ \text{Subject to} \quad & \lambda_1 \leq 1 \qquad \text{and} \\ & k\lambda_1 + \lambda_2 \leq 0 \quad \text{for} \quad k \geq 2 \end{aligned}$$

as the dual problem with the variable $l = (\lambda_1, \lambda_2) \in \mathbb{R}^2$. Prove that $(1, 0, \dots,)$ is the primal optimal plan with $\gamma = 1$ whereas $l = (0, 0)$ is a dual optimal plan with $\beta = 0$.

7. Zero Duality Gap

We turn our attention to a special situation when the infimum γ in the primal problem is equal to the supremum β in the dual problem. In this case, we say that the duality gap is zero or that there is no duality gap. The main objective of this section is to establish a sufficient criterion for the zero duality gap. We are going to use some topology now. Thus we assume that $\langle \rangle_1 : E_1 \times F_1 \longrightarrow \mathbb{R}$ and $\langle \rangle_2 : E_2 \times F_2 \longrightarrow \mathbb{R}$ are dualities of *locally convex* topological vector spaces, that $K_1 \subset E_1$ and $K_2 \subset E_2$ are *closed* convex cones and that $A : E_1 \longrightarrow E_2$ is a *continuous* linear transformation.

We state the criterion in the special case when $K_2 = \{0\}$. The general case, however, can be reduced to this special case; see Problem 1 of Section 7.1 and Problems 3–4 of Section 7.2.

(7.1) Standard and canonical problems. In the context of Section 6.1, let us suppose that $K_2 = \{0\}$, the origin in E_2. Hence $K_2^* = F_2$. To simplify notation, we denote the cone $K_1 \subset E_1$ just by K. Hence we get the problems:

(7.1.1) Primal Problem in the Canonical Form.

$$\begin{aligned} \text{Find} \quad & \gamma = \inf \langle x, c \rangle_1 \\ \text{Subject to} \quad & Ax = b \qquad \text{and} \\ & x \geq_K 0 \end{aligned}$$

in the primal variable $x \in E_1$.

(7.1.2) Dual Problem in the Standard Form.

$$\begin{aligned} \text{Find} \quad & \beta = \sup \langle b, l \rangle_2 \\ \text{Subject to} \quad & A^* l \leq_{K^*} c \end{aligned}$$

in the dual variable $l \in F_2$.

PROBLEMS.

1°. Consider Problem 6.1.1. Let us introduce the new spaces $E = E_1 \oplus E_2$ and $F = F_1 \oplus F_2$ with the duality $\langle\rangle : E \times F \longrightarrow \mathbb{R}$ defined by

$$\big\langle e_1 + e_2,\ f_1 + f_2 \big\rangle = \langle e_1, f_1\rangle_1 + \langle e_2, f_2\rangle_2;$$

cf. Problem 4 of Section 3.2. Let $K = K_1 \times K_2 \subset E$. Let $\hat{c} = (c, 0) \in E$ and let us define a transformation $\hat{A} : E \longrightarrow E_2$ by $\hat{A}(u, v) = Au - v$ for $u \in E_1$ and $v \in E_2$. Show that Problem 6.1.1 is equivalent to the following problem in the canonical form:

$$\begin{aligned} \text{Find} \quad & \gamma = \inf\langle x, \hat{c}\rangle \\ \text{Subject to} \quad & \hat{A}x = b \qquad \text{and} \\ & x \geq_K 0. \end{aligned}$$

Prove that the dual to the above problem is equivalent to Problem 6.1.2.

2°. Let $\langle\rangle : E \times F \longrightarrow \mathbb{R}$ be a duality and let $f_1, \ldots, f_m \in F_1$ be vectors. Let $c \in F_1$ be a vector and let $\beta_1, \ldots, \beta_m$ be real numbers. Let $K \subset E$ be a cone. Consider the linear programming problem:

$$\begin{aligned} \text{Find} \quad & \gamma = \inf\langle x, c\rangle \\ \text{Subject to} \quad & \langle x, f_i\rangle = \beta_i \quad \text{for} \quad i = 1, \ldots, m \qquad \text{and} \\ & x \geq_K 0. \end{aligned}$$

Interpret the problem:

$$\begin{aligned} \text{Find} \quad & \beta = \sup \sum_{i=1}^{m} \beta_i \lambda_i \\ \text{Subject to} \quad & \sum_{i=1}^{d} \lambda_i f_i \leq_{K^*} c \end{aligned}$$

with real variables $(\lambda_1, \ldots, \lambda_m)$ as the dual problem.

Given a duality $\langle\rangle : E \times F \longrightarrow \mathbb{R}$ of topological vector spaces, we can extend it to the duality between $E \oplus \mathbb{R}$ and $F \oplus \mathbb{R}$ by letting

$$\big\langle (e, \alpha_1), (f, \alpha_2) \big\rangle = \langle e, f\rangle + \alpha_1\alpha_2;$$

cf. Problem 4 of Section 4.3. We give $E \oplus \mathbb{R}$ and $F \oplus \mathbb{R}$ the usual topology of the direct product; cf. Section III.2.1.

We are going to establish a sufficient condition for the duality gap in Problems 7.1.1–7.1.2 to be zero. To this end, let us define the linear transformation $\widehat{A} : E_1 \longrightarrow E_2 \oplus \mathbb{R}$ by $\widehat{A}(x) = \big(Ax,\ \langle x, c\rangle_1\big)$. We are interested in the image $\widehat{A}(K)$ of the cone K.

(7.2) Theorem ("zero duality gap"). *Suppose that the cone*

$$\widehat{A}(K) = \Big\{\big(Ax, \ \langle x, c\rangle_1\big) : \quad x \in K\Big\}$$

is closed in $E_2 \oplus \mathbb{R}$ *and that there is a primal feasible plan* x. *Then* $\gamma = \beta$. *If* $\gamma > -\infty$, *then there is a primal optimal plan* x.

Proof. If $\gamma = -\infty$, then there are no dual feasible plans l (see Problem 1, Section 6.2) and hence $\beta = -\infty$. Therefore, we can assume that $\gamma > -\infty$.

In the space $E_2 \oplus \mathbb{R}$, let us consider the straight line $L = (b, \tau)$, where $-\infty < \tau < +\infty$. Then the intersection $L \cap \widehat{A}(K)$ is a closed set of points $\big(b, \langle x, c\rangle_1\big)$, where x is a primal feasible plan; see Figure 50. Since there are primal feasible plans and the objective function is bounded from below, this set is a closed bounded interval or a closed ray bounded from below. Therefore, there is a primal feasible plan x such that $\langle x, c\rangle_1 = \gamma$. Such an x will be a primal optimal plan.

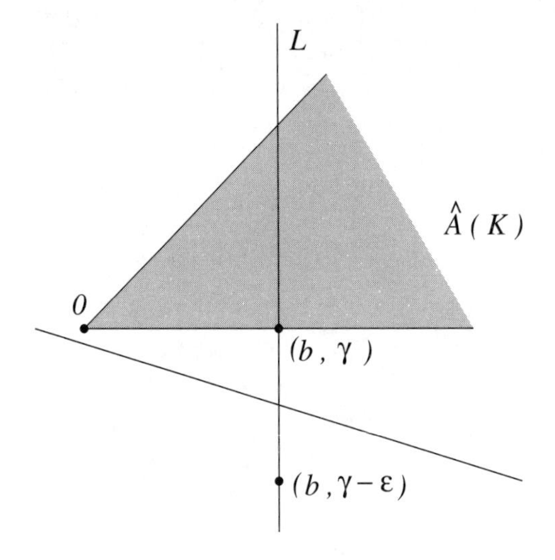

Figure 50

By Part 1 of Theorem 6.2, $\gamma \geq \beta$. Let us prove that for any $\epsilon > 0$ there is a dual feasible plan l such that $\langle b, l\rangle_2 \geq \gamma - \epsilon$. This would imply that $\beta \geq \gamma - \epsilon$, and, therefore, $\beta = \gamma$. We have

$$(b, \gamma - \epsilon) \notin \widehat{A}(K).$$

Since the cone $\widehat{A}(K)$ is closed, the point $(b, \gamma - \epsilon)$ can be strictly separated from $\widehat{A}(K)$ by a closed hyperplane (Theorem III.3.4). In other words, there exists a pair $(l, \sigma) \in F_2 \oplus \mathbb{R}$ and a number α such that

$$\langle b, l\rangle_2 + \sigma(\gamma - \epsilon) > \alpha$$

and

$$\langle Ax, l\rangle_2 + \sigma\langle x, c\rangle_1 < \alpha$$

for all $x \in K$. Choosing $x = 0$, we conclude that $\alpha > 0$. Suppose that for some $x \in K$ we have

$$\langle Ax, l\rangle_2 + \sigma\langle x, c\rangle_1 > 0.$$

Since K is a cone, choosing a sufficiently large $\lambda > 0$, we conclude that the inequality

$$\langle Ax, l\rangle_2 + \sigma\langle x, c\rangle_1 < \alpha$$

is violated for some $x' = \lambda x \in K$. Thus we must have that

$$\langle b, l\rangle_2 + \sigma(\gamma - \epsilon) > 0$$

and

$$\langle Ax, l\rangle_2 + \sigma\langle x, c\rangle_1 \leq 0$$

for all $x \in K$. In particular, if x_0 is a primal optimal plan then $\langle x_0, c\rangle_1 = \gamma$ and $Ax_0 = b$, so

$$\langle b, l\rangle_2 + \sigma\gamma \leq 0.$$

Therefore, $\sigma < 0$ and, by scaling (l, σ), if necessary, we can assume that $\sigma = -1$. Thus we have

$$\langle b, l\rangle_2 - (\gamma - \epsilon) > 0$$

and

$$\langle Ax, l\rangle_2 - \langle x, c\rangle_1 = \langle x, A^*l\rangle_1 - \langle x, c\rangle_1 = \langle x, A^*l - c\rangle_1 \leq 0$$

for all $x \in K$. Therefore, $c - A^*l \in K^*$, that is, $A^*l \leq_{K^*} c$. Hence we conclude that l is a dual feasible plan and that $\langle b, l\rangle_2 > \gamma - \epsilon$. □

Modifying and relaxing some of the conditions of Theorem 7.2, one can get some other useful criteria for the zero duality gap, as stated in the problems below. Problems 1–2 concern linear programming problems 7.1 in the standard and canonical forms whereas problems 3–4 deal with the general linear programming problems 6.1.

PROBLEMS.

1. Suppose that there is a primal feasible plan, that $\gamma > -\infty$ and that for every $\epsilon > 0$ there is a neighborhood $U \subset E_2 \oplus \mathbb{R}$ of $(b, \gamma - \epsilon)$ such that $U \cap \widehat{A}(K) = \emptyset$; see Figure 51. Prove that there is no duality gap. Sometimes, this condition is referred

to as the *stability condition.*

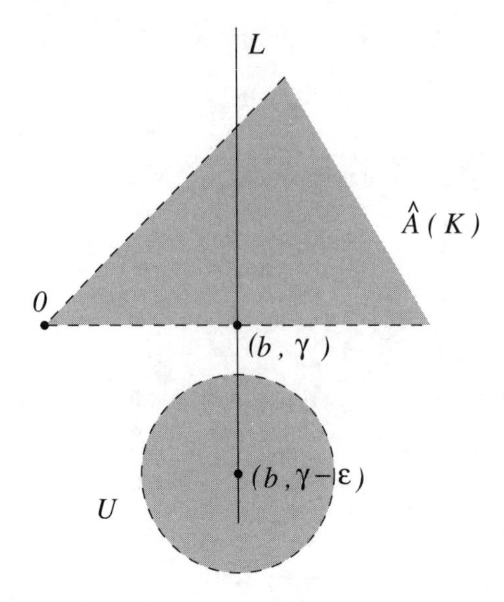

Figure 51

2. Suppose that there is a primal feasible plan x_0 and a neighborhood W of the origin in E_1 such that $x_0 + W \subset K_1$ and $U = AW = \{Aw : w \in W\}$ is a neighborhood of the origin in E_2. Prove that there is no duality gap. This condition is known as the *interior point* or *Kuhn-Tucker* condition.

Hint: Use Problem 1. Without loss of generality, we may suppose that $\gamma > -\infty$. For an $\epsilon > 0$, choose a point $(b, \gamma - \epsilon) \notin \widehat{A}(K)$. Without loss of generality, we may suppose that for some number s we have $|\langle x, c\rangle_1| < s$ for all $x \in x_0 + W$. Choose $\delta > 0$ such that $\delta(|s| + |\gamma|) < \epsilon/2$. Choose a neighborhood $W_0 \subset W$ of the origin such that $W_0 = -W_0$ and $\delta^{-1}W_0 \subset W$. Let $U_0 = AW_0$.

Together with (7.1.1), for an arbitrary $u \in U_0$ consider two problems:

$$\begin{array}{rll} \text{Find} & \gamma_+ = \inf\langle x, c\rangle_1 & \\ \text{Subject to} & Ax = b + u & \text{and} \\ & x \geq_K 0 & \end{array}$$

and

$$\begin{array}{rll} \text{Find} & \gamma_- = \inf\langle x, c\rangle_1 & \\ \text{Subject to} & Ax = b - u & \text{and} \\ & x \geq_K 0. & \end{array}$$

Prove that $\gamma \leq (\gamma_+ + \gamma_-)/2$ and that $\gamma_-, \gamma_+ \leq \gamma + \epsilon/2$. Deduce that $\gamma_+, \gamma_- \geq \gamma - \epsilon/2$ and construct a neighborhood of $(b, \gamma - \epsilon)$ which does not intersect $\widehat{A}(K)$.

The next two exercises address the general linear programming problems 6.1.

In Problems 6.1, let us suppose that $\langle\rangle_1$ and $\langle\rangle_2$ are dualities of locally convex topological vector spaces, that K_1 and K_2 are closed convex cones and that $A : E_1 \longrightarrow E_2$ is a continuous linear transformation.

3. Let us consider the cone $K \subset E_2 \oplus \mathbb{R}$,

$$K = \Big\{ \Big(Ax - y, \quad \langle x, c \rangle_1 \Big) : \quad x \in K_1, y \in K_2 \Big\}.$$

Prove that if K is closed in $E_2 \oplus \mathbb{R}$ and if there is a primal feasible plan, then $\gamma = \beta$. Prove that if $\gamma > -\infty$, then there is a primal optimal plan.

Hint: Use Problem 1 of Section 7.1.

4. Suppose that there is a primal feasible plan $x_0 \in \operatorname{int} K_1$ such that $Ax_0 - b \in \operatorname{int} K_2$ (this condition is known as *Slater's condition*). Suppose that there is a primal optimal plan. Prove that there is no duality gap.

Hint: Use Problem 1 of Section 7.1 and Problem 3 above.

If there is a non-zero duality gap, one can try to choose a stronger topology of E_1 and E_2 and use larger spaces F_1 and F_2 hoping that the duality gap disappears as the spaces grow bigger. Ultimately, we can choose the topology of algebraically open sets in E_1 and E_2; see Problems 2-3 of Section III.3.1. In this case, F_1 and F_2 become the spaces of *all* linear functionals on E_1 and E_2, respectively. Unfortunately, if the spaces F_1 and F_2 become "too large", the absence of the duality gap becomes much less of interest than for "reasonable" F_1 and F_2. Another possibility for eliminating the duality gap is to modify the cones. In the finite-dimensional situation, one can often enforce the interior point condition (see Problems 2 and 4 above) by replacing the cones with their appropriate faces. This trick is much less viable in the infinite-dimensional situation though.

To be able to use Theorem 7.2, we should be able to prove that the image of a certain cone K under a linear transformation is closed. Theorem I.9.2 provides us with one important example, that is, when K is a polyhedral cone. To state another useful result in this direction, we need to recall the definition of a base of a cone; see Definition II.8.3.

(7.3) Lemma. *Let V and W be topological vector spaces, let $K \subset V$ be a cone with a compact convex base and let $T : V \longrightarrow W$ be a continuous linear transformation such that* $(\ker T) \cap K = \{0\}$. *Then $T(K)$ is a closed convex cone in W.*

Proof. Let B be the base of K and let $C = T(B)$. Thus C is a compact convex set in V and $0 \notin C$. Moreover, $T(K) = \operatorname{co}(C)$. Hence by Lemma III.2.10, $T(K)$ is a closed convex cone. □

PROBLEMS.

1. Construct an example of a closed convex cone $K \subset \mathbb{R}^3$ with a compact base and a linear transformation $T : \mathbb{R}^3 \longrightarrow \mathbb{R}^2$ such that $T(K)$ is not closed.

2. Let $C[0,1]$ be the space of continuous functions on the interval $[0,1]$ and let $K = \{f : f(\tau) \geq 0 \text{ for all } \tau \in [0,1]\}$ be the cone of non-negative functions. Let us consider the linear transformation $T : C[0,1] \longrightarrow \mathbb{R}^2$ defined by

$$T(f) = \Big(f(0), \quad \int_0^1 f(\tau)\, d\tau\Big).$$

Prove that K is a closed cone, that $(\ker T) \cap K = \{0\}$ and that $T(K)$ is not closed.

3*. This problem assumes some knowledge of the Banach space theory; see [**Co90**].

Let V and W be Banach spaces, let $K \subset V$ be a closed convex cone and let $T : V \longrightarrow W$ be a continuous linear transformation such that $\dim(\ker T) < \infty$, $\operatorname{im} T \subset W$ is a closed subspace and $\operatorname{codim}(\operatorname{im} T) < \infty$. Suppose further that $K \cap (\ker T) = \{0\}$. Prove that $T(K) \subset W$ is a closed convex cone.

4. Let V be a topological vector space and let $v_1, \ldots, v_n \in V$ be a finite set of points. Prove that $\operatorname{co}(v_1, \ldots, v_n)$ is a closed convex cone.

5. Construct an example of two closed convex cones $K_1, K_2 \subset \mathbb{R}^3$ without straight lines such that $K_1 + K_2$ is not closed.

8. Polyhedral Linear Programming

In this section, we consider what is known as "classical" linear programming in Euclidean space. We call it "polyhedral" linear programming since it deals with orders defined by polyhedral cones.

(8.1) Problems. We consider problems in canonical/standard forms 7.1. Let $E_1 = F_1 = \mathbb{R}^n$ and let $E_2 = F_2 = \mathbb{R}^m$. We consider the standard dualities $\langle\rangle_1$ and $\langle\rangle_2$ of Example 3.2.1, which we denote simply by $\langle\rangle$ as the usual scalar product in Euclidean space. Let $K = \mathbb{R}^n_+$ (see Example 5.2.1). Then $K^* = \mathbb{R}^n_+$ and instead of writing $x \geq_K 0$ and $x \geq_{K^*} 0$, we simply write $x \geq 0$. Let $A : \mathbb{R}^n \longrightarrow \mathbb{R}^m$ be a linear transformation. We can think of A as an $m \times n$ matrix. Then $A^* : \mathbb{R}^m \longrightarrow \mathbb{R}^n$ is represented by the transposed $n \times m$ matrix. Letting $x = (\xi_1, \ldots, \xi_n)$, $c = (\gamma_1, \ldots, \gamma_n)$, $A = (\alpha_{ij})$, $i = 1, \ldots, m$, $j = 1, \ldots, n$, and $b = (\beta_1, \ldots, \beta_m)$, we get the following pair of linear programming problems:

(8.1.1) Primal Problem in the Canonical Form.

$$\text{Find} \quad \gamma = \inf \sum_{j=1}^n \gamma_j \xi_j$$

$$\text{Subject to} \quad \sum_{j=1}^n \alpha_{ij}\xi_j = \beta_i \quad \text{for} \quad i = 1, \ldots, m \quad \text{and}$$

$$\xi_j \geq 0 \quad \text{for} \quad j = 1, \ldots, n$$

in the primal variables $x = (\xi_1, \ldots, \xi_n) \in \mathbb{R}^n$.

(8.1.2) Dual Problem in the Standard Form.

$$\text{Find} \quad \beta = \sup \sum_{i=1}^{m} \beta_i \lambda_i$$

$$\text{Subject to} \quad \sum_{i=1}^{m} \alpha_{ij} \lambda_i \le \gamma_j \quad \text{for} \quad j = 1, \dots, n$$

in the dual variables $l = (\lambda_1, \dots, \lambda_m) \in \mathbb{R}^m$.

Our main result provides a sufficient criterion for zero duality gap and existence of optimal plans in Problems 8.1.1-8.1.2.

(8.2) Theorem ("strong duality"). *Suppose that there exists a primal feasible plan. Then $\gamma = \beta$. If, in addition, $\gamma > -\infty$, then there exist a primal optimal plan and a dual optimal plan.*

Proof. The cone $K = \mathbb{R}^n_+$ is a polyhedron. Therefore, by Theorem I.9.2 the image $\widehat{A}(K) = \{(Ax, \langle c, x \rangle) : x \in K\}$ is a polyhedron and hence is closed. Theorem 7.2 implies that $\gamma = \beta$ and that if $\gamma > -\infty$, then there is a primal optimal plan.

Let us prove that there is a dual optimal plan. One way to show that is to bring the dual problem into canonical form, cf. Problem 1 of Section 6.1 and Problem 3 of Section 7.2. Let us introduce "slack" vectors $y \in \mathbb{R}^n$ and $q_1, q_2 \in \mathbb{R}^m$. Since every vector $l \in \mathbb{R}^m$ can be written as a difference $q_1 - q_2$ for some $q_1, q_2 \in \mathbb{R}^m_+$ and inequality $A^* l \le c$ is equivalent to $A^* l + y = c$ and $y \ge 0$, we can construct a linear programming problem in the canonical form, which is equivalent to Problem 8.1.2:

$$\begin{aligned} \text{Find} \quad & \beta = \sup \langle b, q_1 - q_2 \rangle = -\inf \langle -b, q_1 - q_2 \rangle \\ \text{Subject to} \quad & A^*(q_1 - q_2) + y = c \qquad \text{and} \\ & q_1, q_2, y \ge 0. \end{aligned}$$

As we proved above, for linear programs in the canonical form there is an optimal plan provided $-\infty < \beta < +\infty$. Thus there exists an optimal solution (q_1, q_2, y) to the problem. Therefore, $l = q_1 - q_2$ is an optimal plan in Problem 8.1.2. $\square$

PROBLEMS.

1. Suppose that there exists a dual feasible plan. Prove that $\beta = \gamma$. Suppose, in addition, that $\beta < +\infty$. Prove that there exist a dual optimal plan and a primal optimal plan.

2. Construct an example of Problems 8.1.1 and 8.1.2 such that neither has a feasible plan.

3. Consider a pair of linear programming problems in the form:

$$\text{Find} \quad \gamma = \inf \sum_{j=1}^{n} \gamma_j \xi_j$$

$$\begin{aligned} \text{Subject to} \quad & \sum_{j=1}^{n} \alpha_{ij}\xi_j \geq \beta_i \quad \text{for} \quad i \in I_+ \\ & \sum_{j=1}^{n} \alpha_{ij}\xi_j = \beta_i \quad \text{for} \quad i \in I_0 \quad \text{and} \\ & \xi_j \geq 0 \quad \text{for} \quad j \in J_+ \end{aligned}$$

in the primal variables $x = (\xi_1, \ldots, \xi_n) \in \mathbb{R}^n$, where $I_0 \cup I_+ = \{1, \ldots, m\}$ and $J_+ \subset \{1, \ldots, n\}$

and

$$\text{Find} \quad \beta = \sup \sum_{i=1}^{m} \beta_i \lambda_i$$

$$\begin{aligned} \text{Subject to} \quad & \sum_{i=1}^{m} \alpha_{ij}\lambda_i \leq \gamma_j \quad \text{for} \quad j \in J_+ \\ & \sum_{i=1}^{m} \alpha_{ij}\lambda_i = \gamma_j \quad \text{for} \quad j \notin J_+ \quad \text{and} \\ & \lambda_i \geq 0 \quad \text{for} \quad i \in I_+ \end{aligned}$$

in the dual variables $l = (\lambda_1, \ldots, \lambda_m) \in \mathbb{R}^m$.

Prove that if there is a feasible plan in one of the problems, then $\gamma = \beta$ and if, in addition, $-\infty < \beta = \gamma < +\infty$, then there exist primal and dual optimal plans.

4. Let $c_0, \ldots, c_m \in \mathbb{R}^d$ be vectors and let $\langle c_i, x \rangle \leq 0$ for $i = 0, \ldots, m$ be a system of linear inequalities in $\mathbb{R}^d$. The inequality $\langle c_0, x \rangle \leq 0$ is called *active* provided there is an $x \in \mathbb{R}^d$, such that $\langle c_i, x \rangle \leq 0$ for $i = 1, \ldots, m$ and $\langle c_0, x \rangle > 0$. Prove *Farkas Lemma*: the inequality $\langle c_0, x \rangle \leq 0$ is not active if and only if $c_0 = \lambda_1 c_1 + \ldots + \lambda_m c_m$ for some non-negative $\lambda_1, \ldots, \lambda_m$.

Next, we establish polyhedral versions of the optimality criterion and complementary slackness conditions; cf. Theorem 6.2.

(8.3) Corollary. *Suppose there exist a primal feasible plan and a dual feasible plan. Then there exist a primal optimal plan and a dual optimal plan. Moreover,*

1. *if $x = (\xi_1, \ldots, \xi_n)$ is a primal feasible plan and $l = (\lambda_1, \ldots, \lambda_m)$ is a dual feasible plan and*

$$\xi_j > 0 \quad \textit{implies} \quad \sum_{i=1}^{m} \alpha_{ij}\lambda_i = \gamma_j,$$

then x is a primal optimal plan and l is a dual optimal plan ("optimality criterion");

2. *if $x = (\xi_1, \dots, \xi_n)$ is a primal optimal plan and $l = (\lambda_1, \dots, \lambda_m)$ is a dual optimal plan, then*

$$\xi_j > 0 \quad \textit{implies} \quad \sum_{i=1}^{m} \alpha_{ij}\lambda_i = \gamma_j$$

("complementary slackness").

Proof. Follows by Theorem 8.2 and Theorem 6.2. □

PROBLEM.

1. Consider the linear programming problems of Problem 3, Section 8.2. Suppose that $x = (\xi_1, \dots, \xi_n)$ is a primal feasible plan and that $l = (\lambda_1, \dots, \lambda_m)$ is a dual feasible plan. Prove that x is a primal optimal plan and l is the dual optimal plan if and only if

$$\xi_j > 0 \quad \text{implies} \quad \sum_{i=1}^{m} \alpha_{ij}\lambda_i = \gamma_j \quad \text{for} \quad j \in J_+ \quad \text{and}$$

$$\lambda_i > 0 \quad \text{implies} \quad \sum_{j=1}^{n} \alpha_{ij}\xi_j = \beta_i \quad \text{for} \quad i \in I_+.$$

(8.4) Example. The Diet Problem. As an illustration of the developed theory, let us return to the Diet Problem of Example II.4.4. In Problem 8.1.1, we interpret γ_j as the unit price of the j-th ingredient, α_{ij} as the content of the i-th nutrient in the j-th ingredient and β_i as the target quantity of the i-th nutrient. We are seeking to find the quantity ξ_j of the j-th ingredient so as to get a balanced diet of the minimal possible cost γ.

The dual variable λ_i, $i = 1, \dots, m$, in Problem 8.1.2 can be interpreted as the unit price of the i-th nutrient. Hence the dual problem can be interpreted as the problem of assigning prices to the nutrients in a "consistent" way (so that each ingredient costs at least as much as the nutrients it contains) and the total cost of all involved nutrients is maximized. Note that the prices λ_i are allowed to be negative ("customer incentives" or "bonuses"). Problem 8.1.2 may be interpreted as the problem faced by a manufacturer of vitamin pills who wants to supply the balanced diet in pills containing given nutrients and has to compete with a manufacturer of food. The condition that the package of pills costs not more than the food ingredient it is supposed to substitute means that the pill manufacturer has a chance to compete. Corollary 8.3 implies that for each primal optimal plan $(\xi_1, \dots, \xi_n)$ one can assign costs λ_i of pills in such a way that whenever $\xi_j > 0$ (the diet uses a positive quantity of the j-th ingredient) we have $\sum_{j=1}^{m} \alpha_{ij}\lambda_i = \gamma_j$ (the cost of the j-th ingredient is exactly the sum of the prices of the nutrients contained in the ingredient) and vice versa.

9. An Application: The Transportation Problem

In this section, we consider the Transportation Problem of Section II.7.

We are given a directed graph $G = (V, E)$ with the set $V = \{1, \ldots, n\}$ of vertices and a set E of m edges $i \rightarrow j$. As in Section II.7, we suppose that there are no loops $i \rightarrow i$. To every vertex i a number β_i ("demand" if $\beta_i > 0$, "supply" if $\beta_i < 0$ and "transit" if $\beta_i = 0$) is assigned. To every edge $i \rightarrow j$, a non-negative number γ_{ij} (the cost per unit of transportation) is assigned. The objective is to find a feasible flow ξ_{ij} for all edges $(i \rightarrow j) \in E$ minimizing the total cost of transportation. Hence the problem is stated as follows:

(9.1) Primal (transportation) problem.

$$\text{Find} \quad \gamma = \inf \sum_{(i \rightarrow j) \in E} \gamma_{ij} \xi_{ij}$$

$$\text{Subject to} \quad \sum_{j:(j \rightarrow i) \in E} \xi_{ji} - \sum_{j:(i \rightarrow j) \in E} \xi_{ij} = \beta_i \quad \text{for every vertex} \quad i \in V$$

$$\text{and} \quad \xi_{ij} \geq 0 \quad \text{for all} \quad (i \rightarrow j) \in E.$$

We observe that Problem 9.1 is a primal problem (8.1.1) in the canonical form. Hence we obtain the dual problem:

(9.2) Dual problem.

$$\text{Find} \quad \beta = \sup \sum_{i=1}^{n} \beta_i \lambda_i$$

$$\text{Subject to} \quad \lambda_j - \lambda_i \leq \gamma_{ij} \quad \text{for every edge} \quad (i \rightarrow j) \in E.$$

Problem 9.2 has the following interpretation: the variable λ_i is the price of the commodity at the i-th vertex. A vector $l = (\lambda_1, \ldots, \lambda_n)$ of prices is feasible provided one cannot gain by buying the commodity in one vertex and selling it in another vertex with the transportation costs taken into account. Thus $\{\lambda_i\}$ may be interpreted as the prices of the product if we are to open retail shops at the vertices of the graph. The goal is to assign prices in such a way that the total cost of the commodity over the whole network is maximized.

PROBLEM.

1°. Check that Problem 9.2 is indeed dual to Problem 9.1.

(9.3) Corollary. *Suppose that there is a primal feasible plan (feasible flow). Then there is a primal optimal plan (optimal flow) and there is a dual optimal plan. Let* $x = \big(\xi_{ij} : (i \rightarrow j) \in E\big)$ *be a primal feasible plan and let* $l = (\lambda_1, \ldots, \lambda_n)$ *be a dual feasible plan. Then* x *is a primal optimal plan and* l *is a dual optimal plan if and only if*

$$\xi_{ij} > 0 \quad \textit{implies} \quad \lambda_j - \lambda_i = \gamma_{ij}.$$

Proof. Since the costs γ_{ij} are non-negative, we conclude that $\gamma > -\infty$. Moreover, $l = 0$ is a dual feasible plan. Hence the result follows by Corollary 8.3. □

Corollary 9.3 suggests an algorithm for solving the Transportation Problem. The algorithm turns out to be extremely efficient in practice and constitutes a particular case of a more general *simplex method*; see Chapters 11 and 12 of [**Schr86**].

(9.4) A sketch of the algorithm. Proposition II.7.2 (see also Definition II.7.3) implies that if $x = (\xi_{ij})$ is an extreme point of the set of all feasible flows (transportation polyhedron), then the edges $i \longrightarrow j$, where $\xi_{ij} > 0$, form a forest in G. Suppose that $x = (\xi_{ij})$ is a feasible flow. To make our problem simpler, suppose that the set of edges $i \longrightarrow j$, where $\xi_{ij} > 0$, form a *spanning tree* T in G, that is, a tree such that every vertex i of V is a vertex of the tree. Essentially, this assumption means that the flow is sufficiently non-degenerate.

Let us assign the price λ_1 arbitrarily. Then, if $i \longrightarrow j$ is an edge of T and λ_i is computed, we compute $\lambda_j = \lambda_i + \gamma_{ij}$. Thus we compute the prices λ_i for $i = 1, \ldots, n$. If the vector $l = (\lambda_1, \ldots, \lambda_n)$ is a dual feasible plan, then by Corollary 9.3, x is an optimal flow. Otherwise, there is an edge $(i \longrightarrow j) \in E$, $i \longrightarrow j \notin T$, such that $\lambda_j - \lambda_i > \gamma_{ij}$. This *suggests* to us to use the edge $(i \longrightarrow j)$ in our shipment plan (informal justification: if we buy at i, ship to j and sell at j, we make a profit).

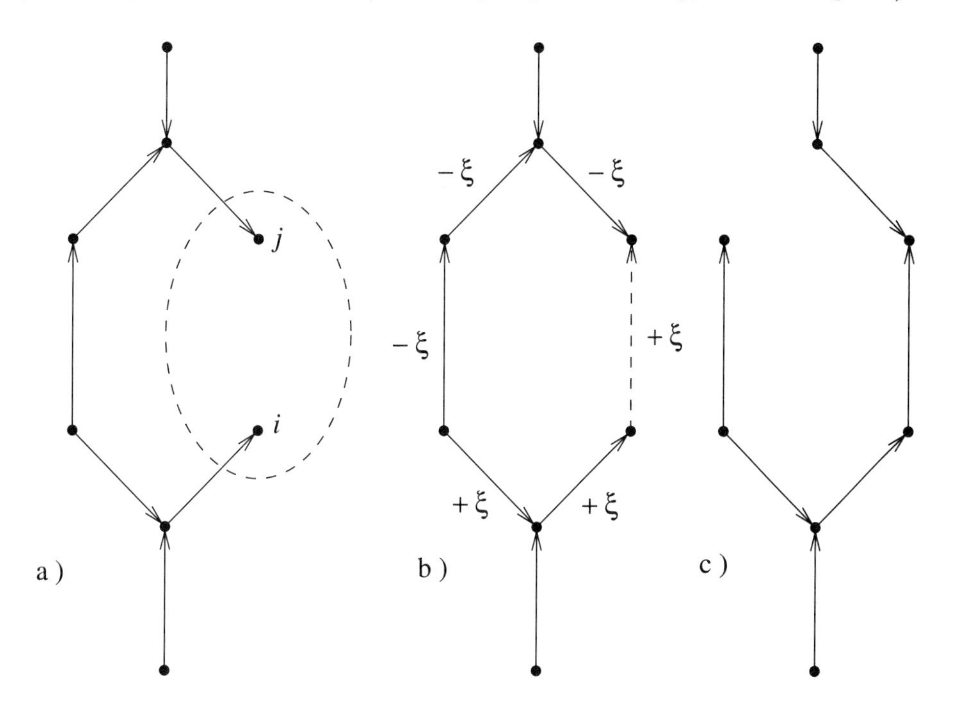

Figure 52. a) locating the edge where $\lambda_j - \lambda_i > \gamma_{ij}$, b) adjusting the flow and c) getting a new flow

Adding the edge to the tree generates a cycle. We change the flow on the edges of the cycle in such a way that the overall balance is preserved: we increase the flow on $i \longrightarrow j$ until the flow on some other edge of the cycle drops to 0 (cf. proof of Proposition II.7.2). We delete the edge and get a new tree T' and a new feasible flow with a smaller total cost (again, we assume that there be a single edge of the cycle that "dries up").

To construct an initial solution x, we modify the network by adding a new vertex 0 with $\beta_0 = 0$ (transit) and introduce new edges $i \longrightarrow 0$ if $\beta_i < 0$ and $0 \longrightarrow i$ if $\beta_i > 0$. We let γ_{0i} and γ_{j0} be very big numbers so that the transportation to/from vertex 0 should be avoided if at all possible. We observe that if the overall balance condition $\sum_{i=1}^{n} \beta_i = 0$ is satisfied, then the initial plan $\xi_{i0} = -\beta_i$, if $\beta_i < 0$, $\xi_{0i} = \beta_i$, if $\beta_i > 0$ and $\xi_{ij} = 0$ otherwise, is a feasible plan. Now the original problem has a feasible plan if the optimal solution in the modified problem has $\xi_{i0} = 0$ and $\xi_{0j} = 0$ for all i and j.

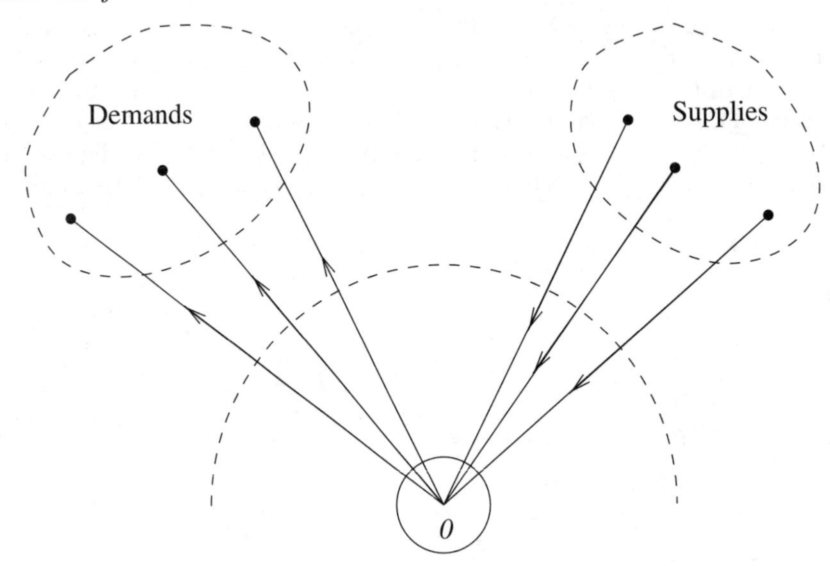

Figure 53. Constructing the initial plan by introducing an extra vertex

PROBLEM.

1. Prove that on every step of the algorithm the cost of the flow indeed decreases.

10. Semidefinite Programming

In this section, we consider linear programming problems in the space Sym_n of $n \times n$ symmetric matrices with respect to the cone $\mathcal{S}_+$ of positive semidefinite matrices;

see Section II.12. This type of linear programming has been known as "semidefinite programming"; see [**VB96**].

(10.1) Problems. We consider problems in the canonical/standard forms of Section 7.1. Let $E_1 = F_1 = \mathrm{Sym}_n$ and let $E_2 = F_2 = \mathbb{R}^m$. We consider the standard dualities (scalar products)

$$\langle A, B \rangle_1 = \mathrm{tr}(AB) = \sum_{i,j=1}^{n} \alpha_{ij}\beta_{ij}$$

for symmetric matrices $A = (\alpha_{ij})$ and $B = (\beta_{ij})$ (cf. Example 3.2.2) and

$$\langle x, y \rangle_2 = \sum_{i=1}^{m} \xi_i \eta_i$$

for vectors $x = (\xi_1, \ldots, \xi_m)$ and $y = (\eta_1, \ldots, \eta_m)$ in $\mathbb{R}^m$ (see Example 3.2.1). To simplify the notation, we denote both scalar products by $\langle \rangle$. We fix $K = \mathcal{S}_+ \subset \mathrm{Sym}_n$ and we write $X \succeq Y$ instead of $X \geq_K Y$ for symmetric matrices X and Y. Hence $X \succeq Y$ means that $X - Y$ is a positive semidefinite matrix.

Recall (Problem 2 of Section 5.3) that $K^* = K$. A linear transformation $\mathrm{Sym}_n \longrightarrow \mathbb{R}^m$ can be written as $X \longmapsto \big(\langle A_1, X \rangle, \ldots, \langle A_m, X \rangle\big)$ for some symmetric matrices $A_1, \ldots, A_m$. The dual linear transformation $\mathbb{R}^m \longrightarrow \mathrm{Sym}_n$ is defined by $(\xi_1, \ldots, \xi_m) \longmapsto \xi_1 A_1 + \ldots + \xi_m A_m$; cf. Problem 3 of Section 3.3.

Let us fix matrices $C, A_1, \ldots, A_m \in \mathrm{Sym}_n$ and a set of real numbers $\beta_1, \ldots, \beta_m$. Hence we arrive at the pair of problems:

(10.1.1) Primal Problem in the Canonical Form.

$$\begin{aligned} \text{Find} \quad & \gamma = \inf \langle C, X \rangle \\ \text{Subject to} \quad & \langle A_i, X \rangle = \beta_i \quad \text{for} \quad i = 1, \ldots, m \quad \text{and} \\ & X \succeq 0 \end{aligned}$$

in the primal matrix variable $X \in \mathrm{Sym}_n$.

(10.1.2) Dual Problem in the Standard Form.

$$\begin{aligned} \text{Find} \quad & \beta = \sup \sum_{i=1}^{m} \beta_i \lambda_i \\ \text{Subject to} \quad & \sum_{i=1}^{m} \lambda_i A_i \preceq C \end{aligned}$$

in the dual variables $(\lambda_1, \ldots, \lambda_m) \in \mathbb{R}^m$.

PROBLEMS.

1. Let $n = 2$, $m = 1$, let $C = \begin{pmatrix} 1 & 0 \\ 0 & 0 \end{pmatrix}$, let $A_1 = \begin{pmatrix} 0 & 1 \\ 1 & 0 \end{pmatrix}$ and let $\beta_1 = 1$. Prove that $\gamma = \beta = 0$, that $\lambda_1 = 0$ is the dual optimal plan and that there is no primal optimal plan.

2. Let $n = 2$, $m = 1$, let $C = \begin{pmatrix} 0 & -0.5 \\ -0.5 & -1 \end{pmatrix}$, let $A_1 = \begin{pmatrix} 1 & 1 \\ 1 & 1 \end{pmatrix}$ and let $\beta_1 = 0$. Prove that $\gamma = 0$, that there exists a primal optimal plan but that there are no dual feasible plans, so $\beta = -\infty$.

3°. Let B be a positive definite $n \times n$ matrix. Prove that if X is an $n \times n$ positive semidefinite matrix such that $\langle B, X \rangle = 0$, then $X = 0$.

Problems 1 and 2 above show that the situation is somewhat different from that of polyhedral linear programming; cf. Section 8. The following result provides a sufficient condition for the existence of the primal optimal plan and for the absence of the duality gap.

(10.2) Proposition. *Suppose that there are real numbers $\alpha_1, \dots, \alpha_m$ and ρ such that*

$$B = \alpha_1 A_1 + \ldots + \alpha_m A_m + \rho C$$

is a positive definite matrix and that there is a primal feasible plan. Then $\gamma = \beta$, and, if $\gamma > -\infty$, there is a primal optimal plan.

Proof. We use Theorem 7.2. Let us consider the linear transformation

$$\widehat{A} : \mathrm{Sym}_n \longrightarrow \mathbb{R}^m, \qquad \widehat{A}(X) = \big(\langle A_1, X \rangle, \dots, \langle A_m, X \rangle, \langle C, X \rangle\big).$$

We claim that $\ker(\widehat{A}) \cap \mathcal{S}_+ = \{0\}$. Indeed, for every $X \in \ker(\widehat{A})$ we have

$$\langle C, X \rangle = \langle A_1, X \rangle = \ldots = \langle A_m, X \rangle = 0$$

and hence $\langle B, X \rangle = 0$. Since B is positive definite and X is positive semidefinite, by Problem 3 of Section 10.1, it follows that $X = 0$. Since $\mathcal{S}_+$ has a compact base (see Problem 4 of Section II.12.2), by Lemma 7.3, we conclude that $\widehat{A}(\mathcal{S}_+)$ is a closed convex cone in $\mathbb{R}^{m+1}$. The result now follows by Theorem 7.2. □

PROBLEMS.

1. Suppose that the matrices $A_1, \dots, A_m$ are linearly independent and that there exists a positive definite matrix X which is a primal feasible plan. Prove that there is no duality gap.

Hint: Use Problem 2 of Section 7.2.

2. Suppose that there is a dual feasible plan, that $\beta < +\infty$ and that there is no dual optimal plan. Prove that there exists a non-zero vector $l = (\lambda_1, \dots, \lambda_m)$ such that

$$\sum_{i=1}^{m} \lambda_i A_i \preceq 0 \quad \text{and} \quad \sum_{i=1}^{m} \beta_i \lambda_i = 0.$$

Hint: Choose a sequence of $\{l_n\}$ of dual feasible plans such that

$$\lim_{n \longrightarrow +\infty} \langle b, l_n \rangle = \beta,$$

where $b = (\beta_1, \dots, \beta_m)$. Show that we must have $\|l_n\| \longrightarrow +\infty$. Consider the sequence $\bar{l}_n = l_n / \|l_n\|$.

3. Suppose that the matrices $A_1, \dots, A_n$ are linearly independent, that there is a positive definite matrix X which is a primal feasible plan and that $\gamma > -\infty$. Prove that there is a dual optimal plan.

Hint: Use Problems 1 and 2 above and Problem 3 of Section 10.1.

Finally, we discuss the positive semidefinite version of the complementary slackness conditions.

(10.3) Corollary.

1. *Suppose that X is a primal feasible plan and $l = (\lambda_1, \dots, \lambda_m)$ is a dual feasible plan. If*

$$\langle X,\ C - \sum_{i=1}^{m} \lambda_i A_i \rangle = 0,$$

then X is a primal optimal plan and l is a dual optimal plan ("optimality criterion").

2. *Suppose that X is a primal optimal plan, $l = (\lambda_1, \dots, \lambda_m)$ is a dual optimal plan and that there is no duality gap. Then*

$$\langle X,\ C - \sum_{i=1}^{m} \lambda_i A_i \rangle = 0$$

("complementary slackness").

Proof. Follows by Theorem 6.2. □

PROBLEMS.

1. Let X and Y be $n \times n$ positive semidefinite matrices such that $\langle X, Y \rangle = 0$. Prove that $\operatorname{rank} X + \operatorname{rank} Y \leq n$.

2. Let X be a primal optimal plan and let $l = (\lambda_1, \dots, \lambda_m)$ be a dual optimal plan. Suppose that there is no duality gap. Prove that

$$\operatorname{rank}(X) + \operatorname{rank}\Big(C - \sum_{i=1}^{m} \lambda_i A_i\Big) \leq n.$$

3. Let us call an $n \times n$ matrix $A = (\alpha_{ij})$ r-diagonal if $\alpha_{ij} = 0$ unless $|i - j| < r$; cf. Problem 5 of Section II.14.3. Suppose that the matrices $A_1, \dots, A_m$ are r-diagonal and there exists a positive semidefinite matrix X such that $\langle A_i, X \rangle = \beta_i$ for $i = 1, \dots, m$. Prove that there exists a positive semidefinite matrix X such that $\langle A_i, X \rangle = \beta_i$ for $i = 1, \dots, m$, and, additionally, $\operatorname{rank} X \leq r$.

Hint: Choose an $n \times n$ positive definite matrix $C = (\gamma_{ij})$ such that

$$\gamma_{ij} = \begin{cases} 1 & \text{if } i = j, \\ \epsilon > 0 & \text{if } |i - j| = r, \\ 0 & \text{elsewhere} \end{cases}$$

in Problem 10.1.1. Prove that there exists a primal optimal plan X^* and that there is no duality gap. Use Problem 2 to show that if there is a dual optimal plan, then $\operatorname{rank} X^* \leq r$. Deduce the general case from that particular case; see [**Barv95**].

11. An Application: The Clique and Chromatic Numbers of a Graph

We discuss an application of semidefinite programming to a combinatorial problem.

(11.1) Cliques and colorings. Let $G = (V, E)$ be a graph with the set of vertices $V = \{1, \ldots, n\}$ and the set of edges E. We assume that the edges are undirected and that there are no loops or multiple edges. A *clique* of G is a set of vertices $W \subset V$, such that every two distinct vertices $i, j \in W$ are connected by an edge of the graph: $(ij) \in E$. The largest number of vertices in a clique of G is called the *clique number* of G and denoted by $\omega(G)$. A *k-coloring* of G is a map $\phi : V \longrightarrow \{1, \ldots, k\}$, such that $\phi(i) \neq \phi(j)$ if $(ij) \in E$. The smallest k, such that a k-coloring exists, is called the *chromatic number* of G and denoted $\chi(G)$.

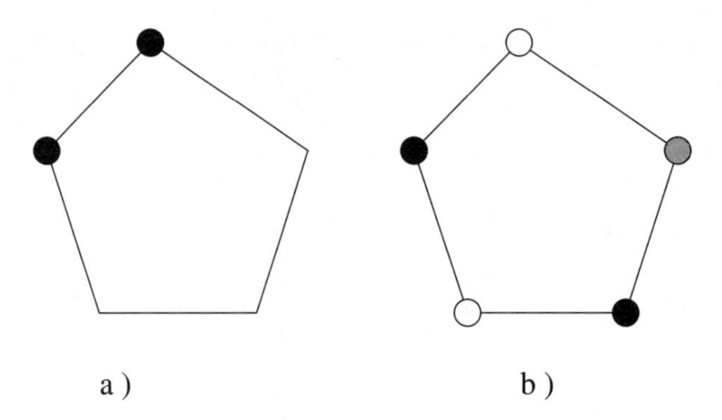

Figure 54. a) a clique, b) a coloring

PROBLEMS.

1°. Prove that $\omega(G) \leq \chi(G)$.

2°. Let G be the pentagon, that is, a graph with five vertices $\{1, 2, 3, 4, 5\}$ and five edges, (12), (23), (34), (45) and (51). Prove that $\omega(G) = 2$ and $\chi(G) = 3$.

Computing or even approximating $\omega(G)$ and $\chi(G)$ for general graphs is computationally hard. Surprisingly, there is a way to compute efficiently a number $\vartheta(G)$ such that $\omega(G) \leq \vartheta(G) \leq \chi(G)$. This number $\vartheta(G)$ was introduced by L. Lovász in 1979 [**Lo79**] and is now called the *Lovász's theta-function.* We do not discuss how to compute $\vartheta(G)$ (see [**GLS93**] and [**Lo86**]). Instead, we show that $\vartheta(G)$ is the optimal value in a certain pair of problems of semidefinite programming related by duality. There is an interesting class of graphs, called perfect graphs, for which we have $\omega(G) = \vartheta(G) = \chi(G)$; see Section 3.2 of [**Lo86**].

(11.2) The primal problem. For $1 \leq i < j \leq n$, let A_{ij} be the $n \times n$ symmetric matrix whose (i,j)-th and (j,i)-th entries are 1's and all other entries are 0. Let J be the $n \times n$ matrix filled by 1's and let I be the $n \times n$ identity matrix. Given a graph $G = (V, E)$, let us consider the following problem of semidefinite programming.

$$\begin{aligned} \text{Find} \quad & \vartheta(G) = \sup \langle J, X \rangle = -\inf \langle -J, X \rangle \\ \text{Subject to} \quad & \langle A_{ij}, X \rangle = 0 \quad \text{for every pair} \quad (ij) \notin E, \\ & \langle I, X \rangle = 1 \qquad \text{and} \\ & X \succeq 0 \end{aligned}$$

in the primal matrix variable X. In other words, we are seeking to maximize the sum of the entries of a positive semidefinite matrix $X = (x_{ij})$ of trace 1, provided $x_{ij} = 0$ if (ij) is not an edge of G.

PROBLEMS.

1°. Let $G_1 = (V, E_1)$ and $G_2 = (V, E_2)$ be graphs with the same vertex set V and such that $E_1 \subset E_2$. Prove that $\omega(G_2) \geq \omega(G_1)$, $\chi(G_2) \geq \chi(G_1)$ and $\vartheta(G_2) \geq \vartheta(G_1)$.

2°. Let $\alpha_1, \ldots, \alpha_k$ be non-negative numbers such that $\sum_{i=1}^k \alpha_i^2 = 1$. Prove that $\sum_{i=1}^k \alpha_i \leq \sqrt{k}$.

It turns out that the value of $\vartheta(G)$ is sandwiched between the clique number and the chromatic number of the graph.

(11.3) Proposition. *We have* $\omega(G) \leq \vartheta(G) \leq \chi(G)$.

Proof. Let $W \subset V$ be a clique and let $|W| = k$. Let $x = (\xi_1, \ldots, \xi_n)$, where

$$\xi_i = \begin{cases} 1 & \text{if } i \in W, \\ 0 & \text{if } i \notin W. \end{cases}$$

Let $X = (x_{ij})$, where $x_{ij} = (\xi_i \xi_j)/k$. Then X is a feasible plan in Problem 11.2 and $\langle J, X \rangle = k$. Therefore, $\vartheta(G) \geq \langle J, X \rangle = k = |W|$. Hence $\vartheta(G) \geq \omega(G)$.

Let $X = (x_{ij})$ be a feasible plan in Problem 11.2. Since X is a positive semidefinite matrix, there exist n vectors $v_1, \ldots, v_n$ in $\mathbb{R}^n$ for which X is the Gram matrix, that is, $\langle v_i, v_j \rangle = x_{ij}$. In particular, $\langle J, X \rangle = \|v_1 + \ldots + v_n\|^2$. Let us choose a coloring $\phi : V \longrightarrow \{1, \ldots, k\}$ of G. We observe that if $\phi(i) = \phi(j)$, then $x_{ij} = 0$ and hence $\langle v_i, v_j \rangle = 0$ (henceforth, such an n-tuple of vectors $v_1, \ldots, v_n$ is called an *orthogonal labeling* of the graph G). Since $\operatorname{tr} X = 1$, we conclude that

$$\sum_{i=1}^n \|v_i\|^2 = 1.$$

For a number ("color") $1 \leq j \leq k$, let us define w_j by

$$w_j = \sum_{i:\phi(i)=j} v_i.$$

Thus w_j is the sum of pairwise orthogonal vectors and we have

$$\|w_j\|^2 = \sum_{i:\phi(i)=j} \|v_i\|^2 \quad \text{and, therefore,} \quad \sum_{j=1}^k \|w_j\|^2 = 1.$$

Also, we observe that

$$\sum_{j=1}^k w_j = \sum_{i=1}^n v_i \quad \text{and hence} \quad \langle J, X\rangle = \Big\|\sum_{j=1}^k w_j\Big\|^2.$$

Now

$$\langle J, X\rangle = \Big\|\sum_{j=1}^k w_j\Big\|^2 \leq \Big(\sum_{j=1}^k \|w_j\|\Big)^2.$$

Denoting $\alpha_j = \|w_j\|$, we conclude that

$$\langle J, X\rangle \leq \Big(\sum_{j=1}^k \alpha_j\Big)^2 \quad \text{for some} \quad \alpha_j \geq 0 \quad \text{such that} \quad \sum_{j=1}^k \alpha_j^2 = 1.$$

Using Problem 2 of Section 11.2, we conclude that $\langle J, X\rangle \leq k$ and hence $\vartheta(G) \leq \chi(G)$. □

Writing the dual to Problem 11.2, we get

(11.4) The dual problem.

$$\begin{aligned} \text{Find} \quad & -\sup \lambda_0 = \inf -\lambda_0 \\ \text{Subject to} \quad & \lambda_0 I + \sum_{(ij)\notin E} \lambda_{ij} A_{ij} \preceq -J \end{aligned}$$

in the dual real variables λ_0 and λ_{ij} with $(ij) \notin E$. It is convenient to make a substitution $\tau = -\lambda_0$ and rewrite the problem as

$$\begin{aligned} \text{Find} \quad & \beta = \inf \tau \\ \text{Subject to} \quad & \tau I - J - \sum_{(ij)\notin E} \lambda_{ij} A_{ij} \succeq 0. \end{aligned}$$

We note that the matrix $\tau I - \Big(J + \sum_{(ij)\notin E} \lambda_{ij} A_{ij}\Big)$ is positive semidefinite if and only if the largest eigenvalue of the matrix $\Big(J + \sum_{(ij)\notin E} \lambda_{ij} A_{ij}\Big)$ does not exceed τ. Hence the dual problem is equivalent to finding the infimum β of the maximum eigenvalue of a symmetric matrix $Y = (\eta_{ij})$ such that $\eta_{ij} = 1$ if $(ij) \in E$ or $i = j$ and all other entries of Y are arbitrary.

(11.5) Proposition. *We have $\beta = \vartheta(G)$.*

Proof. By Proposition 10.2, we conclude that there is no duality gap as soon as some linear combination of the constraint matrices in the primal problem is positive definite. Since the identity matrix I is one of the constraint matrices in Problem 11.2, the result follows. □

PROBLEMS.

1. Prove that there exist a primal optimal plan and a dual optimal plan.

2. Let G be the pentagon; see Problem 2 of Section 11.1. Prove that $\vartheta(G) = \sqrt{5}$.

3. Let $G_1 = (V_1, E_1)$ and $G_2 = (V_2, E_2)$ be two graphs. Assuming that $V_1 \cap V_2 = \emptyset$, let us define the direct sum $G = (V, E)$ as the graph with $V = V_1 \cup V_2$ and $E = E_1 \cup E_2$. Prove that $\vartheta(G) = \vartheta(G_1) + \vartheta(G_2)$.

4. Let $G_1 = (V_1, E_1)$ and $G_2 = (V_2, E_2)$ be two graphs. Let us define their direct product as the graph $G = (V, E)$ with $V = V_1 \times V_2$ such that $\big((i_1, i_2), (j_1, j_2)\big)$ is an edge of G if and only if (i_1, j_1) is an edge of G_1 and (i_2, j_2) is an edge of G_2 or $i_1 = j_1$ and (i_2, j_2) is an edge of G_2 or $i_2 = j_2$ and (i_1, j_1) is an edge of G_1. Prove that $\vartheta(G) = \vartheta(G_1)\vartheta(G_2)$.

12. Linear Programming in L^∞

In this section, we discuss our first instance of infinite-dimensional linear programming. We consider the following optimization problem.

(12.1) Primal problem.

$$\begin{aligned} &\text{Find} \quad \gamma = \inf \int_0^1 g(\tau) u(\tau) \, d\tau \\ &\text{Subject to} \quad \int_0^1 f_i(\tau) u(\tau) \, d\tau = \beta_i \quad \text{for} \quad i = 1, \ldots, m \qquad \text{and} \\ &\qquad 0 \leq u(\tau) \leq 1 \quad \text{for almost all} \quad \tau \in [0, 1]. \end{aligned}$$

Here $g, f, f_1, \ldots, f_m \in L^1[0,1]$ are given functions, $\beta_1, \ldots, \beta_m$ are given real numbers and $u \in L^\infty[0,1]$ is a variable. As we discussed in Section III.6, certain problems of linear optimal control can be stated in the form of Problem 12.1.

There are various ways to write Problem 12.1 as a linear programming problem. We discuss one of them below.

Let $E_1 = L^\infty[0,1] \oplus L^\infty[0,1]$. Hence E_1 consists of ordered pairs (u, v) of functions from $L^\infty[0,1]$. Let $F_1 = L^1[0,1] \oplus L^1[0,1]$. We define the duality $\langle\rangle_1 : E_1 \times F_1 \longrightarrow \mathbb{R}$ by

$$\big\langle (u, v), \ (a, b) \big\rangle_1 = \int_0^1 u(\tau)a(\tau) + v(\tau)b(\tau) \, d\tau;$$

cf. Example 3.2.3. Let us choose $E_2 = F_2 = \mathbb{R}^{m+1}$ with the standard duality of Example 3.2.1. Let us define the cone $K \subset E_1$ as follows:

$$K = \Big\{(u,v): \quad u(\tau), v(\tau) \geq 0 \quad \text{for almost all} \quad \tau \qquad \text{and} \tag{12.1.1}$$
$$u(\tau) + v(\tau) = \text{constant function for almost all} \quad \tau\Big\}.$$

We define a linear transformation $A : E_1 \longrightarrow F_1$ by

$$A(u,v) = \left(\int_0^1 f_1(\tau)u(\tau)\, d\tau, \dots, \int_0^1 f_m(\tau)u(\tau)\, d\tau, \int_0^1 u(\tau) + v(\tau)\, d\tau\right).$$

Letting

$$c = (g, 0) \quad \text{and} \quad b = (\beta_1, \dots, \beta_m, 1),$$

we restate Problem 12.1 as a primal problem in the canonical form:

$$\begin{aligned} \text{Find} \quad & \gamma = \inf\langle x, c\rangle_1 \\ \text{Subject to} \quad & Ax = b \quad \text{and} \\ & x \geq_K 0 \end{aligned}$$

in the primal variable $x = (u,v) \in L^\infty[0,1] \oplus L^\infty[0,1]$.

To interpret the dual problem,

$$\begin{aligned} \text{Find} \quad & \beta = \sup\langle b, p\rangle \\ \text{Subject to} \quad & A^*p \leq_{K^*} c, \end{aligned}$$

we need to find the dual cone K^* and the dual transformation A^*.

(12.2) Lemma. *We have*

$$K^* = \Big\{\big(x+h,\ y+h\big): \quad x(\tau), y(\tau) \geq 0 \quad \textit{for almost all} \quad \tau \qquad \textit{and}$$
$$\int_0^1 h(\tau)\, d\tau \geq 0\Big\},$$

where $x, y, h \in L^1[0,1]$.

Proof. Suppose that $x(\tau), y(\tau) \geq 0$ for almost all τ and that $\int_0^1 h(\tau)\, d\tau \geq 0$. Let us pick a point $(u,v) \in K$. Thus $u(\tau), v(\tau) \geq 0$ and for some constant $\delta \geq 0$ we have $u(\tau) + v(\tau) = \delta$ for almost all τ. Then

$$\big\langle (x+h, y+h),\ (u,v)\big\rangle_1 = \int_0^1 x(\tau)u(\tau) + y(\tau)v(\tau) + \delta h(\tau)\, d\tau \geq 0,$$

from which we conclude that $(x+h, y+h) \in K^*$.

On the other hand, suppose that $(x_1, y_1) \in K^*$. Let

$$h(\tau) = \min\{x_1(\tau), y_1(\tau)\} \quad \text{and let} \quad x = x_1 - h \quad \text{and} \quad y = y_1 - h.$$

Clearly, $x(\tau), y(\tau) \geq 0$ and $x_1 = x + h, y_1 = y + h$. Thus it remains to show that

$$\begin{aligned}\int_0^1 h(\tau)\, d\tau &= \int_0^1 \min\{x_1(\tau), y_1(\tau)\} d\tau \\ &= \int_{\tau : x_1(\tau) \leq y_1(\tau)} x_1(\tau)\, d\tau + \int_{\tau : y_1(\tau) > x_1(\tau)} y_1(\tau)\, d\tau \geq 0. \qquad (12.2.1)\end{aligned}$$

Since $(x_1, y_1) \in K^*$, we must have

$$\int_0^1 x_1(\tau) u(\tau)\, d\tau + \int_0^1 y_1(\tau) v(\tau)\, d\tau \geq 0$$

for any two functions $0 \leq u(\tau), v(\tau) \leq 1$ such that $u(\tau) + v(\tau) = 1$. Choosing an arbitrary $0 \leq u(\tau) \leq 1$, we get

$$\begin{aligned}&\int_0^1 x_1(\tau) u(\tau)\, d\tau + \int_0^1 y_1(\tau) \big(1 - u(\tau)\big)\, d\tau \\ &\quad = \int_0^1 \big(x_1(\tau) - y_1(\tau)\big) u(\tau)\, d\tau + \int_0^1 y_1(\tau)\, d\tau \geq 0.\end{aligned}$$

Let us choose

$$u(\tau) = \begin{cases} 1 & \text{if } x_1(\tau) \leq y_1(\tau), \\ 0 & \text{if } x_1(\tau) > y_1(\tau). \end{cases}$$

Then

$$\int_{\tau : x_1(\tau) \leq y_1(\tau)} x_1(\tau) - y_1(\tau)\, d\tau + \int_0^1 y_1(\tau)\, d\tau \geq 0,$$

which is equivalent to (12.2.1). □

PROBLEM.

1. Prove that the dual transformation $A^* : \mathbb{R}^{m+1} \longrightarrow L^1[0,1] \oplus L^1[0,1]$ is defined by

$$A(\lambda_1, \ldots, \lambda_m, \rho) = \Big(\rho + \sum_{i=1}^m \lambda_i f_i, \quad \rho\Big).$$

From Lemma 12.2 and Problem 1 of Section 12.2, we conclude that the dual problem is

$$\text{Find} \quad \beta = \left(\sup \rho + \sum_{i=1}^m \beta_i \lambda_i\right)$$

$$\begin{aligned}\text{Subject to} \quad & g(\tau) - \rho - \sum_{i=1}^m \lambda_i f_i(\tau) \geq h(\tau) \quad \text{for almost all} \quad \tau \\ & -\rho \geq h(\tau) \quad \text{for almost all} \quad \tau \qquad \text{and} \\ & \int_0^1 h(\tau)\, d\tau \geq 0,\end{aligned}$$

where $h \in L^1[0,1]$ is a function and $\lambda_1, \dots, \lambda_m$ and ρ are dual variables.

Next, we observe that if $(\lambda_1, \dots, \lambda_m, \rho)$ is a feasible plan in the above problem and $\int_0^1 h(\tau)\, d\tau = \epsilon > 0$, we can get a new feasible plan with a better value of the objective function by modifying $h := h - \epsilon$ and $\rho := \rho + \epsilon$. Hence the last condition can be replaced by $\int_0^1 h(\tau)\, d\tau = 0$. Denoting $p(\tau) = -\rho - h(\tau)$, we come to the following

(12.3) Dual problem.

$$\text{Find} \quad \beta = \sup\left(\sum_{i=1}^m \lambda_i \beta_i - \int_0^1 p(\tau)\, d\tau\right)$$

$$\text{Subject to} \quad -p(\tau) + \sum_{i=1}^m \lambda_i f_i(\tau) \leq g(\tau) \quad \text{for almost all} \quad \tau \qquad \text{and}$$

$$p(\tau) \geq 0 \quad \text{for almost all} \quad \tau$$

in dual variables $\lambda_1, \dots, \lambda_m \in \mathbb{R}$ and $p \in L^1[0,1]$.

PROBLEM.

1. Let us consider a discrete version of Problem 12.1:

$$\text{Find} \quad \gamma = \inf \sum_{k=1}^N g(\tau_k) u(\tau_k)$$

$$\text{Subject to} \quad \sum_{k=1}^N f_i(\tau_k) u(\tau_k) = \beta_i \quad \text{for} \quad i = 1, \dots, m$$

$$-u(\tau_k) \geq -1 \quad \text{for} \quad k = 1, \dots N \qquad \text{and}$$

$$u(\tau_k) \geq 0 \quad \text{for} \quad k = 1, \dots N$$

in the primal variables $u(\tau_1), \dots, u(\tau_N) \in \mathbb{R}$. Here $g(\tau_k)$ and $f_i(\tau_k)$ are real numbers for $k = 1, \dots, N$ and $i = 1, \dots, m$.

Using Problem 3 of Section 8.2, show that the dual problem is

$$\text{Find} \quad \beta = \sup\left(\sum_{k=1}^m \lambda_i \beta_i - \sum_{k=1}^N p(\tau_k)\right)$$

$$\text{Subject to} \quad -p(\tau_k) + \sum_{i=1}^m \lambda_i f_i(\tau_k) \leq g(\tau_k) \qquad \text{and}$$

$$p(\tau_k) \geq 0 \quad \text{for} \quad k = 1, \dots, N$$

in the dual variables $\lambda_1, \dots, \lambda_m$ and $p(\tau_1), \dots, p(\tau_N)$.

Next, we compare optimal values in Problems 12.1 and 12.3.

(12.4) Proposition. *Suppose that there is a primal feasible plan $u(\tau)$ in Problem 12.1. Then $\gamma = \beta$ (there is no duality gap) and there is a primal optimal plan.*

Proof. We use Theorem 7.2 and Lemma 7.3. Let us introduce the weak topology of the duality in the spaces $E_1 = L^\infty[0,1] \oplus L^\infty[0,1]$ and $F_1 = L^1[0,1] \oplus L^1[0,1]$. From Proposition III.5.2 it follows that the cone $K \subset E_1$ defined by (12.1.1) has a compact base consisting of the pairs (u, v), where $u(\tau), v(\tau) \geq 0$ and $u(\tau)+v(\tau) = 1$ for almost all τ. Since for $(u, v) \in K$ we have

$$\int_0^1 u(\tau) + v(\tau)\, d\tau = 0 \quad \text{implies} \quad u(\tau) = v(\tau) = 0 \quad \text{for almost all} \quad \tau,$$

we conclude that $\ker(\widehat{A}) \cap K = \{0\}$. Therefore, by Lemma 7.3, the image $\widehat{A}(K)$ is closed in $\mathbb{R}^{m+2}$ and hence by Theorem 7.2 there is no duality gap.

Since

$$\gamma \geq -\int_0^1 |g(\tau)|\, d\tau > -\infty,$$

we conclude that there is a primal optimal plan. □

An interesting feature of Problem 12.3 is that it is, essentially, *finite dimensional.* In fact, the only "true" variables are the real variables $\lambda_1, \dots, \lambda_m$. Once their values are fixed, the best possible $p(\tau)$ is easy to find.

(12.5) Proposition ("the maximum principle"). *Let us fix some real numbers $\lambda_1, \dots, \lambda_m$ and define p by*

$$p(\tau) = \max\Big\{0, -g(\tau) + \sum_{i=1}^m \lambda_i f_i(\tau)\Big\}.$$

Then $(\lambda_1, \dots, \lambda_m; p)$ is a dual feasible plan with the largest possible value of the objective function with the given $\lambda_1, \dots, \lambda_m$. In other words, for any $p_1 \in L^1[0,1]$ such that $(\lambda_1, \dots, \lambda_m; p_1)$ is a dual feasible plan, we have

$$\sum_{i=1}^m \lambda_i \beta_i - \int_0^1 p(\tau)\, d\tau \geq \sum_{i=1}^m \lambda_i \beta_i - \int_0^1 p_1(\tau)\, d\tau$$

and the inequality is strict unless $p_1(\tau) = p(\tau)$ for almost all τ.

Proof. Every feasible function p in Problem 12.3 must satisfy the inequalities

$$p(\tau) \geq -g(\tau) + \sum_{i=1}^m \lambda_i f_i(\tau) \quad \text{and} \quad p(\tau) \geq 0.$$

Hence to choose a feasible p with the smallest value of $\displaystyle\int_0^1 p(\tau)\, d\tau$, we must choose

$$p(\tau) = \max\Big\{0, -g(\tau) + \sum_{i=1}^m \lambda_i f_i(\tau)\Big\},$$

as claimed. □

Often, Proposition 12.5 allows us to find the dual optimal plan if there is any. Then the complementary slackness conditions allow us to find the primal optimal plan.

(12.6) Proposition. *Let $u(\tau)$ be a primal optimal plan and let $(\lambda_1, \dots, \lambda_m; p)$ be a dual optimal plan. Suppose further that the set of roots of the function*

$$g(\tau) - \sum_{i=1}^{m} \lambda_i f_i(\tau)$$

has measure zero. Then

$$u(\tau) = \begin{cases} 1 & \text{if } p(\tau) > 0, \\ 0 & \text{if } p(\tau) = 0 \end{cases}$$

for almost all $\tau \in [0, 1]$.

Proof. We use the "complementary slackness" conditions of Part 3, Theorem 6.2. We observe that in this particular situation, the conditions can be written as

$$\int_0^1 p(\tau)\big(1 - u(\tau)\big)\, d\tau + \int_0^1 \Big(g(\tau) + p(\tau) - \sum_{i=1}^{m} \lambda_i f_i(\tau)\Big) u(\tau)\, d\tau = 0.$$

Both integrals are non-negative, hence we conclude that for almost all τ such that $p(\tau) > 0$ we must have $u(\tau) = 1$ (otherwise, the first integral is positive). From Proposition 12.5 we have

$$p(\tau) = 0 \quad \text{implies} \quad g(\tau) - \sum_{i=1}^{m} \lambda_i f_i(\tau) \geq 0$$

for almost all τ. Thus if $p(\tau) = 0$, we must have $u(\tau) = 0$ for almost all τ (otherwise, the second integral is positive). □

PROBLEM.

1. Consider the primal problem

$$\text{Find} \quad \gamma = \inf \int_0^1 \tau u(\tau)\, d\tau$$

$$\text{Subject to} \quad \int_0^1 u(\tau)\, d\tau = 1/2 \quad \text{and}$$

$$0 \leq u(\tau) \leq 1$$

and the dual problem

$$\text{Find} \quad \beta = \sup \Big(\lambda/2 - \int_0^1 p(\tau)\, d\tau \Big)$$

$$\text{Subject to} \quad -p(\tau) + \lambda \leq \tau \quad \text{and}$$

$$p(\tau) \geq 0.$$

Find the dual optimal plan and the primal optimal plan.

13. Uniform Approximation as a Linear Programming Problem

Let us recall from Section I.6 the problem of the uniform (Chebyshev) approximation: suppose we are given continuous functions $f_1, \ldots, f_m$ on the interval $[0,1]$ and yet another continuous function g on $[0,1]$ (instead of the interval $[0,1]$ we can consider an arbitrary compact metric space X). We want to find a linear combination $\xi_1 f_1 + \ldots + \xi_m f_m$, such that the maximum deviation

$$\xi_0 = \max_{0 \le \tau \le 1} \left| \xi_1 f_1(\tau) + \ldots + \xi_m f_m(\tau) - g(\tau) \right|$$

is the smallest possible.

Hence we can write the problem in the following form.

(13.1) Primal problem.

$$\text{Find} \quad \gamma = \inf \xi_0$$

$$\text{Subject to} \quad \xi_0 + \sum_{i=1}^m \xi_i f_i(\tau) \ge g(\tau) \quad \text{for all} \quad \tau \in [0,1] \qquad \text{and}$$

$$\xi_0 - \sum_{i=1}^m \xi_i f_i(\tau) \ge -g(\tau) \quad \text{for all} \quad \tau \in [0,1]$$

in the real variables $\xi_0, \xi_1, \ldots, \xi_m$.

To interpret Problem 13.1 as a general linear programming problem 6.1.1, we choose

- $E_1 = F_1 = \mathbb{R}^{m+1}$ with the standard duality $\langle\rangle_1$ of Example 3.2.1;
- $E_2 = C[0,1] \oplus C[0,1]$ and $F_2 = V[0,1] \oplus V[0,1]$ with the duality

$$\big\langle (h_1, h_2),\ (\mu_1, \mu_2) \big\rangle_2 = \int_0^1 h_1 \, d\mu_1 + \int_0^1 h_2 \, d\mu_2$$

(cf. Example 3.2.4 and Problem 4 of Section 3.2);

- the linear transformation $A : E_1 \longrightarrow F_1$ defined by

$$A(\xi_0, \xi_1, \ldots, \xi_m) = \Big(\xi_0 + \sum_{i=1}^m \xi_i f_i, \quad \xi_0 - \sum_{i=1}^m \xi_i f_i \Big);$$

- cones $K_1 = \{0\} \subset E_1$ and

$$K_2 = \Big\{ (h_1, h_2) : \ h_1(\tau), h_2(\tau) \ge 0 \quad \text{for all} \quad 0 \le \tau \le 1 \Big\} \subset E_2$$

(cf. Example 5.2.4);

- vectors $b = (g, -g) \in E_2$ and $c = (1, 0, \dots, 0) \in E_1$.

Then the primal problem can be written in the usual form

$$\begin{aligned} \text{Find} \quad & \gamma = \langle x, c \rangle_1 \\ \text{Subject to} \quad & Ax \geq_{K_2} b \qquad \text{and} \\ & x \geq_{K_1} 0. \end{aligned}$$

To write the dual problem, we need to find A^*, K_1^* and K_2^*.

PROBLEMS.

1°. Prove that $A^* : V[0,1] \oplus V[0,1] \longrightarrow \mathbb{R}^{m+1}$ is defined by

$$A^*(\mu_1, \mu_2) =$$
$$\left(\int_0^1 d\mu_1 + \int_0^1 d\mu_2, \ \int_0^1 f_1 \, d\mu_1 - \int_0^1 f_1 \, d\mu_2, \ \dots, \ \int_0^1 f_m \, d\mu_1 - \int_0^1 f_m \, d\mu_2 \right).$$

2°. Prove that $K_1^* = F_1 = \mathbb{R}^m$ and that

$$K_2^* = \{(\mu_1, \mu_2) : \quad \mu_1, \mu_2 \in V_+\};$$

cf. Example 5.2.4 and Problem 2 of Section 5.3.

Let us write $\mu \geq 0$ instead of $\mu \in V_+$.

Summarizing, we conclude that the dual problem

$$\begin{aligned} \text{Find} \quad & \beta = \sup \langle b, l \rangle_2 \\ \text{Subject to} \quad & A^* l \leq_{K_1^*} c \qquad \text{and} \\ & l \geq_{K_2^*} 0 \end{aligned}$$

is stated as follows:

(13.2) Dual problem.

$$\text{Find} \quad \beta = \sup \left(\int_0^1 g \, d\mu_1 - \int_0^1 g \, d\mu_2 \right)$$

$$\begin{aligned} \text{Subject to} \quad & \int_0^1 d\mu_1 + \int_0^1 d\mu_2 = 1, \\ & \int_0^1 f_i \, d\mu_1 - \int_0^1 f_i \, d\mu_2 = 0 \quad \text{for} \quad i = 1, \dots, m \qquad \text{and} \\ & \mu_1, \mu_2 \geq 0, \end{aligned}$$

where the dual variables μ_1 and μ_2 are Borel measures on $[0, 1]$.

PROBLEMS.

1. Let us consider a discrete version of Problem 13.1:

$$\text{Find} \quad \gamma = \inf \xi_0$$

$$\text{Subject to} \quad \xi_0 + \sum_{i=1}^{m} \xi_i f_i(\tau_j) \geq g(\tau_j) \quad \text{for} \quad j = 1, \ldots, N \qquad \text{and}$$

$$\xi_0 - \sum_{i=1}^{m} \xi_i f_i(\tau_j) \geq -g(\tau_j) \quad \text{for} \quad j = 1, \ldots, N,$$

where $\tau_1, \ldots, \tau_N \in [0, 1]$ are some numbers.

Prove that the dual problem is

$$\text{Find} \quad \beta = \sup \sum_{j=1}^{N} (\lambda_j^+ - \lambda_j^-) g(\tau_j)$$

$$\text{Subject to} \quad \sum_{j=1}^{N} (\lambda_j^+ + \lambda_j^-) = 1,$$

$$\sum_{j=1}^{N} (\lambda_j^+ - \lambda_j^-) f_i(\tau_j) = 0 \quad \text{for} \quad i = 1, \ldots, m \qquad \text{and}$$

$$\lambda_j^+, \lambda_j^- \geq 0 \quad \text{for} \quad j = 1, \ldots, N$$

in real variables λ_j^+ and λ_j^- for $j = 1, \ldots, N$. Moreover, prove that the latter problem is equivalent to the following optimization problem:

$$\text{Find} \quad \beta = \sup \sum_{j=1}^{N} \sigma_j g(\tau_j)$$

$$\text{Subject to} \quad \sum_{j=1}^{N} \sigma_j f_i(\tau_j) = 0 \quad \text{for} \quad i = 1, \ldots, m \qquad \text{and}$$

$$\sum_{j=1}^{N} |\sigma_j| \leq 1$$

in real variables $\sigma_1, \ldots, \sigma_N$.

2. For a $\mu \in V[0, 1]$, let us define

$$\|\mu\| = \sup\left\{ \left| \int_0^1 f \, d\mu \right| : \max_{0 \leq \tau \leq 1} |f(\tau)| \leq 1 \right\};$$

cf. Problem 2 of Section III.8.1. Prove that every (signed) measure μ such that $\|\mu\| \leq 1$ can be written as $\mu = \mu_1 - \mu_2$ for some $\mu_1, \mu_2 \geq 0$ such that

$$\int_0^1 d\mu_1 + \int_0^1 d\mu_2 = 1.$$

Conversely, prove that if $\mu = \mu_1 - \mu_2$ where $\mu_1, \mu_2 \geq 0$ and satisfy the above equation, we have $\|\mu\| \leq 1$.

Problem 2 of Section 13.2 suggests the following

(13.3) Reformulation of the dual problem.

$$\begin{aligned} \text{Find} \quad & \beta = \sup \int_0^1 g \, d\mu \\ \text{Subject to} \quad & \|\mu\| \leq 1 \quad \text{and} \\ & \int_0^1 f_i \, d\mu = 0 \quad \text{for} \quad i = 1, \ldots, m \end{aligned}$$

with variable μ which is a signed Borel measure on $[0, 1]$.

Problem 13.1 has a simple geometric interpretation. In the space $C[0, 1]$, let $L = \operatorname{span}(f_1, \ldots, f_m)$ be the subspace spanned by the functions $f_1, \ldots, f_m$. Given a function $g \in C[0, 1]$, we are looking for a function $f \in L$ closest to g in the uniform metric $\operatorname{dist}(f, g) = \max_{0 \leq \tau \leq 1} |f(\tau) - g(\tau)|$.

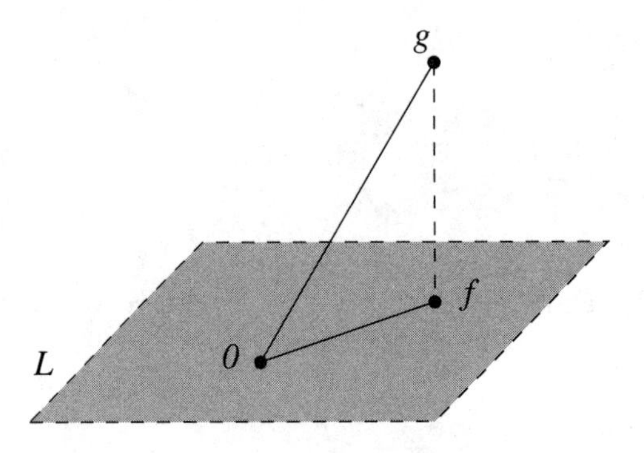

Figure 55

In Problem 13.3 we look for a linear functional μ of the unit norm such that $\mu(h) = 0$ for all $h \in L$ and such that $\mu(g)$ is maximized. If L were a subspace of Euclidean space and the distance $\operatorname{dist}(f, g)$ were the standard Euclidean distance, the largest possible value of $\mu(g)$ should have clearly been equal to the distance from g to L.

It turns out that indeed there is no duality gap in Problems 13.1 and 13.2 even though the metric is different from the Euclidean one.

(13.4) Proposition. *We have $\gamma = \beta$ (there is no duality gap). Moreover, there is a primal optimal plan and a dual optimal plan.*

Sketch of Proof. Problem 13.2 can be considered as a *primal* problem in the canonical form (see Problem 7.1.1) provided we replace sup by $-$ inf of the opposite linear functional. One can check that the dual to that primal problem is equivalent to Problem 13.1, cf. Problem 1 of Section 6.1. Hence we are going to apply Theorem 7.2 to Problem 13.2 as to a *primal* problem. Let us introduce the weak topology of the duality in the spaces E_1, F_1, E_2 and F_2. We observe that there is a feasible plan in Problem 13.2: for example we may choose $\mu_1 = \mu_2 = d\tau/2$. Let $K = K_2^* = \{(\mu_1, \mu_2) : \mu_1, \mu_2 \in V_+\}$. Thus we have to show that the image of $T(K) \subset \mathbb{R}^{m+1}$ is a closed set, where T is defined by

$$T(\mu_1, \mu_2) = \Big(\int_0^1 d\mu_1 + \int_0^1 d\mu_2, \quad \int_0^1 f_1 \, d\mu_1 - \int_0^1 f_1 \, d\mu_2, \quad \ldots, \\ \int_0^1 f_m \, d\mu_1 - \int_0^1 f_m \, d\mu_2, \quad \int_0^1 g \, d\mu_2 - \int_0^1 g \, d\mu_1 \Big).$$

One can observe that $(\ker T) \cap K = \{0\}$ since for $\mu_1, \mu_2 \geq 0$ we have

$$\int_0^1 d\mu_1 + \int_0^1 d\mu_2 = 0 \qquad \text{implies} \qquad \mu_1 = \mu_2 = 0.$$

Moreover, cone K has a compact base

$$B = \Big\{ (\mu_1, \mu_2) : \mu_1, \mu_2 \geq 0 \quad \text{and} \quad \int_0^1 d\mu_1 + \int_0^1 d\mu_2 = 1 \Big\};$$

cf. Section III.8. Hence by Lemma 7.3, $T(K)$ is a closed convex cone and by Theorem 7.2 we conclude that there is no duality gap and that Problem 13.2 has an optimal plan.

To prove that Problem 13.1 has an optimal plan, we use a "brute force" type argument similar to that in the proof of Proposition I.6.3. Without loss of generality, we may assume that $f_1, \ldots, f_m$ are linearly independent. For $x = (\xi_1, \ldots, \xi_m)$, let

$$p(x) = \max_{0 \leq \tau \leq 1} |\xi_1 f_1(\tau) + \ldots + \xi_m f_m(\tau)|.$$

Then $p(x)$ is a continuous function and $p(x) > 0$ for all $x \neq 0$. Let $\mathbb{S}^{m-1} \subset \mathbb{R}^m$ be the unit sphere in the space of all $x = (\xi_1, \ldots, \xi_m)$. Hence there is a $\delta > 0$ such that $p(x) > \delta$ for all $x \in \mathbb{S}^{m-1}$. Therefore $p(x) > \delta R$ for all x such that $\|x\| \geq R$. Let $M = \max_{0 \leq \tau \leq 1} g(\tau)$. If $p(x) > 2M$, then

$$\max_{0 \leq \tau \leq 1} |\xi_1 f_1(\tau) + \ldots + \xi_m f_m(\tau) - g(\tau)| > M$$

and $f \equiv 0$ is a better approximation to g than $\xi_1 f_1 + \ldots + \xi_m f_m$. Thus we conclude that the optimal x is found as a point in the compact set $\{x : \ \|x\| \leq 2M\delta^{-1}\}$ where the minimum value of the continuous function

$$d(x) = \max_{0 \leq \tau \leq 1} |\xi_1 f_1(\tau) + \ldots + \xi_m f_m(\tau) - g(\tau)|$$

is attained. For such an $x = (\xi_1, \ldots, \xi_m)$, the vector $(\xi_0, \xi_1, \ldots, \xi_m)$ with $\xi_0 = d(x)$ is an optimal plan in Problem 13.1. $\square$

PROBLEMS.

The next two problems provide a useful interpretation of the complementary slackness conditions.

1. Prove that in Problem 13.3 there exists an optimal solution μ which is a linear combination of at most $m+1$ delta-measures:

$$\mu = \sum_{i=1}^{k} \sigma_i \delta_{\tau_i} \quad \text{where} \quad k \leq m+1, \quad \tau_i \in [0,1] \quad \text{and} \quad \sum_{i=1}^{k} |\sigma_i| = 1.$$

Hint: Prove that the extreme points of the set $\{\mu : \|\mu\| \leq 1\}$ are the (signed) delta-measures δ_τ and $-\delta_\tau$ for some $\tau \in [0,1]$; cf. Problem 2 of Section III.8.4.

2. Let $x = (\xi_0, \xi_1, \ldots, \xi_m)$ be a feasible plan in Problem 13.1 and let $f = \xi_1 f_1 + \ldots + \xi_m f_m$. Prove that x is an optimal plan if and only if there exist $k \leq m+1$ points $\tau_i \in [0,1]$ and k numbers σ_i such that

$$\sum_{i=1}^{k} |\sigma_i| = 1, \quad \sum_{i=1}^{k} \sigma_i f_j(\tau_i) = 0 \quad \text{for} \quad j = 1, \ldots, m$$

and such that if $\sigma_i > 0$, then $g(\tau_i) - f(\tau_i) = \xi_0$ and if $\sigma_i < 0$, then $g(\tau_i) - f(\tau_i) = -\xi_0$.

Hint: Use Problem 2 and Theorem 6.2

3. In Problems 13.1 and 13.2, let $m = 1$, $f_1(\tau) = 1$ and let $g(\tau) = \tau$. Find a primal optimal plan and a dual optimal plan.

14. The Mass-Transfer Problem

In this section, we discuss an "infinite" version of the Transportation Problem (see Section 9), known as the Mass-Transfer Problem. Historically, it is one of the oldest linear programming problems (although it was recognized as such much later than it was first considered). Apparently, it was considered for the first time by G. Monge in 1781 although it was not until 1942 that L.V. Kantorovich interpreted it as a linear programming problem. We discuss this interesting problem only briefly, leaving much to prove to the problems.

Informally, the problem is described as follows. Suppose we are given an initial mass distribution on the interval $[0,1]$ (or on an arbitrary compact metric space X) and a target mass distribution. For "moving" a unit mass from position $\tau_1 \in [0,1]$ to position $\tau_2 \in [0,1]$ we pay the price of $c(\tau_1, \tau_2)$, where $c : [0,1] \times [0,1] \longrightarrow \mathbb{R}$ is a certain (continuous) function. We are looking for the least expensive way to redistribute the mass.

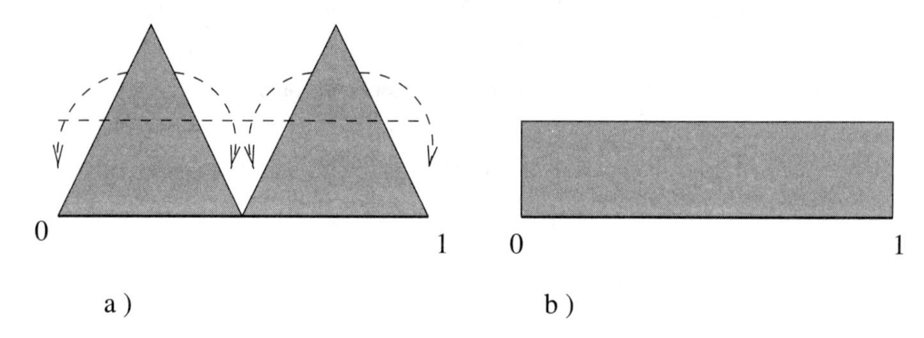

Figure 56. Example: a) the initial mass distribution on the interval $[0, 1]$ and a way to obtain the target mass distribution b)

Let us think of the initial and target mass distributions as non-negative Borel measures $\mu_1, \mu_2 \in V[0, 1]$. Let I_1 and I_2 be two copies of the interval $[0, 1]$. We think of μ_1 as an element of $V(I_1)$ (the space of signed Borel measures on I_1) and of μ_2 as an element of $V(I_2)$ (the space of signed Borel measures on I_2).

Let us describe what "redistributing" means. Let $S = I_1 \times I_2$ be the unit square and let $C(S)$ be the vector space of all real-valued continuous functions $f : S \longrightarrow \mathbb{R}$. We make $C(S)$ a topological vector space in just the same way as we introduced topology in $C[0, 1]$; cf. Section III.8. Namely, we declare a set $U \subset C(S)$ open if for every $f \in U$ there is an $\epsilon > 0$ such that the set

$$U(f, \epsilon) = \Big\{ g \in C(S) : \quad |f(\tau) - g(\tau)| < \epsilon \quad \text{for all} \quad \tau \in S \Big\}$$

is contained in U. Similarly, we introduce the space $V(S)$ of all signed Borel measures on S as the space of all continuous linear functionals $\phi : C(S) \longrightarrow \mathbb{R}$. We say that $f \in C(S)$ is *non-negative* if $f(\tau) \geq 0$ for all $\tau \in S$. Similarly, we say that an element $\mu \in V(S)$ is *positive* (denoted $\mu \geq 0$) provided

$$\int_S f \, d\mu \geq 0 \quad \text{for any non-negative} \quad f \in C(S).$$

For example, the *delta-measure* δ_τ defined for any $\tau \in S$ by

$$\delta_\tau(f) = f(\tau) \quad \text{for all} \quad f \in C(S)$$

is a positive linear functional. Thus we interpret a redistribution of mass as a non-negative measure $\mu \in V(S)$. That redistribution should satisfy initial and target conditions.

Let us define linear transformations $P_1, P_2 : V(S) \longrightarrow V(I_1), V(I_2)$, called *projections*, as follows. Given a function $f \in C(I_1)$, we define its *extension* $F \in C(S)$ by $F(\xi_1, \xi_2) = f(\xi_1)$. Similarly, given a function $f \in C(I_2)$, we define its extension $F \in C(S)$ by $F(\xi_1, \xi_2) = f(\xi_2)$. Given a $\mu \in V(S)$, we define $P_1(\mu) \in V(I_1)$ as the linear functional $C(I_1) \longrightarrow \mathbb{R}$ such that for every $f \in C(I_1)$

$$P_1(\mu)(f) = \int_S F \, d\mu \quad \text{where} \quad F \quad \text{is the extension of} \quad f.$$

Similarly, given a $\mu \in V(S)$, we define $P_2(\mu) \in V(I_2)$ as the linear functional $C(I_2) \longrightarrow \mathbb{R}$ such that for every $f \in C(I_2)$

$$P_2(\mu)(f) = \int_S F \, d\mu \quad \text{where} \quad F \quad \text{is the extension of} \quad f.$$

Now we write the initial condition as $P_1(\mu) = \mu_1$ and the target condition as $P_2(\mu) = \mu_2$.

Finally, we think of a cost function $c(\tau_1, \tau_2)$ as a continuous function $c \in C(S)$ on the square. Hence we arrive at the following optimization problem.

(14.1) Primal problem.

$$\begin{aligned} \text{Find} \quad & \gamma = \inf \int_S c \, d\mu \\ \text{Subject to} \quad & P_1(\mu) = \mu_1, \\ & P_2(\mu) = \mu_2 \qquad \text{and} \\ & \mu \geq 0 \end{aligned}$$

in the primal variable μ which is a Borel measure on the unit square $S = [0,1] \times [0,1]$. Here μ_1 and μ_2 are fixed Borel measures on the interval $[0,1]$ and c is a continuous function on the square S.

To interpret Problem 14.1 as a linear programming problem, we define

- $E_1 = V(S)$, $F_1 = C(S)$ with the duality

$$\langle \mu, f \rangle_1 = \int_S f \, d\mu,$$

- $E_2 = V(I_1) \oplus V(I_2)$, $F_2 = C(I_1) \oplus C(I_2)$ with the duality

$$\big\langle (\mu_1, \mu_2), \ (f_1, f_2) \big\rangle_2 = \int_0^1 f_1 \, d\mu_1 + \int_0^1 f_2 \, d\mu_2,$$

- the linear transformation $A : E_1 \longrightarrow E_2$:

$$A(\mu) = \Big(P_1(\mu), \ P_2(\mu) \Big),$$

- the cone $K \subset E_1$:

$$K = \{\mu \in V(S) : \ \mu \geq 0\},$$

- the vectors $b = (\mu_1, \mu_2) \in E_2$ and $c \in F_1$.

Then Problem 14.1 is stated as the primal linear programming problem in the canonical form:

$$\begin{aligned} &\text{Find} && \langle x, c\rangle_1 \\ &\text{Subject to} && Ax = b \qquad \text{and} \\ & && x \geq_K 0. \end{aligned}$$

To state the dual problems, we need to find the dual transformation A^* and the dual cone K^*. They are found in Problems 3 and 4 below.

PROBLEMS.

1. Let us consider a discrete version of the problem. Let us fix some n distinct points $\tau_1, \dots, \tau_n \in [0,1]$. Suppose that μ_1 and μ_2 are non-negative combinations of delta-measures:

$$\mu_1 = \sum_{i=1}^{n} \alpha_i \delta_{\tau_i} \quad \text{and} \quad \mu_2 = \sum_{i=1}^{n} \beta_i \delta_{\tau_i}$$

for some $\alpha_i, \beta_i \geq 0$. Let us look for the measure $\mu \in V(S)$ which is a linear combination of the delta-measures in the points (τ_i, τ_j), $1 \leq i, j \leq n$:

$$\mu = \sum_{i,j=1}^{n} \xi_{ij} \delta_{(\tau_i, \tau_j)}.$$

Show that Problem 14.1 is equivalent to the following linear programming problem:

$$\begin{aligned} &\text{Find} && \gamma = \inf \sum_{i,j=1}^{n} \gamma_{ij} \xi_{ij} \\ &\text{Subject to} && \sum_{j=1}^{n} \xi_{ij} = \alpha_i \quad \text{for all} \quad i = 1, \dots, n, \\ & && \sum_{i=1}^{n} \xi_{ij} = \beta_j \quad \text{for all} \quad j = 1, \dots, n \qquad \text{and} \\ & && \xi_{ij} \geq 0 \quad \text{for all} \quad i, j. \end{aligned}$$

Interpret the latter problem as a transportation problem of Section 9.

2. Check that the linear transformations $P_1, P_2 : C(S) \longrightarrow C(I_1), C(I_2)$ are well defined.

3. Prove that $A^* : C(I_1) \oplus C(I_2) \longrightarrow C(S)$,

$$A^*\Big((f_1, f_2)\Big) = F \quad \text{where} \quad F(\xi_1, \xi_2) = f_1(\xi_1) + f_2(\xi_2),$$

is the linear transformation dual to A.

4. Prove that

$$K^* = \Big\{f \in C(S) : \ f(\tau) \geq 0 \quad \text{for all} \quad \tau \in S\Big\}.$$

Using Problems 3 and 4 of Section 14.1, we conclude that the dual problem

$$\begin{aligned} \text{Find} \quad & \beta = \sup\langle b, l\rangle_2 \\ \text{Subject to} \quad & A^* l \leq_{K^*} c \end{aligned}$$

is written as follows:

(14.2) Dual problem.

$$\begin{aligned} \text{Find} \quad & \beta = \sup\Big(\int_0^1 l_1 \, d\mu_1 + \int_0^1 l_2 \, d\mu_2\Big) \\ \text{Subject to} \quad & l_1(\tau_1) + l_2(\tau_2) \leq c(\tau_1, \tau_2) \quad \text{for all} \quad \tau_1, \tau_2 \in [0, 1] \end{aligned}$$

in the dual variables l_1 and l_2 which are continuous functions on the interval $[0, 1]$.

PROBLEMS.

1. Suppose that μ_1 and μ_2 are non-negative combinations of delta-measures as in Problem 1 of Section 14.1. Check that Problem 14.2 is written as

$$\begin{aligned} \text{Find} \quad & \beta = \sup \sum_{i=1}^{n} \Big(\lambda_i' \alpha_i + \lambda_i'' \beta_i\Big) \\ \text{Subject to} \quad & \lambda_i' + \lambda_j'' \leq \gamma_{ij} \quad \text{for all} \quad i, j = 1, \ldots, n \end{aligned}$$

in the dual variables λ_i', λ_i'' for $i = 1, \ldots, n$. Show that the latter problem is equivalent to Problem 9.2 for the relevant transportation network.

2. In Problems 14.1 and 14.2, let us choose $\mu_1 = \tau \, d\tau$, $\mu_2 = (1 - \tau) \, d\tau$ and $c(\tau_1, \tau_2) = |\tau_1 - \tau_2|$; see Figure 57. Find a primal optimal plan and a dual optimal plan.

Hint: Guess a primal optimal plan and a dual optimal plan and show that the corresponding values of objective functions are equal.

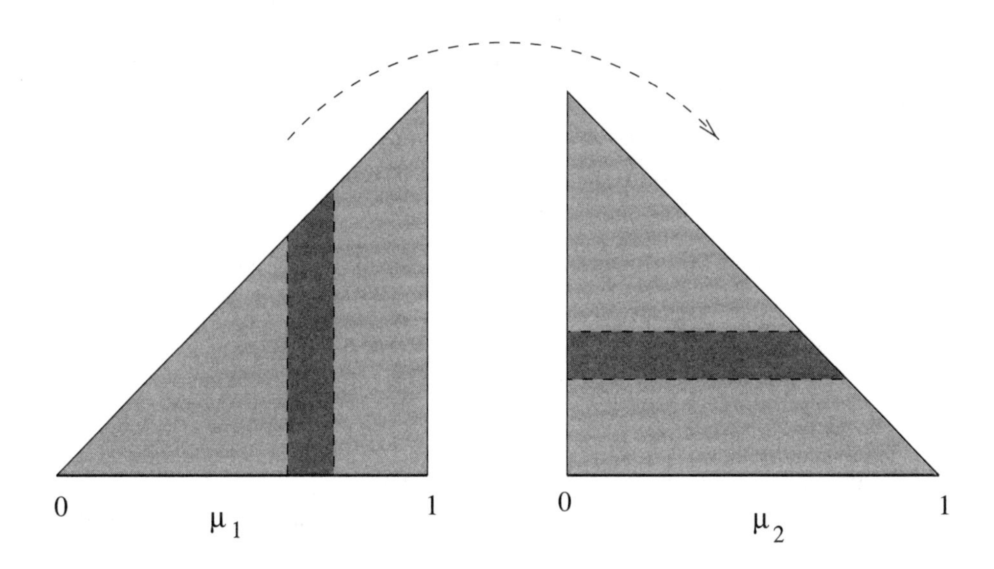

Figure 57. What is the best way to go from the distribution μ_1 to the distribution μ_2 on the interval $[0, 1]$?

3°. Show that a dual feasible plan always exists.

4. Show that a primal feasible plan exists if and only if $\mu_1, \mu_2 \geq 0$ and

$$\int_0^1 d\mu_1 = \int_0^1 d\mu_2.$$

Hint: Prove that there exists a measure $\mu \in V(S)$ (denoted $\mu_1 \times \mu_2$) such that for any $f_1 \in C(I_1)$, $f_2 \in C(I_2)$ and $F(\xi_1, \xi_2) = f_1(\xi_1)f_2(\xi_2)$ one has

$$\int_S F \, d\mu = \Big(\int_0^1 f_1 \, d\mu_1\Big)\Big(\int_0^1 f_2 \, d\mu_2\Big).$$

5. Prove that if $P_1(\mu) = 0$ for $\mu \geq 0$, then $\mu = 0$.

6. Deduce from the Alaoglu Theorem (Theorem III.2.9) that the set

$$B = \Big\{\mu \in V(S) : \quad \mu(S) \geq 0 \quad \text{and} \quad \int_S d\mu = 1\Big\}$$

is compact in the weak* topology of $V(S)$.

(14.3) Proposition. *Suppose that there exists a primal feasible plan. Then $\gamma = \beta$ (no duality gap) and there exists a primal optimal plan.*

Sketch of Proof. We use Theorem 7.2 and Lemma 7.3. Let us introduce the weak topology of the duality (see Section 4) in all relevant spaces E_1, F_1, E_2 and F_2. Let us consider the image $\widehat{A}(K)$ of the cone $K = \{\mu \in V(S) : \mu \geq 0\}$ under the linear transformation $\widehat{A} : V(S) \longrightarrow V(I_1) \oplus V(I_2) \oplus \mathbb{R}$,

$$\mu \longmapsto \left(P_1(\mu), \quad P_2(\mu), \quad \int_S c \, d\mu \right).$$

Problem 5 of Section 14.2 implies that $(\ker \widehat{A}) \cap K = \{0\}$. Moreover, cone K has a compact base

$$B = \left\{ \mu \in V(S) : \quad \mu(S) \geq 0 \quad \text{and} \quad \int_S d\mu = 1 \right\};$$

cf. Problem 6 of Section 14.2. One can show that $\widehat{A}$ is a continuous linear transformation. Therefore, by Lemma 7.3, the image $\widehat{A}(K)$ is closed. Hence by Theorem 7.2 there is no duality gap. Since by Problem 3 of Section 14.2, there is a dual feasible plan, we have $\gamma = \beta > -\infty$ and, by Theorem 7.2, there is a primal optimal plan. □

PROBLEMS.

1. Check that $\widehat{A}$ is indeed continuous.

2*. Prove that there is a dual optimal plan.

15. Remarks

Our reference for polarity in Euclidean space is [**W94**]. The moment cone is described in detail in [**KS66**] and [**KN77**]. For the general concept of duality of (topological) vector spaces, see [**Bou87**]. Our exposition of linear programming is based on [**AN87**] and some original papers [**VT68**] and [**Ve70**]. For the polyhedral linear programming, see, for example, [**PS98**], [**V01**] and [**Schr86**]. A comprehensive survey of semidefinite programming can be found in [**VB96**]. Our general reference in control theory is [**BH75**].

Chapter V

Convex Bodies and Ellipsoids

We explore the metric structure of convex bodies in Euclidean space. We introduce ellipsoids and prove that up to a factor depending on the dimension of the ambient space, any convex body looks like an ellipsoid. Next, we discuss how well a convex body can be approximated by a polynomial hypersurface (ellipsoids correspond to quadratic polynomials). We discuss the Ellipsoid Method for solving systems of linear inequalities. Using the technique of measure concentration for the Gaussian measure in Euclidean space, we obtain new results on existence of low rank approximate solutions to systems of linear equations in positive semidefinite matrices and apply our results to problems of graph realizability with a distortion. Then we briefly discuss the measure and metric on the unit sphere. Exercises address ellipsoidal approximations of particularly interesting convex sets (such as the Traveling Salesman Polytope and the set of non-negative multivariate polynomials), various volume inequalities and some results related to the measure concentration technique.

1. Ellipsoids

In this section, we introduce ellipsoids, which are very important convex sets in Euclidean space $\mathbb{R}^d$.

(1.1) Definition. Let $B = \{x \in \mathbb{R}^d : \ \|x\| \leq 1\}$ be the unit ball, let $a \in \mathbb{R}^d$ be a vector and let $T : \mathbb{R}^d \longrightarrow \mathbb{R}^d$ be an invertible linear transformation. The set

$$E = T(B) + a$$

is called an *ellipsoid* and a is called its *center*; see Figure 58.

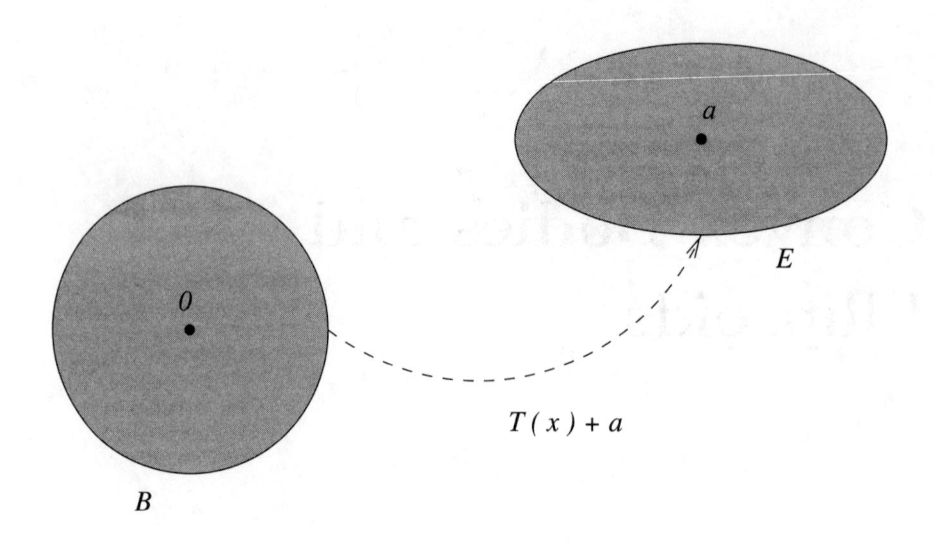

Figure 58

We have $\dim E = d$ and we can write

$$E = \Big\{x \in \mathbb{R}^d : \quad \langle T^{-1}(x-a),\ T^{-1}(x-a)\rangle \le 1\Big\}$$
$$= \Big\{x \in \mathbb{R}^d : \quad \langle Q(x-a),\ x-a\rangle \le 1\Big\},$$

where $Q = (TT^*)^{-1}$ is a positive definite matrix. If we choose a basis of $\mathbb{R}^d$ consisting of the unit eigenvectors of Q, we can define E as

$$E = \Big\{(\xi_1, \dots, \xi_d) : \quad \lambda_1(\xi_1 - \alpha_1)^2 + \ldots + \lambda_d(\xi_d - \alpha_d)^2 \le 1\Big\},$$

where $\lambda_1, \dots, \lambda_d$ are the eigenvalues of Q and $a = (\alpha_1, \dots, \alpha_d)$.

Consequently, for the volume of E, we have

$$\operatorname{vol} E = |\det T| \operatorname{vol} B = \frac{\operatorname{vol} B}{\sqrt{\det Q}}.$$

PROBLEMS.

1°. Prove that any ellipsoid $E \subset \mathbb{R}^d$ centered at a can be written as

$$E = \Big\{x : \quad \langle Q(x-a),\ x-a\rangle \le 1\Big\}$$

for some positive definite matrix Q.

2. Let $E = \{x \in \mathbb{R}^d : \langle Qx, x \rangle \leq 1\}$, where Q is a positive definite matrix, be an ellipsoid. Prove that the polar E° is the ellipsoid defined by

$$E^\circ = \{x \in \mathbb{R}^d : \langle Q^{-1}x, x \rangle \leq 1\}.$$

Deduce that $(\operatorname{vol} E)(\operatorname{vol} E^\circ) = (\operatorname{vol} B)^2$.

3. Let $B \subset \mathbb{R}^d$ be a unit ball and let $T : \mathbb{R}^d \longrightarrow \mathbb{R}^k$ be a linear transformation onto $\mathbb{R}^k$ for some $k \leq d$. Prove that $T(B)$ is an ellipsoid in $\mathbb{R}^k$ centered at the origin.

4°. Let $E \subset \mathbb{R}^d$ be an ellipsoid and let $T : \mathbb{R}^d \longrightarrow \mathbb{R}^d$ be an invertible linear transformation. Prove that $T(E)$ is an ellipsoid.

In a certain sense, ellipsoids are more natural objects than, say, balls. If we choose a different scalar product in Euclidean space, ellipsoids will remain ellipsoids although balls may cease to be balls. In fact, we could have defined ellipsoids without using any Euclidean structure at all: let V be a finite-dimensional real vector space. Then $E \subset V$ is an ellipsoid provided $E = \{x \in V : q(x-a) \leq 1\}$ for some positive definite quadratic form $q : V \longrightarrow \mathbb{R}$ and some $a \in V$.

Last, we will need a couple of technical results.

(1.2) Lemma. *Let $B \subset \mathbb{R}^d$ be a unit ball and let $E \subset \mathbb{R}^d$ be an ellipsoid. Then $E = S(B) + a$ for some $a \in \mathbb{R}^d$ and a linear transformation S whose matrix is positive definite.*

Proof. We observe that $U(B) = B$ for every orthogonal transformation U. Since every square matrix T can be written as $T = SU$, where U is orthogonal and S is positive definite (the polar decomposition), the proof follows by Definition 1.1. □

(1.3) Lemma. *Let X and Y be $d \times d$ positive definite matrices. Then*

$$\det\Big(\frac{X+Y}{2}\Big) \geq \sqrt{\det(X)\det(Y)}.$$

Moreover, the equality holds if and only if $X = Y$.

Proof. Let U be a $d \times d$ orthogonal matrix and let $X' = U^t X U$ and $Y' = U^t Y U$. We observe that the inequality is satisfied (with equality) for X and Y if and only if it is satisfied (with equality) for X' and Y'. By choosing an appropriate orthogonal matrix U, we may assume that X is a diagonal matrix.

Hence we assume that $X = \operatorname{diag}(\lambda_1, \ldots, \lambda_d)$, where $\lambda_i > 0$ for $i = 1, \ldots, d$. Let $T = \operatorname{diag}(\sqrt{\lambda_1}, \ldots, \sqrt{\lambda_d})$, so $X = T^2$. Letting $X' = I$ and $Y' = T^{-1}YT^{-1}$, we observe that the inequality holds (with equality) for X and Y if and only if it holds (with equality) for X' and Y'. Therefore, without loss of generality we may assume that X is the identity matrix I.

Hence we assume that $X = I$ is the identity matrix. Then, by choosing an orthogonal matrix U and letting $Y' = U^t Y U$, we may assume that Y is a diagonal matrix.

Finally, if $X = I$ is the identity matrix and $Y = \operatorname{diag}(\lambda_1, \dots, \lambda_d)$ for some positive $\lambda_1, \dots, \lambda_d$, the inequality reduces to the inequality between the geometric and arithmetic means, that is,

$$\frac{1+\lambda_i}{2} \geq \sqrt{\lambda_i} \qquad \text{for} \quad i = 1, \dots, d,$$

which is satisfied as equality if and only if $\lambda_i = 1$, that is, if and only if $X = Y$. □

Lemma 1.3 leads to a series of fascinating inequalities, some of which are stated in the problems below.

PROBLEMS.

1. Let X and Y be $d \times d$ positive definite matrices and let α and β be positive numbers such that $\alpha + \beta = 1$. Prove that

$$\ln \det\bigl(\alpha X + \beta Y\bigr) \geq \alpha \ln \det X + \beta \ln \det Y.$$

2* (The functional Brunn-Minkowski inequality). Let us fix $\alpha, \beta \geq 0$ such that $\alpha + \beta = 1$. Suppose that f, g and h are non-negative measurable functions on $\mathbb{R}^d$ such that

$$h(\alpha x + \beta y) \geq f^{\alpha}(x) g^{\beta}(y) \quad \text{for all} \quad x, y \in \mathbb{R}^d.$$

Prove that

$$\int_{\mathbb{R}^d} h \, dx \geq \left(\int_{\mathbb{R}^d} f \, dx\right)^{\alpha} \left(\int_{\mathbb{R}^d} g \, dx\right)^{\beta}.$$

The inequality is also known as the Prékopa-Leindler inequality.

Hint: Use induction on the dimension d. For $d = 1$ use the following trick of the "transportation of measure": we may assume that f, g and h are positive and continuous and that

$$\int_{-\infty}^{\infty} f \, dx = \int_{-\infty}^{+\infty} g \, dx = 1.$$

Introduce functions $u, v : [0,1] \longrightarrow \mathbb{R}$ by

$$\int_{-\infty}^{u(t)} f(x) \, dx = t \quad \text{and} \quad \int_{-\infty}^{v(t)} g(x) \, dx = t$$

and let $w(t) = \alpha u(t) + \beta v(t)$. Use that

$$\int_{-\infty}^{\infty} h(x) \, dx = \int_0^1 h\bigl(w(t)\bigr) w'(t) \, dt \quad \text{and} \quad u'(t) f\bigl(u(t)\bigr) = v'(t) g\bigl(v(t)\bigr) = 1;$$

see Section 2.2 of [**Le01**].

3 (The Brunn-Minkowski inequality). Let $A, B \subset \mathbb{R}^d$ be compact convex sets and let $\alpha, \beta \geq 0$ be numbers such that $\alpha + \beta = 1$. Deduce from Problem 2 that

$$\ln \operatorname{vol}\bigl(\alpha A + \beta B\bigr) \geq \alpha \ln \operatorname{vol} A + \beta \ln \operatorname{vol} B.$$

4. Let $A \subset \mathbb{R}^d$ be a compact convex set containing the origin in its interior. Let us define its support function $h_A : \mathbb{R}^d \longrightarrow \mathbb{R}$ by $h_A(x) = \max_{y \in A} \langle x, y \rangle$ (cf. Problem 3 of Section I.8.3). Prove the following formula for the volume of the polar of A:

$$\operatorname{vol} A^\circ = \frac{1}{d!} \int_{\mathbb{R}^d} \exp\{-h_A(x)\} dx;$$

cf. Section IV.1 for the "dual volume".

5 (Firey's inequality). Let $A, B \subset \mathbb{R}^d$ be compact convex sets containing the origin in their interiors and let α and β be positive numbers such that $\alpha + \beta = 1$. Deduce from the formula of Problem 4 that

$$\ln \operatorname{vol}(\alpha A + \beta B)^\circ \leq \alpha \ln \operatorname{vol} A^\circ + \beta \ln \operatorname{vol} B^\circ.$$

6. Let $A \subset \mathbb{R}^d$ be a compact convex set containing the origin in its interior. Let us define a function $f : \operatorname{int} A \longrightarrow \mathbb{R}$ by $f(x) = \ln \operatorname{vol}(A - x)^\circ$. Deduce from Problem 5 that f is a convex function, that is, $f(\alpha x + \beta y) \leq \alpha f(x) + \beta f(y)$ for all $x, y \in \operatorname{int} A$ and all $\alpha, \beta \geq 0$ such that $\alpha + \beta = 1$. Prove that $f(x) \longrightarrow +\infty$ as x approaches ∂A.

Remark: The function f is called the *volumetric barrier* and plays an important role in interior-point methods, a powerful class of methods for solving problems of linear programming; see [**NN94**].

7°. Check that for $d = 1$ the inequalities of Problems 1, 3 and 5 are equivalent to the concavity of $\ln x$ for $x > 0$.

2. The Maximum Volume Ellipsoid of a Convex Body

We define the class of sets we are interested in.

(2.1) Definition. A compact convex set $K \subset \mathbb{R}^d$ with a non-empty interior is called a *convex body*. A convex body K is *symmetric about the origin* provided for every point $x \in \mathbb{R}^d$ we have $x \in K$ if and only if $-x \in K$.

The main result of this section is that for each convex body $K \subset \mathbb{R}^d$ there is a unique ellipsoid $E \subset K$ of the largest volume and that E in some sense "approximates" K. Moreover, the convex bodies that are symmetric about the origin are essentially better approximated than general convex bodies.

We prove the first main result of this section.

(2.2) Theorem. *Let $K \subset \mathbb{R}^d$ be a convex body. Among all ellipsoids E contained in K, there exists a unique ellipsoid of the maximum volume.*

Proof. Let B be the unit ball in $\mathbb{R}^d$. Let us consider the set $\mathcal{X}$ of all pairs (S, a), where S is a linear transformation of $\mathbb{R}^d$ with a positive semidefinite matrix and $a \in \mathbb{R}^d$ is a vector such that $S(B) + a \subset K$. By Lemma 1.2, for each ellipsoid $E \subset K$ there is a pair $(S, a) \in \mathcal{X}$ such that $E = S(B) + a$. Moreover, $\operatorname{vol} E = (\det S) \operatorname{vol} B$.

Since K is compact, there is a number ρ such that $\|x\| \leq \rho$ for all $x \in K$. Therefore, $\|a\| \leq \rho$ and $\|Sx\| \leq 2\rho$ for all $x \in B$ and all $(S, a) \in \mathcal{X}$. In particular, $\mathcal{X}$ is bounded as a subset of $\operatorname{Sym}_d \times \mathbb{R}$; cf. Section II.12. One can see that $\mathcal{X}$ is also closed (see Problem 1 below) and hence compact.

The function $(S, a) \longmapsto \det S$ attains its maximum on $\mathcal{X}$ at a certain point (S_0, a_0). Since K has a non-empty interior, we must have $\det S_0 > 0$, so $E_0 = S_0(B) + a_0$ is an ellipsoid having the largest volume among all ellipsoids inscribed in K.

Let us prove that the inscribed ellipsoid of the largest volume is unique. Since K is convex, one can see that the set $\mathcal{X}$ is also convex. Suppose that $E_1 = S_1(B) + a_1$ and $E_2 = S_2(B) + a_2$ are two ellipsoids of the largest volume among all contained in K. Thus we have $(S_1, a_1) \in \mathcal{X}$ and $(S_2, a_2) \in \mathcal{X}$. Letting $S = (S_1 + S_2)/2$ and $a = (a_1 + a_2)/2$, we observe that $E = S(B) + a \subset K$ and hence $(S, a) \in \mathcal{X}$. Moreover, $\operatorname{vol} E_1 = (\det S_1) \operatorname{vol} B$, $\operatorname{vol} E_2 = (\det S_2) \operatorname{vol} B$ and $\operatorname{vol} E = (\det S) \operatorname{vol} B$. Applying Lemma 1.3, we conclude that unless $S_1 = S_2$, we must have $\operatorname{vol} E > \operatorname{vol} E_1 = \operatorname{vol} E_2$, which is a contradiction. Hence $S_1 = S_2$.

Let us prove that $a_1 = a_2$. Applying a linear transformation and translating, if necessary, we may assume that

$$E_1 = \Big\{x \in \mathbb{R}^d : \quad \|x + a\| \leq 1\Big\} \quad \text{and} \quad E_2 = \Big\{x \in \mathbb{R}^d : \quad \|x - a\| \leq 1\Big\},$$

where $a = (0, \ldots, 0, \alpha)$ for some $\alpha \geq 0$. Let

$$E = \left\{x \in \mathbb{R}^d : \quad \xi_1^2 + \ldots + \xi_{d-1}^2 + \frac{\xi_d^2}{(1 + \alpha)^2} \leq 1\right\}.$$

We observe that $E \subset \operatorname{conv}(E_1 \cup E_2)$.

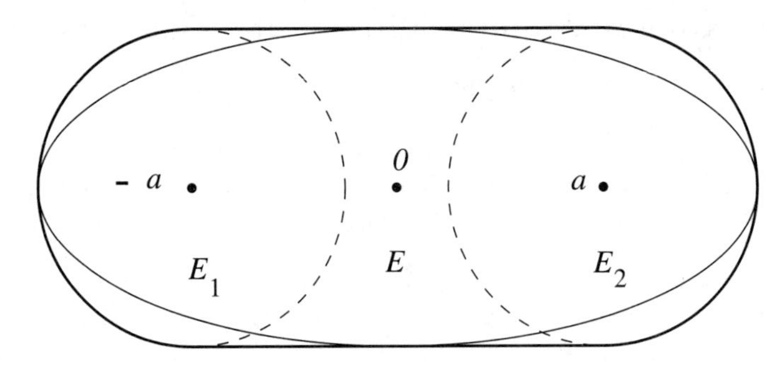

Figure 59

Indeed, let us choose an $x \in E$, $x = (\xi_1, \ldots, \xi_d)$. If $|\xi_d| \leq \alpha$, then x belongs to the cylinder $\xi_1^2 + \ldots + \xi_{d-1}^2 \leq 1$, $|\xi_d| \leq \alpha$, which is a part of $\operatorname{conv}(E_1 \cup E_2)$. If $\xi_d > \alpha$, then $\xi_d \leq 1 + \alpha$ and $(\xi_d - \alpha)^2 \leq \xi_d^2/(1 + \alpha)^2$ (substituting $\tau = \xi_d - \alpha$, we reduce the last inequality to $(1 + \alpha)^2 \leq (1 + \alpha/\tau)^2$ for $0 \leq \tau \leq 1$). Therefore, $x \in E_2$. Similarly, if $\xi_d < -\alpha$, then $x \in E_1$.

Finally, we note that $\operatorname{vol} E = (\operatorname{vol} B)(1+\alpha)$ whereas $\operatorname{vol} E_1 = \operatorname{vol} E_2 = \operatorname{vol} B$. Hence we must have $\alpha = 0$ and the ellipsoid $E \subset K$ of the largest volume is unique. □

(2.3) Definition. Given a convex body $K \subset \mathbb{R}^d$, the unique ellipsoid $E \subset K$ of the maximum volume is called the *maximum volume ellipsoid* of K.

The maximum volume ellipsoid is also known as the Löwner-John ellipsoid or just the John ellipsoid.

PROBLEMS.

1°. Check that the set $\mathcal{X}$ in the proof of Theorem 2.2 is indeed closed.

2. Let $K \subset \mathbb{R}^d$ be a convex body and let E be its maximum volume ellipsoid. For a vector $x = (\xi_1, \ldots, \xi_d)$, let $x \otimes x$ denote the $d \times d$ matrix whose (i,j)-th entry is $\xi_i \xi_j$. Prove that E is the unit ball $B = \{x : \|x\| \leq 1\}$ if and only if there exist unit vectors $u_1, \ldots, u_m \in \partial K$ and positive numbers $\lambda_1, \ldots, \lambda_m$ summing up to 1 such that

$$\sum_{i=1}^m \lambda_i u_i = 0 \quad \text{and} \quad \sum_{i=1}^m \lambda_i (u_i \otimes u_i) = \frac{1}{d} I,$$

where I is the identity matrix. The last condition can be rewritten as

$$\sum_{i=1}^m \lambda_i \langle u_i, x \rangle^2 = \frac{1}{d}\|x\|^2 \quad \text{for all} \quad x \in \mathbb{R}^d.$$

Prove that one can choose $m \leq (d^2 + 3d)/2$.

Hint: Let $X = \partial K \cap B$, let $n = d(d+3)/2$ and let us consider the map

$$\phi : X \longrightarrow \mathbb{R}^n, \quad \text{where} \quad \phi(x) = (x \otimes x, x).$$

If $\left(\frac{1}{d}I, 0\right) \notin \operatorname{conv}(X)$, then $\left(\frac{1}{d}I, 0\right)$ can be separated from $\operatorname{conv}(X)$ by a hyperplane. Use the hyperplane to inscribe an ellipsoid of a bigger volume; see [**B97**].

3. Let E_K denote the maximum volume ellipsoid of a convex body K. Prove that for any two convex bodies A and B in $\mathbb{R}^d$ and any two positive numbers α and β such that $\alpha + \beta = 1$, one has

$$\ln \operatorname{vol} E_{\alpha A + \beta B} \geq \alpha \ln \operatorname{vol} E_A + \beta \ln \operatorname{vol} E_B.$$

4. Let $K \subset \mathbb{R}^d$ be a convex body. Prove that there exists a unique ellipsoid $E \supset K$ of the minimum volume. The ellipsoid E is called the minimum volume ellipsoid (also known as the Löwner-John ellipsoid, the Löwner-Behrend-John ellipsoid or just the Löwner ellipsoid of K).

Although we used the Euclidean structure a lot in the proof of Theorem 2.2, the maximum volume ellipsoid E of a convex body $K \subset \mathbb{R}^d$ does not depend on the Euclidean structure of $\mathbb{R}^d$ at all and is well defined for any convex body in a finite-dimensional vector space V. Indeed, if we change a scalar product in V, the volumes of all measurable sets will be scaled by a positive constant and the maximum volume ellipsoid will remain such.

It is worth noting that the definition of a convex body as well can be made independent of the Euclidean structure. Given a finite-dimensional real vector space V, we call a set $K \subset V$ a *convex body* if K is a convex set not contained in any affine hyperplane and such that the intersection of K with every straight line is a closed bounded interval.

Our goal is to prove that the maximum volume ellipsoid of K approximates K up to a certain factor depending on the dimension alone. Again, the statement of the result is independent of any Euclidean structure although the proof heavily relies on such a structure.

(2.4) Theorem. *Let $K \subset \mathbb{R}^d$ be a convex body and let $E \subset K$ be its maximum volume ellipsoid. Suppose that E is centered at the origin. Then $K \subset dE$.*

Proof. Applying a linear transformation if necessary, we can assume that E is the unit ball B. Suppose there is a point $x \in K$ such that $\|x\| > d$. Let $C = \operatorname{conv}\bigl(B \cup \{x\}\bigr)$. Since K is convex, $C \subset K$. Our goal is to inscribe an ellipsoid E_1 in C such that $\operatorname{vol} E_1 > \operatorname{vol} B$ to obtain a contradiction.

Without loss of generality, we assume that $x = (\rho, 0, \ldots, 0)$ for some $\rho > d$. We look for E_1 in the form:

$$E_1 = \Bigl\{x : \quad \frac{(\xi_1 - \tau)^2}{\alpha^2} + \frac{1}{\beta^2} \sum_{i=2}^{d} \xi_i^2 \leq 1\Bigr\}, \tag{2.4.1}$$

where $\tau > 0$ is sufficiently small. Because of the symmetry, if we find τ, α and β such that E_1 fits inside C for $d = 2$, then E_1 fits inside C for any $d \geq 2$. Hence we assume for a moment that $d = 2$.

The convex hull C is bounded by the two straight line intervals and an arc of ∂B. We inscribe E_1 in C in such a way that E_1 touches B at $(-1, 0)$ and touches the two straight line intervals bounding C as shown on Figure 60. Since E_1 touches B at $(-1, 0)$, we have $\alpha = \tau + 1$. The equation of the tangent line to ∂E_1 at the point $z = (\zeta_1, \zeta_2)$ is

$$\frac{\zeta_1 - \tau}{\alpha^2}(\xi_1 - \tau) + \frac{\zeta_2}{\beta^2}\xi_2 = 1.$$

Since this line passes through $(\rho, 0)$, we get

$$\frac{(\zeta_1 - \tau)^2}{\alpha^2} = \frac{\alpha^2}{(\rho - \tau)^2}.$$

The slope of that line is $-1/\sqrt{\rho^2 - 1}$, hence we deduce the equation

$$\frac{\beta^2}{\zeta_2} = \frac{1}{\sqrt{\rho^2 - 1}} \frac{\alpha^2}{\zeta_1 - \tau} = \frac{\rho - \tau}{\sqrt{\rho^2 - 1}}.$$

Therefore,

$$\frac{\zeta_2^2}{\beta^2} = \frac{\beta^2(\rho^2-1)}{(\rho-\tau)^2}.$$

Since

$$\frac{(\zeta_1-\tau)^2}{\alpha^2} + \frac{\zeta_2^2}{\beta^2} = 1,$$

we get

$$\frac{\alpha^2}{(\rho-\tau)^2} + \frac{\beta^2(\rho^2-1)}{(\rho-\tau)^2} = 1$$

and

$$\beta^2 = \frac{(\rho-\tau)^2-\alpha^2}{\rho^2-1} = \frac{(\rho-\tau)^2-(\tau+1)^2}{\rho^2-1}.$$

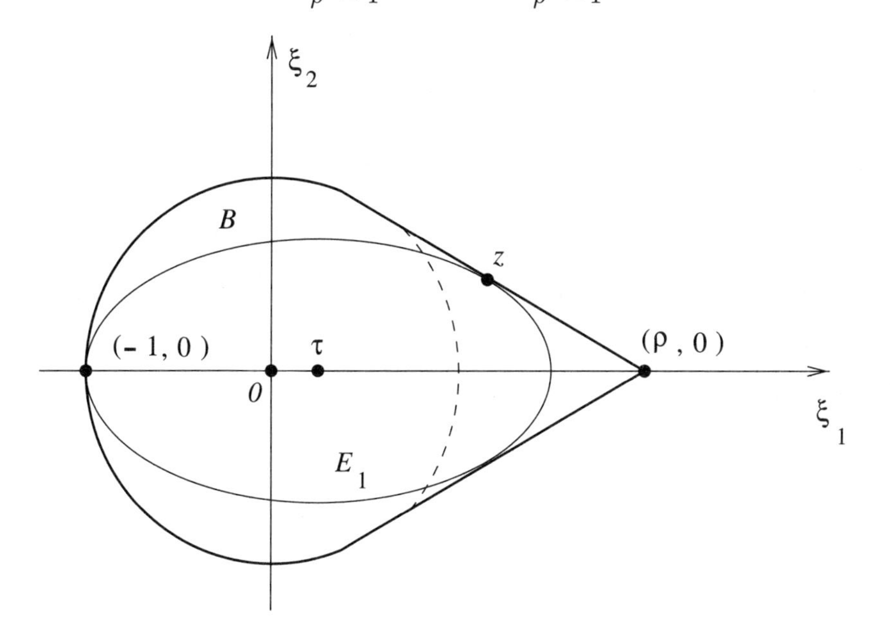

Figure 60. If ρ is large enough, $\operatorname{vol} E_1 > \operatorname{vol} B$.

Summarizing, we conclude that for any $d \geq 2$ and for $0 \leq \tau < (\rho-1)/2$, the ellipsoid E_1 defined by (2.4.1) with

$$\alpha = \tau+1 \quad \text{and} \quad \beta^2 = \frac{(\rho-\tau)^2-(\tau+1)^2}{\rho^2-1}$$

is contained in C. Now we have

$$\ln\frac{\operatorname{vol} E_1}{\operatorname{vol} B} = (d-1)\ln\beta + \ln\alpha = \frac{d-1}{2}\ln\beta^2 + \ln\alpha.$$

Assuming that $\tau > 0$ is small, we obtain

$$\ln \alpha = \tau + O(\tau^2) \quad \text{and} \quad \ln \beta^2 = -\frac{2\tau}{\rho - 1} + O(\tau^2).$$

Therefore, if $\rho > d$, then for a sufficiently small $\tau > 0$, we get $\operatorname{vol} E_1 > \operatorname{vol} B$, which is a contradiction. □

If a convex body K possesses a symmetry, the uniqueness property of the maximum volume ellipsoid E of K implies that E must also possess the symmetry. Sometimes, when the symmetry group of K is rich enough, this may lead to a complete description of E.

PROBLEMS.

1. Let Δ be the standard d-dimensional simplex:

$$\Delta = \Big\{(\xi_1, \ldots, \xi_{d+1}) : \quad \xi_i \geq 0 \quad \text{for} \quad i = 1, \ldots, d+1 \quad \text{and} \quad \xi_1 + \ldots + \xi_{d+1} = 1\Big\}.$$

Consider Δ as a d-dimensional convex body in the affine hyperplane

$$H = \Big\{(\xi_1, \ldots, \xi_{d+1}) : \quad \xi_1 + \ldots + \xi_{d+1} = 1\Big\};$$

cf. Problem 1 of I.2.2. Prove that the maximum volume ellipsoid of Δ is the ball of radius $\dfrac{1}{\sqrt{d(d+1)}}$ centered at $\left(\dfrac{1}{d+1}, \ldots, \dfrac{1}{d+1}\right)$. Show that the constant d in Theorem 2.4 cannot be improved.

2. Let

$$C = \Big\{(\xi_1, \ldots, \xi_d) : \quad |\xi_i| \leq 1 \quad \text{for} \quad i = 1, \ldots, d\Big\}$$

be the cube in $\mathbb{R}^d$; cf. Problem 2 of I.2.2. Prove that the maximum volume ellipsoid of C is the unit ball.

3. Let

$$O = \Big\{(\xi_1, \ldots, \xi_d) : \quad |\xi_1| + \ldots + |\xi_d| \leq 1\Big\}$$

be the standard octahedron in $\mathbb{R}^d$, cf. Problem 3 of I.2.2. Prove that the maximum volume ellipsoid of O is the ball of radius $d^{-1/2}$ centered at the origin.

4°. Suppose that a convex body K is symmetric about the origin. Prove that the maximum volume ellipsoid of K is centered at the origin.

5. Let us fix an $n \times n$ matrix $A = (a_{ij})$,

$$A = \begin{pmatrix} 0 & 1 & 0 & 0 & \ldots & 0 & 1 \\ 1 & 0 & 1 & 0 & \ldots & 0 & 0 \\ 0 & 1 & 0 & 1 & 0 & \ldots & 0 \\ \ldots & \ldots & \ldots & \ldots & \ldots & \ldots & \ldots \\ \ldots & \ldots & \ldots & \ldots & \ldots & \ldots & \ldots \\ 0 & \ldots & \ldots & 0 & 1 & 0 & 1 \\ 1 & 0 & 0 & \ldots & 0 & 1 & 0 \end{pmatrix},$$

i.e.,

$$a_{ij} = \begin{cases} 1 & \text{if } |i-j| \equiv 1 \mod n, \\ 0 & \text{otherwise.} \end{cases}$$

Let P the the convex hull of all matrices obtained from A by a simultaneous permutation of rows and columns: $a_{ij} \longmapsto a_{\sigma^{-1}(i)\sigma^{-1}(j)}$ for a permutation σ of $\{1, \ldots, n\}$. Then P is called the *Symmetric Traveling Salesman Polytope.*

a) Prove that $\dim P = (n^2 - 3n)/2$.

b) Prove that if an affine subspace L in the affine hull of P is invariant under the action of the symmetric group S_n: $x_{ij} \longmapsto x_{\sigma^{-1}(i)\sigma^{-1}(j)}$, then L is either a point $x_{ij} = 2/n$ or the whole affine hull of P (that is, the representation of S_n is irreducible).

c) Prove that the maximum volume ellipsoid of P (where P is considered as a convex body in its affine hull) is a ball centered at $x_{ij} = 2/n$.

d) Prove that the ball of radius $\sqrt{8n-16}/(n^2-3n)$ centered at $x_{ij} = 2/n$ is contained in P.

Remark: See [**BS96**] and [**YKK84**] for information about the combinatorial structure of the Traveling Salesman and other interesting polytopes.

6. Let $K \subset \mathbb{R}^d$ be a convex body and suppose that the points $a_1, \ldots, a_{d+1} \in K$ are chosen in such a way that the volume of $S = \operatorname{conv}(a_1, \ldots, a_{d+1})$ is the maximum possible. Suppose that $a_1 + \ldots + a_{d+1} = 0$. Prove that $K \subset -dS$; cf. Problem 1 of Section I.5.2.

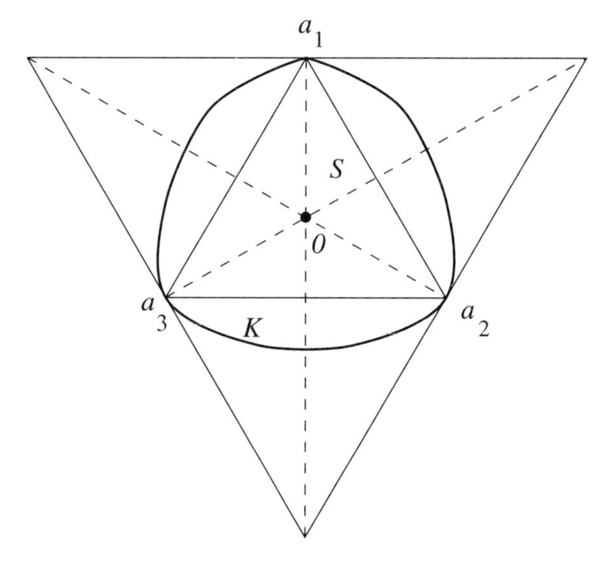

Figure 61

7. In the Euclidean space Sym_n of symmetric $n \times n$ matrices X with the scalar product $\langle A, B \rangle = \operatorname{tr}(AB)$, let $\mathcal{C}$ be the set of all positive semidefinite matrices of

trace 1. Let us consider $\mathcal{C}$ as a convex body in the affine hyperplane $\operatorname{tr} X = 1$ (cf. Section II.12). Prove that the maximum volume ellipsoid of $\mathcal{C}$ is the ball of radius $\dfrac{1}{\sqrt{n(n-1)}}$ centered at $\dfrac{1}{n}I$, where I is the identity matrix. Prove that $\mathcal{C}$ is contained in the ball of radius $\sqrt{\dfrac{n-1}{n}}$ centered at $\dfrac{1}{n}I$.

Remark: It turns out that $\mathcal{C}$ is a counterexample to the famous *Borsuk conjecture* (see Section 31.1 of [**DL97**]) which asserted that every d-dimensional convex body can be partitioned into $d+1$ subsets of a strictly smaller diameter (the diameter of a convex body is the largest distance between some two points of the body).

8* (G. Blekherman). Let $H_{2k,n}$ be the vector space of all real homogeneous polynomials p of degree $2k$ in n variables $x = (\xi_1, \ldots, \xi_n)$; cf. Section I.3. Check that $\dim H_{2k,n} = \binom{n+2k-1}{2k}$. Let us make $H_{2k,n}$ Euclidean space by introducing the scalar product

$$\langle f, g \rangle = \int_{\mathbb{S}^{n-1}} f(x)g(x)\, dx \quad \text{for} \quad f, g \in H_{2k,n},$$

where $\mathbb{S}^{n-1}$ is the unit sphere in $\mathbb{R}^n$ and dx is the rotation invariant probability measure on $\mathbb{S}^{n-1}$.

Let

$$C = \Big\{ f \in H_{2k,n} : \quad f(x) \geq 0 \quad \text{for all} \quad x \in \mathbb{R}^n \quad \text{and} \quad \int_{\mathbb{S}^{n-1}} f(x)\, dx = 1 \Big\}.$$

Considering C as a convex body in its affine hull, prove that the maximum volume ellipsoid of C is the ball B of radius $\sqrt{\dfrac{1}{\dim H_{2k,n} - 1}}$ centered at $p(x) = \|x\|^{2k} = (\xi_1^2 + \ldots + \xi_n^2)^k$.

9* (G. Blekherman). Let C be the convex body of Problem 8 above and let $p = \|x\|^{2k}$, $p \in C$. Let

$$\alpha = \frac{1}{\binom{n+k-1}{k} - 1}.$$

Prove that

$$\alpha(p - C) + p \subset C$$

and that the inclusion fails for any larger α. In other words, the *coefficient of symmetry* of C with p as the origin is α.

It turns out that the estimate of Theorem 2.4 can be strengthened in one important special case.

(2.5) Theorem. *Suppose that a convex body $K \subset \mathbb{R}^d$ is symmetric about the origin. Let $E \subset K$ be the maximum volume ellipsoid of K. Then E is centered at the origin and $K \subset \sqrt{d}E$.*

Proof. It follows by Problem 4, Section 2.4, that E is centered at the origin. Applying a linear transformation, if necessary, we can assume that E is the unit ball B. Suppose that there is a point $x \in K$ such that $\|x\| > \sqrt{d}$. Since K is symmetric about the origin, we have $-x \in K$. Let $C = \operatorname{conv}\big(B \cup \{x\} \cup \{-x\}\big)$. Our goal is to inscribe an ellipsoid $E_1 \subset C$ such that $\operatorname{vol} E_1 > \operatorname{vol} B$ to obtain a contradiction; cf. Figure 62.

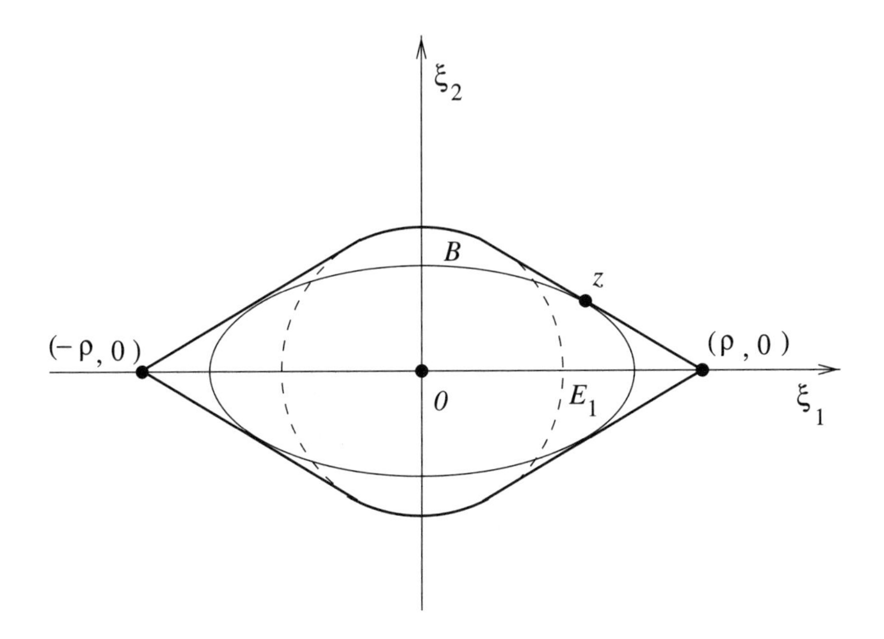

Figure 62. If ρ is large enough, then $\operatorname{vol} E_1 > \operatorname{vol} B$.

The proof essentially follows the proof of Theorem 2.4 with some modifications. Without loss of generality, we assume that $x = (\rho, 0, \ldots, 0)$ for some $\rho > d$. We look for E_1 in the form

$$E_1 = \Big\{(\xi_1, \ldots, \xi_d): \quad \frac{\xi_1^2}{\alpha^2} + \frac{1}{\beta^2} \sum_{i=2}^{d} \xi_i^2 \leq 1\Big\}. \tag{2.5.1}$$

Because of the symmetry, if we find α and β such that E_1 fits inside C for $d = 2$, then E_1 fits inside C for any $d \geq 2$. Hence we assume for a moment that $d = 2$.

The convex hull C is bounded by four straight line intervals and two arcs of ∂B. We inscribe E_1 centered at the origin and in such a way that E_1 touches the intervals bounding C as shown on Figure 62.

The equation of the tangent line to ∂E_1 at the point $z = (\zeta_1, \zeta_2)$ is

$$\frac{\zeta_1}{\alpha^2}\xi_1 + \frac{\zeta_2}{\beta^2}\xi_2 = 1.$$

Since the tangent line passes through $(\rho, 0)$, we get the equation

$$\frac{\zeta_1^2}{\alpha^2} = \frac{\alpha^2}{\rho^2}.$$

The slope of the tangent line is $-1/\sqrt{\rho^2 - 1}$, from which we deduce

$$\frac{\zeta_1 \beta^2}{\zeta_2 \alpha^2} = \frac{1}{\sqrt{\rho^2 - 1}}$$

and

$$\frac{\zeta_2^2}{\beta^2} = \frac{(\rho^2 - 1)\beta^2}{\rho^2}.$$

Since

$$\frac{\zeta_1^2}{\alpha^2} + \frac{\zeta_2^2}{\beta^2} = 1,$$

we get

$$\alpha^2 = \rho^2 - (\rho^2 - 1)\beta^2.$$

Summarizing, we conclude that for any $d \geq 2$ and for any $1 \geq \beta > 0$, the ellipsoid E_1 defined by (2.5.1) with $\alpha^2 = \rho^2 - (\rho^2 - 1)\beta^2$ is contained in C. Now we have

$$\ln \frac{\operatorname{vol} E_1}{\operatorname{vol} B} = (d-1)\ln\beta + \ln\alpha.$$

Let us choose $\beta = 1 - \epsilon$ for some sufficiently small $\epsilon > 0$. Then $\ln\beta = -\epsilon + O(\epsilon^2)$ and $\ln\alpha = \epsilon(\rho^2 - 1) + O(\epsilon^2)$. Therefore, if $\rho > \sqrt{d}$, then for a sufficiently small $\epsilon > 0$, we get $\operatorname{vol} E_1 > \operatorname{vol} B$, which is a contradiction. □

PROBLEMS.

1. Prove that the constant $\sqrt{d}$ in Theorem 2.5 cannot be improved (cf. Problem 3 of Section 2.4).

2. Let $K \subset \mathbb{R}^d$ be a convex body symmetric about the origin and let $E \supset K$ be its minimum volume ellipsoid (see Problem 4, Section 2.3). Prove that $d^{-1/2}E \subset K$.

3. Let $K \subset \mathbb{R}^d$ be a convex body and let $E \supset K$ be its minimum volume ellipsoid. Suppose that E is centered at the origin. Prove that $(1/d)E \subset K$.

3. Norms and Their Approximations

We apply our results to get some information about the structure of norms, an important class of functions on a vector space. We recall the definition of a norm.

(3.1) Definition. Let V be a vector space. A function $p : V \longrightarrow \mathbb{R}$ is called a *norm* if it satisfies the following properties:

- $p(x) \geq 0$ for all $x \in V$ and $p(x) = 0$ if and only if $x = 0$;
- $p(\lambda x) = |\lambda| p(x)$ for all $x \in V$ and all $\lambda \in \mathbb{R}$;
- $p(x + y) \leq p(x) + p(y)$ for all $x, y \in V$.

PROBLEMS.

1°. Prove that the following functions are norms in $\mathbb{R}^d$:

$p(x) = \left(\sum_{i=1}^d \xi_i^2\right)^{1/2}$ (the ℓ^2 norm),

$p(x) = \max_{i=1,\dots,d} |\xi_i|$ (the ℓ^∞ norm) and

$p(x) = \sum_{i=1}^d |\xi_i|$ (the ℓ^1 norm)

for $x = (\xi_1, \dots, \xi_d)$.

2°. Let $C[0,1]$ be the vector space of all continuous functions $f : [0,1] \longrightarrow \mathbb{R}$. Prove that

$$p(f) = \max_{t \in [0,1]} |f(t)|$$

is a norm in $C[0,1]$.

We establish a relationship between norms and convex bodies.

(3.2) Lemma. *Let $p : \mathbb{R}^d \longrightarrow \mathbb{R}$ be a norm. Let*

$$K_p = \bigl\{x \in \mathbb{R}^d : \quad p(x) \leq 1\bigr\}.$$

Then K_p is a convex body symmetric about the origin. Conversely, if $K \subset \mathbb{R}^d$ is a convex body symmetric about the origin, then

$$p_K(x) = \inf\bigl\{\lambda > 0 : \quad x \in \lambda K\bigr\}$$

is a norm in $\mathbb{R}^d$ such that $K = \bigl\{x \in \mathbb{R}^d : \quad p(x) \leq 1\bigr\}$.

Proof. First, we show that K_p is a convex body symmetric about the origin. For any $a, b \in K_p$ and for any $\alpha, \beta \geq 0$ such that $\alpha + \beta = 1$ we have

$$p(\alpha a + \beta b) \leq p(\alpha a) + p(\beta b) = \alpha p(a) + \beta p(b) \leq \alpha + \beta \leq 1,$$

so $\alpha a + \beta b \in K$ and hence K is convex. Since $p(-x) = p(x)$, we conclude that K_p is symmetric about the origin.

Next, we remark that p is a continuous function. Indeed, let $e_1, \dots, e_d$ be the standard basis of $\mathbb{R}^d$, so that $x = \xi_1 e_1 + \ldots + \xi_d e_d$ for $x = (\xi_1, \dots, \xi_d)$. Let

$$\gamma = \max_{i=1,\dots,d} p(e_i).$$

Then

$$p(x) \leq \gamma \sum_{i=1}^d |\xi_i|.$$

Let us choose two points $x = (\xi_1, \dots, \xi_d)$ and $y = (\eta_1, \dots, \eta_d)$ in $\mathbb{R}^d$. Then

$$p(x) = p\bigl(y + (x - y)\bigr) \leq p(y) + p(x - y) \quad \text{and, similarly,}$$
$$p(y) = p\bigl(x + (y - x)\bigr) \leq p(x) + p(y - x).$$

Therefore,

$$|p(x) - p(y)| \leq p(x - y) \leq \gamma \sum_{i=1}^{d} |\eta_i - \xi_i|,$$

from which we conclude that p is continuous.

Since K_p is the inverse image of a closed set $\{\lambda \leq 1\} \subset \mathbb{R}^d$, it is closed. Let $\mathbb{S}^{d-1} = \{x : \|x\| = 1\}$ be the unit sphere in $\mathbb{R}^d$. Since $p(x) > 0$ for every $x \in \mathbb{S}^{d-1}$ and since p is continuous, there exists a number $\delta > 0$ such that $p(x) > \delta$ for all $x \in \mathbb{S}^{d-1}$. It follows then that $\|x\| \leq \delta^{-1}$ for every $x \in K_p$. Hence K_p is bounded and therefore K_p is compact.

It remains to show that K_p contains the origin in its interior. Since p is continuous, there is a number $\beta > 0$ such that $p(x) \leq \beta$ for all $x \in \mathbb{S}^{d-1}$. Then $p(x) \leq 1$ for all x such that $\|x\| \leq 1/\beta$.

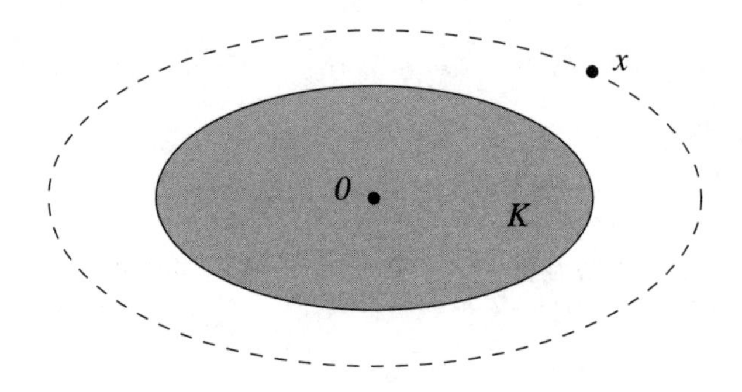

Figure 63. Example: $p_K(x) = 1.5$

Conversely, let $K \subset \mathbb{R}^d$ be a convex body symmetric about the origin and let

$$p_K(x) = \inf\{\lambda > 0 : x \in \lambda K\}.$$

Let us prove that $p_K(x + y) \leq p_K(x) + p_K(y)$ (all the remaining properties are relatively straightforward). Letting $p_K(x) = \lambda_1$ and $p_K(y) = \lambda_2$, we observe that for any $\epsilon > 0$, we have $x \in (\lambda_1 + \epsilon)K$ and $y \in (\lambda_2 + \epsilon)K$. From Problem 4, Section I.1.5, we conclude that $x + y \in (\lambda_1 + \lambda_2 + 2\epsilon)K$. Hence $p_K(x + y) \leq \lambda_1 + \lambda_2 + 2\epsilon$ and since $\epsilon > 0$ was arbitrary, we conclude that $p_K(x+y) \leq \lambda_1 + \lambda_2 = p(x) + p(y)$. □

PROBLEMS.

1. Let V be a vector space and let $p : V \longrightarrow \mathbb{R}$ be a norm. Let

$$K_p = \{x \in V : \quad p(x) \leq 1\}.$$

Prove that K_p is a convex set, symmetric about the origin, which does not contain straight lines and such that $\bigcup_{\lambda \geq 0}(\lambda K_p) = V$. Conversely, let $K \subset V$ be a convex set, symmetric about the origin, which does not contain straight lines and such that $\bigcup_{\lambda>0}(\lambda K) = V$. For $x \in V$ let $p_K(x) = \inf\{\lambda > 0 : x \in \lambda K\}$. Prove that p_K is a norm.

2°. Prove that $p_{K_p} = p$ for any norm $p : V \longrightarrow \mathbb{R}$.

3. Let $E \subset \mathbb{R}^d$ be an ellipsoid centered at the origin. Prove that $p_E(x) = \sqrt{q(x)}$ for some positive definite quadratic form $q : \mathbb{R}^d \longrightarrow \mathbb{R}$.

4. Let $p : \mathbb{R}^d \longrightarrow \mathbb{R}$ be a norm. Prove that there exists a positive definite quadratic form $q : \mathbb{R}^d \longrightarrow \mathbb{R}$ such that $\sqrt{q(x)} \leq p(x) \leq \sqrt{dq(x)}$ for all $x \in \mathbb{R}^d$.

Hint: Combine Theorem 2.5, Lemma 3.2 and Problem 3 above.

5. Let $p : \mathbb{R}^d \longrightarrow \mathbb{R}$ be a norm and let $S = \{x \in \mathbb{R}^d : p(x) = 1\}$. Prove that for any $\delta > 0$ there is a subset $N \subset S$ consisting of not more than $(1+2/\delta)^d$ points such that for every $x \in S$ there exists $y \in N$ such that $p(x-y) \leq \delta$. The set N is called a *δ-net* of p.

Hint: Choose N to be the maximal subset $N \subset S$ with the property that $p(x-y) > \delta$ for all $x, y \in N$.

Problem 4 of Section 3.2 tells us that an arbitrary norm in $\mathbb{R}^d$ can be approximated by a square root of a quadratic form up to a certain factor depending on the dimension d alone (equivalently, a convex body symmetric about the origin can be approximated by an ellipsoid). The question we are going to address now is whether we can get a better approximation by using higher degree polynomials (equivalently, whether we can obtain better approximations of a convex body by using higher degree algebraic hypersurfaces). As a preparation, we need to review some linear algebra.

(3.3) Tensor powers. By the n-th tensor power

$$W = \underbrace{\mathbb{R}^d \otimes \ldots \otimes \mathbb{R}^d}_{n \text{ times}}$$

of Euclidean space $\mathbb{R}^d$ we mean the space of all $d \times \ldots \times d$ arrays (tensors)

$$x = \left(\xi_{i_1 \ldots i_n} : 1 \leq i_1, \ldots, i_n \leq d\right)$$

of real numbers (coordinates) $\xi_{i_1 \ldots i_n}$. Coordinatewise addition of arrays and multiplication of an array by a number make W a d^n-dimensional vector space. In particular, $\mathbb{R}^d \otimes \mathbb{R}^d$ can be thought of as the space of all (real) $d \times d$ matrices. We introduce the scalar product

$$\langle x, y \rangle = \sum_{1 \leq i_1, \ldots, i_n \leq d} \xi_{i_1 \ldots i_n} \eta_{i_1 \ldots i_n} \quad \text{where} \quad x = \left(\xi_{i_1 \ldots i_n}\right) \quad \text{and} \quad y = \left(\eta_{i_1 \ldots i_n}\right)$$

thus making W a Euclidean space which can be identified with $\mathbb{R}^{d^n}$.

Let $x_1, \ldots, x_n$ be vectors from $\mathbb{R}^d$, $x_i = (\xi_{i1}, \ldots, \xi_{id})$. We write $x_1 \otimes \ldots \otimes x_n$ for the tensor

$$c = \left(\gamma_{i_1 \ldots i_n}\right) \quad \text{where} \quad \gamma_{i_1 \ldots i_n} = \xi_{1 i_1} \cdots \xi_{n i_n}.$$

There is an identity

$$\langle x_1 \otimes \ldots \otimes x_n,\ y_1 \otimes \ldots \otimes y_n \rangle = \prod_{i=1}^{n} \langle x_i, y_i \rangle \quad \text{for} \quad x_i, y_i \in \mathbb{R}^d,$$

where the scalar product in the left-hand side is taken in W and the scalar products in the right-hand side are taken in $\mathbb{R}^d$.

We are interested in the subspace $\operatorname{Sym}(W)$ of W consisting of the tensors $x = (\xi_{i_1 \ldots i_n})$ whose coordinates $\xi_{i_1, \ldots, i_n}$ depend only on the multiset $\{i_1, \ldots, i_n\}$ but not on the order of the indices in the sequence $i_1, \ldots, i_n$, that is

$$\xi_{i_1 \ldots i_n} = \xi_{j_1 \ldots j_n}$$

provided $i_1 \ldots i_n$ is a permutation of $j_1 \ldots j_n$. Thus the value of $\xi_{i_1 \ldots i_n}$ depends on how many 1's, 2's, ..., d's are among the indices $i_1, \ldots, i_n$. Hence the dimension of W is equal to the number of non-negative integer solutions of the equation $k_1 + \ldots + k_d = n$, that is,

$$\dim \operatorname{Sym}(W) = \binom{n+d-1}{n}.$$

Finally, we observe that for any $x \in \mathbb{R}^d$, the tensor

$$x^{\otimes n} = \underbrace{x \otimes \ldots \otimes x}_{n \text{ times}}$$

lies in $\operatorname{Sym}(W)$.

PROBLEMS.

1°. Let us identify $\mathbb{R}^d \otimes \mathbb{R}^d$ with the space V of $d \times d$ matrices. Show that the scalar product in $\mathbb{R}^d \otimes \mathbb{R}^d$ is defined by $\langle A, B \rangle = \operatorname{tr}(AB^t)$. Show that $\operatorname{Sym}(V)$ consists of the symmetric $d \times d$ matrices and that $x \otimes x$ are the positive semidefinite matrices whose rank does not exceed 1.

2. Let $u_1, \ldots, u_m$ be points in $\mathbb{R}^d$. Consider the points

$$v_i = u_i \otimes u_i \quad \text{for} \quad i = 1, \ldots, m$$

in $W = \mathbb{R}^d \otimes \mathbb{R}^d$. Let $P = \operatorname{conv}(v_i : i = 1, \ldots, m)$. Let $L \subset U$ be a subspace (hence $0 \in L$) and let $I \subset \{1, \ldots, m\}$ be the set of all i such that $u_i \in L$. Prove that $\operatorname{conv}(v_i : i \in I)$ is a face of P.

3°. Let $B = \{x \in \mathbb{R}^d :\ \|x\| \leq 1\}$ be the unit ball. Prove that

$$\|y\| = \max_{x \in B} \langle y, x \rangle \quad \text{for all} \quad y \in \mathbb{R}^d.$$

Now we can prove that by using higher degree polynomials, we can approximate the norm better.

(3.4) Theorem. *Let $p : \mathbb{R}^d \longrightarrow \mathbb{R}$ be a norm. For any integer $n \geq 1$ there exists a homogeneous polynomial $q : \mathbb{R}^d \longrightarrow \mathbb{R}$ of degree $2n$ such that*

1. *The polynomial q is a sum of squares:*

$$q = \sum_{i \in I} q_i^2,$$

where $q_i : \mathbb{R}^d \longrightarrow \mathbb{R}$ are homogeneous polynomials of degree n. In particular, $q(x) \geq 0$ for all $x \in \mathbb{R}^d$.

2. *For all $x \in \mathbb{R}^d$*

$$q^{\frac{1}{2n}}(x) \leq p(x) \leq \binom{n+d-1}{n}^{\frac{1}{2n}} q^{\frac{1}{2n}}(x).$$

Proof. Let $K = K_p = \{x : p(x) \leq 1\}$. Then, by Lemma 3.2, K is a convex body in $\mathbb{R}^d$ symmetric about the origin. Let K° be the polar of K. Then K° is also a convex body symmetric about the origin. Applying Theorem IV.1.2 (the Bipolar Theorem), we can write

$$K = \Big\{x : \quad \langle x, y \rangle \leq 1 \quad \text{for all} \quad y \in K^\circ\Big\}.$$

By Problem 2 of Section 3.2, we can write

$$\begin{aligned} p(x) &= \inf\big\{\lambda > 0 : x \in \lambda K\big\} \\ &= \inf\big\{\lambda > 0 : \lambda^{-1} x \in K\big\} \\ &= \inf\big\{\lambda > 0 : \lambda^{-1}\langle x, y \rangle \leq 1 \quad \text{for all} \quad y \in K^\circ\big\} \\ &= \inf\big\{\lambda > 0 : \langle x, y \rangle \leq \lambda \quad \text{for all} \quad y \in K^\circ\big\}. \end{aligned}$$

Finally,

$$p(x) = \max_{y \in K^\circ} \langle x, y \rangle = \max_{y \in K^\circ} |\langle x, y \rangle|. \tag{3.4.1}$$

Let

$$W = \underbrace{\mathbb{R}^d \otimes \ldots \otimes \mathbb{R}^d}_{n \text{ times}} = \mathbb{R}^{d^n}.$$

For a vector $x \in \mathbb{R}^d$, let

$$x^{\otimes n} = \underbrace{x \otimes \ldots \otimes x}_{n \text{ times}} \in W.$$

Hence we have $\langle x^{\otimes n}, y^{\otimes n} \rangle = \big(\langle x, y \rangle\big)^n$; cf. Section 3.3. From (3.4.1), we can write

$$p^n(x) = \max_{y \in K^\circ} \langle x^{\otimes n}, y^{\otimes n} \rangle = \max_{y \in K^\circ} |\langle x^{\otimes n}, y^{\otimes n} \rangle|.$$

Let

$$A = \operatorname{conv}\big(y^{\otimes n}, -y^{\otimes n} : \ y \in K^\circ\big) \subset W.$$

Thus

$$p^n(x) = \max_{z \in A} \langle x^{\otimes n}, z \rangle. \tag{3.4.2}$$

Since the map $y \longmapsto y^{\otimes n}$ is continuous and K° is compact, by Corollary I.2.4, the set A is compact. Moreover, A is symmetric about the origin. Since $y^{\otimes n} \in \operatorname{Sym}(W)$ for all $y \in \mathbb{R}^d$ (cf. Section 3.3), we have $A \subset \operatorname{Sym}(W)$. Therefore,

$$\dim A \leq \binom{n+d-1}{n}.$$

Let E be the maximum volume ellipsoid of A in the affine hull of A. Then, by Theorem 2.5,

$$E \subset A \subset \binom{n+d-1}{n}^{1/2} E. \tag{3.4.3}$$

Let us define

$$f(x) = \max_{z \in E} \langle x^{\otimes n}, z \rangle.$$

We claim that $q(x) = f^2(x)$ is a polynomial satisfying Parts 1 and 2 of the theorem. Indeed, let B be the standard unit ball in W:

$$B = \Big\{ (\gamma_{i_1 \ldots i_n}) : \sum_{1 \leq i_1, \ldots, i_n \leq d} \gamma^2_{i_1 \ldots i_n} \leq 1 \Big\}.$$

Since E is an ellipsoid (in the affine hull of A), there is a (non-invertible) linear transformation T of W such that $T(B) = E$. Then

$$\begin{aligned} f(x) &= \max_{z \in E} \langle x^{\otimes n}, z \rangle = \max_{u \in B} \langle x^{\otimes n}, T(u) \rangle \\ &= \max_{u \in B} \langle T^*(x^{\otimes n}), u \rangle = \| T^*(x^{\otimes n}) \|, \end{aligned}$$

where $\| \cdot \|$ is the standard Euclidean norm in W; cf. Problem 3 of Section 3.3. Now the coordinates of $x^{\otimes n}$ are the monomials of degree n in the coordinates of $x \in \mathbb{R}^d$ and hence the coordinates of $T^*(x^{\otimes n})$ are homogeneous polynomials $q_{i_1 \ldots i_n}(x)$ of degree n in the coordinates of x. Hence Part 1 follows. Part 2 follows by (3.4.2) and (3.4.3). □

Choosing, for example, $n = d$ in Theorem 3.4, we conclude that any norm $p : \mathbb{R}^d \longrightarrow \mathbb{R}$ can be approximated by the $2d$-th root of a polynomial q of degree $2d$ within a factor of $\binom{2d-1}{d}^{1/2d} \leq 2$.

PROBLEMS.

1°. Show that for any $\epsilon > 0$ there exists an even $m = m(\epsilon)$ such that for any sufficiently large $d \geq d(\epsilon)$ and any norm $p : \mathbb{R}^d \longrightarrow \mathbb{R}$ there exists a homogeneous polynomial $q : \mathbb{R}^d \longrightarrow \mathbb{R}$ of degree m such that $q(x) \geq 0$ for all $x \in \mathbb{R}^d$ and

$$q^{1/m}(x) \leq p(x) \leq \epsilon \sqrt{d} \cdot q^{1/m}(x) \quad \text{for all} \quad x \in \mathbb{R}^d.$$

2. Let $p : \mathbb{R}^d \longrightarrow \mathbb{R}$ be the ℓ^1 norm,

$$p(\xi_1, \ldots, \xi_d) = |\xi_1| + \ldots + |\xi_d|,$$

and let $q : \mathbb{R}^d \longrightarrow \mathbb{R}$ be a homogeneous polynomial of degree 6 such that $q(x) \geq 0$ for all $x \in \mathbb{R}^d$ and

$$q^{1/6}(x) \leq p(x) \leq C(d) q^{1/6}(x) \quad \text{for all} \quad x \in \mathbb{R}^d,$$

where $C(d)$ is some constant. Prove that $C(d) \geq c\sqrt{d}$ for some absolute constant $c > 0$.

3°. Show that for any $\alpha > 1$ there exists $\beta = \beta(\alpha) > 0$ such that for any norm $p : \mathbb{R}^d \longrightarrow \mathbb{R}$ there exists a homogeneous polynomial $q : \mathbb{R}^d \longrightarrow \mathbb{R}$ such that $m = \deg q \leq \beta d$, $q(x) \geq 0$ for all $x \in \mathbb{R}^d$ and

$$q^{1/m}(x) \leq p(x) \leq \alpha q^{1/m}(x) \quad \text{for all} \quad x \in \mathbb{R}^d.$$

4°. In the context of Theorem 3.4, assume that p satisfies all the requirements of Definition 3.1 except, perhaps, that $p(x) = 0$ implies $x = 0$. Let $K = \{x : p(x) \leq 1\}$ and let K° be the polar of K. Let $D = \dim \operatorname{span}(x^{\otimes n} : x \in K^\circ)$. Check that

$$q^{\frac{1}{2n}}(x) \leq p(x) \leq D^{\frac{1}{2n}} q^{\frac{1}{2n}}(x) \quad \text{for all} \quad x \in \mathbb{R}^d.$$

5. Let $H_{k,d}$ be the vector space of all homogeneous real polynomials of degree k in d variables $x = (\xi_1, \ldots, \xi_d)$, so $\dim H_{k,d} = \binom{k+d-1}{k}$. Let $K \subset \mathbb{R}^d$ be a compact set symmetric about the origin. Let $p : H_{k,d} \longrightarrow \mathbb{R}$ be defined by

$$p(f) = \max_{x \in K} |f(x)| \quad \text{for} \quad f \in H_{k,d}.$$

Prove that for any integer $n \geq 1$ there is a homogeneous polynomial $q : H_{k,d} \longrightarrow \mathbb{R}$ of degree $2n$ such that $q(f) \geq 0$ for all $f \in H_{k,d}$ and

$$q^{\frac{1}{2n}}(f) \leq p(f) \leq \binom{kn+d-1}{d-1}^{\frac{1}{2n}} q^{\frac{1}{2n}}(f) \quad \text{for all} \quad f \in H_{k,d}.$$

Hint: Use Problem 4 above.

6°. Let $p : \mathbb{R}^d \longrightarrow \mathbb{R}$ be the ℓ^∞ norm,

$$p(\xi_1, \ldots, \xi_d) = \max_{i=1,\ldots,d} |\xi_i|.$$

For a positive integer n, let us define a polynomial $q : \mathbb{R}^d \longrightarrow \mathbb{R}$ by

$$q(\xi_1, \ldots, \xi_d) = \sum_{i=1}^d \xi_i^{2n}.$$

Prove that

$$q^{\frac{1}{2n}}(x) \leq p(x) \leq d^{\frac{1}{2n}} q^{\frac{1}{2n}}(x) \quad \text{for all} \quad x \in \mathbb{R}^d.$$

7. Let $H_{2,d}$ be the vector space of all quadratic forms $q : \mathbb{R}^d \longrightarrow \mathbb{R}$ and let $\mathbb{S}^{d-1} \subset \mathbb{R}^d$ be the unit sphere. Let $p : H_{2,d} \longrightarrow \mathbb{R}$ be defined by

$$p(f) = \max_{x \in \mathbb{S}^{d-1}} |f(x)| \quad \text{for} \quad f \in H_{2,d}.$$

Prove that for any positive integer n there exists a homogeneous polynomial $q : H_{2,d} \longrightarrow \mathbb{R}$ of degree $2n$ such that $q(f) \geq 0$ for all $f \in H_{2,d}$ and

$$q^{\frac{1}{2n}}(f) \leq p(f) \leq d^{\frac{1}{2n}} q^{\frac{1}{2n}}(f) \quad \text{for all} \quad f \in H_{2,d}.$$

The next two problems require some representation theory.

8. Let G be a compact subgroup of the group of orthogonal transformations of $\mathbb{R}^d$. Let $v \in \mathbb{R}^d$ be a vector and let

$$\mathcal{O}_v = \{g(v) : \quad g \in G\}$$

be the orbit of v. Let us define a function $p : \mathbb{R}^d \longrightarrow \mathbb{R}$ by

$$p(x) = \max_{g \in G} \left| \langle x, \ g(v) \rangle \right|.$$

Let dg be the Haar probability measure on G and let us define the quadratic form $q : \mathbb{R}^d \longrightarrow \mathbb{R}$ by

$$q(x) = \int_G \langle x, \ g(v) \rangle^2 \ dg.$$

Prove that

$$\sqrt{q(x)} \leq p(x) \leq \sqrt{d}\sqrt{q(x)} \quad \text{for all} \quad x \in \mathbb{R}^d$$

and that, more precisely,

$$\sqrt{q(x)} \leq p(x) \leq \sqrt{\dim \operatorname{span}(\mathcal{O}_v)}\sqrt{q(x)} \quad \text{for all} \quad x \in \mathbb{R}^d.$$

Hint: Note that the eigenspaces of q are G-invariant subspaces of V.

9. In Problem 8 above, for an integer $n > 0$, let us define a polynomial $q : \mathbb{R}^d \longrightarrow \mathbb{R}$ by

$$q(x) = \int_G \langle x, \ g(v) \rangle^{2n} \ dg$$

(thus $n = 1$ in Problem 8). Deduce from Problem 8 that

$$q^{\frac{1}{2n}}(x) \leq p(x) \leq \binom{d+n-1}{n}^{\frac{1}{2n}} q^{\frac{1}{2n}}(x) \quad \text{for all} \quad x \in \mathbb{R}^d,$$

and that, more precisely,

$$q^{\frac{1}{2n}}(x) \leq p(x) \leq \left(\dim \operatorname{span} \mathcal{O}_{v^{\otimes n}}\right)^{\frac{1}{2n}} q^{\frac{1}{2n}}(x) \quad \text{for all} \quad x \in \mathbb{R}^d,$$

where $\mathcal{O}_{v^{\otimes n}} = \left\{ \left(g(v)\right)^{\otimes n} : \ g \in G \right\}$ is the orbit of $v^{\otimes n} \in (\mathbb{R}^d)^{\otimes n}$.

10. In Problem 5 above, let

$$K = \mathbb{S}^{d-1} = \left\{ x \in \mathbb{R}^d : \quad \|x\| = 1 \right\}$$

be the unit sphere. Using Problem 9, show that one can choose

$$q(f) = \int_{\mathbb{S}^{d-1}} f^{2n}(x) \ dx,$$

where dx is the rotation invariant probability measure on $\mathbb{S}^{d-1}$.

4. The Ellipsoid Method

In this section, we briefly describe a method for finding a solution to a system of linear inequalities. The method, known as the *Ellipsoid Method*, was first developed by N.Z. Shor, A.S. Nemirovskii and L.G. Khachiyan. In 1979, L.G. Khachiyan applied it to solving linear inequalities and problems of *linear programming*. It resulted in the first *polynomial time* algorithm in linear programming, see [**Schr86**], [**PS98**], [**GLS93**].

(4.1) Systems of linear inequalities. Suppose we are given a system of linear inequalities

$$\langle a_i, x \rangle < \beta_i, \quad i = 1, \ldots, m,$$

where $a_i \in \mathbb{R}^d$ are given real vectors and β_i are given real numbers. We do not discuss how numbers and vectors may be "given" (cf. Chapter 1 of [**GLS93**]); for example, we assume that β_i and all the coordinates of a_i are rational numbers. Also, the method can be modified for systems of non-strict inequalities, but we don't discuss it here.

Our goal is to find a vector $x \in \mathbb{R}^d$ satisfying (4.1) or to show that none exists. We would like to be able to do that reasonably fast. The method is based on the following geometric result.

(4.2) Lemma. *Let*

$$B^+ = \left\{ (\xi_1, \ldots, \xi_d) \in \mathbb{R}^d : \quad \xi_1^2 + \ldots + \xi_d^2 \leq 1 \quad \textit{and} \quad \xi_d \geq 0 \right\}$$

be the "upper half" of the unit ball B. Let

$$E = \left\{ (\xi_1, \ldots, \xi_d) \in \mathbb{R}^d : \ \frac{d^2-1}{d^2}\xi_1^2 + \ldots + \frac{d^2-1}{d^2}\xi_{d-1}^2 + \frac{(d+1)^2}{d^2}\left(\xi_d - \frac{1}{d+1}\right)^2 \leq 1 \right\}$$

be an ellipsoid. Then

1. $B^+ \subset E$;
2. $\dfrac{\operatorname{vol} E}{\operatorname{vol} B} \leq \exp\left\{ -\dfrac{1}{2(d+1)} \right\}$.

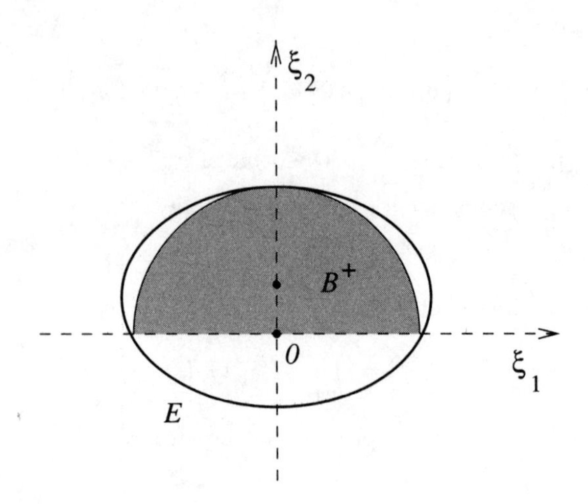

Figure 64

Proof. Let us choose an $x \in B^+$, $x = (\xi_1, \ldots, \xi_d)$. Then $\xi_1^2 + \ldots + \xi_{d-1}^2 \leq 1 - \xi_d^2$, so

$$\frac{d^2-1}{d^2}\Big(\xi_1^2 + \ldots + \xi_{d-1}^2\Big) \leq \frac{d^2-1}{d^2} - \frac{d^2-1}{d^2}\xi_d^2.$$

Furthermore,

$$\frac{(d+1)^2}{d^2}\Big(\xi_d - \frac{1}{d+1}\Big)^2 = \frac{(d+1)^2}{d^2}\xi_d^2 - \frac{2(d+1)}{d^2}\xi_d + \frac{1}{d^2}.$$

Adding the two inequalities together, we get

$$\frac{d^2-1}{d^2}\xi_1^2 + \ldots + \frac{d^2-1}{d^2}\xi_{d-1}^2 + \frac{(d+1)^2}{d^2}\Big(\xi_d - \frac{1}{d+1}\Big)^2 \leq 1 + \frac{2d+2}{d^2}\Big(\xi_d^2 - \xi_d\Big) \leq 1,$$

since $\xi_d^2 - \xi_d \leq 0$ for $0 \leq \xi_d \leq 1$. Hence Part 1 is proven.

To prove Part 2, we note that

$$\frac{\operatorname{vol} E}{\operatorname{vol} B} = \frac{d}{d+1}\Big(\frac{d^2}{d^2-1}\Big)^{(d-1)/2}.$$

Using the inequality $1 + x \leq \exp\{x\}$ for $x = -1/(d+1)$ and for $x = 1/(d^2-1)$, we get

$$\frac{\operatorname{vol} E}{\operatorname{vol} B} \leq \exp\Big\{-\frac{1}{d+1}\Big\}\exp\Big\{\frac{d-1}{2}\cdot\frac{1}{d^2-1}\Big\} = \exp\Big\{-\frac{1}{2(d+1)}\Big\}.$$

□

PROBLEMS.

1°. Check that the boundary of the ellipsoid E constructed in Lemma 4.2 contains the "great circle" $\big\{(\xi_1, \ldots, \xi_{d-1}, 0) : \xi_1^2 + \ldots + \xi_{d-1}^2 = 1\big\}$ and the point $(0, \ldots, 0, 1)$.

2°. Let $E \subset \mathbb{R}^d$ be an ellipsoid centered at $c \in \mathbb{R}^d$ and let $a \in \mathbb{R}^d$ be a non-zero vector. Consider the "half-ellipsoid" $E^- = \big\{x \in E : \langle a, x\rangle \leq \langle a, c\rangle\big\}$. Construct an ellipsoid $E_1 \subset \mathbb{R}^d$ such that $E^- \subset E_1$ and

$$\frac{\operatorname{vol} E_1}{\operatorname{vol} E} \leq \exp\Big\{-\frac{1}{2(d+1)}\Big\}.$$

(4.3) The description of the method. Let $P = \big\{x : \langle a_i, x\rangle < \beta_i, i = 1, \ldots, m\big\}$ be the set of all solutions to the system of (4.1). Suppose we know a pair of real numbers $R > r > 0$ with the following property:

If P is non-empty, then for some (hitherto unknown) point $x_0 \in P$ we have $B(x_0, r) \subset P \cap B(0, R)$, where $B(b, \rho) = \{x : \|x - b\| < \rho\}$ is the (open) ball of radius ρ centered at a point b.

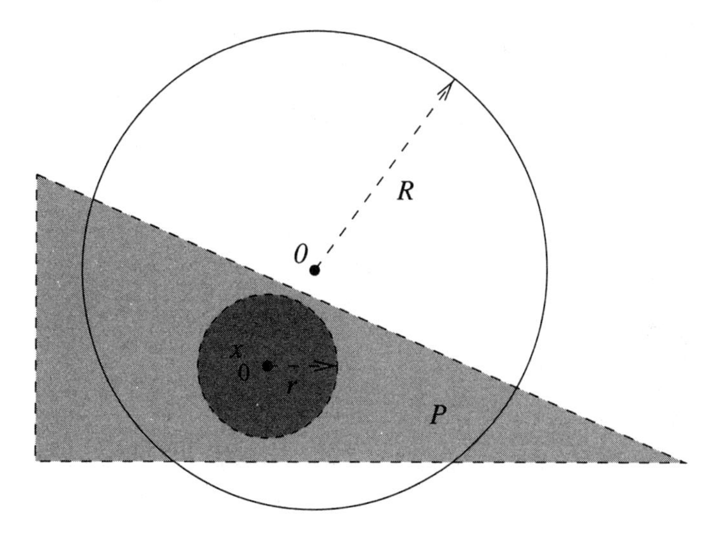

Figure 65

In other words, if there is a solution, then there is a solution in a sufficiently large ball and the set of solutions is sufficiently "thick". The numbers R and r can be determined from the numerical data (see Problems below). The ratio R/r can be thought of as the "condition number" of the problem (the problems with a large ratio R/r are "ill-conditioned").

We construct a sequence of ellipsoids $E_0, \ldots, E_k, \ldots$ which satisfy the properties:

1. If P is non-empty, then $B(x_0, r) \subset E_k$ for all k and

2. $$\frac{\operatorname{vol} E_{k-1}}{\operatorname{vol} E_k} \leq \exp\Big\{-\frac{1}{2(d+1)}\Big\} \quad \text{for all} \quad k \geq 1.$$

We let $E_0 = B(0, R)$. If E_k is constructed, we check whether its center c_k satisfies the system (4.1). If it does, then a feasible point $x = c_k$ is constructed. If not, we pick an inequality violated on c_k, say, $\langle a_{i_0}, c_k \rangle \geq \beta_{i_0}$. Let

$$E_k^- = \Big\{x \in E_k : \quad \langle a_{i_0}, x \rangle \leq \langle a_{i_0}, c_k \rangle\Big\}$$

be the "half ellipsoid" which contains solutions if they exist. Using Problem 2 of Section 4.2, we construct an ellipsoid E_{k+1} containing E_k^- and such that

$$\frac{\operatorname{vol} E_{k+1}}{\operatorname{vol} E_k} \leq \exp\Big\{-\frac{1}{2(d+1)}\Big\}.$$

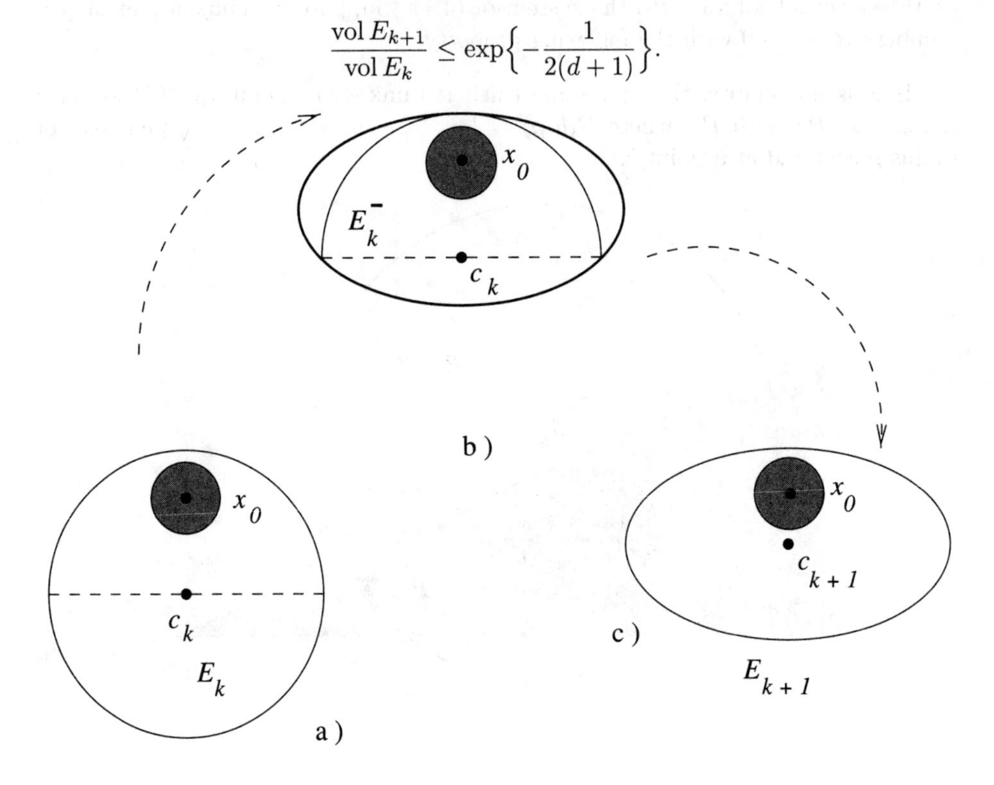

Figure 66. a) If the center c_k of the current ellipsoid E_k does not satisfy the linear inequalities, we construct the half ellipsoid E_k^- which contains the solutions. b) We construct the ellipsoid E_{k+1} containing E_k^-. c) We check whether the center c_{k+1} of the ellipsoid E_{k+1} satisfies the linear inequalities and proceed as above with E_k replaced by E_{k+1}.

We continue until we either hit a solution or $\operatorname{vol} E_k < \operatorname{vol} B(x_0, r)$, in which case we conclude that there are no solutions. Since each time the volume of the ellipsoid decreases by a factor of $\exp\{1/(2d+2)\}$, the total number of the constructed ellipsoids does not exceed

$$(2d+2)\ln\frac{\operatorname{vol} B(x_0, R)}{B(x_0, r)} = 2d(d+1)\ln\frac{R}{r}.$$

Hence the number of iterations is quadratic in the dimension d and linear in the *logarithm* of the "condition number" R/r. From the computational complexity point of view, this means that the running time of the algorithm is polynomial in the "size of the input".

PROBLEMS.

In the problems below, for a matrix $A = (\alpha_{ij})$, we let $L_A = \max_{ij} |\alpha_{ij}|$ and for a vector $b = (\beta_1, \dots, \beta_m)$, we let $L_b = \max_k |\beta_k|$. We do not aim for the best possible bounds here but want to convey a general flavor of the estimates.

1. Let $A = (\alpha_{ij})$ be an invertible $n \times n$ integer matrix and let $b = (\beta_1, \dots, \beta_n)$ be an integer n-vector. Let $x = (\xi_1, \dots, \xi_n)$ be the (necessarily unique) solution to the system of linear equations

$$\sum_{i=1}^{n} \alpha_{ij}\xi_j = \beta_i \quad \text{for} \quad i = 1, \dots, n.$$

Prove that $\xi_j = p_j/q$ are rational numbers, where p_j and q are integers with $1 \leq q \leq n^{n/2} L_A^n$ and $|p_j| \leq n^{n/2} L_b L_A^{n-1}$ for $j = 1, \dots, n$.

Hint: Use Cramer's rule and Hadamard's inequality for determinants.

2. Let $A = (\alpha_{ij})$ be an $m \times d$ real matrix and let $b = (\beta_1, \dots, \beta_m)$ be a real m-vector. Prove that a solution $x = (\xi_1, \dots, \xi_d)$ to the system

$$\sum_{j=1}^{d} \alpha_{ij}\xi_j \leq \beta_i \quad \text{for} \quad i = 1, \dots, m \tag{4.3.1}$$

of linear inequalities exists if and only if there exists a solution $u = (\eta_1', \dots, \eta_d'; \eta_1'', \dots, \eta_d''; \zeta_1, \dots, \zeta_m)$ to the system

$$\begin{aligned} \sum_{j=1}^{d} \alpha_{ij}(\eta_j' - \eta_j'') = \beta_i - \zeta_i \quad &\text{for} \quad i = 1, \dots, m, \\ \eta_j', \eta_j'' \geq 0 \quad &\text{for} \quad j = 1, \dots, d, \\ \zeta_i \geq 0 \quad &\text{for} \quad i = 1, \dots, m \end{aligned} \tag{4.3.2}$$

of linear inequalities. Prove that if the set U of solutions u to the system (4.3.2) is non-empty, then U has an extreme point.

Hint: Use Lemma II.3.5.

3. Let $A = (\alpha_{ij})$ be a (non-zero) $m \times d$ integer matrix and let $b = (\beta_1, \dots, \beta_m)$ be an integer vector. Suppose that the set of solutions U to the system (4.3.2) is

non-empty and let $u = (\eta'_1, \dots, \eta'_d; \eta''_1, \dots, \eta''_d; \zeta_1, \dots, \zeta_m)$ be an extreme point of U. Prove that each coordinate of u is a rational number which can be represented in the form p_j/q, where p_j and q are integers with $1 \leq q \leq (2d+m)^{d+m/2} L_A^{2d+m}$ and $|p_j| \leq (2d+m)^{d+m/2} L_b L_A^{2d+m-1}$ for $j = 1, \dots, 2d+m$.

Hint: Use Problem 1 and Theorem II.4.2.

4. Let $A = (\alpha_{ij})$ be a (non-zero) $m \times d$ integer matrix and let $b = (\beta_1, \dots, \beta_m)$ be an integer vector. Suppose that the system (4.3.1) of Problem 2 has a solution.

Prove that there exists a solution $(\xi_1, \dots, \xi_d)$ with $\xi_j = p_j/q$, where p_j and q are integers such that $1 \leq q \leq (2d+m)^{d+m/2} L_A^{2d+m}$ and $|p_j| \leq 2(2d+m)^{d+m/2} L_b L_A^{2d+m}$ for $j = 1, \dots, d$.

Hint: Use Problems 2 and 3.

5. Let $A = (\alpha_{ij})$ be a $k \times d$ integer matrix. Suppose that the system

$$\sum_{j=1}^{d} \alpha_{ij} \xi_j < 0 \quad \text{for} \quad i = 1, \dots, k \tag{4.3.3}$$

has a solution $x = (\xi_1, \dots, \xi_d)$. Prove that there exists a solution $(\xi_1, \dots, \xi_d)$ with $\xi_j = p_j/q$, where p_j and q are integers such that $1 \leq q \leq (2d+k)^{(d+k/2)k} L_A^{(2d+k)k}$ and $|p_j| \leq 2k(2d+k)^{(d+k/2)k} L_A^{(2d+k)k}$ for $j = 1, \dots, d$.

Hint: For $i_0 = 1, \dots, i_k$ consider the modified system

$$\sum_{j=1}^{d} \alpha_{ij} \xi_j \leq 0 \quad \text{for} \quad i \neq i_0; \qquad \sum_{j=1}^{d} \alpha_{i_0 j} \xi_j = -1.$$

Let x_{i_0} be its solution that exists in view of Problem 4 and let x be the sum of x_{i_0} for $i_0 = 1, \dots, k$.

6. Let $A = (\alpha_{ij})$ be an $m \times d$ integer matrix and let $b = (\beta_1, \dots, \beta_m)$ be an integer vector. Suppose that the system

$$\sum_{j=1}^{d} \alpha_{ij} \xi_j < \beta_j \quad \text{for} \quad i = 1, \dots, m \tag{4.3.4}$$

of strict linear inequalities has a solution $x = (\xi_1, \dots, \xi_d)$ and $\xi_j = p_j/q$, where $1 \leq q$ and p_j are integers. Prove that the set of solutions contains an (open) ball of radius $(qL_A\sqrt{d})^{-1}$ centered at x.

7. Suppose that the system (4.3.4) has a solution. Prove that there exists a solution $(\xi_1, \dots, \xi_j)$ with $\xi_j = p_j/q$, where p_j and q are integers such that

$$1 \leq q \leq 4dm(2d+m)^{(d+m/2)(2m+1)} L_A^{(2d+m)(2m+1)+1} \qquad \text{and}$$
$$|p_j| \leq 16dm(2d+m)^{(d+m/2)(2m+1)} L_b L_A^{(2d+m)(2m+1)+1}$$

for $j = 1, \dots, d$.

Hint: Let $x_0 = (\xi_1, \dots, \xi_d)$ be the solution to (4.3.1) whose existence is asserted by Problem 4 and let $I = \left\{i : \sum_{j=1}^{d} \alpha_{ij}\xi_j = \beta_i\right\}$. Let $y = (\eta_1, \dots, \eta_d)$ be the solution to the system

$$\sum_{j=1}^{d} \alpha_{ij}\eta_j < 0, \quad i \in I,$$

whose existence is asserted by Problem 5. Consider $x = x_0 + \epsilon y$ for a suitable $\epsilon > 0$; see Figure 67.

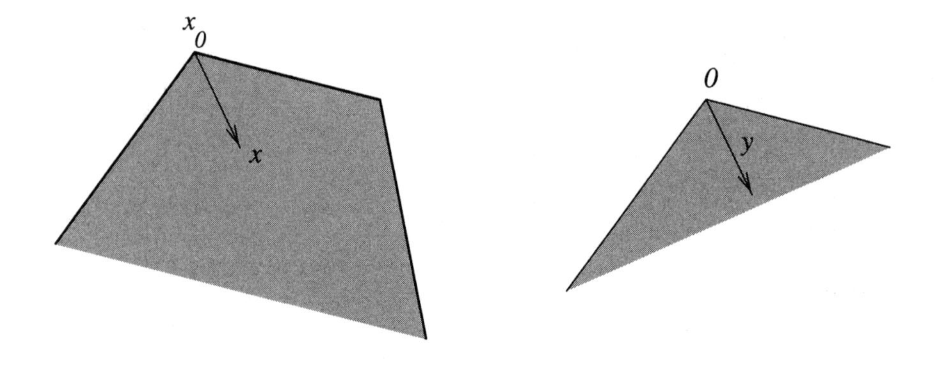

Figure 67

8. Let A be an $m \times d$ integer matrix and let b be an integer vector. Show that to solve the system (4.3.4) of linear inequalities by the Ellipsoid Method 4.3, one can choose

$$r = \frac{1}{4}d^{-3/2}m^{-1}(2d+m)^{-(d+m/2)(2m+1)}L_A^{-(2d+m)(2m+1)-2} \quad \text{and}$$
$$R = 4\sqrt{d}(2d+m)^{d+m/2}L_b L_A^{2d+m}.$$

Hint: Use Problems 4, 6 and 7.

We note that $\ln R$ is bounded by a polynomial in $\ln L_A$, $\ln L_b$, d and m whereas $\ln r^{-1}$ is bounded by a polynomial in $\ln L_A$, d and m only, thus being independent of b.

One more problem.

9*. Let $A_1, \dots, A_k$ be $n \times n$ symmetric matrices and let $\beta_1, \dots, \beta_k$ be numbers. Develop a version of the Ellipsoid Method for the problem of finding an $n \times n$ positive definite matrix X such that $\langle A_i, X \rangle = \beta_i$ for $i = 1, \dots, k$.

5. The Gaussian Measure on Euclidean Space

The purpose of the next three sections is to introduce the technique of *measure concentration*, which turns out to be quite useful in exploring metric properties of convex bodies. We start with the Gaussian measure on Euclidean space.

(5.1) The standard Gaussian measure on $\mathbb{R}$. Let us consider the *standard univariate Gaussian density*

$$\frac{1}{\sqrt{2\pi}} \exp\{-\xi^2/2\} \quad \text{for} \quad \xi \in \mathbb{R}.$$

As is known,

$$\frac{1}{\sqrt{2\pi}} \int_{-\infty}^{+\infty} \exp\{-\xi^2/2\}\, d\xi = 1,$$

hence we can define a probability measure γ, called the *standard Gaussian measure*, on the real line $\mathbb{R}$:

$$\gamma(A) = \frac{1}{\sqrt{2\pi}} \int_A \exp\{-\xi^2/2\}\, d\xi$$

for a Borel set $A \subset \mathbb{R}$.

PROBLEMS.

1°. Prove that

$$\frac{1}{\sqrt{2\pi}} \int_{-\infty}^{+\infty} \exp\{-\lambda\xi^2\}\, d\xi = \lambda^{-1/2} \quad \text{for any} \quad \lambda > 0.$$

2°. Prove that

$$\frac{1}{\sqrt{2\pi}} \int_{-\infty}^{+\infty} \xi^2 \exp\{-\xi^2/2\}\, d\xi = 1.$$

3°. Prove that

$$\frac{1}{\sqrt{2\pi}} \int_{-\infty}^{+\infty} \exp\{\lambda\xi - \xi^2/2\}\, d\xi = \exp\{\lambda^2/2\} \quad \text{for any} \quad \lambda.$$

Hint: Substitute $\xi = (\xi' + \lambda)$.

4°. Prove that $\gamma(A) = \gamma(-A)$ for any Borel set $A \subset \mathbb{R}$.

The following lemma, while providing us with a simple and useful estimate, introduces a general and powerful technique for estimating "tails", which we will use several times in this section.

(5.2) Lemma. *For $\tau \geq 0$, let*

$$A_\tau = \Bigl\{\xi \in \mathbb{R} : \quad \xi \geq \tau\Bigr\} \quad \textit{and} \quad B_\tau = \Bigl\{\xi \in \mathbb{R} : \quad \xi \leq -\tau\Bigr\}.$$

Then

$$\gamma(A_\tau) = \gamma(B_\tau) \leq \exp\{-\tau^2/2\}.$$

Proof. Let us choose $\lambda \geq 0$ (to be adjusted later). Then

$$\xi \geq \tau \quad \text{implies} \quad \exp\{\lambda\xi\} \geq \exp\{\lambda\tau\}.$$

Therefore,

$$\begin{aligned} &\frac{1}{\sqrt{2\pi}} \int_{-\infty}^{+\infty} \exp\{\lambda\xi\} \exp\{-\xi^2/2\}\, d\xi \\ &\geq \frac{1}{\sqrt{2\pi}} \int_{A_\tau} \exp\{\lambda\xi\} \exp\{-\xi^2/2\}\, d\xi \\ &\geq \exp\{\lambda\tau\} \cdot \frac{1}{\sqrt{2\pi}} \int_{A_\tau} \exp\{-\xi^2/2\}\, d\xi = \gamma(A_\tau) \cdot \exp\{\lambda\tau\}. \end{aligned} \tag{5.2.1}$$

By Problem 3 of Section 5.1,

$$\frac{1}{\sqrt{2\pi}} \int_{-\infty}^{+\infty} \exp\{\lambda\xi\} \exp\{-\xi^2/2\}\, d\xi = \exp\{\lambda^2/2\}.$$

Therefore, from (5.2.1) we conclude that

$$\gamma(A_\tau) \leq \exp\{\lambda^2/2\} \cdot \exp\{-\lambda\tau\} = \exp\{\lambda^2/2 - \lambda\tau\}.$$

Substituting $\lambda = \tau$, we obtain the desired bound for $\gamma(A_\tau)$. Since $B_\tau = -A_\tau$, by Problem 4 of Section 5.1 we obtain $\gamma(B_\tau) = \gamma(A_\tau)$. □

(5.3) The standard Gaussian measure on $\mathbb{R}^n$. Let us consider the *standard Gaussian density* in $\mathbb{R}^n$:

$$(2\pi)^{-n/2} \exp\{-\|x\|^2/2\} = \prod_{i=1}^{n} \frac{1}{\sqrt{2\pi}} \exp\{-\xi_i^2/2\} \quad \text{for} \quad x = (\xi_1, \ldots, \xi_n) \in \mathbb{R}^n.$$

Thus

$$(2\pi)^{-n/2} \int_{\mathbb{R}^n} \exp\{-\|x\|^2/2\}\, dx = 1.$$

Hence we can define a probability measure γ_n on $\mathbb{R}^n$:

$$\gamma_n(A) = (2\pi)^{-n/2} \int_A \exp\{-\|x\|^2/2\}\, dx$$

for a Borel set $A \subset \mathbb{R}^n$. The measure γ_n is called the *standard Gaussian measure* on $\mathbb{R}^n$.

PROBLEMS.

1°. Let Q be an $n \times n$ symmetric matrix and let $q : \mathbb{R}^n \longrightarrow \mathbb{R}$ be the corresponding quadratic form, $q(x) = \langle Qx, x \rangle$ for $x \in \mathbb{R}^n$. Prove that

$$(2\pi)^{-n/2} \int_{\mathbb{R}^n} q(x) \exp\{-\|x\|^2/2\}\, dx = \operatorname{tr}(Q).$$

Hint: Use Problem 2 of Section 5.1.

2°. Let Q be an $n \times n$ positive definite matrix and let $q : \mathbb{R}^n \longrightarrow \mathbb{R}$ be the corresponding quadratic form, $q(x) = \langle Qx, x \rangle$ for $x \in \mathbb{R}^n$. Prove that

$$(2\pi)^{-n/2} \int_{\mathbb{R}^n} \exp\{-q(x)/2\}\, dx = \left(\det Q\right)^{-1/2}.$$

Hint: Applying an orthogonal transformation of the coordinates, reduce to the case of a diagonal matrix Q and use Problem 1 of Section 5.1.

The measure γ_n has some interesting properties; see Sections 4.1–4.3 of [**Bo98**] for Problems 3 and 4 below.

3. Let $A, B \subset \mathbb{R}^n$ be convex bodies and let $\alpha, \beta \geq 0$ be numbers such that $\alpha + \beta = 1$. From Problem 2 of Section 1.3, deduce the Brunn-Minkowski inequality:

$$\ln \gamma_n(\alpha A + \beta B) \geq \alpha \ln \gamma_n(A) + \beta \gamma_n(B).$$

4*. Let $B = \{x \in \mathbb{R}^n : \|x\| \leq 1\}$ be the unit ball. Let $A \subset \mathbb{R}^n$ be a closed set and let $H \subset \mathbb{R}^n$ be a halfspace such that $\gamma_n(A) = \gamma_n(H)$. Prove the isoperimetric inequality: $\gamma_n(A + \rho B) \geq \gamma_n(H + \rho B)$ for all $\rho \geq 0$.

Equivalently, if

$$\gamma_n(A) = \frac{1}{\sqrt{2\pi}} \int_{-\infty}^{\alpha} \exp\{-\xi^2/2\}\, d\xi \quad \text{for a suitable} \quad \alpha \in \mathbb{R},$$

then

$$\gamma_n(A + \rho B) \geq \frac{1}{\sqrt{2\pi}} \int_{-\infty}^{\alpha+\rho} \exp\{-\xi^2/2\}\, d\xi.$$

5°. Prove that $\gamma_n(A) = \gamma\big(U(A)\big)$ for any Borel set $A \subset \mathbb{R}^n$ and any orthogonal transformation $U : \mathbb{R}^n \longrightarrow \mathbb{R}^n$.

Now we prove that if the dimension n is large, then "almost all" measure γ_n is concentrated in the vicinity of the sphere $\|x\| = \sqrt{n}$. The method of the proof is similar to that of the proof of Lemma 5.2.

(5.4) Proposition.

1. *For any* $\delta \geq 0$

$$\gamma_n\Big\{x \in \mathbb{R}^n : \quad \|x\|^2 \geq n + \delta\Big\} \leq \left(\frac{n}{n+\delta}\right)^{-n/2} \exp\{-\delta/2\}.$$

2. *For any* $0 < \delta \leq n$

$$\gamma_n\Big\{x \in \mathbb{R}^n : \quad \|x\|^2 \leq n - \delta\Big\} \leq \left(\frac{n}{n-\delta}\right)^{-n/2} \exp\{\delta/2\}.$$

Proof. To prove Part 1, let us choose $\lambda \in [0,1]$ (to be adjusted later). Then

$$\|x\|^2 \geq n+\delta \quad \text{implies} \quad \exp\{\lambda\|x\|^2/2\} \geq \exp\{\lambda(n+\delta)/2\}.$$

We observe that

$$\begin{aligned} &(2\pi)^{-n/2}\int_{\mathbb{R}^n} \exp\{\lambda\|x\|^2/2\}\exp\{-\|x\|^2/2\}\,dx \\ &\geq \gamma_n\Big\{x\in\mathbb{R}^n: \quad \exp\{\lambda\|x\|^2\} \geq \exp\{\lambda(n+\delta)/2\}\Big\}\cdot\exp\{\lambda(n+\delta)/2\} \\ &\geq \gamma_n\Big\{x\in\mathbb{R}^n: \quad \|x\|^2 \geq n+\delta\Big\}\cdot\exp\{\lambda(n+\delta)/2\}. \end{aligned} \tag{5.4.1}$$

Evaluating the integral in the left-hand side, we obtain

$$\begin{aligned} &(2\pi)^{-n/2}\int_{\mathbb{R}^n} \exp\{\lambda\|x\|^2/2\}\exp\{-\|x\|^2/2\}\,dx \\ &= \left(\frac{1}{\sqrt{2\pi}}\int_{-\infty}^{+\infty}\exp\{-(1-\lambda)\xi^2/2\}\,d\xi\right)^n = (1-\lambda)^{-n/2}; \end{aligned}$$

cf. Problem 1 of Section 5.1.

Hence from (5.4.1) we conclude that

$$\gamma_n\Big\{x\in\mathbb{R}^n: \quad \|x\|^2 \geq n+\delta\Big\} \leq (1-\lambda)^{-n/2}\exp\{-\lambda(n+\delta)/2\}.$$

Now we choose $\lambda = \delta/(n+\delta)$ and Part 1 follows.

To prove Part 2, let us choose $\lambda > 0$ (to be adjusted later). Then

$$\|x\|^2 \leq n-\delta \quad \text{implies} \quad \exp\{-\lambda\|x\|^2/2\} \geq \exp\{-\lambda(n-\delta)/2\}.$$

We observe that

$$\begin{aligned} &(2\pi)^{-n/2}\int_{\mathbb{R}^n} \exp\{-\lambda\|x\|^2/2\}\exp\{-\|x\|^2/2\}\,dx \\ &\geq \gamma_n\Big\{x\in\mathbb{R}^n: \quad \|x\|^2 \geq n-\delta\Big\}\cdot\exp\{-\lambda(n-\delta)/2\}. \end{aligned} \tag{5.4.2}$$

The right-hand side integral of (5.4.2) evaluates to $(1+\lambda)^{-n/2}$. Hence from (5.4.2) we get:

$$\gamma_n\Big\{x\in\mathbb{R}^n: \quad \|x\|^2 \geq n-\delta\Big\} \leq (1+\lambda)^{-n/2}\exp\{\lambda(n-\delta)/2\}.$$

Substituting $\lambda = \delta/(n-\delta)$, we complete the proof of Part 2. $\square$

For practical purposes, the following weaker estimates are more convenient.

(5.5) Corollary. *For any* $0 < \epsilon < 1$

$$\gamma_n\Big\{x \in \mathbb{R}^n: \quad \|x\|^2 \geq \frac{n}{1-\epsilon}\Big\} \leq \exp\{-\epsilon^2 n/4\},$$
$$\gamma_n\Big\{x \in \mathbb{R}^n: \quad \|x\|^2 \leq (1-\epsilon)n\Big\} \leq \exp\{-\epsilon^2 n/4\}.$$

Proof. Let us choose $\delta = \epsilon n/(1-\epsilon)$ in Part 1 of Proposition 5.4. Then

$$\left(\frac{n}{n+\delta}\right)^{-n/2} \exp\{-\delta/2\} = \exp\Big\{-\frac{n}{2}\ln(1-\epsilon) - \frac{n}{2}\frac{\epsilon}{1-\epsilon}\Big\}.$$

Expanding

$$\ln(1-\epsilon) = -\epsilon - \epsilon^2/2 - \epsilon^3/3 - \ldots \quad \text{and} \quad \frac{1}{1-\epsilon} = 1 + \epsilon + \epsilon^2 + \ldots,$$

from Part 1 of Proposition 5.4, we get the desired bound

$$\gamma_n\Big\{x \in \mathbb{R}^n: \quad \|x\|^2 \geq \frac{n}{1-\epsilon}\Big\} \leq \exp\{-\epsilon^2 n/4\}.$$

Let us choose $\delta = \epsilon n$ in Part 2 of Proposition 5.4. Then

$$\left(\frac{n}{n-\delta}\right)^{-n/2} \exp\{\delta/2\} = \exp\Big\{\frac{n}{2}\ln(1-\epsilon) + \frac{n\epsilon}{2}\Big\}.$$

Expanding

$$\ln(1-\epsilon) = -\epsilon - \epsilon^2/2 - \epsilon^3/3 - \ldots,$$

we obtain the desired estimate

$$\gamma_n\Big\{x \in \mathbb{R}^n: \quad \|x\|^2 \leq (1-\epsilon)n\Big\} \leq \exp\{-\epsilon^2 n/4\}.$$

□

Corollary 5.5 implies that for any sequence $\rho_n \longrightarrow +\infty$, $n = 1, 2, \ldots$, we have

$$\gamma_n\Big\{x \in \mathbb{R}^n: \quad \sqrt{n} - \rho_n \leq \|x\| \leq \sqrt{n} + \rho_n\Big\} \longrightarrow 1.$$

This is a *concentration property* of the Gaussian measure. Some other related concentration properties are discussed in problems below.

PROBLEMS.

1. Let $A \subset \mathbb{R}^n$ be a closed set such that $\gamma_n(A) = 1/2$ and let $B = \big\{x \in \mathbb{R}^n: \|x\| \leq 1\big\}$. Deduce from Lemma 5.2 and Problem 4 of Section 5.3 the following *concentration inequality* for the Gaussian measure γ_n on $\mathbb{R}^n$:

$$\gamma_n(A + \rho B) \geq 1 - \exp\{-\rho^2/2\}.$$

2. Let $f : \mathbb{R}^n \longrightarrow \mathbb{R}$ be a function such that

$$|f(x) - f(y)| \leq \|x - y\| \quad \text{for all} \quad x, y \in \mathbb{R}^n.$$

Let α be the *median* of f, that is, α is a number such that

$$\gamma_n\Big\{x \in \mathbb{R}^n : \quad f(x) \leq \alpha\Big\} \geq \frac{1}{2} \quad \text{and} \quad \gamma_n\Big\{x \in \mathbb{R}^n : \quad f(x) \geq \alpha\Big\} \geq \frac{1}{2}$$

(check that such an α indeed exists). Deduce from Problem 1 above the *concentration inequality for Lipschitz functions*:

$$\gamma_n\Big\{x \in \mathbb{R}^n : \quad |f(x) - \alpha| \leq \rho\Big\} \geq 1 - 2\exp\{-\rho^2/2\} \quad \text{for any} \quad \rho \geq 0.$$

Remark: For this and related inequalities, see [**Bo98**] and [**Le01**].

We conclude this section with a concentration inequality for positive semidefinite quadratic forms. The following result extends Proposition 5.4 and the proof is very similar to that of Proposition 5.4.

(5.6) Proposition. *Let $q : \mathbb{R}^n \longrightarrow \mathbb{R}$ be a positive semidefinite quadratic form, $q(x) = \langle x, Qx\rangle$ for an $n \times n$ positive semidefinite matrix Q.*

Let $\mu_1, \ldots, \mu_n$ be the eigenvalues of Q and let

$$\|Q\| = \sqrt{\operatorname{tr}(Q^2)} = \sqrt{\mu_1^2 + \ldots + \mu_n^2}.$$

Then

1. *for any $\tau \geq 0$ we have*

$$\gamma_n\Big\{x \in \mathbb{R}^n : \quad q(x) < \operatorname{tr}(Q) - \tau\|Q\|\Big\} \leq \exp\{-\tau^2/4\};$$

2. *for any $\tau \geq 0$ such that*

$$\tau\mu_i \leq \|Q\| \quad \text{for} \quad i = 1, \ldots, n$$

we have

$$\gamma_n\Big\{x \in \mathbb{R}^n : \quad q(x) > \operatorname{tr}(Q) + \tau\|Q\|\Big\} \leq \exp\{-\tau^2/8\}.$$

Proof. Clearly, we may assume that $Q \neq 0$.

To prove Part 1, let us choose $\lambda \geq 0$ (to be adjusted later). Then

$$q(x) < \operatorname{tr}(Q) - \tau\|Q\| \quad \text{implies} \quad \exp\{-\lambda q(x)/2\} \geq \exp\{-\lambda \operatorname{tr}(Q)/2 + \lambda\tau\|Q\|/2\}.$$

We observe that

$$(5.6.1)\quad \begin{aligned}&(2\pi)^{-n/2}\int_{\mathbb{R}^n} \exp\{-\lambda q(x)/2\} \exp\{-\|x\|^2/2\}\, dx \\ &\geq \gamma_n\Big\{x \in \mathbb{R}^n : \quad q(x) < \operatorname{tr}(Q) - \tau\|Q\|\Big\} \cdot \exp\{-\lambda \operatorname{tr}(Q)/2 + \lambda\tau\|Q\|/2\}.\end{aligned}$$

By Problem 2 of Section 5.3, the integral in (5.6.1) evaluates to

$$\det^{-1/2}(I + \lambda Q) = \prod_{i=1}^{n}(1 + \lambda\mu_i)^{-1/2},$$

where I is the identity matrix. Hence we deduce from (5.6.1) that

$$\begin{aligned}\gamma_n\Big\{x \in \mathbb{R}^n : \quad & q(x) < \operatorname{tr}(Q) - \tau\|Q\|\Big\} \\ \leq \quad & \det^{-1/2}(I + \lambda Q) \cdot \exp\{\lambda \operatorname{tr}(Q)/2 - \lambda\tau\|Q\|/2\} \\ = \exp\Big\{ & -\frac{1}{2}\sum_{i=1}^{n} \ln(1 + \lambda\mu_i) + \lambda \operatorname{tr}(Q)/2 - \lambda\tau\|Q\|/2\Big\}.\end{aligned}$$

Now we use that $\ln(1 + \lambda\mu_i) \geq \lambda\mu_i - \lambda^2\mu_i^2/2$, that $\operatorname{tr}(Q) = \mu_1 + \ldots + \mu_n$ and that $\|Q\|^2 = \mu_1^2 + \ldots + \mu_n^2$. Thus

$$\begin{aligned}&\gamma_n\Big\{x \in \mathbb{R}^n : \quad q(x) < \operatorname{tr}(Q) - \tau\|Q\|\Big\} \\ &\leq \exp\Big\{\frac{1}{4}\sum_{i=1}^{n} \lambda^2\mu_i^2 - \lambda\tau\|Q\|/2\Big\} = \exp\Big\{\lambda^2\|Q\|^2/4 - \lambda\tau\|Q\|/2\Big\}.\end{aligned}$$

Substituting $\lambda = \tau/\|Q\|$, we complete the proof of Part 1.

To prove Part 2, let us choose $\lambda \geq 0$ (to be adjusted later). Then

$$q(x) > \operatorname{tr}(Q) + \tau\|Q\| \quad \text{implies} \quad \exp\{\lambda q(x)/2\} \geq \exp\{\lambda \operatorname{tr}(Q)/2 + \lambda\tau\|Q\|/2\}.$$

We observe that

$$(5.6.2)\quad \begin{aligned}&(2\pi)^{-n/2}\int_{\mathbb{R}^n} \exp\{\lambda q(x)/2\} \exp\{-\|x\|^2/2\}\, dx \\ &\geq \gamma_n\Big\{x \in \mathbb{R}^n : \quad q(x) \geq \operatorname{tr}(Q) + \tau\|Q\|\Big\} \cdot \exp\{\lambda \operatorname{tr}(Q)/2 + \lambda\tau\|Q\|/2\}.\end{aligned}$$

By Problem 2 of Section 5.3, the integral in (5.6.2) evaluates to

$$\det^{-1/2}(I - \lambda Q) = \prod_{i=1}^{n}(1 - \lambda\mu_i),$$

where I is the identity matrix. Hence we deduce from (5.6.2) that

$$\begin{aligned}\gamma_n\Big\{x \in \mathbb{R}^n: \quad q(x) > \operatorname{tr}(Q) + \tau\|Q\|\Big\} \\ \leq \quad \det^{-1/2}(I - \lambda Q) \cdot \exp\{-\lambda \operatorname{tr}(Q)/2 - \lambda\tau\|Q\|/2\} \\ = \exp\Big\{-\frac{1}{2}\sum_{i=1}^{n} \ln(1-\lambda\mu_i) - \lambda \operatorname{tr}(Q)/2 - \lambda\tau\|Q\|/2\Big\}.\end{aligned}$$

Now we use that $\ln(1-\lambda\mu_i) \geq -\lambda\mu_i - \lambda^2\mu_i^2$ provided $\lambda\mu_i \leq 1/2$, that $\operatorname{tr}(Q) = \mu_1 + \ldots + \mu_n$ and that $\|Q\|^2 = \mu_1^2 + \ldots + \mu_n^2$. Thus

$$\begin{aligned}\gamma_n\Big\{x \in \mathbb{R}^n: \quad q(x) > \operatorname{tr}(Q) + \tau\|Q\|\Big\} \\ \leq \exp\Big\{\frac{1}{2}\sum_{i=1}^{n} \lambda^2\mu_i^2 - \lambda\tau\|Q\|/2\Big\} = \exp\Big\{\lambda^2\|Q\|^2/2 - \lambda\tau\|Q\|/2\Big\},\end{aligned}$$

provided $\lambda\mu_i \leq 1/2$ for $i = 1, \ldots, n$. Substituting $\lambda = \tau/(2\|Q\|)$ and noting that

$$\lambda\mu_i = \frac{\tau\mu_i}{2\|Q\|} \leq \frac{1}{2} \quad \text{for} \quad i = 1, \ldots, n,$$

we complete the proof. $\square$

PROBLEMS.

1. Let $q: \mathbb{R}^n \longrightarrow \mathbb{R}$ be a quadratic form, $q(x) = \langle x, Qx\rangle$ for an $n \times n$ symmetric matrix Q. Let $\mu_1, \ldots, \mu_n$ be the eigenvalues of Q and let $\|Q\| = \sqrt{\mu_1^2 + \ldots + \mu_n^2}$. Prove that for any $\tau \geq 0$ such that $|\tau\mu_i| \leq \|Q\|$ for $i = 1, \ldots, n$, we have

$$\begin{aligned}\gamma_n\Big\{x \in \mathbb{R}^n: \quad q(x) < \operatorname{tr}(Q) - \tau\|Q\|\Big\} \leq \exp\{-\tau^2/8\} \quad \text{and} \\ \gamma_n\Big\{x \in \mathbb{R}^n: \quad q(x) > \operatorname{tr}(Q) + \tau\|Q\|\Big\} \leq \exp\{-\tau^2/8\}.\end{aligned}$$

2°. Prove that $\operatorname{tr}(Q) \geq \|Q\|$ for any positive semidefinite matrix Q.

3°. Let A be an $n \times n$ positive semidefinite matrix and let Q be the $mn \times mn$ matrix consisting of m diagonal blocks A. Prove that $\operatorname{tr}(Q) = m\operatorname{tr}(A)$, $\|Q\| = \sqrt{m}\|A\|$ and $\mu \leq m^{-1/2}\|Q\|$ for every eigenvalue μ of Q.

4. Prove the following version of Part 2 of Proposition 5.6:

For every $\epsilon > 0$ there exists $\delta > 0$ such that

$$\gamma_n\Big\{x \in \mathbb{R}^n: \quad q(x) > \operatorname{tr}(Q) + \tau\|Q\|\Big\} \leq \exp\Big\{-(1-\epsilon)\tau^2/4\Big\}$$

provided $\tau \geq 0$ and $\tau\mu_i \leq \delta\|Q\|$ for $i = 1, \ldots, n$.

6. Applications to Low Rank Approximations of Matrices

In this section, we apply our methods to obtain new results regarding systems of linear equations in positive semidefinite matrices, which we considered in Sections II.13–15. We use the notation of Sections II.13–15. Thus $\langle A, B\rangle = \operatorname{tr}(AB)$ for symmetric matrices A and B; $X \succeq 0$ means that X is positive semidefinite and $x \otimes x$ denotes the matrix X whose (i,j)-th entry is $\xi_i \xi_j$, where $x = (\xi_1, \ldots, \xi_n)$.

Proposition II.13.1 asserts, roughly, that if a system of k equations $\langle A_i, X\rangle = \alpha_i$ has a solution $X \succeq 0$, then there is a solution $X_0 \succeq 0$ such that $\operatorname{rank} X_0 = O(\sqrt{k})$. Now we show that if we are willing to settle for an *approximate* solution, we can make $\operatorname{rank} X_0 = O(\ln k)$. To state what "approximate" means, we assume that the matrices A_i are positive semidefinite and so the α_i are non-negative.

(6.1) Proposition. *Let us fix k positive semidefinite $n \times n$ matrices $A_1, \ldots, A_k$, k non-negative numbers $\alpha_1, \ldots, \alpha_k$ and $0 < \epsilon < 1$.*

Suppose that there is a matrix $X \succeq 0$ such that

$$\langle A_i, X\rangle = \alpha_i \quad \textit{for} \quad i = 1, \ldots, k.$$

Let m be a positive integer such that

$$m \geq \frac{8}{\epsilon^2} \ln(4k).$$

Then there is a matrix $X_0 \succeq 0$ such that

$$\alpha_i(1-\epsilon) \leq \langle A_i, X_0\rangle \leq \alpha_i(1+\epsilon) \quad \textit{for} \quad i = 1, \ldots, k$$

and

$$\operatorname{rank} X_0 \leq m.$$

Proof. First, we show that without loss of generality we may assume that $X = I$, the identity matrix. Indeed, suppose that $X \succeq 0$ satisfies the system

$$\langle A_i, X\rangle = \alpha_i \quad \text{for} \quad i = 1, \ldots, k.$$

Let us write $X = TT^*$ for an $n \times n$ matrix T and let $B_i = TA_iT^*$. Then the B_i are positive semidefinite matrices and

$$\alpha_i = \langle A_i, X\rangle = \operatorname{tr}(A_i X) = \operatorname{tr}(A_i TT^*) = \operatorname{tr}(TA_iT^*) = \operatorname{tr}(B_i) = \langle B_i, I\rangle,$$

so I satisfies the system $\langle B_i, I\rangle = \alpha_i$ for $i = 1, \ldots, k$.

Moreover, if $Y_0 \succeq 0$ satisfies the inequalities

$$(1-\epsilon)\alpha_i \leq \langle B_i, Y_0\rangle \leq (1+\epsilon)\alpha_i \quad \text{for} \quad i = 1, \ldots, k,$$

then for $X_0 = T^* Y_0 T$ we have

$$\langle A_i, X_0\rangle = \operatorname{tr}(A_i T^* Y_0 T) = \operatorname{tr}(TA_iT^* Y_0) = \operatorname{tr}(B_i Y_0) = \langle B_i, Y_0\rangle$$

and hence

$$(1-\epsilon)\alpha_i \leq \langle A_i, X_0\rangle \leq (1+\epsilon)\alpha_i \quad \text{for} \quad i=1,\dots,k.$$

In addition, $X_0 \succeq 0$ and $\operatorname{rank} X_0 \leq \operatorname{rank} Y_0$.

Thus we assume that $X = I$ and so

$$\alpha_i = \langle A_i, I\rangle = \operatorname{tr}(A_i) \quad \text{for} \quad i=1,\dots,k.$$

Let $d = mn$ and let us consider the direct sum of m copies of $\mathbb{R}^n$:

$$\mathbb{R}^d = \underbrace{\mathbb{R}^n \oplus \ldots \oplus \mathbb{R}^n}_{m \text{ times}}.$$

Thus a vector $x \in \mathbb{R}^d$ is identified with an m-tuple, $x = (x_1, \dots, x_m)$, where $x_j \in \mathbb{R}^n$ for $j = 1, \dots, m$.

For $i = 1, \dots, k$, we define a quadratic form $q_i : \mathbb{R}^d \longrightarrow \mathbb{R}$ by

$$q_i(x_1,\dots,x_m) = \frac{1}{m}\sum_{j=1}^m \langle A_i x_j,\ x_j\rangle = \frac{1}{m}\sum_{j=1}^m \langle A_i,\ x_j \otimes x_j\rangle, \tag{6.1.1}$$

where $x_1, \dots, x_m \in \mathbb{R}^n$. Thus the matrix Q_i of q_i consists of m diagonal blocks $\frac{1}{m}A_i$:

$$Q_i = \begin{pmatrix} \frac{1}{m}A_i & 0 & 0 & \dots & 0 \\ 0 & \frac{1}{m}A_i & 0 & \dots & 0 \\ \dots & \dots & \dots & \dots & \dots \\ 0 & 0 & \dots & 0 & \frac{1}{m}A_i \end{pmatrix}.$$

Let us consider the standard Gaussian measure γ_d in $\mathbb{R}^d$. We apply Proposition 5.6 to the forms q_i. We have

$$\operatorname{tr}(Q_i) = \operatorname{tr}(A_i) = \alpha_i, \quad \|Q_i\| = \frac{\|A_i\|}{\sqrt{m}} \leq \frac{\operatorname{tr}(A_i)}{\sqrt{m}} = \frac{\alpha_i}{\sqrt{m}} \quad \text{and} \quad \mu \leq \frac{\|Q_i\|}{\sqrt{m}}$$

for every eigenvalue μ of Q_i; cf. Problems 2–3 of Section 5.6.

Let us choose $\tau = \epsilon\sqrt{m}$ in Proposition 5.6. Then

$$\gamma_d\Big\{x \in \mathbb{R}^d : \quad |q_i(x) - \alpha_i| > \epsilon\alpha_i\Big\} \leq 2\exp\big\{-\epsilon^2 m/8\big\} \leq \frac{1}{2k} \quad \text{for} \quad i=1,\dots,k.$$

Therefore,

$$\gamma_d\Big\{x \in \mathbb{R}^d : \quad |q_i(x) - \alpha_i| \leq \epsilon\alpha_i \quad \text{for} \quad i=1,\dots,k\Big\} \geq 1 - k\cdot\frac{1}{2k} = \frac{1}{2}.$$

In particular, there exists $x = (x_1, \dots, x_m) \in \mathbb{R}^d$ such that $|q_i(x) - \alpha_i| \leq \epsilon\alpha_i$ for $i = 1, \dots, k$. Let

$$X_0 = \frac{1}{m}\sum_{j=1}^m x_j \otimes x_j.$$

Then, by (6.1.1),

$$\big|\langle A_i, X_0\rangle - \alpha_i\big| \leq \epsilon\alpha_i \quad \text{for} \quad i = 1, \ldots, k$$

and the result follows. □

The proof of Proposition 6.1 suggests a simple recipe to construct the desired matrix X_0. First, we find a matrix T such that $X = TT^*$. Then we sample m vectors $y_1, \ldots, y_m$ at random from the standard Gaussian distribution in $\mathbb{R}^n$ and we let $Y_0 = \frac{1}{m}\sum_{i=1}^m y_i \otimes y_m$. Finally, we let $X_0 = T^* Y_0 T$. With probability at least 1/2, the matrix X_0 satisfies the constraints of Proposition 6.1.

PROBLEMS.

1. Let $A_1, \ldots, A_k$ be symmetric matrices and let $\alpha_1, \ldots, \alpha_k$ be numbers. Suppose that the eigenvalues of every matrix A_i do not exceed 1 in absolute value. Suppose further that there is a matrix $X \succeq 0$ such that

$$\langle A_i, X\rangle = \alpha_i \quad \text{for} \quad i = 1, \ldots, k$$

and $\operatorname{tr}(X) = 1$. Let $\epsilon > 0$ and let m be a positive integer such that

$$m \geq 32\epsilon^{-2}\ln(4k + 4).$$

Prove that there exists a matrix $X_0 \succeq 0$ such that

$$\alpha_i - \epsilon \leq \langle A_i, X_0\rangle \leq \alpha_i + \epsilon \quad \text{for} \quad i = 1, \ldots, k$$

and

$$\operatorname{rank} X_0 \leq m.$$

2. Let $\mathbb{S}^{n-1} = \big\{x \in \mathbb{R}^n : \ \|x\| = 1\big\}$ be the unit sphere and let $q_i : \mathbb{S}^{n-1} \longrightarrow \mathbb{R}$ be quadratic forms such that $|q_i(x)| \leq 1$ for all $x \in \mathbb{S}^{n-1}$ and $i = 1, \ldots, k$. Consider the map $\phi : \mathbb{S}^{n-1} \longrightarrow \mathbb{R}^k$, $\phi(x) = \big(q_1(x), \ldots, q_k(x)\big)$. Let $\epsilon > 0$ and let m be a positive integer such that $m \geq 32\epsilon^{-2}\ln(4k+4)$.

Prove that for any point $a = (\alpha_1, \ldots, \alpha_k) \in \operatorname{conv}\Big(\phi\big(\mathbb{S}^{n-1}\big)\Big)$ there exists a point $b = (\beta_1, \ldots, \beta_k)$ such that

$$|\alpha_i - \beta_i| \leq \epsilon \quad \text{for} \quad i = 1, \ldots, k$$

and b is a convex combination of m points of $\phi\big(\mathbb{S}^{n-1}\big)$.

3. Prove the following version of Proposition 6.1:

For every $\delta > 0$ there exists $\epsilon_0 > 0$ such that for any positive semidefinite matrices $A_1, \ldots, A_k$, for any non-negative numbers $\alpha_1, \ldots, \alpha_k$, for any positive $\epsilon < \epsilon_0$ and for any positive integer

$$m \geq \frac{4+\delta}{\epsilon^2}\ln(2k)$$

there exists a matrix $X_0 \succeq 0$ such that

$$\alpha_i(1-\epsilon) \le \langle A_i, X_0\rangle \le \alpha_i(1+\epsilon) \quad \text{for} \quad i=1,\dots,k$$

and

$$\operatorname{rank} X_0 \le m$$

provided there exists a matrix $X \succeq 0$ such that

$$\langle A_i, X_0\rangle = \alpha_i \quad \text{for} \quad i=1,\dots,k.$$

Hint: Use Problem 4 of Section 5.6.

4* (A. Frieze and R. Kannan). Let $A = (a_{ij})$ be an $m \times n$ matrix such that $|a_{ij}| \le 1$ for all i, j. Prove that for any $\epsilon \in (0,1)$ there exists an $m \times n$ matrix D such that $\operatorname{rank} D = O(1/\epsilon^2)$ and the sum of the elements in every submatrix of $A - D$ (among 2^{m+n} submatrices) does not exceed ϵmn in absolute value.

Remark: See [**FK99**].

As in Chapter II, we apply our results to the graph realization problem; cf. Section II.15.

(6.2) Realization with a distortion. Suppose we are given a weighted graph $G = (V, E; \rho)$, where $V = \{v_1, \dots, v_n\}$ is the set of vertices, E is the set of edges and $\rho : E \longrightarrow \mathbb{R}_+$ is a function, which assigns to every edge $(i,j) \in E$ a non-negative number ("length") ρ_{ij}. Recall that G is realizable if one can place the vertices $v_1, \dots, v_n$ in $\mathbb{R}^d$ for some d in such a way that

$$\|v_i - v_j\| = \rho_{ij} \quad \text{for every edge} \quad \{i,j\} \in E.$$

Suppose that we are willing to permit a certain *distortion.* We obtain the following result.

(6.3) Corollary. *Suppose that a graph G with k edges is realizable. Then, for any $0 < \epsilon < 1$ and any $m \ge 8\epsilon^{-2}\ln(4k)$, one can place the vertices $v_1, \dots, v_n$ in $\mathbb{R}^m$ so that*

$$(1-\epsilon)\rho_{ij}^2 \le \|v_i - v_j\|^2 \le (1+\epsilon)\rho_{ij}^2 \quad \textit{for every edge} \quad \{i,j\} \in E.$$

Proof. As in the proof of Proposition II.15.4, we reduce the problem to a system of linear equations in positive semidefinite matrices.

For $(i,j) \in E$, let A_{ij} be the $n \times n$ matrix such that $\langle A_{ij}, X\rangle = x_{ii} - 2x_{ij} + x_{jj}$ for any $n \times n$ matrix X. Since G is realizable, there exists $X \succeq 0$ such that

$$\langle A_{ij}, X\rangle = \rho_{ij}^2 \quad \text{for all} \quad (i,j) \in E$$

(we choose X to be the Gram matrix of the vectors $v_1, \dots, v_n$ in a realization of G). Now we note that $A_{ij} \succeq 0$ and apply Proposition 6.1. □

Corollary 6.3 with different constants and a slightly different method of proof is due to W.B. Johnson and J. Lindenstrauss [**JL84**]; see also Problem 2 of Section 7.1.

PROBLEM.

1. Prove the following version of Corollary 6.3:

For every $\delta > 0$ there exists $\epsilon_0 > 0$ such that for every positive $\epsilon < \epsilon_0$, for every k and every positive integer $m \geq (1+\delta)\epsilon^{-2}\ln(2k)$, if a graph $G=(V,E;\rho)$ with k edges is realizable, one can place the vertices $v_1, \ldots, v_n$ of G in $\mathbb{R}^m$ in such a way that

$$(1-\epsilon)\rho_{ij} \leq \|v_i - v_j\| \leq (1+\epsilon)\rho_{ij} \quad \text{for every edge} \quad \{i,j\} \in E.$$

Hint: Use Problem 3 of Section 6.1.

7. The Measure and Metric on the Unit Sphere

Let $\mathbb{S}^{n-1} \subset \mathbb{R}^n$ be the unit sphere:

$$\mathbb{S}^{n-1} = \Big\{ x \in \mathbb{R}^n : \quad \|x\| = 1 \Big\}.$$

Let ν_{n-1} be the Haar probability measure on $\mathbb{S}^{n-1}$. That is, ν_{n-1} is the unique measure defined on Borel sets $A \subset \mathbb{S}^{n-1}$ such that $\nu_{n-1}(\mathbb{S}^{n-1}) = 1$ and $\nu_{n-1}(U(A)) = \nu_{n-1}(A)$ for every orthogonal transformation U and every Borel set $A \subset \mathbb{S}^{n-1}$. Since the standard Gaussian measure γ_n is also rotation invariant, there is a simple relationship between ν_{n-1} and γ_n: for $A \subset \mathbb{S}^{n-1}$, let

$$\overline{A} = \Big\{ \lambda x \in \mathbb{R}^n : \quad x \in A \quad \text{and} \quad \lambda \geq 0 \Big\}.$$

Then

$$\nu_{n-1}(A) = \gamma_n(\overline{A});$$

see Figure 68.

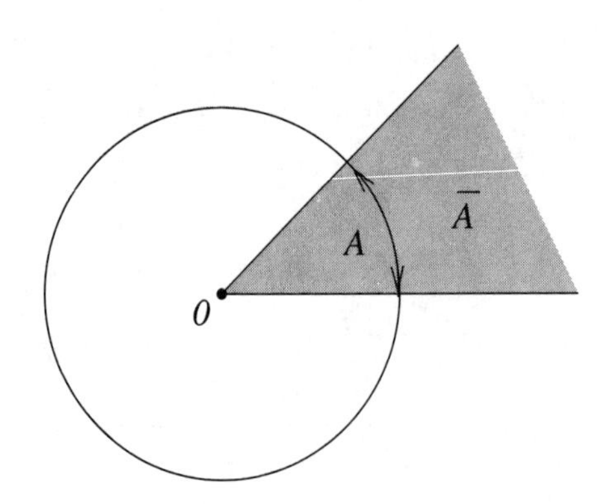

Figure 68. A set $A \subset \mathbb{S}^{n-1}$ and the corresponding set $\overline{A} \subset \mathbb{R}^n$

The following estimate will be used in Section VI.8 (the bound is far from the best possible but works fine for our purposes).

(7.1) Lemma. *Let us choose $y \in \mathbb{S}^{n-1}$ and $\epsilon > 0$. Then*

$$\nu_{n-1}\Big\{ x \in \mathbb{S}^{n-1} : \quad \langle x, y \rangle \geq \epsilon \Big\} \leq 2\exp\big\{-\epsilon^2 n/16\big\}.$$

Proof. Without loss of generality we may assume that $y = (1, 0, \dots, 0)$. Hence our goal is to estimate $\nu_{n-1}(A)$, where

$$A = \Big\{(\xi_1, \dots, \xi_n) \in \mathbb{S}^{n-1} : \quad \xi_1 \geq \epsilon\Big\}.$$

We have $\nu_{n-1}(A) = \gamma_n(\overline{A})$, where

$$\overline{A} = \Big\{x = (\xi_1, \dots, \xi_n) \in \mathbb{R}^n : \quad \xi_1 \geq \epsilon\|x\|\Big\}.$$

Let

$$B = \Big\{(\xi_1, \dots, \xi_n) \in \mathbb{R}^n : \quad \sum_{i=1}^n \xi_i^2 \geq (1-\epsilon)n \quad \text{and} \quad \xi_1^2 \leq \epsilon^2(1-\epsilon)n\Big\}.$$

By Corollary 5.5 and Lemma 5.2,

$$\gamma_n(B) \geq 1 - \exp\big\{-\epsilon^2 n/4\big\} - \exp\big\{-\epsilon^2(1-\epsilon)n/2\big\}.$$

Thus

$$\gamma_n(B) \geq 1 - 2\exp\big\{-\epsilon^2 n/4\big\} \quad \text{provided} \quad \epsilon \leq \frac{1}{2}.$$

Then $\overline{A} \subset \mathbb{R}^n \setminus B$ and hence

$$\gamma_n(\overline{A}) \leq 1 - \gamma_n(B) \leq 2\exp\big\{-\epsilon^2 n/4\big\} \quad \text{provided} \quad \epsilon \leq \frac{1}{2}.$$

We observe that $\nu_{n-1}(A)$ decreases as ϵ grows and that A is empty for $\epsilon > 1$. Hence

$$\nu_{n-1}(A) \leq 2\exp\big\{-n/16\big\} \quad \text{for} \quad \epsilon \geq \frac{1}{2}$$

and, in any case,

$$\nu_{n-1}(A) \leq 2\exp\big\{-\epsilon^2 n/16\big\}.$$

□

Lemma 7.1 implies that if the dimension n is large, then with high probability, for any two randomly chosen vectors $x, y \in \mathbb{S}^{n-1}$, we have $|\langle x, y\rangle| = O(n^{-1/2})$, so x and y are "almost orthogonal". This is an example of a *concentration property* for the uniform probability measure on the unit sphere. Some other related concentration properties are discussed in problems below.

PROBLEMS.

1. Let $L \subset \mathbb{R}^n$ be a subspace and let $k = \dim L$. For $x \in \mathbb{S}^{n-1}$, let x_L denote the orthogonal projection of x onto L. Prove that for any $0 < \epsilon < 1$

$$\nu_{n-1}\left\{x \in \mathbb{S}^{n-1} : \quad (1-\epsilon)\sqrt{\frac{k}{n}} \leq \|x_L\| \leq (1-\epsilon)^{-1}\sqrt{\frac{k}{n}}\right\} \geq 1 - 4\exp\big\{-\epsilon^2 k/4\big\}.$$

Thus if k is large, then the length of the projection x_L for "almost any" vector $x \in \mathbb{S}^{n-1}$ is close to $\sqrt{k/n}$.

2. Instead of fixing a subspace $L \subset \mathbb{R}^n$, we can fix a vector $x \in \mathbb{S}^{n-1}$ and choose a random k-dimensional subspace L. More precisely, let $G_k(\mathbb{R}^n)$ be the Grassmannian of all k-dimensional subspaces $L \subset \mathbb{R}^n$ and let $\nu_{n,k}$ be the rotation invariant probability measure on $G_k(\mathbb{R}^n)$; cf. Section I.8.3. Let us fix $x \in \mathbb{S}^{n-1}$. Prove that for any $0 < \epsilon < 1$

$$\nu_{n,k}\left\{L \in G_k(\mathbb{R}^n): \quad (1-\epsilon)\sqrt{\frac{k}{n}} \leq \|x_L\| \leq (1-\epsilon)^{-1}\sqrt{\frac{k}{n}}\right\} \geq 1 - 4\exp\{-\epsilon^2 k/4\}.$$

Hence if k is large, then the length of the projection x_L for "almost any" k-dimensional subspace $L \subset \mathbb{R}^n$ is close to $\sqrt{k/n}$.

Remark: The original proof of Corollary 6.3 in [**JL84**] uses the following construction. Let us choose a realization $v_1, \ldots, v_n \in \mathbb{R}^n$ of G and let $L \subset \mathbb{R}^n$ be a randomly chosen m-dimensional subspace. Let u_i be the orthogonal projection of v_i onto L for $i = 1, \ldots, m$. Then, with high probability, the points $(\sqrt{n/k})u_i$ provide the desired m-realization of G.

3. Let $\phi_n : \mathbb{S}^{n-1} \longrightarrow \mathbb{R}$,

$$\phi_n(\xi_1, \ldots, \xi_n) = \xi_1\sqrt{n},$$

be a map. Prove that

$$\gamma(A) = \lim_{n \longrightarrow +\infty} \nu_{n-1}\big(\phi_n^{-1}(A)\big)$$

for any Borel set $A \subset \mathbb{R}$. Hence the standard Gaussian measure can be obtained as a limit of the appropriately scaled projection of the uniform measure ν_{n-1} on the sphere.

4*. Let us make $\mathbb{S}^{n-1}$ a metric space by letting $\operatorname{dist}(x,y) = \arccos\langle x, y\rangle$ for all $x, y \in \mathbb{S}^{n-1}$. Thus $0 \leq \operatorname{dist}(x,y) \leq \pi$ is the angle between x and y. For a closed set $A \subset \mathbb{S}^{n-1}$ and $\rho > 0$, let

$$A_\rho = \Big\{x \in \mathbb{S}^{n-1}: \quad \operatorname{dist}(x,y) \leq \rho \quad \text{for some} \quad y \in A\Big\}$$

be the ρ-neighborhood of A. The ρ-neighborhood of a point $y \in \mathbb{S}^{n-1}$ is called the *spherical cap* of radius ρ centered at y and denoted $C(y,\rho)$. Let $A \subset \mathbb{S}^{n-1}$ be a closed set and let $C(y,\rho)$ be a spherical cap such that

$$\nu_{n-1}(A) = \nu_{n-1}\big(C(y,\rho)\big).$$

Prove the *isoperimetric inequality* for the unit sphere: for any $\epsilon > 0$

$$\nu_{n-1}(A_\epsilon) \geq \nu_{n-1}\big(C(y, \rho + \epsilon)\big).$$

Remark: See [**FLM77**].

5. Deduce the isoperimetric inequality for the Gaussian measure (cf. Problem 4 of Section 5.3) from the isoperimetric inequality for the unit sphere and Problem 3 above.

6. Let $A \subset \mathbb{S}^{d-1}$ be a closed set such that $\nu_{n-1}(A) = 1/2$. Deduce from Problem 4 above and Lemma 7.1 a *concentration inequality* for the unit sphere:

$$\nu_{n-1}(A_\epsilon) \geq 1 - 2\exp\bigl\{-c\epsilon^2 n\bigr\}$$

for some absolute constant $c > 0$ and all $\epsilon > 0$.

7. Let $f : \mathbb{S}^{n-1} \longrightarrow \mathbb{R}$ be a function such that

$$|f(x) - f(y)| \leq \operatorname{dist}(x, y) \quad \text{for all} \quad x, y \in \mathbb{S}^{n-1};$$

cf. Problem 4 above. Let α be the *median* of f, that is, α is a number such that

$$\nu_{n-1}\Bigl\{x \in \mathbb{S}^{n-1} : \quad f(x) \leq \alpha\Bigr\} \geq \frac{1}{2} \quad \text{and} \quad \nu_{n-1}\Bigl\{x \in \mathbb{S}^{n-1} : \quad f(x) \geq \alpha\Bigr\} \geq \frac{1}{2};$$

cf. Problem 2 of Section 5.5. Deduce from Problem 6 above a *concentration inequality for Lipschitz functions*:

$$\nu_{n-1}\Bigl\{x \in \mathbb{S}^{n-1} : \quad |f(x) - \alpha| \leq \epsilon\Bigr\} \geq 1 - 4\exp\bigl\{-c\epsilon^2 n\bigr\}$$

for some absolute constant $c > 0$ and all $\epsilon > 0$.

Concentration inequalities are used to prove the existence of "almost ellipsoidal" sections and projections of convex bodies. Those results are beyond the scope of this book (see [**MiS86**] and [**P94**]), but we state some of them as problems.

8* (Dvoretzky's Theorem). Prove that for any $\epsilon > 0$ and for any positive integer k there exists a positive integer $m = m(k, \epsilon)$ (one can choose m about $\exp\bigl\{O(\epsilon^{-2}k)\bigr\}$) such that for any $n \geq m$ and for any convex body $K \subset \mathbb{R}^n$ symmetric about the origin there exists a k-dimensional subspace $L \subset \mathbb{R}^n$ and an ellipsoid $E \subset L$ centered at the origin such that

$$(1 - \epsilon)E \ \subset \ K \cap L \ \subset \ (1 + \epsilon)E.$$

Remark: The two main ingredients of the proof from [**FLM77**] are the results of Problem 7 above and Problem 5 of Section 3.2.

9. Here is the dual form of Dvoretzky's Theorem: For any $\epsilon > 0$ and for any positive integer k there exists a positive integer $m = m(k, \epsilon)$ (one can choose m about $\exp\bigl\{O(\epsilon^{-2}k)\bigr\}$) such that for any $n \geq m$ and for any convex body $K \subset \mathbb{R}^n$ symmetric about the origin there exists a k-dimensional subspace $L \subset \mathbb{R}^n$ and an ellipsoid $E \subset L$ centered at the origin such that

$$(1 - \epsilon)E \ \subset \ K_L \ \subset \ (1 + \epsilon)E,$$

where K_L is the orthogonal projection of K onto L. Deduce the equivalence of the statements of Problems 8 and 9.

10* (Milman's QS Theorem). Prove that for any $0 < \epsilon < 1$ there exists $\alpha = \alpha(\epsilon) > 0$ such that for any convex body $K \subset \mathbb{R}^n$ symmetric about the origin there exist subspaces $L_2 \subset L_1 \subset \mathbb{R}^n$ with $\dim L_2 \geq (1-\epsilon)n$ and an ellipsoid $E \subset L_2$ such that

$$E \subset (K \cap L_1)_{L_2} \subset \alpha E,$$

where $(K \cap L_1)_{L_2}$ is the orthogonal projection of the section $K \cap L_1$ onto L_2.

Remark: "QS" stands for "quotient of subspace". In the dual (equivalent) form of the theorem we take the projection first and then the intersection.

11* (The "Volume Ratio" Theorem). Let $K \subset \mathbb{R}^n$ be a convex body symmetric about the origin and let $E \subset K$ be the maximum volume ellipsoid; see Section 2. Let

$$\alpha = \left(\frac{\operatorname{vol} K}{\operatorname{vol} E}\right)^{1/n}.$$

Prove that for $k = 1, \ldots, n-1$ there exists a subspace $L \subset \mathbb{R}^n$ with $\dim L = k$ such that

$$K \cap L \subset (4\pi\alpha)^{\frac{n}{n-k}} E.$$

8. Remarks

For an accessible introduction to metric convex geometry including approximating ellipsoids and Dvoretzky's Theorem, see [**B97**]. For more advanced texts, see [**P94**] and [**MiS86**]. We did not discuss other interesting and important ellipsoids, such as Milman's ellipsoid and the inertia ellipsoid, associated with a (symmetric) convex body; see [**P94**], [**B97**]. The volume inequalities and the Brunn-Minkowski Theory are discussed in detail in [**Sc93**]. First counterexamples to the Borsuk conjecture (see the remark after Problem 7 of Section 2.4) were constructed by J. Kahn and G. Kalai; see [**K95**]. Results of Problems 8 and 9 of Section 2.4 regarding the metric structure of the set of non-negative polynomials are due to [**Bl02**]. For the Ellipsoid Method, see [**Lo86**], [**GLS93**] and [**PS98**]. A comprehensive reference for the measure concentration techniques is [**Le01**]. For Gaussian measures, see [**Bo98**].

Chapter VI

Faces of Polytopes

We explore the combinatorial structure of polytopes. We discuss the number of faces of a given dimension that a polytope can have, how the faces fit together and what is the facial structure of some particularly interesting polytopes, such as the permutation polytope and the cyclic polytope. Our approach is based on considering a sufficiently generic linear function on a polytope and using some combinatorial (counting) or metric arguments.

1. Polytopes and Polarity

Recall (Definition I.2.2) that a polytope is the convex hull of a finite set of points in $\mathbb{R}^d$. We proved that a polytope is a polyhedron (Corollary IV.1.3) and that a bounded polyhedron is a polytope (Corollary II.4.3). Recall that an extreme point of a polyhedron (polytope) is called a vertex (cf. Definition II.4.1). In this section, we apply polarity (see Chapter IV) to obtain some general results about the facial structure of polytopes.

To warm up, we prove that a face of a polytope is the convex hull of the vertices of the polytope that belong to the face.

(1.1) Lemma. *Let $P = \operatorname{conv}(v_1, \ldots, v_m) \subset \mathbb{R}^d$ be a polytope and let $F \subset P$ be a face. Then $F = \operatorname{conv}\big(v_i : v_i \in F\big)$. In particular, a face of a polytope is a polytope and the number of faces of a polytope is finite.*

Proof. Since F is a face of P, there exists a linear functional $f : \mathbb{R}^d \longrightarrow \mathbb{R}$ and a number α such that $f(x) \leq \alpha$ for all $x \in P$, and $f(x) = \alpha$ if and only if $x \in F$ (see Definition II.2.6). Let $I = \big\{i : f(v_i) = \alpha\big\}$, so $f(v_i) < \alpha$ for all $i \notin I$. Obviously, $\operatorname{conv}(v_i : i \in I) \subset F$. It remains to show that $F \subset \operatorname{conv}(v_i : i \in I)$. Let us choose an $x \in F$. Since $x \in P$, we have

$$x = \sum_{i=1}^{m} \lambda_i v_i, \quad \text{where} \quad \lambda_i \geq 0 \quad \text{for} \quad i = 1, \ldots, m \quad \text{and} \quad \sum_{i=1}^{m} \lambda_i = 1.$$

If $\lambda_i > 0$ for some $i \notin I$, then $f(\lambda_i v_i) = \lambda_i f(v_i) < \lambda_i \alpha$ and

$$f(x) = \sum_{i=1}^{m} \lambda_i f(v_i) < \alpha \sum_{i=1}^{m} \lambda_i = \alpha,$$

which is a contradiction. Hence $\lambda_i = 0$ for $i \notin I$ and $x \in \operatorname{conv}(v_i : i \in I)$. □

As an immediate corollary, we conclude that the faces of a polytope fit together nicer than in the case of a general convex body; cf. Problem 6, Section II.2.6.

(1.2) Corollary. *Let P be a polytope, let F be a face of P and let G be a face of F. Then G is a face of P.*

Proof. Suppose that

$$P = \operatorname{conv}\big(v_1, \ldots, v_m\big), \quad F = \operatorname{conv}\big(v_i : i \in I\big) \quad \text{and} \quad G = \operatorname{conv}\big(v_j : j \in J \subset I\big).$$

There exists a linear functional $f : \mathbb{R}^d \longrightarrow \mathbb{R}$ and a number α such that $f(v_i) = \alpha$ for all $i \in I$ and $f(v_i) < \alpha$ for all $i \notin I$, and there exists a linear functional $g : \mathbb{R}^d \longrightarrow \mathbb{R}$ and a number β such that $g(v_i) = \beta$ for $i \in J$ and $g(v_i) < \beta$ for $i \in I \setminus J$. Then, for a sufficiently small $\epsilon > 0$, the functional $h = f + \epsilon g$ has the following property: $h(v_i) = \alpha + \epsilon\beta$ for $i \in J$ and $h(v_i) < \alpha + \epsilon\beta$ for $i \notin J$. Hence G is a face of P. □

PROBLEMS.

1°. Prove that a vertex of a polytope is a 0-dimensional face of the polytope.

2°. Prove that a face of a polyhedron is a polyhedron.

3. Let $P = \big\{x \in \mathbb{R}^d : \ \langle c_i, x \rangle \leq \beta_i \text{ for } i = 1, \ldots, m\big\}$ be a polyhedron and let $u \in P$ be a point. Let $I_u = \{i : \ \langle c_i, u \rangle = \beta_i\}$. Let $F = \big\{x \in P : \ \langle c_i, x \rangle = \beta_i$ for $i \in I_u\big\}$. Prove that F is the smallest face of P containing u (we agree here that P is a face of itself).

4. Prove that a polyhedron has finitely many faces.

5. Let $P \subset \mathbb{R}^d$ be a polyhedron, let $F \subset P$ be a face of P and let $G \subset F$ be a face of F. Prove that G is a face of P.

6°. Prove that the intersection of two polytopes is a polytope.

Now we present the main result of this section, which establishes an inclusion reversing correspondence between the faces of a polytope and the faces of its polar. It will serve us as a "translation device", which, in the general spirit of duality, allows us to obtain some results "for free".

(1.3) Theorem. *Let $P \subset \mathbb{R}^d$ be a polytope, containing the origin in its interior and let $Q = P^\circ$. Then Q is a polytope. For a face F of P, let*

$$\widehat{F} = \Big\{x \in Q : \quad \langle x, y \rangle = 1 \quad \textit{for each} \quad y \in F\Big\}$$

(we agree that $\widehat{\emptyset} = Q$ *and that* $\widehat{P} = \emptyset$*). Then*

1. *the set* $\widehat{F}$ *is a face of* Q *and*

$$\dim F + \dim \widehat{F} = d - 1;$$

2. *if* $F \subset G$ *are faces of* P*, then* $\widehat{G} \subset \widehat{F}$*;*
3. *let* G *be a face of* Q *and let*

$$F = \Big\{ y \in P : \quad \langle x, y \rangle = 1 \quad \textit{for each} \quad x \in G \Big\}$$

(we agree that $F = P$ *for* $G = \emptyset$ *and that* $F = \emptyset$ *for* $G = Q$*). Then*

$$\widehat{F} = G.$$

Proof. The proof that Q is a polytope is incorporated in the proof of Corollary IV.1.3. Suppose that $P = \operatorname{conv}\big(v_1, \ldots, v_m\big)$. By Lemma 1.1, $F = \operatorname{conv}\big(v_i : i \in I\big)$ for some set of indices I. Let $v = \dfrac{1}{|I|} \sum\limits_{i \in I} v_i$, where $|I|$ is the cardinality of I. Hence $v \in F$. We claim that

$$\widehat{F} = \big\{ x \in Q : \quad \langle x, v \rangle = 1 \big\}. \tag{1.3.1}$$

Indeed, $v \in F$, so $\langle x, v \rangle = 1$ for any $x \in \widehat{F}$. On the other hand, for any $x \in Q$ and any v_i, we have $\langle x, v_i \rangle \leq 1$. Therefore, if $x \in Q$ and $\langle x, v \rangle = 1$, then $\langle x, v_i \rangle = 1$ for $i \in I$ and hence $x \in \widehat{F}$. By (1.3.1), we conclude that $\widehat{F}$ is a face of Q.

Let us prove that $\dim F + \dim \widehat{F} = d - 1$. Since F is a face, there is a vector $c \in \mathbb{R}^d$ and a number α, such that

$$\langle c, v_i \rangle = \alpha \quad \text{for} \quad i \in I \qquad \text{and} \qquad \langle c, v_i \rangle < \alpha \quad \text{for} \quad i \notin I.$$

Since $0 \in \operatorname{int} P$, $\alpha > 0$. Scaling, if necessary, we can assume that $\alpha = 1$. Then $c \in \widehat{F}$. Suppose that $\dim F = k$. For any $x \in \widehat{F}$, we must have $\langle x, v_i \rangle = 1$ for $i \in I$, or, in other words, $\langle x - c, v_i \rangle = 0$ for $i \in I$. Since $\dim \operatorname{span}(F) = k + 1$, we get that $\dim \widehat{F} \leq d - k - 1$. On the other hand, if $y \in \mathbb{R}^d$ is a vector such that $\langle y, v_i \rangle = 0$ for $i \in I$, then $x = c + \epsilon y \in \widehat{F}$ for a sufficiently small $\epsilon > 0$. Hence $\dim \widehat{F} = d - k - 1$. Part 1 is proved.

Let us prove Part 2 of the theorem. For any $x \in \widehat{G}$, we have $\langle x, y \rangle = 1$ for any $y \in G$. Since $F \subset G$, we have $\langle x, y \rangle = 1$ for any $y \in F$. Therefore, $x \in \widehat{F}$ and the proof of Part 2 follows.

By Theorem IV.1.2, $Q^\circ = P$. Then, by Part 1, F is a face of P and it is clear that $G \subset \widehat{F}$. Let $c \in \mathbb{R}^d$ be a vector such that $\langle c, x \rangle = 1$ for all $x \in G$ and $\langle c, x \rangle < 1$ for $x \notin Q \setminus G$ (see the proof of Part 2 above). Then $c \in F$. Therefore, for all $x \in \widehat{F}$ we have $\langle x, c \rangle = 1$, so $x \in G$. Therefore, $\widehat{F} \subset G$. Hence $G = \widehat{F}$ and the proof of Part 3 follows. □

Figure 69 presents an example of the polarity correspondence between faces of the cube and octahedron in $\mathbb{R}^3$. It should be noted that the figure reflects only some general combinatorial features of the correspondence since each polytope is pictured in its own copy of the ambient space $\mathbb{R}^3$.

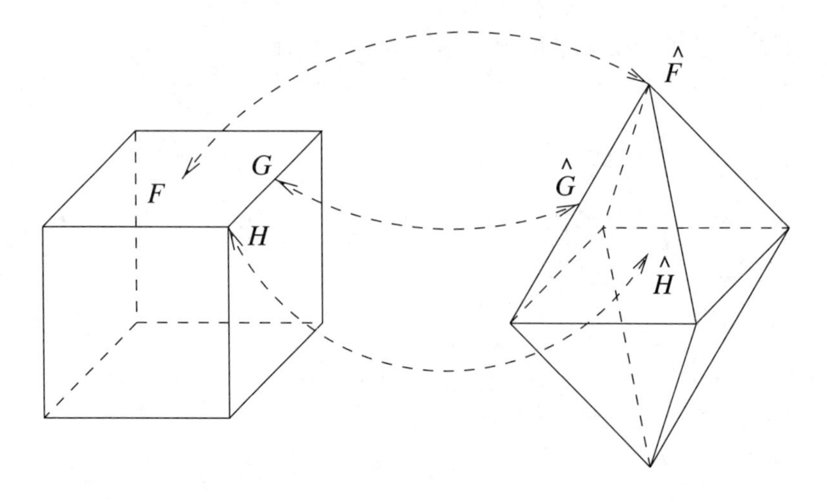

Figure 69. The polarity correspondence between faces

Faces of certain dimensions have special names.

(1.4) Definitions. A 0-dimensional face of a polytope is called a *vertex.* A 1-dimensional face of a polytope is called an *edge.* A $(d-1)$-dimensional face of a d-dimensional polytope is called a *facet.* A $(d-2)$-dimensional face of a d-dimensional polytope is called a *ridge.* Vertices v and u of a polytope are called *neighbors* if the interval $[u, v]$ is an edge of the polytope.

PROBLEMS.

1. Prove that every d-dimensional polytope has a facet. Deduce that a d-dimensional polytope has a k-dimensional face for each $0 \leq k \leq d-1$.

2. Let $P \subset \mathbb{R}^d$ be a d-dimensional polytope. Prove that every ridge of P belongs to precisely two facets.

Hint: What is the dual statement?

3. Prove the following *diamond property.* Let P be a polytope and $G \subset F$ be two faces of P such that $\dim F - \dim G = 2$. Then there are precisely two faces H_1, H_2, such that $G \subset H_1, H_2 \subset F$ (all inclusions are proper).

4. Let $P \subset \mathbb{R}^d$ be a d-dimensional polytope containing the origin in its interior, let $F \subset P$ be a k-dimensional face of P and let $\widehat{F} \subset P^\circ$ be the face from Theorem 1.3. Hence F is a $(d-k-1)$-dimensional polytope. We consider F as a full-dimensional polytope in its affine hull and choose the origin to be in its interior.

Let $R = F^{\circ}$. Hence R is a $(d-k-1)$-dimensional polytope. Establish an inclusion-preserving bijection ϕ between the faces of R and the faces H of P containing F. The (facial structure of) the polytope R is called the *face figure* of F in P and denoted P/F.

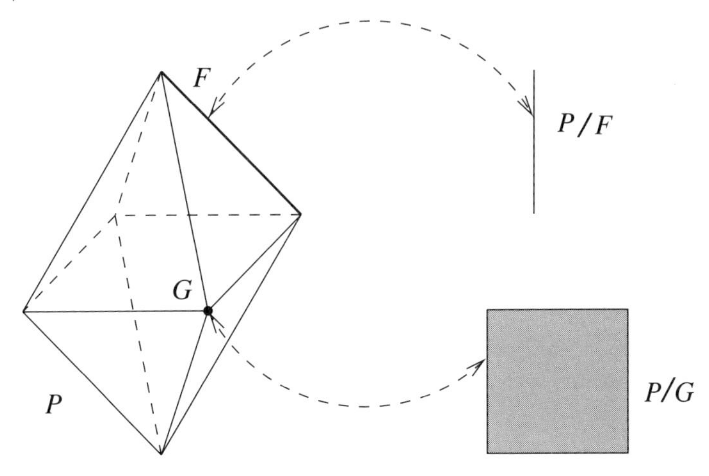

Figure 70. A polytope P and its face figures

Remark: The picture of the face figure P/F can be obtained by intersecting P with an appropriate affine subspace of dimension $\dim P - \dim F - 1$.

5. Let $K \subset \mathbb{R}^d$ be a compact convex set containing the origin in its interior. Let $Q_0 = K^{\circ}$, let $p \in \operatorname{int} Q_0$ be a point and let Q_p be the polar of K with the origin moved to p:

$$Q_p = \Big\{x \in \mathbb{R}^d : \quad \langle x - p,\ y - p \rangle \leq 1 \quad \text{for all} \quad y \in K\Big\}.$$

Prove that Q_p is the image of Q_0 under the *projective transformation*:

$$x \longmapsto p + \frac{x}{1 - \langle x, p \rangle}.$$

We will need the following useful result, which can be considered as a sharpening of Corollary IV.1.3.

(1.5) Lemma. *Let $P \subset \mathbb{R}^d$ be a d-dimensional polytope. Then P can be represented in the form*

$$P = \Big\{x \in \mathbb{R}^d : \quad f_i(x) \leq \alpha_i \quad for \quad i = 1, \ldots, m\Big\},$$

where $f_i : \mathbb{R}^d \longrightarrow \mathbb{R}$ are linear functionals, $\alpha_i \in \mathbb{R}$ are numbers and the sets

$$F_i = \Big\{x \in P : \quad f_i(x) = \alpha_i\Big\}$$

are the facets of P for $i = 1, \ldots, m$.

Proof. The proof follows the proof of Corollary IV.1.3 with some modifications. Let us choose the origin in the interior of P and let $Q = P^\circ$ be the polar of P. We can write $Q = \operatorname{conv}(v_i : i = 1, \ldots, m)$, where v_i are the vertices of Q. Then

$$P = \{x : \quad \langle v_i, x \rangle \leq 1 \quad \text{for} \quad i = 1, \ldots, m\}.$$

Let us choose $f_i(x) = \langle v_i, x \rangle$ and $\alpha_i = 1$. By Theorem 1.3, $F_i = \widehat{v_i}$ are the facets of P. □

2. The Facial Structure of the Permutation Polytope

In this section, we describe the facial structure of a particular polytope. Recall (see Definition II.6.1) that the permutation polytope $P(a)$, where $a = (\alpha_1, \ldots, \alpha_n)$, $a \in \mathbb{R}^n$, is the convex hull of the points $\sigma(a)$ obtained from a by permutations of the coordinates. In this section, we describe the facial structure of $P(a)$ assuming that the coordinates $\alpha_1, \ldots, \alpha_n$ are distinct. We need a simple and useful result.

(2.1) Lemma. *Let $x = (\xi_1, \ldots, \xi_n)$ and $y = (\eta_1, \ldots, \eta_n)$ be n-vectors. Suppose that $\xi_i > \xi_j$ and $\eta_i < \eta_j$ for some pair of indices $i \neq j$. Let $\overline{y}$ be the vector obtained from y by swapping η_i and η_j. Then*

$$\langle x, \overline{y} \rangle > \langle x, y \rangle.$$

Proof. We have

$$\langle x, \overline{y} \rangle - \langle x, y \rangle = \xi_i \eta_j + \xi_j \eta_i - \xi_i \eta_i - \xi_j \eta_j = (\xi_i - \xi_j)(\eta_j - \eta_i) > 0.$$

□

PROBLEM.

The intuitive meaning of Lemma 2.1 is conveyed by the problem below.

1°. Suppose that there are six boxes in front of you. The first is stuffed with \$100 bills, the second with \$50 bills, the third with \$20 bills, the fourth with \$10 bills, the fifth with \$5 bills and the sixth with \$1 bills. You are allowed to take from each box a number of bills with the only condition that the (unordered) set of the numbers of bills taken is the (unordered) set $\{3, 5, 10, 2, 15, 8\}$ (that is, you should take three bills from some box, then five bills from some other box and so forth). How many bills should you take from each box to maximize the total amount taken?

Surprising as it may seem, the reasoning behind Problem 1 of Section 2.1 is powerful enough to lead to the complete description of the facial structure of the permutation polytope $P(a)$; cf. Figure 16.

(2.2) Proposition. *Let $a = (\alpha_1, \ldots, \alpha_n)$ be a point such that $\alpha_1 > \alpha_2 > \ldots > \alpha_n$ and let $P = P(a)$ be the corresponding permutation polytope. For a number $1 \leq k \leq n$, let $\mathcal{S}$ be a partition of the set $\{1, \ldots, n\}$ into k pairwise disjoint non-empty subsets $S_1, \ldots, S_k$. Let $s_i = |S_i|$ be the cardinality of the i-th subset for $i = 1, \ldots, k$, let $t_i = \sum_{j=1}^{i} s_j$ for $i = 1, \ldots, k$ and let us define sets $A_1 = \{\alpha_j : 1 \leq j \leq s_1\}$ and $A_i = \{\alpha_j : t_{i-1} \leq j \leq t_i\}$ for $i = 2, \ldots, k$.*

Let $F_{\mathcal{S}}$ be the convex hull of the points $b = \sigma(a)$, $b = (\beta_1, \ldots, \beta_n)$ such that $\{\beta_j : j \in S_i\} = A_i$ for all $i = 1, \ldots, k$. In words: we permute the first s_1 numbers $\alpha_1, \ldots, \alpha_{s_1}$ in the coordinate positions prescribed by $S_1 \subset \{1, \ldots, n\}$, the second s_2 numbers $\alpha_{s_1+1}, \ldots, \alpha_{s_1+s_2}$ in the coordinate positions prescribed by $S_2 \subset \{1, \ldots, n\}$ and so forth, and take the convex hull $F_{\mathcal{S}}$ of all resulting points.

Then $F_{\mathcal{S}}$ is a face of P, $\dim F_{\mathcal{S}} = n - k$ and for every face F of P we have $F = F_{\mathcal{S}}$ for some partition $\mathcal{S}$.

Example. Let $n = 10$, $a = (20, 16, 15, 10, 9, 7, 5, 4, 2, 1)$, $S_1 = \{4, 5, 6, 7\}$, $S_2 = \{1, 2, 3\}$ and $S_3 = \{8, 9, 10\}$. Hence $A_1 = \{20, 16, 15, 10\}$, $A_2 = \{9, 7, 5\}$ and $A_3 = \{4, 2, 1\}$. The face $F_{\mathcal{S}}$ is the convex hull of all points $b = \sigma(a)$ of the type

$$\Big(\underbrace{\bullet \quad \bullet \quad \bullet}_{\text{permutation of } 9,7,5} \qquad \underbrace{\bullet \quad \bullet \quad \bullet \quad \bullet}_{\text{permutation of } 20,16,15,10} \qquad \underbrace{\bullet \quad \bullet \quad \bullet}_{\text{permutation of } 4,2,1} \Big)$$

and $\dim F_{\mathcal{S}} = 7$.

Proof of Proposition 2.2. Let us describe all the faces F of P containing a.

Let $c = (\gamma_1, \ldots, \gamma_n)$ be a vector and λ be a number such that $\langle c, x \rangle \leq \lambda$ for all $x \in P$ and $\langle c, x \rangle = \lambda$ if and only if $x \in F$. Since $a \in F$, we have $\langle c, a \rangle = \lambda$. Lemma 2.1 implies that we must have $\gamma_1 \geq \gamma_2 \geq \ldots \geq \gamma_n$, since if for some $i < j$ we had $\gamma_i < \gamma_j$, we would have obtained $\langle c, \tau(a) \rangle > \langle c, a \rangle$ for the transposition τ that swaps α_i and α_j.

Let us split the sequence $\gamma_1 \geq \gamma_2 \geq \ldots \geq \gamma_n$ into the subintervals $S_1, \ldots, S_k$ for which the γ's do not change:

$$\underbrace{\gamma_1 = \ldots = \gamma_{t_1}}_{S_1 = \{1, \ldots, t_1\}} > \underbrace{\gamma_{t_1+1} = \ldots = \gamma_{t_2}}_{S_2 = \{t_1+1, \ldots, t_2\}} > \gamma_{t_2+1} = \ldots > \underbrace{\gamma_{t_{k-1}+1} = \ldots = \gamma_n}_{S_k = \{t_{k-1}+1, \ldots, n\}}.$$

Hence $S_1 = \{j : \gamma_j = \gamma_1\}$, $s_1 = t_1 = |S_1|$ and $S_i = \{j : \gamma_j = \gamma_{t_{i-1}+1}\}$, $s_i = |S_i|$ and $t_i = t_{i-1} + s_i$ for $i = 2, \ldots, k$.

We observe that for $b = \sigma(a)$, $b = (\beta_1, \ldots, \beta_n)$, we have $\langle b, c \rangle = \langle a, c \rangle$ if and only if $(\beta_1, \ldots, \beta_{t_1})$ is a permutation of $(\alpha_1, \ldots, \alpha_{t_1})$, $(\beta_{t_1+1}, \ldots, \beta_{t_2})$ is a permutation of $(\alpha_{t_1+1}, \ldots, \alpha_{t_2})$, and so forth, so that $(\beta_{t_{k-1}+1}, \ldots, \beta_n)$ is a permutation of $(\alpha_{t_{k-1}+1}, \ldots, \alpha_n)$. Applying Lemma 1.1, we conclude that $F = F_{\mathcal{S}}$ for the partition $\mathcal{S} = \{S_1, \ldots, S_k\}$. In this case, we have $A_1 = \{\alpha_1, \ldots, \alpha_{s_1}\}$ and $A_i = \{\alpha_{t_{i-1}+1}, \ldots, \alpha_{t_i}\}$ for $i = 2, \ldots, k$.

Vice versa, every vector $c = (\gamma_1, \ldots, \gamma_n)$ with $\gamma_1 \geq \gamma_2 \geq \ldots \geq \gamma_n$ gives rise to a face $F_{\mathcal{S}}$ containing a, where $\mathcal{S} = S_1 \cup S_2 \cup \ldots \cup S_k$ is the partition of $\{1, \ldots, n\}$ into the subintervals on which γ's do not change.

Let $a_1 = (\alpha_1, \ldots, \alpha_{s_1}) \in \mathbb{R}^{s_1}$ and let $a_i = (\alpha_{t_{i-1}+1}, \ldots, \alpha_{t_i}) \in \mathbb{R}^{s_i}$ for $i = 2, \ldots, k$. Geometrically, the face $F_{\mathcal{S}}$ is the direct product

$$F_{\mathcal{S}} = P(a_1) \times \ldots \times P(a_k)$$

of the permutation polytopes $P(a_i) \subset \mathbb{R}^{s_i}$. Since a has distinct coordinates, the coordinates of each a_i are distinct as well, so by Problem 4, Section II.6.1, we have $\dim P(a_i) = s_i - 1$ for $i = 1, \ldots, k$. Therefore,

$$\dim F_{\mathcal{S}} = \sum_{i=1}^{k} \dim P(a_i) = \sum_{i=1}^{k} (s_i - 1) = \left(\sum_{i=1}^{k} s_i\right) - k = n - k.$$

Hence we have described the faces F of P containing a. Let $\sigma \in S_n$ be a permutation and let $\sigma(x) = y$ for $x = (\xi_1, \ldots, \xi_n)$ and $y = (\eta_1, \ldots, \eta_n)$ provided $\eta_i = \xi_{\sigma^{-1}(i)}$. Then the action $x \longmapsto \sigma(x)$ is an orthogonal transformation of $\mathbb{R}^n$. Hence we conclude that F is a face of P if and only if for some permutation σ, the set $\sigma(F)$ is a face of P containing a. If $\sigma(F) = F_{\mathcal{S}}$ for $\mathcal{S} = S_1 \cup \ldots \cup S_k$, then $F = F_{\mathcal{S}'}$, where $\mathcal{S}' = \sigma^{-1}(S_1) \cup \ldots \cup \sigma^{-1}(S_k)$. This completes the proof. □

PROBLEMS.

1°. Let $a = (\alpha_1, \ldots, \alpha_n)$ be a vector such that $\alpha_1 > \alpha_2 > \ldots > \alpha_n$ and let $P(a)$ be the corresponding permutation polytope. Let $b = (\beta_1, \ldots, \beta_n)$ be a vertex of P. Let b' be another vertex of P. Prove that the interval $[b, b']$ is an edge of P if and only if b' is obtained from b by swapping two values β_i and β_j such that $\beta_i = \alpha_k$ and $\beta_j = \alpha_{k+1}$ for some $k = 1, \ldots, n-1$.

2°. Prove that the facets of the permutation polytope can be described as follows. Let us choose a partition $\mathcal{S}$ of the set $\{1, \ldots, n\}$ into two non-empty disjoint subsets S_1 and S_2. Let $s_1 = |S_1|$ be the cardinality of S_1 and let $s_2 = |S_2|$ be the cardinality of S_2, so that $s_1 + s_2 = n$. Let $F_{\mathcal{S}}$ be the convex hull of all points b obtained by permuting the numbers $\alpha_1, \ldots, \alpha_{s_1}$ in the coordinate positions prescribed by $S_1 \subset \{1, \ldots, n\}$ and permuting independently the numbers $\alpha_{s_1+1}, \ldots, \alpha_n$ in the coordinate positions prescribed by $S_2 \subset \{1, \ldots, n\}$. Prove that $F_{\mathcal{S}}$ is a facet of $P(a)$ and that every facet of $P(a)$ has the form $F_{\mathcal{S}}$ for some partition $\mathcal{S} = S_1 \cup S_2$.

3. Let $a \in \mathbb{R}^n$ be a vector with distinct components and let $P(a)$ be the corresponding permutation polytope. Let f_k be the number of k-dimensional faces of P. Prove that

$$f_k = \sum_{\substack{m_1 + \ldots + m_{n-k} = n \\ m_1, \ldots, m_{n-k} \text{ are positive integers}}} \frac{n!}{m_1! \cdots m_{n-k}!}.$$

4*. Let us choose $a = (n-1, n-2, \ldots, 1, 0)$, $a \in \mathbb{R}^n$. The polytope $P_{n-1} = P(a)$ is called the *permutohedron*. Prove that the permutohedron is a *zonotope*, that is, the Minkowski sum of finitely many (namely, $\binom{n}{2}$) straight line intervals.

Hint: Let $e_1, \ldots, e_n$ be the standard basis of $\mathbb{R}^n$. For $i < j$ let

$$I_{ij} = \operatorname{conv}\Big(\frac{1}{2}e_i - \frac{1}{2}e_j, \quad \frac{1}{2}e_j - \frac{1}{2}e_i\Big).$$

Then

$$P_{n-1} = \frac{n-1}{2}(1, \ldots, 1) + \sum_{i<j} I_{ij},$$

where "$\sum$" stands for the Minkowski sum; see Example 7.15 of [**Z95**].

5. Let $a \in \mathbb{R}^n$ be a vector with distinct components and let $P(a)$ be the corresponding permutation polytope. Let us consider $P(a)$ as a full-dimensional polytope in its affine hull and let us choose the origin in the interior of $P(a)$. Describe the combinatorial structure of the polar Q to $P(a)$.

Hint: Combinatorially, Q is the *first barycentric subdivision* of the simplex. Namely, we start with the standard $(d-1)$-dimensional simplex Δ (see Problem 1 of Section I.2.2) considered as a convex body in its affine hull, choose a point p_F in the interior of every face F of Δ and slightly push p_F "outward" by the distance $\epsilon^{d-\dim F}$ for a sufficiently small $\epsilon > 0$.

As a corollary of Proposition 2.2, we obtain a theorem of R. Rado (1952), which gives us a useful criterion for checking whether a given point belongs to a given permutation polytope.

(2.3) Rado's Theorem. *Let $a = (\alpha_1, \ldots, \alpha_n)$ be a point, where $\alpha_1 \geq \alpha_2 \geq \ldots \geq \alpha_n$ and let $P(a) \subset \mathbb{R}^n$ be the corresponding permutation polytope. Let $b \in \mathbb{R}^n$ be a point and let $\beta_1 \geq \beta_2 \geq \ldots \geq \beta_n$ be the ordering of the coordinates of b. Then $b \in P(a)$ if and only if*

$$\sum_{i=1}^{k} \beta_i \leq \sum_{i=1}^{k} \alpha_i \quad \text{for} \quad k = 1, \ldots, n-1$$

and

$$\sum_{i=1}^{n} \beta_i = \sum_{i=1}^{n} \alpha_i.$$

Proof. We prove the result assuming that $\alpha_1 > \alpha_2 > \ldots > \alpha_n$. The proof in the general case follows then by a continuity argument.

We know that $\dim P(a) = n - 1$ and that $P(a)$ lies in the affine hyperplane

$$H = \Big\{(\xi_1, \ldots, \xi_n) : \quad \xi_1 + \ldots + \xi_n = \alpha_1 + \ldots + \alpha_n\Big\}$$

(see Problem 3, Section II.6.1). Problem 2 of Section 2.2 implies that the facets $F_{\mathcal{S}}$ of P are indexed by the partitions $\mathcal{S} = S_1 \cup S_2$ of the set $\{1, \ldots, n\}$ into non-empty disjoint subsets. Suppose that $|S_1| = s_1$ and $|S_2| = s_2$, so $s_1, s_2 > 0$ and $s_1 + s_2 = n$. The facet F_S is a convex hull of the points $x = \sigma(a)$, $x = (\xi_1, \ldots, \xi_n)$ such that

$$\{\xi_i : i \in S_1\} = \{\alpha_1, \ldots, \alpha_{s_1}\} \quad \text{and} \quad \{\xi_i : i \in S_2\} = \{\alpha_{s_1+1}, \ldots, \alpha_{s_n}\}.$$

It follows then that in H the facet F_S is defined by the inequality

$$\sum_{i \in S_1} \xi_i \leq \sum_{i=1}^{s_1} \alpha_i \tag{2.3.1}$$

since (2.3.1) is trivially satisfied on all vertices of P and is sharp on the vertices from F_S. Lemma 1.5 implies that $b \in P(a)$ if and only if $b \in H$ and b satisfies the inequality (2.3.1) for every facet F_S of P. Since for every subset $S_1 \subset \{1, \ldots, n\}$ we have

$$\sum_{i \in S_1} \beta_i \leq \sum_{i=1}^{s_1} \beta_i,$$

the result follows. □

3. The Euler-Poincaré Formula

In this section, we prove a classical relation for the number of faces of an arbitrary polytope. Throughout this section, we use the Euler characteristic χ (see Section I.7). For a set $A \subset \mathbb{R}^d$, we write simply $\chi(A)$ instead of $\chi([A])$. In what follows, we agree that a polytope P is a face of itself. By convention, the dimension of the empty face is -1.

(3.1) Lemma. *Let $P \subset \mathbb{R}^d$ be a d-dimensional polytope, let ∂P be the boundary of P and let* int P *be the interior of P. Then*

$$\chi(\partial P) = 1 + (-1)^{d-1} \quad \textit{and} \quad \chi(\operatorname{int} P) = (-1)^d.$$

Proof. Since ∂P is the union of faces of P, the indicator function $[\partial P]$ belongs to the algebra $\mathcal{K}(\mathbb{R}^d)$ of compact convex sets, so $\chi(\partial P)$ is defined. We prove the first identity by induction on d. If $d = 1$, then P is an interval, so ∂P consists of two points and $\chi(\partial P) = 2$.

Let us introduce a family of hyperplanes $H_\tau = \{(\xi_1, \ldots, \xi_d) : \xi_d = \tau\}$ for $\tau \in \mathbb{R}$. By Lemma I.7.5,

$$\chi(\partial P) = \sum_{\tau \in \mathbb{R}} \Big(\chi(\partial P \cap H_\tau) - \lim_{\epsilon \longrightarrow +0} \chi(\partial P \cap H_{\tau - \epsilon}) \Big).$$

Let

$$\tau_{\min} = \min_{x \in P} \xi_d \quad \text{and} \quad \tau_{\max} = \max_{x \in P} \xi_d.$$

For any $\tau_{\min} < \tau < \tau_{\max}$, the intersection $P \cap H_\tau$ is a $(d-1)$-dimensional polytope and $\partial(P \cap H_\tau) = \partial P \cap H_\tau$. Therefore, $\chi(\partial P \cap H_\tau) = 1 + (-1)^{d-2}$ for all

$\tau \in [\tau_{\min}, \tau_{\max}]$. Furthermore, for $\tau = \tau_{\min}$ and $\tau = \tau_{\max}$, the intersection $P \cap H_\tau$ is a face of P lying in ∂P. Therefore, $\chi(\partial P \cap H_\tau) = 1$ for $\tau = \tau_{\max}$ or $\tau = \tau_{\min}$. Finally, for $\tau > \tau_{\max}$ and $\tau < \tau_{\min}$, the intersection $P \cap H_\tau$ is empty, so $\chi(P \cap H_\tau) = 0$.

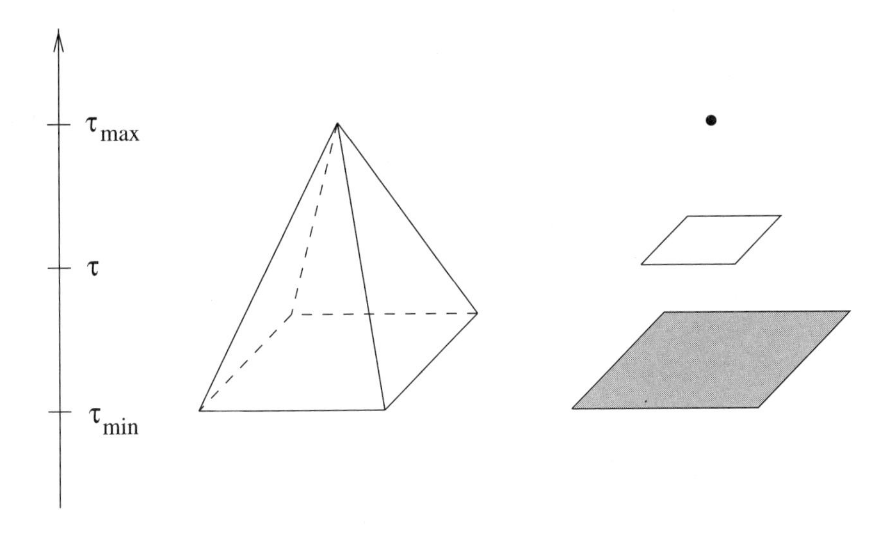

Figure 71. The boundary of a polytope and its "slices"

Summarizing,

$$\chi(\partial P) = 1 - (1 + (-1)^{d-2}) + 1 = 1 + (-1)^{d-1},$$

as claimed. Now we observe that $[\operatorname{int} P] = [P] - [\partial P]$ and hence

$$\chi(\operatorname{int} P) = \chi(P) - \chi(\partial P) = 1 - 1 - (-1)^{d-1} = (-1)^d.$$

□

Now we are ready to obtain the Euler-Poincaré Formula for polytopes.

(3.2) Corollary (The Euler-Poincaré Formula).

Let $P \subset \mathbb{R}^d$ be a d-dimensional polytope and let $f_i(P)$ be the number of i-dimensional faces of P. Then

$$\sum_{i=0}^{d-1} (-1)^i f_i(P) = 1 + (-1)^{d-1}.$$

Proof. For every (non-empty) face F of P, including P itself, let $\operatorname{int} F$ be the interior of F considered in its affine hull (in particular, if F is a vertex of P, then $\operatorname{int} F = F$). Then we can write:

$$[P] = \sum_F [\operatorname{int} F],$$

where the sum is taken over all non-empty faces F of P including P; cf. Figure 72.

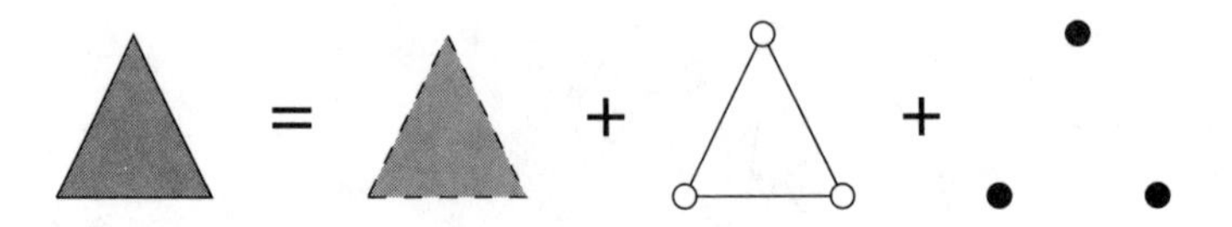

Figure 72. Writing a polytope as the sum of the interiors of the faces

Let us apply the Euler characteristic χ to the both sides of the identity. By Lemma 3.1, $\chi[\operatorname{int} F] = (-1)^{\dim F}$ and the result follows. □

The first complete proof of Corollary 3.2 was, apparently, obtained by H. Poincaré in 1893 (using methods of algebraic topology). A somewhat incomplete proof was given by L. Schläfli in 1852. The formula for $d \leq 3$ was known to L. Euler.

(3.3) Definition. Let P be a d-dimensional polytope. The d-tuple $\big(f_0(P), \dots, f_{d-1}(P)\big)$, where $f_i(P)$ is the number of i-dimensional faces of P, is called the *f-vector* of P. By convention, $f_{-1}(P) = f_d(P) = 1$.

PROBLEMS.

1. Let $P \subset \mathbb{R}^d$ be a d-dimensional polytope and let $H \subset \mathbb{R}^d$ be an affine hyperplane passing through an interior point of P and not containing any of the vertices of P. Let H_+ be an open halfspace bounded by H and let f_i^+ be the number of i-dimensional faces contained in H_+. Prove that

$$\sum_{i=0}^{d-1} (-1)^i f_i^+ = 1.$$

2. Let $P \subset \mathbb{R}^d$ be a d-dimensional polytope and let $\ell : \mathbb{R}^d \longrightarrow \mathbb{R}$ be a linear functional, which takes different values on different vertices of P. For a vertex v of P, let us denote by f_i^v the number of i-dimensional faces F of P such that $\ell(v) = \max\{\ell(x) : x \in F\}$. Prove that

$$\sum_{i=0}^{d-1} (-1)^i f_i^v = \begin{cases} 1 & \text{if } \ell(v) \text{ is the minimum of } \ell \text{ on } P, \\ (-1)^{d-1} & \text{if } \ell(v) \text{ is the maximum of } \ell \text{ on } P, \\ 0 & \text{otherwise.} \end{cases}$$

3. Let P be a d-dimensional polytope and let $F \subset P$ be its k-dimensional face. Let $f_j(F, P)$ denote the number of j-dimensional faces of P that contain F. Prove that

$$\sum_{j=k}^{d-1} (-1)^j f_j(F, P) = (-1)^{d-1}.$$

4. Let $P \subset \mathbb{R}^d$ be an unbounded d-dimensional polyhedron, which does not contain a straight line. Let $f_i^0(P)$ be the number of bounded i-dimensional faces of P, let f_i^∞ be the number of unbounded i-dimensional faces of P and let $f_i(P) = f_i^0(P) + f_i^\infty(P)$ be the total number of i-dimensional faces of P. Prove that

$$\sum_{i=0}^{d-1} (-1)^i f_i^0(P) = 1, \quad \sum_{i=1}^{d} (-1)^{i+1} f_i^\infty(P) = 1 \quad \text{and} \quad \sum_{i=0}^{d} (-1)^i f_i(P) = 0.$$

5. Let $P \subset \mathbb{R}^d$ be a 3-dimensional polytope. Let us define the *curvature* $\kappa(v)$ at a vertex v of P as follows. Let $F_1, \ldots, F_m$ be the facets containing v. Then F_i is a planar polygon and let α_i be the angle at the vertex v of F_i. Let

$$\kappa(v) = 2\pi - \sum_{i=1}^{m} \alpha_i.$$

Prove the *Gauss-Bonnet Formula*:

$$\sum_v \kappa(v) = 4\pi,$$

where the sum is taken over all vertices v of P.

Remark: The general Gauss-Bonnet Formula asserts that the value of the similar sum taken over the vertices of an oriented polyhedral 2-dimensional surface S is $2\pi\chi(S)$, where χ is the Euler characteristic. For smooth surfaces the sum is replaced by an integral.

6. Let us fix $d \geq 1$. In $\mathbb{R}^d$, consider the set of all f-vectors of d-dimensional polytopes. Prove that the affine hull of this set is the hyperplane $\sum_{i=0}^{d-1} (-1)^i f_i = 1 + (-1)^{d-1}$. In other words, prove that there are no linear relations for the numbers f_i of i-dimensional faces other than the Euler-Poincaré Formula, which would hold for all d-dimensional polytopes.

7. A finite set $\mathcal{P} = \{P_i \subset \mathbb{R}^d, i \in I\}$ of distinct polytopes in $\mathbb{R}^d$ is called a *polytopal complex* provided the following two conditions are satisfied: for any two polytopes $P_i, P_j \in \mathcal{P}$, the intersection $P_i \cap P_j$ is a face of both P_i and P_j and if $P_i \in \mathcal{P}$ and F is a face of P_i, then $F \in \mathcal{P}$. The union

$$|\mathcal{P}| = \bigcup_{i \in I} P_i$$

is called the *support* of $\mathcal{P}$.

Let $\mathcal{P}$ be a polytopal complex, let $|\mathcal{P}| \subset \mathbb{R}^d$ be its support and let $f_i(\mathcal{P})$ be the number of i-dimensional polytopes in $\mathcal{P}$. Prove that

$$\chi(|\mathcal{P}|) = \sum_{i=0}^{d} (-1)^i f_i(\mathcal{P}).$$

8. Let $P \subset \mathbb{R}^d$ be a d-dimensional polytope. Prove that

$$(-1)^d [\operatorname{int} P] = \sum_F (-1)^{\dim F} [F],$$

where the sum is taken over all faces F of P, including P.

4. Polytopes with Many Faces: Cyclic Polytopes

Our goal is to construct polytopes with many faces. To this end, we modify the definition of the moment curve; see Section II.9.1.

(4.1) Definition. Let us fix an interval, say, $[0,1] \subset \mathbb{R}^1$. Let

$$q(\tau) = (\tau, \tau^2, \ldots, \tau^d): \quad 0 \leq \tau \leq 1$$

be a curve in $\mathbb{R}^d$. Let us pick n distinct points $0 < \tau_1 < \tau_2 < \ldots < \tau_n < 1$ and let $v_i = q(\tau_i)$ for $i = 1, \ldots, n$. The polytope

$$C(d,n) = \operatorname{conv}(v_1, \ldots, v_n)$$

is called the *cyclic polytope*.

Although the construction depends on how the points $\tau_1 < \tau_2 < \ldots < \tau_n$ are chosen, we will see soon that the facial structure of $C(d,n)$ is independent of the choice of the points as long as their number n is fixed.

PROBLEMS.

1°. Prove that each affine hyperplane $H \subset \mathbb{R}^d$ intersects the moment curve $q(\tau)$ in at most d points.

2°. Prove that $\dim C(d,n) = d$ for $n \geq d+1$.

It turns out that any subset of at most $d/2$ vertices of $C(d,n)$ is the set of vertices of some face of $C(d,n)$. For example, every two vertices of the 4-dimensional polytope $C(4,n)$ are neighbors (the endpoints of an edge).

(4.2) Proposition. *Let $C(d,n) = \operatorname{conv}(v_1, \ldots, v_n)$ be a cyclic polytope. Let $I \subset \{1, \ldots, n\}$ be a set such that $|I| \leq d/2$. Then $F = \operatorname{conv}\big(v_i : i \in I\big)$ is a face of $C(d,n)$.*

Proof. We will find a non-zero vector $c = (\gamma_1, \ldots, \gamma_d)$ and a number α such that

$$\langle c, v_i \rangle = \alpha \quad \text{for} \quad i \in I \qquad \text{and} \qquad \langle c, v_i \rangle < \alpha \quad \text{for} \quad i \notin I.$$

This would clearly imply that F is a face of $C(d, n)$.

Let $I = \{i_1, i_2, \ldots, i_k\}$, hence $k \leq d/2$. Let us consider a polynomial p of degree d in τ:

$$p(\tau) = -\tau^{d-2k}(\tau - \tau_{i_1})^2 \cdots (\tau - \tau_{i_k})^2.$$

Hence $p(\tau_i) = 0$ for $i \in I$ and $p(\tau_i) < 0$ for $i \in \{1, \ldots, n\} \setminus I$. Let us write

$$p(\tau) = \gamma_d \tau^d + \gamma_{d-1}\tau^{d-1} + \ldots + \gamma_1 \tau_1 - \alpha$$

and let $c = (\gamma_1, \ldots, \gamma_d)$. Then

$$\big\langle c,\ q(\tau) \big\rangle = \gamma_1 \tau + \gamma_2 \tau^2 + \ldots + \gamma_d \tau^d = p(\tau) + \alpha.$$

Hence $\langle c, v_i \rangle = \alpha$ for $i \in I$ and $\langle c, v_i \rangle < \alpha$ for $i \notin I$ and the result follows. □

PROBLEMS.

1. Let $P = \operatorname{conv}(v_1, \ldots, v_n) \subset \mathbb{R}^d$ be a d-dimensional polytope with n vertices and let $k > d/2$. Prove that if every k vertices $v_{i_1}, \ldots, v_{i_k}$ are the vertices of a face of P, then $n = d + 1$.

Hint: Use Radon's Theorem (see Theorem I.4.1).

2. Let $I \subset \{1, \ldots, n\}$ be a set, $|I| = d$. Prove that $F = \operatorname{conv}(v_i : i \in I)$ is a facet of $C(d, n)$ if and only if *Gale's evenness condition* is satisfied: every two elements i, j which are not in I are separated by an even number of elements from I.

Figure 73. Example: white dots are not in I; black dots are in I. Gale's condition is satisfied.

3. Describe the faces of the polytope $C(4, n)$.

4*. Prove that for $1 \leq k \leq d - 1$, the number f_k of k-dimensional faces of $C(d, n)$ is

$$f_k = \begin{cases} \sum_{j=1}^{(d+1)/2} \frac{n}{n-j} \binom{n-j}{j} \binom{j}{k+1-j} & \text{if } d \text{ is odd,} \\ \sum_{j=0}^{d/2} \frac{k+2}{n-j} \binom{n-j}{j+1} \binom{j+1}{k+1-j} & \text{if } d \text{ is even.} \end{cases}$$

5. Let us define a closed curve

$$\phi(\tau) = \Big(\sin \tau, \cos \tau, \sin 2\tau, \cos 2\tau, \ldots, \sin k\tau, \cos k\tau\Big), \quad 0 \leq \tau \leq 2\pi$$

in $\mathbb{R}^d$, $d = 2k$. Let $0 < \tau_1 < \tau_2 < \ldots < \tau_n < 2\pi$ and let $v_i = \phi(\tau_i)$. Let $P = \operatorname{conv}(v_1, \ldots, v_n)$. Prove that P and $C(d,n)$ have the same facial structure. In particular, the facial structure of P does not depend on the choice of $0 < \tau_1 < \ldots < \tau_n < 2\pi$ and the cyclic permutation $v_1 \longmapsto v_2 \longmapsto \ldots \longmapsto v_n \longmapsto v_1$ gives rise to a permutation of faces of $C(d,n)$.

The cyclic polytope $C(d,n)$ has the following important property: among all polytopes in $\mathbb{R}^d$ with n vertices, the polytope $C(d,n)$ has the largest number of faces of each dimension $k = 1, \ldots, d-1$. Our next goal is to sketch a proof of this result.

5. Simple Polytopes

In this section, we introduce an important class of polytopes.

(5.1) Definitions. A d-dimensional polytope P is called *simple* if every vertex v of P belongs to exactly d facets of P. In particular, suppose that a polytope $P \subset \mathbb{R}^d$ with $\dim P = d$ is defined as a d-dimensional bounded polyhedron,

$$P = \Big\{x \in \mathbb{R}^d: \ \langle c_i, x\rangle \leq \beta_i \quad \text{for} \quad i = 1, \ldots, m\Big\}.$$

Then P is simple provided for every vertex v of P the set $I(v) = \{i : \langle c_i, v\rangle = \beta_i\}$ of inequalities that are active on v consists of precisely d elements (cf. Theorem II.4.2).

PROBLEMS.

1°. Prove that the cube $I = \big\{x \in \mathbb{R}^d: \ -1 \leq \xi_i \leq 1 \text{ for } i = 1, \ldots, d\big\}$ is a simple polytope and that the (hyper)octahedron $O = \big\{x \in \mathbb{R}^d : |\xi_1| + \ldots + |\xi_d| \leq 1\big\}$ is not simple for $d > 2$.

2. A polytope which is the convex hull of affinely independent points is called a *simplex*. Let $P \subset \mathbb{R}^d$ be a polytope such that $0 \in \operatorname{int} P$. Prove that P is simple if and only if every facet of P° is a simplex (such polytopes are called *simplicial*).

3. Prove that if a polytope is both simple and simplicial (see Problem 2 above), then it is either a simplex or a polygon.

4°. Let us fix a vector $a = (\alpha_1, \ldots, \alpha_d)$ with distinct coordinates and let $P(a)$ be the corresponding permutation polytope (see Section 2). Prove that $P(a)$ is a simple $(d-1)$-dimensional polytope.

5. Prove that the cyclic polytope $C(d,n)$ is simplicial.

6. Let $P \subset \mathbb{R}^d$ be a polytope, let v be a vertex of P and let $u_1, \ldots, u_n$ be the neighbors of P. Suppose that $\ell : \mathbb{R}^d \longrightarrow \mathbb{R}$ is a linear functional such that $\ell(v) > \ell(u_i)$ for $i = 1, \ldots, n$. Prove that the maximum of ℓ on P is attained at v.

Locally, in the neighborhood of every vertex, a simple d-dimensional polytope P looks like the standard non-negative orthant

$$\mathbb{R}^d_+ = \Big\{(\xi_1, \ldots, \xi_d): \quad \xi_i \geq 0 \quad \text{for} \quad i = 1, \ldots, d\Big\}.$$

The following result summarizes some useful properties of simple polytopes.

(5.2) Proposition. *Let $P \subset \mathbb{R}^d$ be a d-dimensional simple polytope. Then*

1. *Every vertex v of P has precisely d neighbors $u_1, \ldots, u_d$.*
2. *For every vertex v of P and for every $k < d$ neighbors $u_{i_1}, \ldots, u_{i_k}$ of v there exists a unique k-dimensional face F of P containing v and $u_{i_1}, \ldots, u_{i_k}$.*
3. *The intersection of any $k \leq d$ facets of P containing v is a $(d-k)$-dimensional face of P.*
4. *Let $\ell : \mathbb{R}^d \longrightarrow \mathbb{R}$ be a linear functional such that $\ell(u_i) < \ell(v)$ for all neighbors u_i of some vertex v of P. Then the maximum of ℓ on P is attained at v.*
5. *Every face of P is a simple polytope.*

Proof. Let

$$P = \Big\{ x \in \mathbb{R}^d : \quad \langle c_i, x \rangle \leq \beta_i \quad \text{for} \quad i = 1, \ldots, m \Big\}$$

and let $v \in P$ be a vertex. Translating P, if necessary, we may assume that $v = 0$ is the origin, and changing the basis, if necessary, we may assume that the inequalities that are active on v are $\xi_i \geq 0$ for $i = 1, \ldots, d$. Thus P is the intersection of the non-negative orthant $\mathbb{R}^d_+ = \big\{ (\xi_1, \ldots, \xi_d) : \ \xi_i \geq 0 \text{ for } i = 1, \ldots, d \big\}$ with finitely many halfspaces containing the origin in their interior.

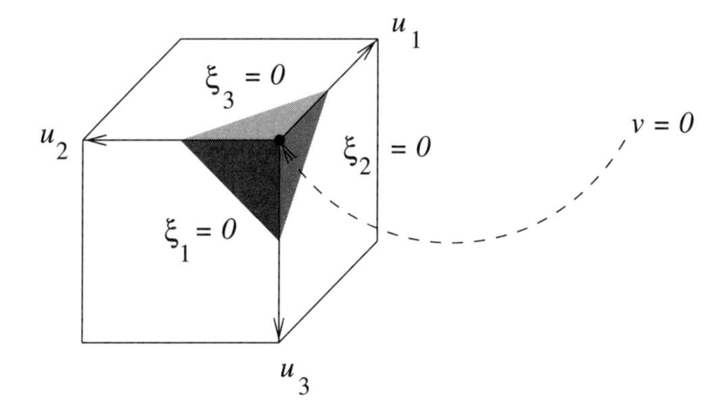

Figure 74

To prove Part 1, note that v belongs to exactly d edges; the i-th edge is the intersection of P with the hyperplane $\sum_{j \neq i} \xi_j = 0$. The other endpoints of these edges are the neighbors $u_1, \ldots, u_d$ of v. Thus all but the i-th coordinate of u_i are 0 and the i-th coordinate of u_i is positive. Let us prove that there are no other edges with the endpoint $v = 0$. Indeed, suppose that $c = (\gamma_1, \ldots, \gamma_d)$ is a vector such that $0 = \langle c, v \rangle \geq \langle c, x \rangle$ for all $x \in P$, so that $F = \big\{ x : \langle c, x \rangle = 0 \big\}$ is a face of P containing v. Substituting $x = u_i$, we get that $\gamma_i \leq 0$ for $i = 1, \ldots, d$. Then $F = \big\{ x \in P : \sum_{i: \gamma_i < 0} \xi_i = 0 \big\}$, which is an edge of P if and only if all but one γ_i are equal to 0.

To prove Part 2, note that F is the intersection of P with the hyperplane $\sum_{j \neq i_1, \ldots, i_k} \xi_j = 0$.

To prove Part 3, we observe that the i-th facet is defined by the equation $\xi_i = 0$. Therefore, the intersection of the facets with the indices $i_1, i_2, \ldots, i_k$ is the face F defined by the equation $\sum_{j=i_1,\ldots,i_k} \xi_j = 0$. Clearly $\dim F = d-k$, since F contains a point where $\xi_j > 0$ for $j \neq i_1, \ldots, i_k$.

To prove Part 4, assume that $\ell(x) = \gamma_1\xi_1 + \ldots + \gamma_d\xi_d$. Hence we have $\ell(u_i) < \ell(v) = 0$, so all γ_i's are negative. Therefore $\ell(x) \leq 0$ for any $x \in \mathbb{R}^d_+$ and $\ell(x) = 0$ if and only if $x = v = 0$.

To prove Part 5, it suffices to prove that if F is a facet of P, then F is a $(d-1)$-dimensional simple polytope. Let us choose a vertex v of F. Then v is a vertex of P and there are exactly d facets $G_1, \ldots, G_d$ of P containing v. Without loss of generality, we may assume that $F = G_d$. As above, we may assume that $v = 0$ and that G_i is defined by the equation $\xi_i = 0$. Then $F \cap G_i$, $i = 1, \ldots, d-1$, are the facets of F containing v (cf. Part 3). □

Note that Part 4 of Proposition 5.2 holds for all (not necessarily simple) polytopes; cf. Problem 6 of Section 5.1.

PROBLEMS.

1*. Let $P \subset \mathbb{R}^d$ be a d-dimensional polytope with n facets. Construct a simple d-dimensional polytope $\tilde{P} \subset \mathbb{R}^d$ with n facets such that $f_i(P) \leq f_i(\tilde{P})$ for $i = 0, \ldots, d-2$, where f_i is the number of i-dimensional faces.

Hint: Suppose that

$$P = \{x : \quad \langle c_i, x\rangle \leq \beta_i \quad \text{for} \quad i = 1, \ldots, n\}.$$

Let

$$\tilde{P} = \{x : \quad \langle c_i, x\rangle \leq \tilde{\beta}_i \quad \text{for} \quad i = 1, \ldots, n\},$$

where $\tilde{\beta}_i$ are generic small perturbations of β_i.

2°. Prove that every face of a simplicial polytope is a simplex (see Problem 2, Section 5.1).

3. Prove that a d-dimensional polytope is simple if every vertex of the polytope has precisely d neighbors.

4. Let P be a simple d-dimensional polytope. Prove that $df_0(P) = 2f_1(P)$.

5. Let P be a 3-dimensional simple polytope. Then the facets of P are polygons and let p_k be the number of k-gons among its facets. Prove that $3p_3 + 2p_4 + p_5 = 12 + \sum_{k\geq 7}(k-6)p_k$.

6*. Let p_k for $k > 2, k \neq 6$ be non-negative integers such that $3p_3 + 2p_4 + p_5 = 12 + \sum_{k\geq 7}(k-6)p_k$. Prove that for some p_6 there exists a simple 3-dimensional polytope whose facets consist of p_k k-gons for $k = 3, 4, 5, \ldots$.

Remark: This is Eberhard's Theorem; see Section 13.3 of [**Gr67**].

6. The h-vector of a Simple Polytope. Dehn-Sommerville Equations

The information about the number of faces of a simple polytope is best encoded by its *h-vector.*

(6.1) Definition. Let P be a d-dimensional simple polytope and let $f_i(P)$ be the number of the i-dimensional faces of P (we agree that $f_d(P) = 1$). Let

$$h_k(P) = \sum_{i=k}^{d} (-1)^{i-k} \binom{i}{k} f_i(P) \quad \text{for} \quad k = 0, \dots, d,$$

where $\displaystyle\binom{m}{n} = \frac{m!}{n!(m-n)!}$ is the binomial coefficient.

The $(d+1)$-tuple $\bigl(h_0(P), \dots, h_d(P)\bigr)$ is called the *h-vector* of P.

(6.2) Lemma. *Let P be a d-dimensional simple polytope. Then*

$$f_i(P) = \sum_{k=i}^{d} \binom{k}{i} h_k(P) \quad \textit{for} \quad i = 0, \dots, d.$$

Furthermore, the numbers $h_k(P)$ are uniquely determined by the above equations.

Proof. Let us introduce two polynomials in one variable τ:

$$\mathbf{f}(\tau) = \sum_{i=0}^{d} f_i(P)\tau^i \quad \text{and} \quad \mathbf{h}(\tau) = \sum_{k=0}^{d} h_k(P)\tau^k.$$

Then the equations of Definition 6.1 are equivalent to the identity $\mathbf{f}(\tau - 1) = \mathbf{h}(\tau)$, whereas the equations of Lemma 6.2 are equivalent to the identity $\mathbf{f}(\tau) = \mathbf{h}(\tau + 1)$. The result now follows. □

We observe that the expession of $f_i(P)$ in terms of $h_k(P)$ is a linear combination with non-negative coefficients. Therefore, a bound for $h_k(P)$ for $k = 0, \dots, d$ easily implies a bound for $f_i(P)$ for $i = 0, \dots, d$. The following is the main result of this section.

(6.3) Theorem. *Let $P \subset \mathbb{R}^d$ be a simple d-dimensional polytope. Let $\ell : \mathbb{R}^d \longrightarrow \mathbb{R}$ be a linear functional such that $\ell(v_i) \neq \ell(v_j)$ for every pair of neighbors v_i, v_j. For a vertex v of P let us define the index of v with respect to ℓ as the number of neighbors v_i of v such that $\ell(v_i) < \ell(v)$. Then for every $0 \leq k \leq d$ the number of vertices of P with index k with respect to ℓ is equal to $h_k(P)$. In particular, this number does not depend on the functional ℓ.*

Proof. Let us count the i-dimensional faces F of P. On one hand, we know that this number is $f_i(P)$. On the other hand, on every i-dimensional face F let us mark the unique vertex $v \in F$ where the maximum of ℓ on F is attained: $\ell(v) = \max_{x \in F} \ell(x)$ (cf. Part 4 of Proposition 5.2). Let us see how many times a given vertex $v \in P$ gets marked.

Suppose that the index of v is equal to $k \geq i$. Then there are exactly k neighbors v_j of v for which $\ell(v_j) < \ell(v)$ (we call them the *lower neighbors* of v as opposed to the $d-k$ neighbors v_i for which $\ell(v_i) > \ell(v)$ and which we call the *upper neighbors* of v). By Part 2 of Proposition 5.2, for any i of the lower neighbors there exists a unique i-dimensional face F containing them and by Part 4 of Proposition 5.2, ℓ attains its maximum on F at v. Conversely, if F is an i-dimensional face such that the maximum of ℓ on F is attained at v, then v should have i neighbors v_j (see Parts 5 and 1 of Proposition 5.2) and for every such neighbor v_j we must have $\ell(v_j) < \ell(v)$. In particular, the index of v with respect to ℓ must be at least i.

Summarizing, we observe that every vertex v of index $k \geq i$ is marked precisely $\binom{k}{i}$ times and that the vertices whose indices are smaller than i are not marked at all. Denoting for a moment by $h_k(P, \ell)$ the number of vertices of P whose index with respect to ℓ is k, we conclude that

$$f_i(P) = \sum_{k=i}^{d} \binom{k}{i} h_k(P, \ell) \quad \text{for} \quad i = 0, \dots, d.$$

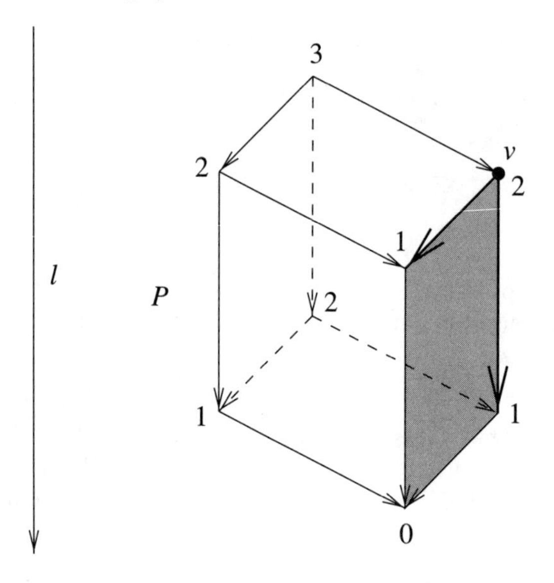

Figure 75. Example: a polytope P, a linear function ℓ and a vertex v. The numbers show the index of the corresponding vertex. The arrows show the direction along which ℓ decreases. We have $h_3(P) = 1$, $h_2(P) = 3$, $h_1(P) = 3$ and $h_0(P) = 1$. There are two edges (bold) and one 2-dimensional face (shaded) with the maximum of ℓ attained at v.

By Lemma 6.2, the numbers $h_k(P, \ell)$ are uniquely determined by the above identities and must coincide with $h_k(P)$. $\square$

(6.4) Corollary (Dehn - Sommerville Equations). *Let P be a simple d-dimensional polytope. Then*

$$h_k(P) = h_{d-k}(P) \quad \textit{for} \quad k = 0, \ldots, d.$$

Proof. Let us choose a linear functional $\ell : \mathbb{R}^d \longrightarrow \mathbb{R}$ such that $\ell(v_i) \neq \ell(v_j)$ for any two neighboring vertices v_i, v_j of P. Then $h_k(P)$ is the number of vertices of P that have index k with respect to ℓ (Theorem 6.3). Applying Theorem 6.3 with the functional $-\ell$, we conclude that the number of vertices of P with index $d - k$ with respect to $-\ell$ is equal to $h_{d-k}(P)$. The proof is completed by the observation that the index of a vertex with respect to ℓ is k if and only if the index of the vertex with respect to $-\ell$ is $d - k$. $\square$

The relations (in a different form) were found by M. Dehn in 1905 for $d \leq 5$ and for an arbitrary d by D. Sommerville in 1927. The idea of the above proof is due to P. McMullen; see [**MSh71**].

PROBLEMS.

1°. Prove that $\sum_{k=0}^{d} h_k(P) = f_0(P)$.

2°. Let P be a simple d-dimensional polytope. Check that the equation $h_0(P) = h_d(P)$ is the Euler-Poincaré Formula for P.

3°. Prove that $h_k(P) \geq 0$ for every simple polytope P.

4. Compute the h-vector of a d-dimensional simplex (the convex hull of $d + 1$ affinely independent points in $\mathbb{R}^d$).

5. Let P be a simple 3-dimensional polytope. Prove that the Dehn-Sommerville equations for P are equivalent to $f_0 - f_1 + f_2 = 2$ and $3f_0 - 2f_1 = 0$. In particular, the number of faces of a simple 3-dimensional polytope is determined by the number of vertices.

6. Let P be a simple 4-dimensional polytope. Prove that the Dehn-Sommerville equations for P are equivalent to $f_0 - f_1 + f_2 - f_3 = 0$ and $f_1 = 2f_0$.

7. Let $I = \{(\xi_1, \ldots, \xi_d) : \; 0 \leq \xi_i \leq 1\}$ be a d-dimensional cube. Prove that $h_k(I) = \binom{d}{k}$ for $k = 0, \ldots, d$.

8. For a permutation σ of the set $\{1, \ldots, n\}$, let us define a *descent* as a number $i = 2, \ldots, n$ such that $\sigma(i) < \sigma(i-1)$. Let $E(n, k)$ be the number of permutations having precisely $k - 1$ descents, $k = 1, \ldots, n$. Let $a = (\alpha_1, \ldots, \alpha_n)$ be a point with distinct coordinates $\alpha_1 > \ldots > \alpha_n$ and let $P = P(a)$ be the corresponding permutation polytope (see Section 2). Then $P(a)$ is a simple $(n-1)$-dimensional polytope (see Problem 4, Section 5.1) and one can define $h_k(P)$ for $k = 0, \ldots, n-1$. Prove that $h_k(P) = E(n, k+1)$ and that $E(n, k) = E(n, n-k+1)$ for $k = 1, \ldots, n$. The numbers $E(n, k)$ are called *Eulerian numbers*.

7. The Upper Bound Theorem

In this section, we sketch a proof of the Upper Bound Theorem, which tells us the maximum number of faces that a polytope of a given dimension and with a given number of vertices (equivalently, facets) may have. The proof below belongs to P. McMullen (1970). More precisely, McMullen gave a detailed proof which is, in a sense, dual to our approach, but he also remarked about what the dual (to his proof, that is, our proof) looks like.

(7.1) Lemma. *Let P be a d-dimensional simple polytope and let F be a facet. Then F is a $(d-1)$-dimensional simple polytope and*

$$h_k(F) \leq h_k(P) \quad \textit{for} \quad k = 0, \ldots, d-1.$$

Moreover, if the intersection of every $k+1$ facets of P is non-empty, then $h_k(F) = h_k(P)$.

Proof. Part 5 of Proposition 5.2 implies that F is a simple polytope. Since F is a facet of P, there is a linear functional ℓ and a number α such that $\ell(x) \geq \alpha$ for any $x \in P$ and $\ell(x) = \alpha$ if and only if $x \in F$. In particular, $\ell(v) > \alpha$ for any vertex v of P that is not in F. Let us perturb $\ell \longmapsto \tilde{\ell}$ slightly so that $\tilde{\ell}$ has different values on different vertices of P and $\tilde{\ell}(v) > \tilde{\ell}(u)$ for every pair of vertices u and v of P, where $v \notin F$ and $u \in F$.

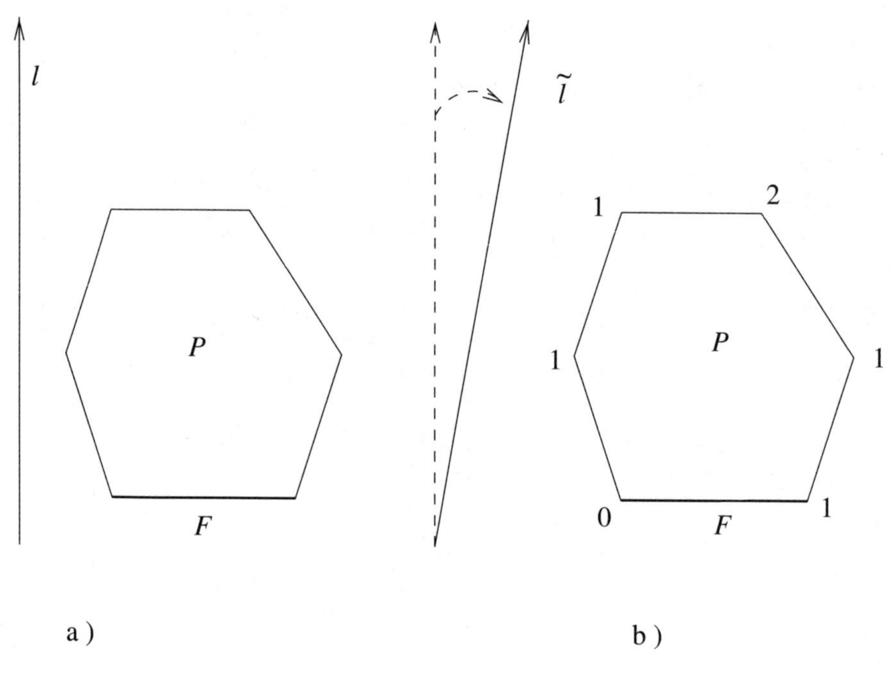

Figure 76. a) A polytope P, a facet F of P and a linear function ℓ attaining its minimum on F, b) a perturbed linear function $\tilde{\ell}$. Now every vertex of the polytope acquires an index.

Theorem 6.3 asserts that $h_k(P)$ (resp. $h_k(F)$) is the number of vertices of P (resp. of F) that have index k with respect to $\tilde{\ell}$. Let v be a vertex of F of index k. Then v has $d-1$ neighbors $u_1, \ldots, u_{d-1}$ in F and precisely k of them produce a smaller value of $\tilde{\ell}$. As a vertex of P, v has one additional neighbor $u_d \notin F$, where we have $\tilde{\ell}(u_d) > \tilde{\ell}(v)$. Therefore, the index of v in P is also k. Hence we proved that $h_k(F) \leq h_k(P)$.

Suppose that the intersection of every k facets of P is non-empty. Suppose that v is a vertex of P of index k and v is *not* in F. There are $d-k$ upper neighbors, say, $u_{k+1}, \ldots, u_d$ of v such that $\tilde{\ell}(u_i) > \tilde{\ell}(v)$ for all $i = k+1, \ldots, d$. Then, by Part 2 of Proposition 5.2, there is $d-k$-dimensional face G containing v and $u_{k+1}, \ldots, u_d$ and by Part 4 it follows that $\tilde{\ell}$ achieves its minimum value on G at v. Moreover, using Part 3 of Proposition 5.2, we can represent G as the intersection of k facets, $G = F_1 \cap \ldots \cap F_k$, where F_i is a facet of P containing v and all but u_i of its lower neighbors. Then the intersection $F \cap F_1 \cap \ldots \cap F_k = F \cap G$ is non-empty, and hence we must have $\tilde{\ell}(x) > \tilde{\ell}(v)$ for some $x \in F$, which is a contradiction; see Figure 76.

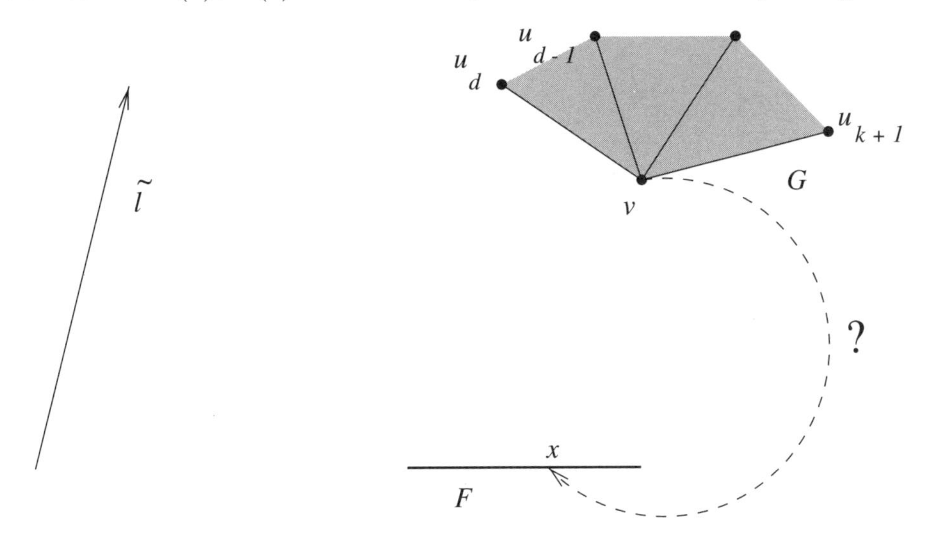

Figure 77. A vertex v, its upper neighbors $u_{k+1}, \ldots, u_d$ and a face G containing them all. For $x \in G \cap F$ we must have $\tilde{\ell}(x) > \tilde{\ell}(v)$, which is a contradiction.

The contradiction shows that there are no points of index k outside of F and hence $h_k(F) = h_k(P)$. □

(7.2) Lemma. *Let P be a simple d-dimensional polytope. Then*

$$\sum_{F \text{ is a facet of } P} h_k(F) = (d-k)h_k(P) + (k+1)h_{k+1}(P).$$

Proof. To prove the formula, let us choose a linear functional ℓ, which attains different values at neighbors in P. By Theorem 6.3, the left-hand side of the formula counts the vertices on the facets of P that have index k relative to the facet. Let us see how many times a given vertex v of P gets counted. There are exactly d facets of P containing v and each such facet F of P contains all but one neighbor of v (see Part 2 of Proposition 5.2).

If the index of v in P with respect to ℓ is smaller than k, then there are fewer than k lower neighbors u_i of v with $\ell(u_i) < \ell(v)$, so v is not counted at all in the left-hand side.

Similarly, if the index of v is greater than $k+1$, then every facet F contains at least $k+1$ lower neighbors u_i of v with $\ell(u_i) < \ell(v)$, and so v is not counted in the left-hand side.

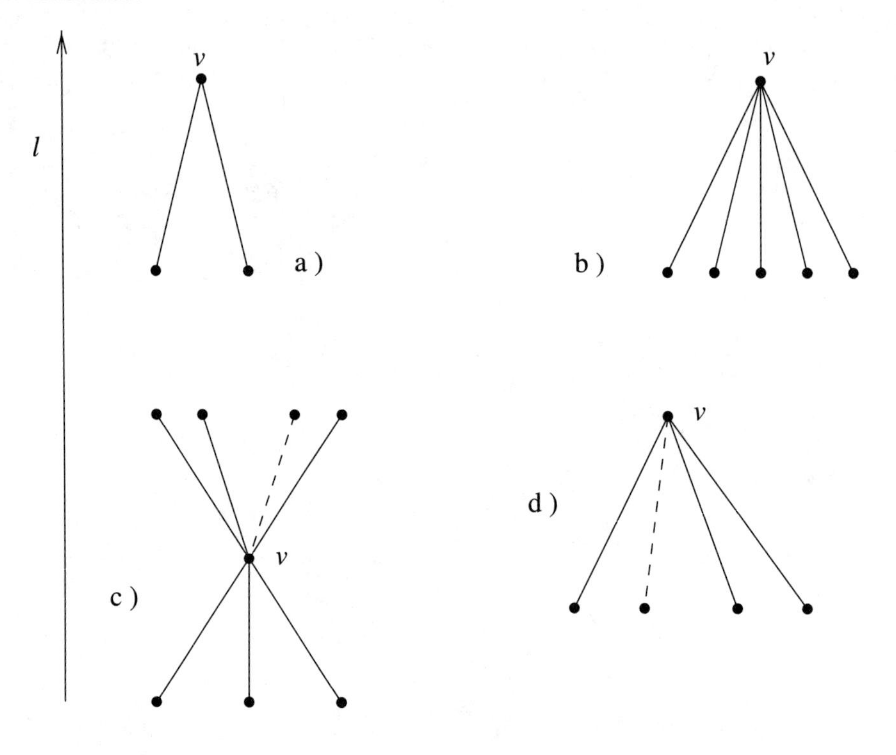

Figure 78. Example: let $k = 3$. In how many ways can we choose a facet F so that the index of v relative to F is k? a) If there are fewer than k lower neighbors, there is no F. b) If there are more than $k+1$ lower neighbors, there is no F. c) If there are k lower neighbors, we can choose F in $d-k$ ways by choosing an upper neighbor not in F. d) If there are $k+1$ lower neighbors, we can choose F in $k+1$ ways by choosing a lower neighbor not in F.

If the index of v in P is k, then there are precisely k lower neighbors $u_1, \ldots, u_k$ of v with $\ell(u_i) < \ell(v)$. Then there are precisely $d-k$ facets F of P that contain v

and all its neighbors except one of $d-k$ upper neighbors. Hence every vertex v of P of the index k is counted $d-k$ times.

Finally, suppose that the index of v in P is $k+1$. Thus v has $k+1$ lower neighbors. Then there are exactly $k+1$ facets F of P such that the index of v relative to the facet is k. The i-th such facet F is determined uniquely by the condition that it contains v and all the neighbors of v except one lower neighbor u_i. Hence a vertex v of P of index $k+1$ is counted $k+1$ times. The proof now follows. □

Summarizing, we get

(7.3) Theorem. *Let P be a d-dimensional simple polytope with n facets. Then*

$$h_k(P) \leq \binom{n-d+k-1}{k} \quad \text{for} \quad k = 0, \ldots, d.$$

Moreover, if the intersection of every k facets of P is non-empty, then

$$h_k(P) = \binom{n-d+k-1}{k}.$$

Proof. Combining Lemma 7.1 and Lemma 7.2, we get

$$nh_i(P) \geq (d-i)h_i(P) + (i+1)h_{i+1}(P)$$

or

$$h_{i+1}(P) \leq \frac{n-d+i}{i+1} h_i(P) \quad \text{for} \quad i = 0, \ldots, d,$$

and the equality holds if every $i+1$ facets of P intersect.

Combining it with $h_0(P) = h_d(P) = 1$, we get the desired inequality. □

Let $C(d,n) \subset \mathbb{R}^d$ be the cyclic polytope and let $Q(d,n)$ be the polar of $C(d,n)$ with the origin chosen in the interior of $C(d,n)$. Hence $Q(d,n)$ is a d-dimensional polytope with n facets and for every $k \leq d/2$, the intersection of any k facets of $Q(d,n)$ is a $(d-k)$-dimensional face of $Q(d,n)$; cf. Proposition 4.2 and Theorem 1.3. Now we prove the celebrated Upper Bound Theorem which asserts that $Q(d,n)$ has the largest number of faces of any dimension among all d-dimensional polytopes with n facets.

(7.4) The Upper Bound Theorem. *Let P be a d-dimensional polytope with n facets and let Q be a d-dimensional simple polytope with n facets, such that for every $k \leq d/2$, the intersection of k facets of Q is a $(d-k)$-dimensional face of Q. Then*

$$f_i(P) \leq f_i(Q) \quad \text{for all} \quad i = 0, \ldots, d-1.$$

Moreover, for $i \leq \lfloor d/2 \rfloor$,

$$f_i(Q) = \sum_{k=i}^{\lfloor d/2 \rfloor} \binom{k}{i}\binom{n-d+k-1}{k} + \sum_{k=\lfloor d/2 \rfloor+1}^{d} \binom{k}{i}\binom{n-k-1}{d-k} \tag{7.4.1}$$

and for $i > \lfloor d/2 \rfloor$,

$$f_i(Q) = \binom{n}{d-i}. \tag{7.4.2}$$

Proof. The equation (7.4.2) is obvious. By Theorem 7.3,

$$h_k(Q) = \binom{n-d+k-1}{k} \quad \text{for} \quad k = 0, \ldots, \lfloor d/2 \rfloor. \tag{7.4.3}$$

Combining this with the Dehn-Sommerville equations $h_k(Q) = h_{d-k}(Q)$ (Corollary 6.4) for $k > d/2$, we get (7.4.1) from Lemma 6.2.

If P is a simple polytope, then from Theorem 7.3 and (7.4.3), we have $h_k(P) \leq h_k(Q)$ for $k \leq d/2$. Combining this with the Dehn-Sommerville equations, we get $h_k(P) \leq h_k(Q)$ for all $k = 0, \ldots, d$. Then by Lemma 6.2, $f_i(P) \leq f_i(Q)$ for all $i = 1, \ldots, d-1$.

The case of an arbitrary polytope P reduces to that of a simple polytope P by Problem 1, Section 5.2. □

PROBLEMS.

1. Let us fix the dimension d. Prove that the number of faces of a polytope $P \subset \mathbb{R}^d$ with n facets is $O(n^{\lfloor d/2 \rfloor})$.

Remark: A simple proof of this fact due to R. Seidel [**Se95**] goes as follows. It suffices to bound the number of vertices of a simple polytope P. Let us choose a generic linear functional ℓ. Then every vertex v of P has either at most $d/2$ lower neighbors or at most $d/2$ upper neighbors with respect to ℓ. Consequently, v is either the highest point or the lowest point of some face F of P with $\dim F \leq d/2$. Therefore, the number of vertices is at most twice as big as the number of faces F with $\dim F \leq d/2$.

2. Deduce the Upper Bound Theorem for d-dimensional polytopes with n vertices.

3. Write the inequalities of the Upper Bound Theorem for a 4-dimensional polytope with n facets and for a 4-dimensional polytope with n vertices.

4*. Let P be a simple d-dimensional polytope. Prove the Unimodality Theorem: $h_0(P) \leq h_1(P) \leq h_2(P) \leq \ldots \leq h_{\lfloor d/2 \rfloor}(P)$.

Remark: The proof was first obtained by R. Stanley using a technique from algebraic geometry [**St80**]. A "convex geometry" proof was later found by P. McMullen in [**Mc93b**] and [**Mc96**].

8. Centrally Symmetric Polytopes

Let $P \subset \mathbb{R}^d$ be a d-dimensional polytope. Suppose that P is symmetric about the origin (or *centrally symmetric*), that is, $P = -P$.

In this section, we prove that a centrally symmetric polytope must have either many vertices or many facets or both. Apart from the result itself, the method of proof is very interesting. For the first time in this chapter we rely on metric and measure arguments.

The following result was obtained by T. Figiel, J. Lindenstrauss and V.D. Milman [**FLM77**].

(8.1) Proposition. *There exists a constant $\gamma > 0$ such that for any centrally symmetric d-dimensional polytope P with $|V|$ vertices and $|F|$ facets one has*

$$\ln|V| \cdot \ln|F| \geq \gamma d.$$

Proof. Let P be a d-dimensional centrally symmetric polytope. Applying a suitable linear transformation, we may assume that the maximum volume ellipsoid of P is the unit ball $B = \{x : \|x\| \leq 1\}$; see Section V.2. Hence, by Theorem V.2.5, $B \subset P \subset \sqrt{d}B$.

Let $\mathbb{S}^{d-1} = \{x \in \mathbb{R}^d : \|x\| = 1\}$ be the unit sphere and let ν be the Haar probability measure on $\mathbb{S}^{d-1}$; see Section V.7.

For $x \neq 0$ let us denote by $\overline{x} = x/\|x\|$ the radial projection of x onto the unit sphere $\mathbb{S}^{d-1}$. Let us choose some $\epsilon > 0$ and consider the set $A_\epsilon \subset \mathbb{S}^{d-1}$ defined as follows:

$$A_\epsilon = \Big\{c \in \mathbb{S}^{d-1} : \langle c, \overline{v} \rangle \geq \epsilon \quad \text{for some} \quad v \in V\Big\}.$$

By Lemma V.7.1, we have

$$\nu(A_\epsilon) \leq 2|V| \exp\big\{-\epsilon^2 d/16\big\}.$$

We observe that for every $c \in \mathbb{S}^{d-1} \setminus A_\epsilon$

$$\max\{\langle c, x \rangle : x \in P\} = \max\{\langle c, v \rangle : v \in V\} \leq \sqrt{d} \max\{\langle c, \overline{v} \rangle : v \in V\} \leq \epsilon\sqrt{d}.$$

We want to choose an $\epsilon > 0$ so that $\nu(A_\epsilon) \leq 1/3$. We can take, say,

$$\epsilon = \gamma_1 d^{-1/2} \ln^{1/2} |V|,$$

where $\gamma_1 > 0$ is a sufficiently large number. Then

$$\max\{\langle c, x \rangle : x \in P\} \leq \gamma_1 \ln^{\frac{1}{2}} |V| \quad \text{for every} \quad c \in \mathbb{S}^{d-1} \setminus A_\epsilon. \tag{8.1.1}$$

Suppose that P is given by a system of linear inequalities

$$P = \Big\{x \in \mathbb{R}^d : \langle a_i, x \rangle \leq \alpha_i \quad \text{for} \quad i = 1, \ldots, |F|\Big\}$$

for some $a_i \in \mathbb{R}^d$. Since P contains the unit ball we must have $\alpha_i > 0$ and, therefore, using a proper scaling, we can assume that $\alpha_i = 1$ for all i. Since the point $a_i/\|a_i\| = \overline{a_i}$ is in P, we must have $\|a_i\| \leq 1$ for $i = 1, \ldots, |F|$. Thus

$$P = \Big\{x \in \mathbb{R}^{d-1} : \langle a_i, x \rangle \leq 1 \quad \text{for} \quad i = 1, \ldots, |F|\Big\}$$
$$\text{where} \quad \|a_i\| \leq 1 \quad \text{for} \quad i = 1, \ldots, |F|.$$

Let us choose some $\delta > 0$ and consider the set $B_\delta \subset \mathbb{S}^{d-1}$ defined as follows:

$$B_\delta = \Big\{c \in \mathbb{S}^{d-1} : \ \langle c, \overline{a_i} \rangle \geq \delta \quad \text{for some} \quad i = 1, \ldots, |F|\Big\}.$$

By Lemma V.7.1, we get

$$\nu(B_\delta) \leq 2|F| \exp\{-\delta^2 d/16\}.$$

Let us choose a $c \in \mathbb{S}^{d-1} \setminus B_\delta$. Since $\|a_i\| \leq 1$, we conclude that $\langle c, a_i \rangle \leq \delta$ for $i = 1, \ldots, |F|$. Therefore, the vector $x = \delta^{-1} c$ satisfies the system of linear inequalities $\langle a_i, x \rangle \leq 1$ and hence belongs to the polytope P. Hence for every $c \in \mathbb{S}^{d-1} \setminus B_\delta$ we have

$$\max\{\langle c, x \rangle : \ x \in P\} \geq \delta^{-1}.$$

Now we want to choose a $\delta > 0$ so that $\nu(B_\delta) \leq 1/3$. We can take, say,

$$\delta = \gamma_2 d^{-1/2} \ln^{1/2} |F|,$$

where $\gamma_2 > 0$ is a sufficiently large number. Then

$$\max\{\langle c, x \rangle : \ x \in P\} \geq \gamma_2^{-1} d^{\frac{1}{2}} \ln^{-\frac{1}{2}} |F| \quad \text{for every} \quad c \in \mathbb{S}^{d-1} \setminus B_\delta. \tag{8.1.2}$$

Since $\nu(A_\epsilon) \leq 1/3$ and $\nu(B_\delta) \leq 1/3$, there is a point $c \notin A_\epsilon \cup B_\delta$ and for such c both (8.1.1) and (8.1.2) are satisfied. Thus we have

$$\gamma_2^{-1} d^{\frac{1}{2}} \ln^{-\frac{1}{2}} |F| \leq \gamma_1 \ln^{\frac{1}{2}} |V|,$$

or, in other words,

$$\ln |V| \cdot \ln |F| \geq \gamma d,$$

where $\gamma = (\gamma_1 \gamma_2)^{-2}$. □

PROBLEMS.

1. Prove that there exists $\gamma > 0$ such that for any $k \leq l \leq d$ and for any centrally symmetric d-dimensional polytope P we have $\ln f_l \cdot \ln f_k \geq \gamma(l-k)$, where f_i is the number of i-dimensional faces of P.

2. Let $B \subset \mathbb{R}^d$ be the unit ball centered at the origin. Prove that there exists an absolute constant $\gamma > 0$ such that for any polytope $P \subset \mathbb{R}^d$ with the property that $B \subset P \subset \rho B$ one has

$$\ln |V| \cdot \ln |F| \geq \gamma d^2 / \rho^2,$$

where $|V|$ is the number of vertices of P and $|F|$ is the number of facets of P.

Here are a couple of open problems.

3*. Is it true that any d-dimensional centrally symmetric polytope has at least 3^d faces?

4*. What is the maximal number of edges that a centrally symmetric 4-dimensional polytope can have?

9. Remarks

For combinatorial theory of polytopes, see [**Gr67**], [**MSh71**], [**Brø83**], [**Z95**] and [**YKK84**]. For the structure of the permutation polytope and related polytopes, see [**BS96**] and [**YKK84**]. Rado's Theorem (Theorem 2.3) and its ramifications are discussed in [**YKK84**] and [**MO79**]. A simple self-contained proof of the Euler-Poincaré Formula can be found in [**L97**]. Our discussion of simple polytopes (Sections 5–7) follows [**Brø83**]. The necessary and sufficient algebraic conditions for a d-tuple $(f_0, \ldots, f_{d-1})$ to be the f-vector of a simple (or simplicial) polytope are known. They were conjectured by P. McMullen, the necessity part was established by R.P. Stanley [**St80**] and the sufficiency part was proved by L.J. Billera and C.W. Lee [**BL81**].

Chapter VII

Lattices and Convex Bodies

We discuss some discrete aspects of convexity. We define a lattice in Euclidean space and explore how lattices interact with convex bodies. In this chapter, we focus on metric rather than combinatorial aspects of this interaction. The landmark results of this chapter are the theorems of Minkowski and Minkowski-Hlawka with applications to number theory problems and construction of dense sphere packings (which are related to coding), flatness results and the Lenstra-Lenstra-Lovász basis reduction algorithm. Problems address properties of some particular lattices, such as $\mathbb{Z}^n$, D_n, A_n, E_6, E_7 and E_8; results on enumeration of lattice points in convex bodies, such as Pick's Formula and its extensions; and other interesting results, such as Doignon's lattice version of Helly's Theorem.

1. Lattices

We define the main object of this chapter.

(1.1) Definitions. A set $\Lambda \subset \mathbb{R}^d$ is called an (additive) *subgroup* of $\mathbb{R}^d$ provided

- $0 \in \Lambda$,
- $x + y \in \Lambda$ for any two $x, y \in \Lambda$ and
- $-x \in \Lambda$ for any $x \in \Lambda$.

A subgroup $\Lambda \subset \mathbb{R}^d$ is called *discrete* provided there is an $\epsilon > 0$ such that the ball $B(0, \epsilon)$ of radius ϵ centered at the origin does not contain any non-zero lattice point:

$$B(0,\epsilon) \cap \Lambda = \{0\} \quad \text{where} \quad B(0,\epsilon) = \{x : \ \|x\| \leq \epsilon\} \quad \text{for some} \quad \epsilon > 0.$$

A discrete subgroup Λ of $\mathbb{R}^d$ such that $\operatorname{span}(\Lambda) = \mathbb{R}^d$ is called a *lattice*. The dimension d is called the *rank* of Λ and denoted $\operatorname{rank} \Lambda$.

Given a lattice $\Lambda \subset \mathbb{R}^d$, a set of linear independent vectors $u_1, \ldots, u_d \in \Lambda$ is called a *basis* of Λ if every $x \in \Lambda$ can be written in the form $x = \mu_1 u_1 + \ldots + \mu_d u_d$, where $\mu_i \in \mathbb{Z}$ for $i = 1, \ldots, d$.

Here are some examples.

(1.2) Examples.

(1.2.1) Let $\mathbb{Z}^d \subset \mathbb{R}^d$ be the set of points with integer coordinates:

$$\mathbb{Z}^d = \Big\{(\xi_1, \ldots, \xi_d) : \ \xi_i \in \mathbb{Z} \quad \text{for} \quad i = 1, \ldots, d\Big\}.$$

The lattice $\mathbb{Z}^d$ is called the *standard integer lattice.*

(1.2.2) Let us identify $\mathbb{R}^d$ with the hyperplane

$$H = \Big\{(\xi_1, \ldots, \xi_{d+1}) \in \mathbb{R}^{d+1} : \ \xi_1 + \ldots + \xi_{d+1} = 0\Big\}$$

in $\mathbb{R}^{d+1}$. Let $A_d = \mathbb{Z}^{d+1} \cap H$, so $A_d \subset \mathbb{R}^d$ is a lattice.

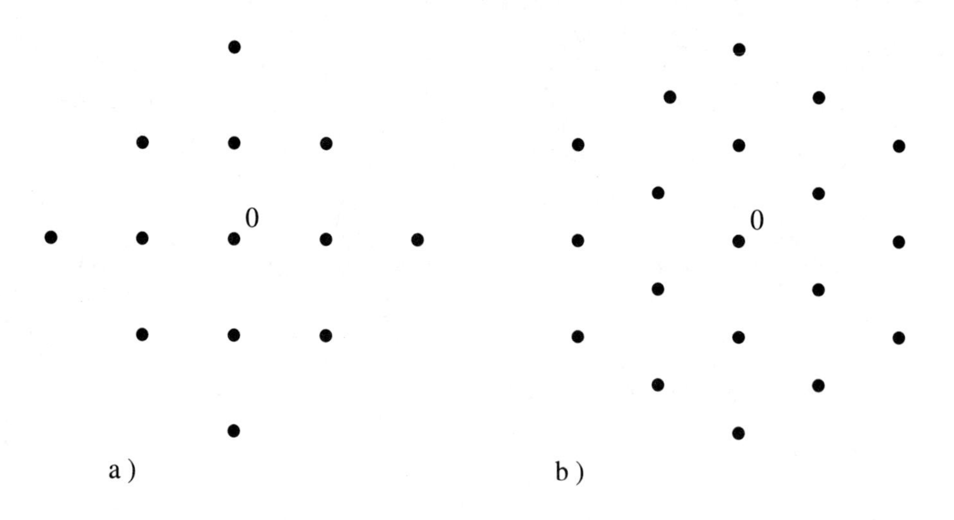

Figure 79. Lattices a) $\mathbb{Z}^2$ and b) A_2

(1.2.3) Let us define $D_n \subset \mathbb{Z}^n$,

$$D_n = \Big\{(\xi_1, \ldots, \xi_n) : \quad \xi_i \in \mathbb{Z} \quad \text{for} \quad i = 1, \ldots, n \quad \text{and} \\ \xi_1 + \ldots + \xi_n \quad \text{is an even integer}\Big\}.$$

(1.2.4) Let n be an even number and let $x_0 = (1/2, \ldots, 1/2) \in \mathbb{R}^n$. Let us define $D_n^+ = D_n \cup (D_n + x_0)$, where $D_n + x_0 = \{x + x_0 : \ x \in D_n\}$ is a shift of D_n (check that if n is odd, then the set D_n^+ so defined is *not* a lattice). The lattice D_8^+ is special and is called E_8.

(1.2.5) Let us identify $\mathbb{R}^7$ with the hyperplane

$$H = \Big\{(\xi_1, \ldots, \xi_8) : \quad \xi_1 + \ldots + \xi_8 = 0\Big\}$$

in $\mathbb{R}^8$. We define $E_7 = E_8 \cap H$, so E_7 is a lattice in $\mathbb{R}^7$.

(1.2.6) Let us identify $\mathbb{R}^6$ with the subspace $L \subset \mathbb{R}^8$,

$$L = \Big\{(\xi_1, \ldots, \xi_8) : \quad \xi_1 + \xi_8 = \xi_2 + \ldots + \xi_7 = 0\Big\}.$$

We define $E_6 = E_8 \cap L$, so E_6 is a lattice in $\mathbb{R}^6$.

PROBLEMS.

1°. Prove that a set $\Lambda \subset \mathbb{R}^n$ is a subgroup if and only if Λ is non-empty and $x - y \in \Lambda$ for any two $x, y \in \Lambda$.

2°. Let $B(x_0, \epsilon) = \{x : \|x - x_0\| \leq \epsilon\}$ be the ball centered at x_0 of radius ϵ. Let $\Lambda \subset \mathbb{R}^d$ be a lattice. Prove that if $B(0, \epsilon) \cap \Lambda = \{0\}$, then $B(x_0, \epsilon) \cap \Lambda = \{x_0\}$ for any $x_0 \in \Lambda$.

3°. Let $\Lambda \subset \mathbb{R}^d$ be a lattice and let $B(x_0, \rho) = \{x : \|x - x_0\| \leq \rho\}$ be a ball. Prove that $B \cap \Lambda$ is a finite set.

4°. Construct a subgroup $\Lambda \subset \mathbb{R}^m$ which is not discrete.

5. Let $u_1, \ldots, u_d \in \mathbb{R}^d$ be linearly independent vectors. Let

$$\Lambda = \Big\{\mu_1 u_1 + \ldots + \mu_d u_d : \quad \mu_i \in \mathbb{Z} \quad \text{for} \quad i = 1, \ldots, d\Big\}.$$

Prove that Λ is a lattice.

Hint: Construct an invertible linear transformation

$$T : \mathbb{R}^d \longrightarrow \mathbb{R}^d$$

such that $T(\mathbb{Z}^d) = \Lambda$. Then T^{-1} is an invertible linear transformation such that $T^{-1}(\Lambda) = \mathbb{Z}^d$. In particular, T^{-1} is continuous.

6. Check that $\mathbb{Z}^d$, A_d, D_n, D_n^+, E_8, E_7 and E_6 are indeed lattices.

7°. Draw pictures of D_2 and A_2.

8°. Construct a basis of $\mathbb{Z}^d$.

9. Prove that

$$\begin{aligned}
&u_1 = (2,0,0,0,0,0,0,0), \quad u_2 = (-1,1,0,0,0,0,0,0),\\
&u_3 = (0,-1,1,0,0,0,0,0), \quad u_4 = (0,0,-1,1,0,0,0,0),\\
&u_5 = (0,0,0,-1,1,0,0,0), \quad u_6 = (0,0,0,0,-1,1,0,0),\\
&u_7 = (0,0,0,0,0,-1,1,0) \quad \text{and} \quad u_8 = \left(\frac{1}{2},\frac{1}{2},\frac{1}{2},\frac{1}{2},\frac{1}{2},\frac{1}{2},\frac{1}{2},\frac{1}{2}\right)
\end{aligned}$$

is a basis of E_8.

10°. Let $u_1, \ldots, u_d$ be a basis of a lattice $\Lambda \subset \mathbb{R}^d$. Let us choose a pair of indices $i \neq j$ and let $u_i' = u_i + \alpha u_j$, where $\alpha \in \mathbb{Z}$. Prove that $u_1, \ldots, u_{i-1}, u_i', u_{i+1}, \ldots, u_d$ is a basis of Λ.

11. A set $S \subset \mathbb{R}^d$ is called a *semigroup* provided $x+y \in S$ for any two $x, y \in S$. A set $X \subset S$ is called a set of *generators* of S if and only if every $s \in S$ can be represented as a finite sum $s = \sum_X \mu_x x$, where μ_x are non-negative integers. Prove that every semigroup $S \subset \mathbb{Z}$ possesses a finite set of generators and construct a semigroup $S \subset \mathbb{Z}^2$ which does not have a finite set of generators.

Our first goal is to prove that every lattice has a basis. In fact, we will obtain the stronger result that for any set of linearly independent lattice vectors $b_1, \ldots, b_d \in \Lambda$ with $d = \operatorname{rank}\Lambda$ there is a basis $u_1, \ldots, u_d$ of Λ which is "reasonably close" to $b_1, \ldots, b_d$. To this end, let us invoke the distance function in $\mathbb{R}^d$:

$$\operatorname{dist}(x,y) = \|x-y\| \quad \text{for any two points} \quad x, y \in \mathbb{R}^d$$

and let

$$\operatorname{dist}(x,A) = \inf_{y \in A} \operatorname{dist}(x,y) \quad \text{for a point} \quad x \in \mathbb{R}^d \quad \text{and a set} \quad A \subset \mathbb{R}^d.$$

First, we prove that given a subspace spanned by lattice points which does not contain the entire lattice, there is a lattice point with the minimum possible positive distance to the subspace. In fact, the exact structure of the distance function is not important here. It could have come from any norm in $\mathbb{R}^d$; see Section V.3.

We will use the following notation:

For a real number ξ, let $\lfloor \xi \rfloor$ denote the largest integer not exceeding ξ and let $\{\xi\} = \xi - \lfloor \xi \rfloor$, so $0 \leq \{\xi\} < 1$ for all ξ.

(1.3) Lemma. *Let $\Lambda \subset \mathbb{R}^d$ be a lattice and let $b_1, \ldots, b_k \in \Lambda$, $k < d$, be linearly independent points. Let $L = \operatorname{span}(b_1, \ldots, b_k)$. Then there exists a point $v \in \Lambda \setminus L$ and a point $x \in L$ such that*

$$\operatorname{dist}(v,x) \leq \operatorname{dist}(w,y) \quad \textit{for every} \quad w \in \Lambda \setminus L \quad \textit{and every} \quad y \in L.$$

In words: among all lattice points not in L, there exists a point closest to L.

Proof. Let Π be the parallelepiped spanned by $b_1, \ldots, b_k$:

$$\Pi = \Bigl\{\sum_{i=1}^k \alpha_i b_i : \quad 0 \leq \alpha_i \leq 1 \quad \text{for} \quad i = 1, \ldots, k\Bigr\}.$$

Then Π is a compact set. We claim that among all lattice points not in L, there is a point v which is closest to Π.

Indeed, let us choose any $a \in \Lambda \setminus L$ and let $\rho = \operatorname{dist}(a, \Pi) > 0$. Let us consider the ρ-neighborhood of Π:

$$\Pi_\rho = \Bigl\{x \in \mathbb{R}^d : \quad \operatorname{dist}(x, \Pi) \leq \rho\Bigr\}.$$

Then Π_ρ is a bounded set and since Λ is discrete, the intersection $\Pi_\rho \cap \Lambda$ is finite (Problem 3, Section 1.2). Moreover, there are points in $\Pi_\rho \cap \Lambda$ which are not contained in L (such is, for example, the point a).

Let us choose a lattice point $v \in \Pi_\rho \setminus L$ which is closest to Π:

$$\operatorname{dist}(v, \Pi) \leq \operatorname{dist}(w, \Pi) \quad \text{for every lattice point} \quad w \in \Pi_\rho \setminus L.$$

Let $x \in \Pi$ be a point such that

$$\operatorname{dist}(v, x) = \operatorname{dist}(v, \Pi).$$

We claim that v and x satisfy the requirements of the lemma.

Indeed, let us choose any $w \in \Lambda \setminus L$ and any $y \in L$. We can write

$$y = \sum_{i=1}^k \gamma_i b_i \quad \text{for some real} \quad \gamma_i.$$

We observe that

$$z = \sum_{i=1}^k \lfloor \gamma_i \rfloor b_i \quad \text{and} \quad w - z$$

are lattice points, that $w - z \notin L$ and that

$$y - z = \sum_{i=1}^k \{\gamma_i\} b_i$$

is a point from Π. Therefore,

$$\operatorname{dist}(w, y) = \operatorname{dist}\bigl(w - z,\ y - z\bigr) \geq \operatorname{dist}\bigl(w - z,\ \Pi\bigr) \geq \operatorname{dist}\bigl(v,\ \Pi\bigr) = \operatorname{dist}(v, x)$$

and the result follows. □

PROBLEMS.

1. Let $\Lambda \subset \mathbb{R}^d$ be a lattice, let $b_1, \ldots, b_k \in \Lambda$, $k < d+1$, be lattice points and let A be the affine hull of $\{b_1, \ldots, b_k\}$. Prove that there is a lattice point with the minimum possible positive distance to A.

2. Give an example of a straight line $L \subset \mathbb{R}^2$ such that

$$\inf_{x \in \mathbb{Z}^2 \setminus L} \operatorname{dist}(x, L) = 0.$$

3°. Let $\Lambda \subset \mathbb{R}^d$ be a lattice, let $u_1, \ldots, u_m \in \Lambda$ be lattice points and let $L = \operatorname{span}(u_1, \ldots, u_m)$. Let us consider the orthogonal projection $pr : \mathbb{R}^d \longrightarrow L^\perp$ onto the orthogonal complement $L^\perp$ of L. Prove that the image $\Lambda_1 = pr(\Lambda)$ is a lattice in $L^\perp$.

4°. Let $\Lambda \subset \mathbb{R}^d$ be a lattice and let $b_1, \ldots, b_d \in \Lambda$ be linearly independent lattice points. Let $L_0 = \{0\}$ and $L_k = \operatorname{span}(b_1, \ldots, b_k)$ for $k = 1, \ldots, d$. Prove that for $k = 1, \ldots, d$ there exists a lattice point $u_k \in L_k \setminus L_{k-1}$ closest to L_{k-1}.

Now we are ready to prove that every lattice of positive rank has a basis. In fact, we prove a stronger result.

(1.4) Theorem. *Let $d > 0$ and let $\Lambda \subset \mathbb{R}^d$ be a lattice. Let $b_1, \ldots, b_d \in \Lambda$ be linearly independent lattice vectors. Let us define subspaces $\{0\} = L_0 \subset L_1 \subset \ldots \subset L_d \subset \mathbb{R}^d$ by*

$$L_k = \operatorname{span}(b_1, \ldots, b_k) \quad \textit{for} \quad k = 1, \ldots, d.$$

For $k = 1, \ldots, d$ let u_k be a lattice point in $L_k \setminus L_{k-1}$ closest to L_{k-1}. Then $u_1, \ldots, u_d$ is a basis of Λ. In particular, every lattice of positive rank has a basis.

Proof. First, we observe that by Problem 4 of Section 1.3 such points $u_1, \ldots, u_d$ indeed exist. Let $\Lambda_k = \Lambda \cap L_k$. We conclude that Λ_k is a lattice in L_k.

We prove by induction on k that $u_1, \ldots, u_k$ is a basis of Λ_k. For $k = 1$, we have

$$u_1 = \alpha_1 b_1 \quad \text{for some} \quad \alpha_1 \neq 0.$$

Let $v \in \Lambda_1 \setminus \{0\}$ be a point. Then

$$v = \beta b_1 \quad \text{for some} \quad \beta \in \mathbb{R}.$$

We claim that $\mu = \beta/\alpha_1$ is an integer. Otherwise, $0 < \{\mu\} < 1$ and the lattice point

$$u_1' = v - \lfloor \mu \rfloor u_1 = v - \mu u_1 + \{\mu\} u_1 = \{\mu\} u_1 \in \Lambda_1 \setminus \{0\}$$

is closer to the origin than u_1, which is a contradiction. Thus $v = \mu u_1$ for some $\mu \in \mathbb{Z}$ and hence u_1 is a basis of Λ_1.

Suppose that $k > 1$. For a point

$$x = \sum_{i=1}^{k} \gamma_i b_i, \quad x \in L_k,$$

we have

$$\operatorname{dist}(x,\ L_{k-1}) = \operatorname{dist}(\gamma_k b_k,\ L_{k-1}) = |\gamma_k| \operatorname{dist}(b_k,\ L_{k-1}).$$

We have

$$u_k = \sum_{i=1}^{k} \alpha_i b_i \quad \text{for some real} \quad \alpha_i \quad \text{such that} \quad \alpha_k \neq 0.$$

Let $v \in \Lambda_k$ be a point. Then

$$v = \sum_{i=1}^{k} \beta_i b_i \quad \text{for some} \quad \beta_i \in \mathbb{R}.$$

We claim that $\mu = \beta_k / \alpha_k$ is an integer. Otherwise, $0 < \{\mu\} < 1$ and the lattice point

$$u_k' = v - \lfloor \mu \rfloor u_k = v - \mu u_k + \{\mu\} u_k = \{\mu\} \alpha_k b_k + \sum_{i=1}^{k-1} (\beta_i - \lfloor \mu \rfloor \alpha_{ki}) b_i$$

is a point from $\Lambda_k \setminus \Lambda_{k-1}$ which is closer to L_{k-1} than u_k, which is a contradiction. Thus $\mu \in \mathbb{Z}$ and $v - \mu u_k \in \Lambda_{k-1}$. Applying the induction hypothesis, we conclude that v is an integer linear combination of $u_1, \ldots, u_k$ and the result follows. □

PROBLEMS.

1. Construct bases of A_n, D_n, D_n^+, E_6 and E_7 (see Example 1.2).

2°. Let $u_1, \ldots, u_d$ be a basis of a lattice $\Lambda \subset \mathbb{R}^d$, let $L_k = \operatorname{span}(u_1, \ldots, u_k)$ for $k = 1, \ldots, d$ and let $L_0 = \{0\}$. Prove that

$$\|v\| \geq \min_{k=1,\ldots,d} \operatorname{dist}(u_k,\ L_{k-1})$$

for every vector $v \in \Lambda \setminus \{0\}$.

3. Let $\Lambda \subset \mathbb{R}^2$ be a lattice. Prove that there exists a basis u_1, u_2 of Λ such that the angle between u_1 and u_2 is between $60°$ and $90°$.

Hint: Choose u_1 to be a shortest non-zero lattice vector and u_2 to be a shortest lattice vector such that u_1, u_2 is a basis of Λ.

4. Let $\Lambda \subset \mathbb{R}^d$ be a lattice and let $b_1, \ldots, b_d \in \Lambda$ be linearly independent vectors. Prove that there exists a basis $u_1, \ldots, u_d$ of Λ such that

$$u_k = \sum_{i=1}^{k} \alpha_{ki} b_i \quad \text{where} \quad 0 \leq \alpha_{ki} \leq 1 \quad \text{for} \quad i = 1, \ldots, k$$

for $k = 1, \ldots, d$.

5. Let $\Lambda \subset \mathbb{R}^d$ be a lattice and let $b_1, \ldots, b_d \in \Lambda$ be linearly independent vectors. Prove that there exists a basis $u_1, \ldots, u_d$ of Λ such that

$$u_k = \sum_{i=1}^{k} \alpha_{ki} b_i \quad \text{where} \quad 0 \leq \alpha_{kk} \leq 1 \quad \text{and} \quad |\alpha_{ik}| \leq 1/2 \quad \text{for} \quad i = 1, \ldots, k-1$$

for $k = 1, \ldots, d$.

The following convex set plays a crucial role in what follows.

(1.5) Definition. Let $\Lambda \subset \mathbb{R}^d$ be a lattice and let $u_1, \ldots, u_d$ be a basis of Λ. The set

$$\Pi = \left\{ \sum_{i=1}^d \alpha_i u_i : \quad 0 \leq \alpha_i < 1 \quad \text{for} \quad i = 1, \ldots d \right\}$$

is called *the fundamental parallelepiped* of the basis $u_1, \ldots, u_d$ and *a fundamental parallelepiped* of the lattice Λ.

PROBLEM.

1°. Prove that the fundamental parallelepiped is a convex bounded set.

2. The Determinant of a Lattice

We proceed to define an important metric invariant of a lattice.

(2.1) Lemma. *Let $\Lambda \subset \mathbb{R}^d$ be a lattice and let Π be a fundamental parallelepiped of Λ. Then, for each point $x \in \mathbb{R}^d$ there exist unique $v \in \Lambda$ and $y \in \Pi$ such that $x = v + y$.*

Proof. Suppose that Π is the fundamental parallelepiped of a basis $u_1, \ldots, u_d$ of Λ. Then $u_1, \ldots, u_d$ is a basis of $\mathbb{R}^d$ and every point x can be written as $x = \alpha_1 u_1 + \ldots + \alpha_d u_d$ for some real $\alpha_1, \ldots, \alpha_d$. Let

$$v = \sum_{i=1}^d \lfloor \alpha_i \rfloor u_i \quad \text{and} \quad y = \sum_{i=1}^d \{\alpha_i\} u_i,$$

where $\lfloor \cdot \rfloor$ is the integer part and $\{\cdot\}$ is the fractional part of a number. Clearly, $v \in \Lambda$, $y \in \Pi$ and $x = v + y$.

Suppose that there are two decompositions $x = v_1 + y_1$ and $x = v_2 + y_2$, where $v_1, v_2 \in \Lambda$ and

$$y_1 = \sum_{i=1}^d \alpha_i u_i \quad \text{and} \quad y_2 = \sum_{i=1}^d \beta_i u_i, \quad \text{where} \quad 0 \leq \alpha_i, \beta_i < 1 \quad \text{for } i = 1, \ldots, d.$$

Then

$$v_1 - v_2 = y_2 - y_1 = \sum_{i=1}^d \gamma_i u_i, \quad \text{where} \quad \gamma_i = \beta_i - \alpha_i.$$

We observe that $|\gamma_i| < 1$ for $i = 1, \ldots, d$ and that $v_1 - v_2 \in \Lambda$. Since $u_1, \ldots, u_d$ is a basis of Λ, the numbers γ_i must be integers. Since $|\gamma_i| < 1$, we conclude that $\gamma_i = 0$ and $\alpha_i = \beta_i$ for $i = 1, \ldots, d$. Therefore, $y_2 = y_1$ and hence $v_2 = v_1$. □

(2.2) Corollary. *Let $\Lambda \subset \mathbb{R}^d$ be a lattice and let Π be a fundamental parallelepiped of Λ. Then the translates $\{\Pi + u : u \in \Lambda\}$ cover the whole space $\mathbb{R}^d$ without overlapping.* □

Now we are ready to introduce the most important invariant of a lattice.

(2.3) Theorem. *Let $\Lambda \subset \mathbb{R}^d$ be a lattice. Then the volume of the fundamental parallelepiped of a basis of Λ does not depend on the basis. It is called the determinant of Λ and denoted* $\det \Lambda$.

For $\rho > 0$ let $B(\rho) = \{x \in \mathbb{R}^d : \|x\| \leq \rho\}$ be the ball of radius ρ centered at the origin and let $|B(\rho) \cap \Lambda|$ be the number of lattice points in $B(\rho)$. Then

$$\lim_{\rho \longrightarrow +\infty} \frac{\operatorname{vol} B(\rho)}{|B(\rho) \cap \Lambda|} = \det \Lambda.$$

In other words, $\det \Lambda$ *can be interpreted as the "volume per lattice point".*

Proof. Let us choose a fundamental parallelepiped Π of Λ. The parallelepiped Π is a bounded set (see Problem 1 of Section 1.5), so there is a γ such that $\Pi \subset B(\gamma)$. Let us choose a $\rho > \gamma$. Let us consider the union

$$X(\rho) = \bigcup_{u \in B(\rho) \cap \Lambda} (\Pi + u).$$

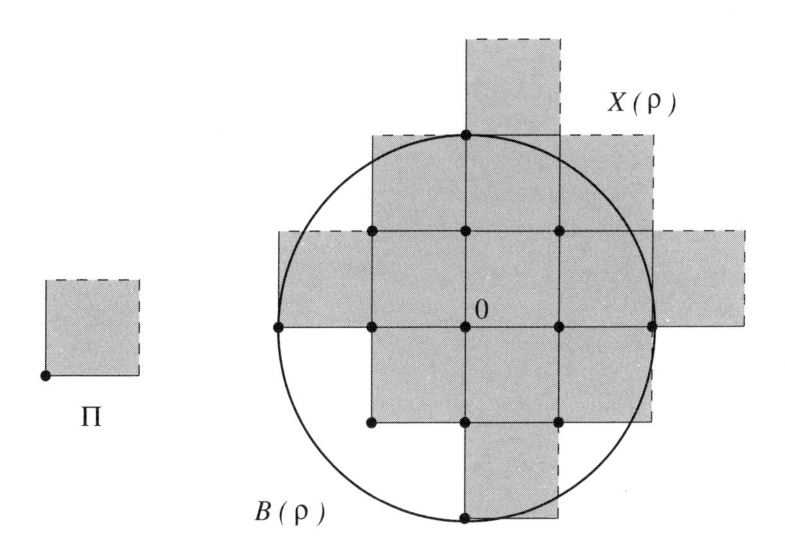

Figure 80. A fundamental parallelepiped Π, a ball $B(\rho)$ and the set $X(\rho)$

By Corollary 2.2, the translates $\Pi + u$ do not overlap, so

$$\operatorname{vol} X(\rho) = |B(\rho) \cap \Lambda| \cdot \big(\operatorname{vol} \Pi\big) \quad \text{and} \quad \big(\operatorname{vol} \Pi\big) = \frac{\operatorname{vol} X(\rho)}{|B(\rho) \cap \Lambda|}. \tag{2.3.1}$$

Since $\Pi \subset B(\gamma)$, we have

$$X(\rho) \subset B(\rho + \gamma) \quad \text{and} \quad \operatorname{vol} X(\rho) \leq \operatorname{vol} B(\rho + \gamma). \tag{2.3.2}$$

Corollary 2.2 implies that for every $x \in B(\rho - \gamma)$ there is a $u \in \Lambda$ such that $x \in \Pi + u$. Then we must have $\|x - u\| \leq \gamma$ and so u must be in $B(\rho)$. Hence we conclude that

(2.3.3) $$B(\rho - \gamma) \subset X(\rho) \quad \text{and} \quad \operatorname{vol} B(\rho - \gamma) \leq \operatorname{vol} X(\rho).$$

Combining (2.3.1)–(2.3.3), we conclude that

$$(\operatorname{vol} \Pi) \cdot \frac{\operatorname{vol} B(\rho)}{\operatorname{vol} B(\rho - \gamma)} \geq \frac{\operatorname{vol} B(\rho)}{|B(\rho) \cap \Lambda|} \geq (\operatorname{vol} \Pi) \cdot \frac{\operatorname{vol} B(\rho)}{\operatorname{vol} B(\rho + \gamma)}.$$

Since $\operatorname{vol} B(\rho \pm \gamma) = (\rho \pm \gamma)^d \operatorname{vol} B(1)$, we have

$$\lim_{\rho \longrightarrow +\infty} \frac{\operatorname{vol} B(\rho)}{\operatorname{vol} B(\rho - \gamma)} = \lim_{\rho \longrightarrow +\infty} \frac{\operatorname{vol} B(\rho)}{\operatorname{vol} B(\rho + \gamma)} = 1,$$

which completes the proof. □

Lattices of determinant 1 are called *unimodular*.

PROBLEM.

1. Let $u_1, \ldots, u_d$ and $v_1, \ldots, v_d$ be two bases of a lattice Λ. Suppose that $u_i = \sum_{j=1}^d \alpha_{ij} v_j$ and $v_i = \sum_{j=1}^d \beta_{ij} u_j$. Let $A = (\alpha_{ij})$ and $B = (\beta_{ij})$ be the $d \times d$ matrices composed of α_{ij}'s and β_{ij}'s correspondingly. Prove that A and B are integer matrices and that $AB = I$ is the identity matrix. Deduce that $|\det A| = |\det B| = 1$. Give an alternative proof of the fact that the volume of a fundamental parallelepiped does not depend on the basis.

(2.4) Definitions. Let $\Lambda \subset \mathbb{R}^d$ be a lattice. A lattice $\Lambda_0 \subset \Lambda$ is called a *sublattice* of Λ. For $x \in \Lambda$ the set $x + \Lambda_0 = \{x + y : y \in \Lambda_0\}$ is called a *coset* of Λ modulo Λ_0. The set of all cosets is denoted Λ/Λ_0. The number of cosets of Λ modulo Λ_0 is called the *index* of Λ_0 in Λ and denoted $|\Lambda/\Lambda_0|$.

PROBLEMS.

1°. Prove that two cosets of Λ modulo Λ_0 either coincide or do not intersect.

2. Let $\Lambda \subset \mathbb{R}^d$ be a lattice and let $\Lambda_0 \subset \Lambda$ be a sublattice. Prove that there exist a basis $u_1, \ldots, u_d$ of Λ and a basis $v_1, \ldots, v_d$ of Λ_0 such that $v_i = \lambda_i u_i$ for some positive integers λ_i for $i = 1, \ldots, d$, where λ_i divides λ_{i+1} for $i = 1, \ldots, d-1$.

Hint: With a basis $U = (u_1, \ldots, u_d)$ of Λ and a basis $V = (v_1, \ldots, , v_d)$ of Λ_0, let us associate an integer $d \times d$ matrix $A = A_{U,V}$, where $A = (\alpha_{ij})$ and

$$v_i = \sum_{j=1}^d \alpha_{ij} u_j \quad \text{for} \quad i = 1, \ldots, d.$$

Let us choose a pair U, V of bases such that $\alpha_{11} > 0$ and the value of α_{11} is the smallest possible among all pairs U, V with positive α_{11}. Prove that all other entries α_{ij} must be divisible by α_{11}. Modify the bases so that $\alpha_{1j} = 0$ and $\alpha_{i1} = 0$ for all $i, j = 2, \ldots, d$. Repeat the same argument with $\alpha_{22}, \ldots, \alpha_{dd}$.

(2.5) Theorem. *Let $\Lambda \subset \mathbb{R}^d$ be a lattice and let $\Lambda_0 \subset \Lambda$ be a sublattice. Let Π_0 be a fundamental parallelepiped of Λ_0. Then*

$$|\Lambda/\Lambda_0| = |\Pi_0 \cap \Lambda| = \frac{\det \Lambda_0}{\det \Lambda}.$$

In particular, the number of points from Λ in a fundamental parallelepiped of Λ_0 does not depend on the parallelepiped.

Proof. By Lemma 2.1, for every $x \in \Lambda$ there is unique representation $x = v + y$, where $v \in \Lambda_0$ and and $y \in \Pi_0$. Since $x \in \Lambda$ and $v \in \Lambda$, we conclude that $y \in \Lambda$. Hence the points of $\Pi_0 \cap \Lambda$ are the coset representatives of Λ/Λ_0, so $|\Lambda/\Lambda_0| = |\Pi_0 \cap \Lambda|$. In particular, we conclude that $|\Lambda/\Lambda_0|$ is finite.

Let $B(\rho) = \{x \in \mathbb{R}^d : \|x\| \leq \rho\}$ be the ball of radius ρ. From Theorem 2.3,

$$\lim_{\rho \longrightarrow +\infty} \frac{|B(\rho) \cap \Lambda|}{\operatorname{vol} B(\rho)} = \frac{1}{\det \Lambda} \quad \text{and} \quad \lim_{\rho \longrightarrow +\infty} \frac{|B(\rho) \cap \Lambda_0|}{\operatorname{vol} B(\rho)} = \frac{1}{\det \Lambda_0}.$$

We claim that for any $x \in \mathbb{R}^d$,

$$\lim_{\rho \longrightarrow +\infty} \frac{|B(\rho) \cap (\Lambda_0 + x)|}{\operatorname{vol} B(\rho)} = \frac{1}{\det \Lambda_0}.$$

Indeed, let $B(-x, \rho)$ be the ball of radius ρ centered at $-x$. Then $B(\rho) \cap (\Lambda_0 + x) = B(-x, \rho) \cap \Lambda_0$. Since $B(0, \rho - \|x\|) \subset B(-x, \rho) \subset B(0, \rho + \|x\|)$, we obtain the desired limit as in the proof of Theorem 2.3. Since $\Lambda = \bigcup_{x \in \Pi_0}(x + \Lambda_0)$ and the cosets do not intersect (cf. Problem 1, Section 2.4), we have

$$|B(\rho) \cap \Lambda| = \sum_{x \in \Pi_0 \cap \Lambda} |B(\rho) \cap (\Lambda_0 + x)|.$$

Summarizing,

$$\frac{1}{\det \Lambda} = \frac{|\Pi_0 \cap \Lambda|}{\det \Lambda_0} \quad \text{and} \quad |\Lambda/\Lambda_0| = \frac{\det \Lambda_0}{\det \Lambda},$$

which completes the proof. □

(2.6) Corollary. *Let $u_1, \ldots, u_d \in \mathbb{Z}^d$ be linearly independent vectors. Then the number of integer points in the "semi-open" parallelepiped*

$$\Pi = \Big\{ \sum_{i=1}^{d} \alpha_i u_i : \quad 0 \leq \alpha_i < 1 \quad \textit{for} \quad i = 1, \ldots, d \Big\}$$

is equal to the volume of Π, that is, the absolute value of the determinant of the matrix with the columns $u_1, \ldots, u_d$.

Proof. Let Λ_0 be the lattice with the basis $u_1, \ldots, u_d$ (cf. Problem 5, Section 1.2). The proof follows by Theorem 2.5 since $\det \mathbb{Z}^d = 1$. □

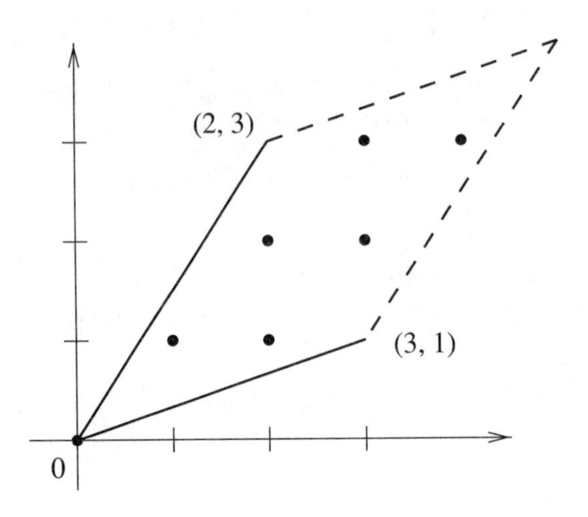

Figure 81. Example: the parallelepiped spanned by $(3,1)$ and $(2,3)$ contains $\left|\det\begin{pmatrix}3 & 2\\ 1 & 3\end{pmatrix}\right| = 7$ integer points.

PROBLEMS.

1°. Prove that $\det \mathbb{Z}^d = 1$, $\det D_n = 2$ and $\det D_n^+ = 1$ (cf. Examples 1.2.1, 1.2.3 and 1.2.4).

2. Prove that $\det A_n = \sqrt{n+1}$ (cf. Example 1.2.2).

3. Prove that $\det E_7 = \sqrt{2}$ (cf. Example 1.2.5) and that $\det E_6 = \sqrt{3}$ (cf. Example 1.2.6).

4. Let $\alpha_1, \ldots, \alpha_{d+1}$ be a set of coprime integers and let us identify $\mathbb{R}^d$ with the hyperplane $H = \big\{(\xi_1, \ldots, \xi_{d+1}) : \ \alpha_1\xi_1 + \ldots + \alpha_{d+1}\xi_{d+1} = 0\big\}$. Let $\Lambda = H \cap \mathbb{Z}^{d+1}$ be a set in $H = \mathbb{R}^d$. Prove that Λ is a lattice in H and that $\det \Lambda = \sqrt{\alpha_1^2 + \ldots + \alpha_{d+1}^2}$.

Hint: Let $a = (\alpha_1, \ldots, \alpha_{d+1}) \in \mathbb{Z}^{d+1}$ and let $n = \alpha_1^2 + \ldots + \alpha_{d+1}^2$. Consider the lattice $\Lambda_1 \subset \mathbb{R}^{d+1}$,

$$\Lambda_1 = \big\{x \in \mathbb{Z}^d : \ \alpha_1\xi_1 + \ldots + \alpha_{d+1}\xi_{d+1} \equiv 0 \mod n\big\}.$$

Prove that $|\mathbb{Z}^{d+1}/\Lambda_1| = n$ and use the fact that if $u_1, \ldots, u_d$ is a basis of Λ, then $u_1, \ldots, u_d, a$ is a basis of Λ_1.

5°. Let $\Lambda \subset \mathbb{R}^d$ be a lattice and let $u_1, \ldots, u_d$ be a set of vectors from Λ such that the volume of the parallelepiped

$$\big\{\alpha_1 u_1 + \ldots + \alpha_d u_d, \quad 0 \leq \alpha_i \leq 1 \quad \text{for} \quad i = 1, \ldots, d\big\}$$

is equal to $\det \Lambda$. Prove that $u_1, \ldots, u_d$ is a basis of Λ.

6. The convex hull of finitely many points in $\mathbb{Z}^2$ is called an *integer polygon.* Following the steps below, prove *Pick's Formula:* the number of integer points in an integer polygon with a non-empty interior is equal to the area of the polygon plus half of the number of integer points on the boundary plus 1.

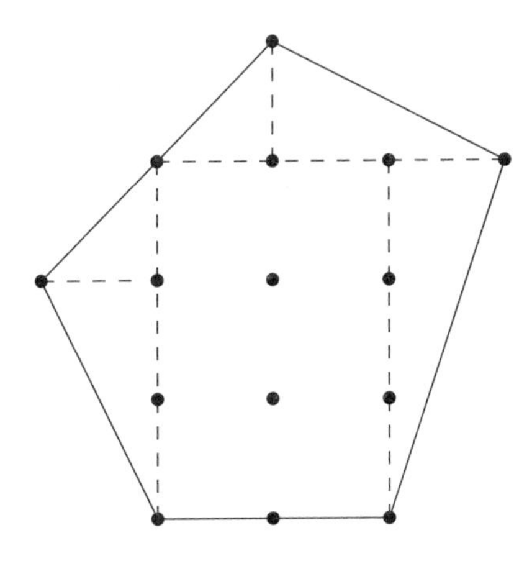

Figure 82. Example: area = 10.5, number of integer points = 15, number of integer points on the boundary = 7; Pick's Formula: 15 = 10.5 + 7/2 + 1

a) Let $a, b \in \mathbb{Z}^2$ be linearly independent vectors and let $c = a + b$. Prove that the number of integer points in the triangle $\operatorname{conv}(0, a, b)$ is equal to the number of integer points in the triangle $\operatorname{conv}(a, b, c)$.

Hint: Consider the transformation $x \longmapsto c - x$.

b) Using Corollary 2.6 and part a), prove Pick's Formula for integer triangles.

c) Prove Pick's Formula in whole generality using part b) and the induction on the number of vertices of the polygon.

7. Prove that linear independent vectors $u, v \in \mathbb{Z}^2$ constitute a basis of $\mathbb{Z}^2$ if and only if the triangle $\operatorname{conv}(0, u, v)$ contains no integer points other than $0, u$ and v.

8. Construct an example of linearly independent vectors $u, v, w \in \mathbb{Z}^3$ such that $\operatorname{conv}(0, u, v, w)$ contains no integer points other than $0, u, v, w$ but u, v, w is not a basis of $\mathbb{Z}^3$.

9. Let $A \subset \mathbb{R}^d$ be a compact convex set. For a point $a \in A$, let us define the *solid angle* $\phi(A, a)$ of A at a by

$$\phi(A, a) = \lim_{\epsilon \longrightarrow 0} \frac{\operatorname{vol}\big(A \cap B(a, \epsilon)\big)}{\operatorname{vol} B(a, \epsilon)}.$$

Let

$$\Phi(A) = \sum_{a \in \mathbb{Z}^d} \phi(A, a).$$

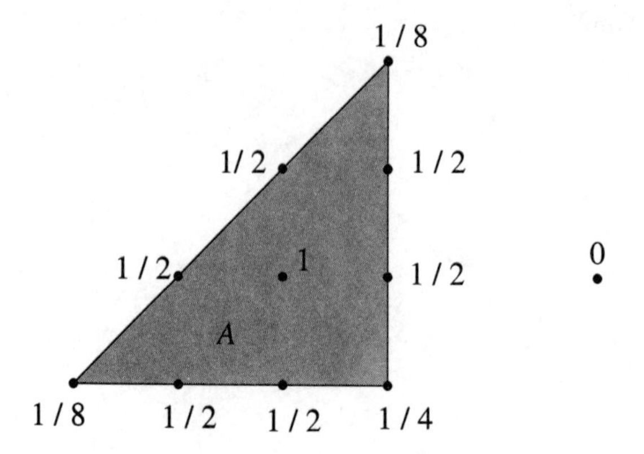

Figure 83. Example: $\Phi(A) = 1 + 6(1/2) + 2(1/8) + 1/4 = 4.5$

a) Let A be an integer polygon (see Problem 6 above). Show that Pick's Formula is equivalent to the identity $\Phi(A) =$ area of A.

b) Prove that Φ gives rise to a valuation (see Section I.7): if $A_1, \ldots, A_m \subset \mathbb{R}^d$ are compact convex sets and $\alpha_1, \ldots, \alpha_m$ are real numbers, then

$$\sum_{i=1}^{m} \alpha_i [A_i] = 0 \quad \text{implies} \quad \sum_{i=1}^{m} \alpha_i \Phi(A_i) = 0.$$

Prove further that $\Phi(A + u) = \Phi(A)$ for all $u \in \mathbb{Z}^d$ and that if $\dim A < d$, then $\Phi(A) = 0$.

c) Let $u_1, \ldots, u_d \in \mathbb{Z}^d$ be linearly independent vectors and let

$$A = \Big\{ \sum_{i=1}^{d} \alpha_i u_i : \quad 0 \leq \alpha_i \leq 1 \quad \text{for} \quad i = 1, \ldots d \Big\}$$

be the parallelepiped spanned by $u_1, \ldots, u_d$. Prove that $\Phi(A + x) = \operatorname{vol} A$ for any vector $x \in \mathbb{R}^d$.

d*) Let $I_i = [u_i, v_i]$, $u_i, v_i \in \mathbb{Z}^d$ for $i = 1, \ldots, m$ be a collection of intervals. Let $A = I_1 + \ldots + I_m$ (such a set A is called a *zonotope*). Prove that $\Phi(A) = \operatorname{vol} A$.

Hint: Show that A can be dissected into a union of parallelepipeds and use Parts b) and c) above.

e) Let $A \subset \mathbb{R}^d$ be a polytope and let $\Lambda \subset \mathbb{Z}^d$ be a lattice such that $\mathbb{R}^d = \bigcup_{u \in \Lambda} (A + u)$ and $\operatorname{int}(A + u_1) \cap \operatorname{int}(A + u_2) = \emptyset$ for distinct $u_1, u_2 \in \Lambda$. Prove that $\Phi(A) = \operatorname{vol} A$.

f*) Let $A \subset \mathbb{R}^d$ be a polytope with integer vertices. Suppose that for every facet F of P there is a point $u_F \in F$ such that $2u_F - F = F$ (in other words, each facet F has a center of symmetry). Prove that $\Phi(A) = \operatorname{vol} A$.

g*) Let $A \subset \mathbb{R}^d$ be a polytope with integer vertices. Prove that for some $\alpha_0(A), \ldots, \alpha_d(A)$

$$\Phi(mA) = \sum_{k=0}^{d} \alpha_k(A) m^k \quad \text{and all positive integers} \quad m.$$

Prove that $\alpha_d(A) = \operatorname{vol} A$, $\alpha_0 = 0$ and $\alpha_k = 0$ if $d - k$ is odd.

Remark: See [**BP99**].

3. Minkowski's Convex Body Theorem

In this section, we prove one of the most elegant and powerful results in convexity, Minkowski's Convex Body Theorem. It stands along with Helly's Theorem (Theorem I.4.2) as one of the most glorious results in finite-dimensional convexity. We start with a result of H.F. Blichfeldt (1914), which states that a set of a sufficiently large volume contains two points that differ by a non-zero lattice point.

(3.1) Blichfeldt's Theorem. *Let $\Lambda \subset \mathbb{R}^d$ be a lattice and let $X \subset \mathbb{R}^d$ be a (Lebesgue measurable) set such that $\operatorname{vol} X > \det \Lambda$. Then there exist two points $x \neq y \in X$ such that $x - y \in \Lambda$.*

Proof. Let Π be a fundamental parallelepiped of Λ. For every lattice point $u \in \Lambda$ let us define a set $X_u \subset \Pi$ as follows:

$$X_u = \{y \in \Pi : \quad y + u \in X\}, \quad \text{that is,} \quad X_u = \big((\Pi + u) \cap X\big) - u;$$

cf. Figure 84. Corollary 2.2 implies that the translates $X_u + u$ cover X without overlapping. Hence

$$\sum_{u \in \Lambda} \operatorname{vol} X_u = \operatorname{vol} X > \operatorname{vol} \Pi = \det \Lambda.$$

We claim that some two subsets X_u and X_v have a non-empty intersection for some lattice points $u \neq v$. Indeed, let $[X_u]$ be the indicator function of X_u (see Definition I.7.1) and let

$$f = \sum_{u \in \Lambda} [X_u].$$

Then

$$\int_\Pi f \, dx = \sum_{u \in \Lambda} \int_\Pi [X_u] \, dx = \sum_{u \in \Lambda} \operatorname{vol} X_u > \operatorname{vol} \Pi.$$

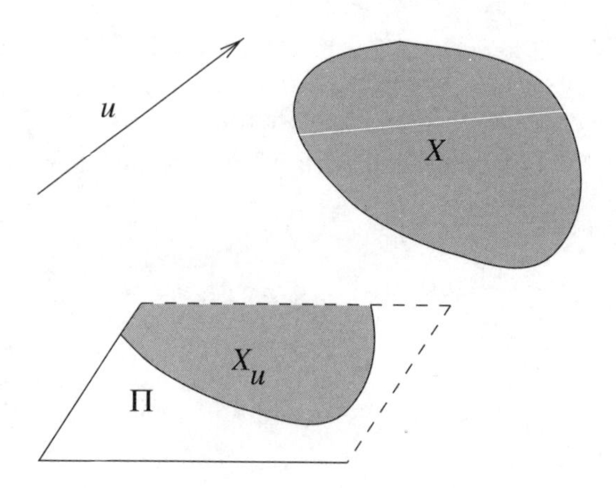

Figure 84. A set X, a fundamental parallelepiped Π, a lattice vector u and the set X_u

Therefore, for some $x \in \Pi$ we have $f(x) > 1$, so $f(x) \geq 2$ and hence $X_u \cap X_v \neq \emptyset$ for some $u \neq v$. Let $z \in X_u \cap X_v$. Then $z + u = x \in X$ and $z + v = y \in X$. We observe that $x - y = u - v \in \Lambda$ and the result follows. □

There are numerous extensions of Blichfeldt's Theorem. Some of them are presented below.

PROBLEMS.

1. Suppose that $\operatorname{vol} X = \det \Lambda$ and X is compact. Prove that there exist two points $x \neq y \in X$ such that $x - y$ is a lattice point.

2. Suppose that $\operatorname{vol} X > m \det \Lambda$ or $\operatorname{vol} X = m \det \Lambda$ and X is compact, where m is a positive integer. Prove that there exist $m+1$ distinct points $x_1, \ldots, x_{m+1} \in X$ such that $x_i - x_j \in \Lambda$ for all i and j.

3. Let f be a non-negative integrable function on $\mathbb{R}^d$ and let $\Lambda \subset \mathbb{R}^d$ be a lattice. Prove that there exists a point $z \in \mathbb{R}^d$ such that

$$\sum_{u \in \Lambda} f(u + z) \geq \frac{1}{\det \Lambda} \int_{\mathbb{R}^d} f(x)\, dx.$$

Now we are ready to prove Minkowski's Theorem.

(3.2) Minkowski's Convex Body Theorem. *Let $\Lambda \subset \mathbb{R}^d$ be a lattice and let $A \subset \mathbb{R}^d$ be a convex set such that $\operatorname{vol} A > 2^d \det \Lambda$ and A is symmetric about the origin. Then A contains a non-zero lattice point u. Furthermore, if A is compact, then the inequality $\operatorname{vol} A > 2^d \det \Lambda$ can be relaxed to $\operatorname{vol} A \geq 2^d \det \Lambda$.*

Proof. Let

$$X = \frac{1}{2}A = \Big\{ \frac{1}{2}x : \quad x \in A \Big\}.$$

Then $\operatorname{vol} X = 2^{-d} \operatorname{vol} A > \det \Lambda$. Therefore, by Theorem 3.1, there exists a pair $x, y \in X$ such that $x - y = u$ is a non-zero lattice vector. Now $2x, 2y \in A$ and since A is symmetric about the origin, $-2y \in A$. Then, since A is convex,

$$u = x - y = \frac{1}{2}(2x) + \frac{1}{2}(-2y) \in A.$$

Hence A contains a non-zero lattice point u.

Suppose now that A is compact and that $\operatorname{vol} A = 2^d \det \Lambda$. Then for any $1 < \rho < 2$ and $\rho A = \big\{ \rho x : x \in A \big\}$ we have $\operatorname{vol}(\rho A) = \rho^d (\operatorname{vol} A) > 2^d \det \Lambda$, so there is a non-zero lattice point $u_\rho \in \rho A$. Since A is compact, the family $\{u_\rho\}$ has a limit point, which has to be a non-zero lattice point from A. □

PROBLEMS.

1°. Construct an example of a convex symmetric non-compact set $A \subset \mathbb{R}^d$, such that $\operatorname{vol} A = 2^d$ but A does not contain a non-zero point from $\mathbb{Z}^d$.

2°. Construct an example of a convex (but not symmetric) set $A \subset \mathbb{R}^2$ of an arbitrarily large volume such that $A \cap \mathbb{Z}^2 = \emptyset$.

3. Let $\Lambda \subset \mathbb{R}^d$ be a lattice and let $A \subset \mathbb{R}^d$ be a centrally symmetric convex set with $\operatorname{vol} A > m 2^n \det \Lambda$ for some positive integer m. Prove that A contains at least m pairs of non-zero lattice points $u_i, -u_i : i = 1, \ldots, m$.

4 (K. Mahler). Let $A \subset \mathbb{R}^d$ be a convex body containing the origin as its interior point. Suppose that for some $\sigma > 0$ we have $-\sigma x \in A$ for all $x \in A$. Prove that if $\operatorname{vol} A > (1 + \sigma^{-1})^d \det \Lambda$, then A contains a non-zero lattice point u.

5* (D.B. Sawyer). Show that the inequality of Problem 4 can be relaxed to $\operatorname{vol} A > (1 + \sigma^{-1})^d (1 - (1 - \sigma)^d) \det \Lambda$.

6* (E. Ehrhart). Prove that if $A \subset \mathbb{R}^2$ is a compact convex set whose center of gravity coincides with the origin and $\operatorname{vol} A > \frac{9}{2} \det \Lambda$, then A contains a non-zero lattice point.

7. Prove the following version of the Minkowski Theorem, suggested by C.L. Siegel. Let $A \subset \mathbb{R}^d$ be a compact centrally symmetric convex body which does not contain a non-zero point from $\mathbb{Z}^d$. Then

$$2^d = \operatorname{vol} A + 4^d \Big(\operatorname{vol} A \Big)^{-1} \sum_{u \in \mathbb{Z}^d \setminus \{0\}} \Big| \int_{\frac{1}{2}A} \exp\{-2\pi i \langle u, x \rangle\} \, dx \Big|^2.$$

Hint: Let

$$\phi(x) = \sum_{u \in \mathbb{Z}^d} [u + (1/2)A],$$

where $[X]$ is the indicator function of the set $X \subset \mathbb{R}^d$; see Definition I.7.1. Then ϕ is a periodic function on $\mathbb{R}^d$ and we may apply Parseval's Formula to ϕ.

8*. Prove the following *Minkowski Extremal Convex Body Theorem.* Let $K \subset \mathbb{R}^d$ be a convex body symmetric about the origin such that $\operatorname{vol} K = 2^d \det \Lambda$ and such that $(\operatorname{int} K) \cap \Lambda = \{0\}$. Prove that K is a polytope with at most $2(2^d - 1)$ facets.

Hint: Through every non-zero lattice point u, let us draw an affine hyperplane H_u isolating K in such a way that H_u and H_{-u} are parallel. Let $\overline{H_u^+}$ be the closed halfspace containing K. Let

$$P = \bigcap_{u \in \Lambda \setminus \{0\}} \overline{H_u^+}.$$

Then P is a symmetric convex body containing K and not containing any non-zero lattice point. We must have $\operatorname{vol} P = 2^d \det \Lambda$ and hence $P = K$. Show that the halfspaces $\overline{H_u^+}$ which correspond to "far away" lattice points u can be discarded. Deduce that $P = K$ is a polyhedron. Show that each facet of K must contain a lattice point in its interior (otherwise, we could increase K). Let $\Lambda_0 = 2\Lambda = \{2u : u \in \Lambda\}$ be a sublattice of Λ. Prove that $|\Lambda/\Lambda_0| = 2^d$. Argue that if the number of facets is greater than $2(2^d - 1)$, then there is a pair of lattice points $x \neq -y$ in different facets of K such that $x - y \in \Lambda_0$. Show that $z = (x - y)/2$ would have been an interior point of K.

9. Let $\Lambda \subset \mathbb{R}^d$ be a lattice. Let

$$K = \Big\{x \in \mathbb{R}^d : \ \operatorname{dist}(x, 0) \leq \operatorname{dist}(x, u) \quad \text{for all} \quad u \in \Lambda\Big\}.$$

Prove that K is a convex body and that $\operatorname{vol} K = \det \Lambda$. Prove that $\operatorname{int}(2K) \cap \Lambda = \{0\}$. Deduce from Problem 8 that K is a polytope.

10. Let $\Lambda = D_4$ (see Example 1.2.3) and let $K \subset \mathbb{R}^4$ be the polytope of Problem 9 above. Prove that up to a change in the coordinates, K is the 24-cell of Problem 6, Section IV.1.3.

11. Let $\Lambda \subset \mathbb{R}^d$ be a lattice and let $A \subset \mathbb{R}^d$ be a convex d-dimensional set symmetric about the origin. Let us define *successive minima* $\lambda_1, \ldots, \lambda_d$ by

$$\lambda_k = \inf\Big\{\tau > 0 : \quad \dim \operatorname{span}\big(\tau A \cap \Lambda\big) = k\Big\}.$$

a°) Check that $\lambda_1 \leq \lambda_2 \leq \ldots \leq \lambda_d$.

b) Check that Minkowski's Convex Body Theorem is equivalent to the inequality $\lambda_1^d \operatorname{vol}(A) \leq 2^d \det \Lambda$.

c*) Prove *Minkowski's Second Theorem:* $\lambda_1 \cdots \lambda_d \operatorname{vol}(A) \leq 2^d \det \Lambda$.

Hint: To describe the idea of a possible approach, let us first sketch a slightly different proof of Theorem 3.2 in the case of $\Lambda = \mathbb{Z}^d$. Let $X \subset \mathbb{R}^d$ be a (Jordan measurable or otherwise "nice") set and let us consider the map $\Phi : X \longrightarrow [0, 1)^d$, $(\xi_1, \ldots, \xi_d) \longmapsto (\{\xi_1\}, \ldots, \{\xi_d\})$, where $\{\xi\}$ is the fractional part of ξ. Prove that Φ preserves volume *locally.* Deduce that if Φ is injective, then $\operatorname{vol} \Phi(X) = \operatorname{vol} X$. Deduce that if $\operatorname{vol} X > 1$, then $X - X$ contains a non-zero lattice point.

Now, let us refine the above reasoning. First, show that it suffices to prove Minkowski's Second Theorem in the case of $\Lambda = \mathbb{Z}^d$. Then show that by changing the coordinates, if necessary, we may assume that for any k and for any $\tau < \lambda_k$, non-zero coordinates of the points in $\tau A \cap \mathbb{Z}^d$ are permitted in the first $k-1$ positions only. Let $X = (1/2)A$. Taking one fractional part at a time, prove that $\operatorname{vol}\big(\Phi(\lambda_d X)\big) = \lambda_1 \cdots \lambda_d \operatorname{vol} X$. Deduce Minkowski's Second Theorem from there.

12. Let $\Lambda \subset \mathbb{R}^d$ be a lattice and let $p : \mathbb{R}^d \longrightarrow \mathbb{R}$ be a norm; see Section V.3. Let $K_p = \{x \in \mathbb{R}^d : p(x) \leq 1\}$. Deduce from Problem 11, c) above and Problem 5 of Section 1.4 that there is a basis $u_1, \ldots, u_d$ of Λ such that

$$\operatorname{vol}(K_p) \prod_{i=1}^{d} p(u_i) \leq (d+1)! \det \Lambda.$$

Remark: General references for Problems 3–9 and 11–12 are [**C97**] and [**GL87**].

Before we proceed with applications of Minkowski's Theorem, we need to do some volume computations.

The volume of the unit ball in $\mathbb{R}^d$. Let $B(\rho) = \{x \in \mathbb{R}^d : \|x\| \leq \rho\}$ be the ball of radius $\rho > 0$ and let $\mathbb{S}^{d-1}(\rho) = \{x \in \mathbb{R}^d : \|x\| = \rho\}$ be the sphere of radius $\rho > 0$.

We will use the integral

$$\int_{-\infty}^{+\infty} e^{-\xi^2} \, d\xi = \sqrt{\pi};$$

cf. Section V.5.1.

(3.3) Definition. The *Gamma function* is defined by the formula

$$\Gamma(x) = \int_0^{+\infty} t^{x-1} e^{-t} \, dt \quad \text{for} \quad x > 0.$$

We recall (without proof) *Stirling's Formula:*

$$\Gamma(x+1) = \sqrt{2\pi x} \left(\frac{x}{e}\right)^x \Big(1 + O(x^{-1})\Big) \quad \text{as} \quad x \longrightarrow +\infty.$$

PROBLEMS.

1. Prove that $\Gamma(x+1) = x\Gamma(x)$.
2. Prove that $\Gamma(x) = (x-1)!$ for positive integers x.
3. Prove that $\Gamma(1/2) = \sqrt{\pi}$.

(3.4) Lemma. *Let β_d be the volume of the unit ball $B = \{x \in \mathbb{R}^d : \|x\| \leq 1\}$. Then*

$$\beta_d = \frac{\pi^{d/2}}{\Gamma(d/2+1)}.$$

Proof. Let κ_{d-1} be the surface area of the unit sphere $\mathbb{S}^{d-1} = \{x \in \mathbb{R}^d : \|x\| = 1\}$ and let

$$\mathbb{S}^{d-1}(\rho) = \{x \in \mathbb{R}^d : \quad \|x\| = \rho\}$$

denote the sphere of radius ρ.

Since $\|x\|^2 = \xi_1^2 + \ldots + \xi_d^2$, we have

$$\int_{\mathbb{R}^d} e^{-\|x\|^2} \, dx = \left(\int_{-\infty}^{+\infty} e^{-\xi^2} \, d\xi \right)^d = \pi^{d/2}.$$

Using polar coordinates, we can write

$$\pi^{d/2} = \int_{\mathbb{R}^d} e^{-\|x\|^2} \, dx = \int_0^{+\infty} e^{-\rho^2} \operatorname{vol}_{d-1} \mathbb{S}^{d-1}(\rho) \, d\rho = \kappa_{d-1} \int_0^{+\infty} \rho^{d-1} e^{-\rho^2} \, d\rho.$$

Substituting $\tau = \rho^2$ in the last integral, we get

$$\pi^{d/2} = \frac{\kappa_{d-1}}{2} \int_0^{+\infty} \tau^{(d-2)/2} e^{-\tau} \, d\tau = \frac{\kappa_{d-1}}{2} \Gamma(d/2).$$

Therefore,

$$\kappa_{d-1} = \frac{2\pi^{d/2}}{\Gamma(d/2)}.$$

Now,

$$\beta_d = \int_0^1 \operatorname{vol}_{d-1} \mathbb{S}^{d-1}(\rho) \, d\rho = \kappa_{d-1} \int_0^1 \rho^{d-1} \, d\rho = \frac{\kappa_{d-1}}{d} = \frac{\pi^{d/2}}{\Gamma(d/2+1)}.$$

□

PROBLEM.

1°. Compute β_1, β_2, β_3 and β_4.

4. Applications: Sums of Squares and Rational Approximations

In this section, we discuss some applications of Minkowski's Convex Body Theorem (Theorem 3.2). Our first goal is to give a proof of Lagrange's result that every positive integer is a sum of four squares of integers. We start with a simple lemma.

(4.1) Lemma. *Let $c_1, \ldots, c_m \in \mathbb{Z}^d$ be integral vectors and let $\gamma_1, \ldots, \gamma_m$ be positive integers. Let*

$$\Lambda = \Big\{x \in \mathbb{Z}^d : \ \langle c_i, x \rangle \equiv 0 \mod \gamma_i \quad \textit{for} \quad i = 1, \ldots, m\Big\}.$$

Then Λ is a lattice and $\det \Lambda \leq \gamma_1 \cdots \gamma_m$.

Proof. Obviously, Λ is a discrete subgroup of $\mathbb{R}^d$. Since Λ contains $(\gamma_1 \cdots \gamma_m)\mathbb{Z}^d$, we conclude that Λ is a lattice. Let us construct the coset representatives of $\mathbb{Z}^d/\Lambda$. For an m-tuple of numbers $b = (\beta_1, \ldots, \beta_m)$, where $0 \leq \beta_i < \gamma_i$, let $x_b \in \mathbb{Z}^d$ be a lattice point, if one exists, such that $\langle c_i, x_b \rangle \equiv \beta_i \mod \gamma_i$ for $i = 1, \ldots, m$. Thus for every $x \in \mathbb{Z}^d$ there is unique $x_b \in \mathbb{Z}^d$ such that

$$\langle c_i, x \rangle \equiv \langle c_i, x_b \rangle \mod \gamma_i \quad \text{for} \quad i = 1, \ldots, m.$$

Therefore, $\mathbb{Z}^d = \bigcup_b (x_b + \Lambda)$ and hence x_b are the coset representatives of $\mathbb{Z}^d/\Lambda$. It follows that the number of possible x_b's does not exceed the product $\gamma_1 \cdots \gamma_m$. By Theorem 2.5, $\det \Lambda = |\mathbb{Z}^d/\Lambda| \det \mathbb{Z}^d = |\mathbb{Z}^d/\Lambda| \leq \gamma_1 \cdots \gamma_m$ and the proof follows.

□

PROBLEM.

1°. Let $\xi_1, \xi_2, \xi_3, \xi_4$ and $\eta_1, \eta_2, \eta_3, \eta_4$ be numbers. Let

$$\zeta_1 = \xi_1\eta_1 - \xi_2\eta_2 - \xi_3\eta_3 - \xi_4\eta_4, \quad \zeta_2 = \xi_1\eta_1 + \xi_2\eta_1 + \xi_3\eta_4 - \xi_4\eta_3,$$
$$\zeta_3 = \xi_1\eta_3 - \xi_2\eta_4 + \xi_3\eta_1 + \xi_4\eta_2 \quad \text{and} \quad \zeta_4 = \xi_1\eta_4 + \xi_2\eta_3 - \xi_3\eta_2 + \xi_4\eta_1.$$

Check that

$$(\xi_1^2 + \xi_2^2 + \xi_3^2 + \xi_4^2)(\eta_1^2 + \eta_2^2 + \eta_3^2 + \eta_4^2) = \zeta_1^2 + \zeta_2^2 + \zeta_3^2 + \zeta_4^2.$$

In particular, the product of two sums of four squares of integers is a sum of four squares of integers.

(4.2) Lagrange's Theorem. *Every positive integer number n can be represented as a sum $n = \xi_1^2 + \xi_2^2 + \xi_3^2 + \xi_4^2$, where ξ_1, ξ_2, ξ_3 and ξ_4 are integers.*

Proof. Problem 1 of Section 4.1 implies that the product of sums of four squares of integers is a sum of four squares of integers. Since every integer is a product of primes, it suffices to prove the result assuming that n is a prime number. Assuming that n is prime, let us show there are numbers α and β such that

$$\alpha^2 + \beta^2 + 1 \equiv 0 \mod n.$$

Indeed, if $n = 2$ we take $\alpha = 1$ and $\beta = 0$. If n is odd, then all the $(n+1)/2$ numbers $\alpha^2 : 0 \leq \alpha < n/2$ have distinct residues mod n. To see this, assume that $\alpha_1^2 \equiv \alpha_2^2 \mod n$ for some $0 \leq \alpha_1, \alpha_2 < n/2$. Then $(\alpha_1 - \alpha_2)(\alpha_1 + \alpha_2) \equiv 0 \mod n$. On the other hand, $0 < \alpha_1 + \alpha_2 < n$ and since n is prime, we must have $\alpha_1 = \alpha_2$. Similarly, the $(n+1)/2$ numbers $-1 - \beta^2 : \ 0 \leq \beta < n/2$ have distinct residues mod n. Therefore there is a pair α, β such that $\alpha^2 = -1 - \beta^2 \mod n$, or, equivalently, $\alpha^2 + \beta^2 + 1 \equiv 0 \mod n$.

Let us consider the lattice

$$\Lambda = \Big\{(\xi_1, \xi_2, \xi_3, \xi_4) \in \mathbb{Z}^4 : \ \xi_1 \equiv \alpha\xi_3 + \beta\xi_4 \mod n \quad \text{and}$$
$$\xi_2 = \beta\xi_3 - \alpha\xi_4 \mod n\Big\}.$$

Lemma 4.1 implies that Λ is a lattice and that $\det \Lambda \leq n^2$. Let us consider an open ball of radius $2n$:

$$B = \Big\{(\xi_1, \xi_2, \xi_3, \xi_4) \in \mathbb{R}^4 : \quad \xi_1^2 + \xi_2^2 + \xi_3^2 + \xi_4^2 < 2n\Big\}.$$

From Lemma 3.4, we have

$$\operatorname{vol} B = 2n^2\pi^2 > 16n^2 \geq 2^4 \det \Lambda.$$

Therefore, by Theorem 3.2, there is a point $x = (\xi_1, \xi_2, \xi_3, \xi_4) \in \Lambda$ such that $0 < \xi_1^2 + \xi_2^2 + \xi_3^2 + \xi_4^2 < 2n$. On the other hand, since $x \in \Lambda$, we get that

$$\xi_1^2 + \xi_2^2 + \xi_3^2 + \xi_4^2 \equiv (\alpha^2 + \beta^2 + 1)\xi_3^2 + (\alpha^2 + \beta^2 + 1)\xi_4^2 \equiv 0 \mod n.$$

Therefore, $\xi_1^2 + \xi_2^2 + \xi_3^2 + \xi_4^2$ is divisible by n and since $0 < \xi_1^2 + \xi_2^2 + \xi_3^2 + \xi_4^2 < 2n$, we must have $\xi_1^2 + \xi_2^2 + \xi_3^2 + \xi_4^2 = n$. □

This proof is due to H. Davenport.

PROBLEMS.

1°. Give an example of a positive integer that cannot be represented as a sum of three squares of integers.

2. Let k be a positive integer. Prove that if there is a solution to the congruence $\xi^2 + 1 \equiv 0 \mod k$, then k is the sum of two squares of integers. Deduce that every prime number $p \equiv 1 \mod 4$ is the sum of two squares of integers.

3*. Prove that the number of integral solutions of the equation $\xi_1^2 + \xi_2^2 + \xi_3^2 + \xi_4^2 = n$ is 8 times the sum of all d such that d divides n and 4 does not divide d (Jacobi's Formula).

Remark: For a short proof, see [**AEZ93**].

Next, we consider how to approximate a given real number by a rational number. Obviously, given a real number α and a positive integer q, we can approximate α by a rational number p/q so that $|\alpha - p/q| \leq 1/2q$. It turns out, we can do better.

(4.3) Theorem. *There exists a constant $0 < C < 1$ such that for any $\alpha \in \mathbb{R}$ one can find an arbitrarily large positive integer q and an integer p such that*

$$\left|\alpha - \frac{p}{q}\right| \leq \frac{C}{q^2}.$$

Proof. Without loss of generality, we may assume that α is irrational. Let us choose a positive integer $Q > 2$ and consider the set $A \subset \mathbb{R}^2$:

$$A = \Big\{(x, y) : \quad |\alpha x - y| \leq \frac{1}{Q}, \ |x| \leq Q\Big\}.$$

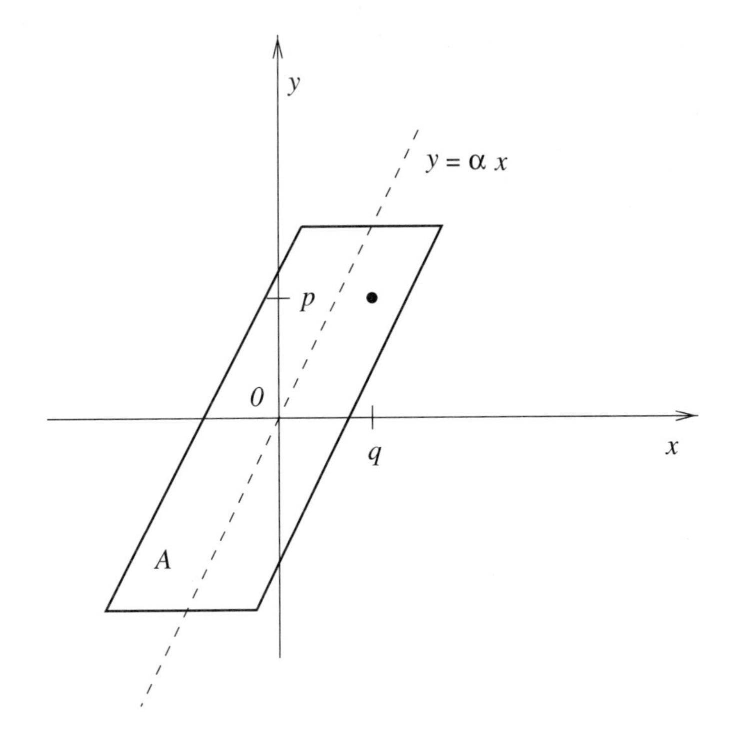

Figure 85

Then A is a compact convex centrally symmetric set (parallelogram). Furthermore, $\operatorname{vol} A = 4$. Therefore, by Theorem 3.2, there is a non-zero integer point $(q, p) \in A$. We observe that $q \neq 0$ since otherwise $p = 0$. Because of the symmetry, we may assume that $q > 0$. It now follows that

(4.3.1) $$\left|\alpha - \frac{p}{q}\right| \leq \frac{1}{qQ} \quad \text{and} \quad 0 < q \leq Q.$$

The inequalities (4.3.1) imply that

(4.3.2) $$\left|\alpha - \frac{p}{q}\right| \leq \frac{1}{q^2}.$$

Our goal is to establish that q can be made arbitrarily large in (4.3.2). Indeed, let M be a natural number. Since α is irrational, we can choose Q in (4.3.1) so big that the inequality (4.3.1) is not satisfied for any $1 < q < M$ (since there are finitely many possibilities to approximate a number within a given error by a fraction whose denominator does not exceed M and not a single such approximation will be precise). Therefore, (4.3.1) and thus (4.3.2) are satisfied with some $q > M$. $\square$

The construction used in the proof of Theorem 4.3 does not provide the best way to get a rational approximation of a given real number. It is known that the best value of C is $C = 1/\sqrt{5}$ (one can take $\alpha = (\sqrt{5}-1)/2$), whereas the proof of Theorem 4.3 gives us only that $C \leq 1$. A better method of rational approximation is via *continued fractions* which are not discussed here (see [**Kh97**]). However, in the case of a *simultaneous approximation* (see Problem 1 below), the approach via Minkowski's Theorem turns out to be quite useful.

PROBLEM.

1. Prove that there exists $C > 0$ such that for any real numbers $\alpha_1, \dots, \alpha_n$ one can find an arbitrarily large positive integer q and integers $p_1, \dots, p_n$ such that

$$\left|\alpha_i - \frac{p_i}{q}\right| \leq \frac{C}{q^{1+\frac{1}{n}}} \quad \text{for } i = 1, \dots, n.$$

5. Sphere Packings

In this section, we discuss the notion of the packing density which is closely related to Blichfeldt's and Minkowski's Theorems (Theorems 3.1 and 3.2).

(5.1) Definitions. For a number $\rho > 0$ and a point $x_0 \in \mathbb{R}^d$ let $B(x_0, \rho) = \{x \in \mathbb{R}^d : \|x - x_0\| < \rho\}$ denote the open ball of radius ρ centered at x_0. Let $\Lambda \subset \mathbb{R}^d$ be a lattice. The *packing radius* of Λ is the largest $\rho > 0$ such that the open balls $B(x, \rho)$ and $B(y, \rho)$ do not intersect for any two distinct points $x, y \in \Lambda$. Let $X = \bigcup_{x \in \Lambda} B(x, \rho)$ be the part of the space $\mathbb{R}^d$ covered by the balls centered at the lattice points, where ρ is the packing radius.

The *packing density* of Λ is the number

$$\sigma(\Lambda) = \lim_{\tau \longrightarrow +\infty} \frac{\operatorname{vol}\Big(X \cap B(0, \tau)\Big)}{\operatorname{vol} B(0, \tau)}.$$

In other words, $\sigma(\Lambda)$ is the "fraction" of the space $\mathbb{R}^d$ filled by the largest congruent non-overlapping balls centered at the lattice points.

It has been known for some time that some lattice packings utilize space better (that is, have a higher packing density) than others; see Figure 86.

PROBLEMS.

1°. Show that the packing radius exists and that it is equal to one half of the minimum length of a non-zero vector from Λ.

2°. Show that the packing density exists and that it is equal to

$$\frac{\beta_d \rho^d}{\det \Lambda} = \frac{\pi^{d/2} \rho^d}{\Gamma(d/2 + 1) \det \Lambda},$$

where ρ is the packing radius and β_d is the volume of the unit ball in $\mathbb{R}^d$; see Lemma 3.4.

3°. Explain why the Minkowski Convex Body Theorem (Theorem 3.2) for a ball centered at the origin is equivalent to the statement that the packing density $\sigma(\Lambda)$ of any lattice Λ does not exceed 1.

Lattices Λ_1 and Λ_2 in $\mathbb{R}^d$ are called *similar* (denoted $\Lambda_1 \sim \Lambda_2$) if one can be obtained from another by a composition of an orthogonal transformation of $\mathbb{R}^d$ and a dilation $x \longmapsto \alpha x$, $\alpha \neq 0$.

4. Prove that the packing densities of similar lattices are equal.
5. Prove that $D_2 \sim \mathbb{Z}^2$, $D_3 \sim A_3$, and $D_4^+ \sim \mathbb{Z}^4$ (cf. Example 1.2).
6. Prove the following identities for the packing radii:

$$\rho(\mathbb{Z}^d) = \frac{1}{2}, \quad \rho(A_n) = \rho(D_n) = \frac{\sqrt{2}}{2} \quad \text{for} \quad n \geq 2,$$

$$\rho(D_n^+) = \frac{\sqrt{2}}{2} \quad \text{for} \quad n \geq 8,$$

$$\rho(D_2^+) = \frac{1}{2\sqrt{2}}, \quad \rho(D_4^+) = \frac{1}{2}, \quad \rho(D_6) = \sqrt{\frac{3}{8}} \quad \text{and}$$

$$\rho(E_6) = \rho(E_7) = \frac{\sqrt{2}}{2}.$$

(cf. Example 1.2).

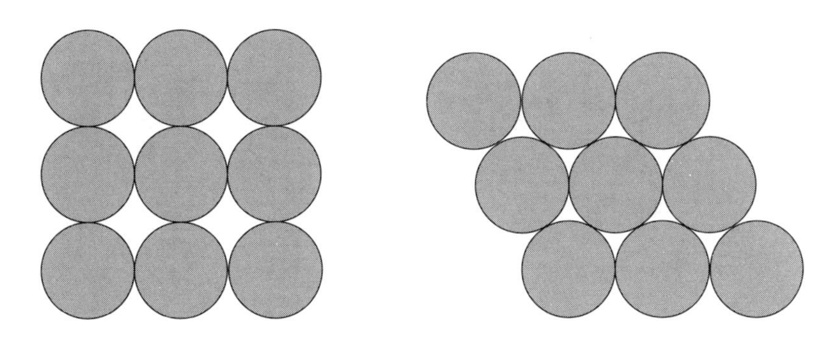

Figure 86. The $\mathbb{Z}^2$ packing and the A_2 packing of discs in $\mathbb{R}^2$. The latter has a higher packing density.

7. Prove the following identities for the packing densities

$$\sigma(\mathbb{Z}) = 1, \quad \sigma(A_2) = \frac{\pi}{\sqrt{12}} \approx 0.9069, \quad \sigma(A_3) = \frac{\pi}{\sqrt{18}} \approx 0.7405,$$

$$\sigma(D_4) = \frac{\pi^2}{16} \approx 0.6169, \quad \sigma(D_5) = \frac{\pi^2}{15\sqrt{2}} \approx 0.4653, \quad \sigma(E_6) = \frac{\pi^3}{48\sqrt{3}} \approx 0.3729,$$

$$\sigma(E_7) = \frac{\pi^3}{105} \approx 0.2953 \quad \text{and} \quad \sigma(E_8) = \frac{\pi^4}{384} \approx 0.2537$$

(cf. Examples 1.2).

8. Let $K \subset \mathbb{R}^d$ be a compact convex set with non-empty interior. The *kissing number* of K is the maximum number of congruent copies of K that can touch K without any two overlapping. Using lattices A_2, A_3, D_4 and E_8, prove that the kissing number of a ball in $\mathbb{R}^d$ is at least 6 for $d=2$, at least 12 for $d=3$, at least 24 for $d=4$ and at least 240 for $d=8$.

9*. Prove that for every dimension d there exists a lattice $\Lambda \subset \mathbb{R}^d$ with the largest possible packing density among all lattices of rank d.

(5.2) Packing densities and signal transmission. Suppose we want to transmit a signal which we interpret as a d-tuple of numbers $(\xi_1, \ldots, \xi_d)$. Hence the space $\mathbb{R}^d$ is interpreted as the space of all signals. If we want to transmit information, some signals may serve as *codes* of transmitted symbols.

Realistically, we can expect that every transmitted signal gets somewhat distorted. Then to *decode* a signal, we need to "round" it to a code, hence there is some error bound $\rho > 0$, such that within the distance less than ρ of the signal there is at most one code and that code can be found more or less efficiently.

If the codes form a lattice in $\mathbb{R}^d$, then ρ is the packing radius and the packing density σ is the "percentage" of all signals that can be decoded. Naturally, we would like to make σ as big as possible. Of course, there are other circumstances which we would like to take into account. For example, we want to make decoding, that is, finding a lattice point closest to a given signal, as simple as possible.

PROBLEMS.

1. Prove the following inequalities for packing densities:

$$\begin{aligned}
&\sigma(A_2) > \sigma(\mathbb{Z}^2),\\
&\sigma(A_3) > \sigma(\mathbb{Z}^3),\\
&\sigma(D_4) > \sigma(A_4) > \sigma(\mathbb{Z}^4),\\
&\sigma(D_5) > \sigma(A_5) > \sigma(\mathbb{Z}^5),\\
&\sigma(E_6) > \sigma(D_6) > \sigma(A_6) > \sigma(\mathbb{Z}^6),\\
&\sigma(E_7) > \sigma(D_7) > \sigma(A_7) > \sigma(\mathbb{Z}^7),\\
&\sigma(E_8) > \sigma(D_8) > \sigma(A_8) > \sigma(\mathbb{Z}^8).
\end{aligned}$$

2. Construct an efficient algorithm of rounding a point $x \in \mathbb{R}^4$ to the nearest point of D_4.

3. Construct an efficient algorithm of rounding a point $x \in \mathbb{R}^8$ to the nearest point of E_8.

Minkowski's Theorem results in the following general estimates.

(5.3) Corollary. *Let $\Lambda \subset \mathbb{R}^d$ be a lattice. Then for the packing radius $\rho(\Lambda)$ we have*

$$\rho(\Lambda) \leq \frac{\Gamma^{1/d}(d/2+1)}{\sqrt{\pi}} \Big(\det \Lambda\Big)^{1/d} = \sqrt{\frac{d}{2\pi e}} \Big(1 + O(1/d)\Big) \Big(\det \Lambda\Big)^{1/d}.$$

Proof. By Problem 1, Section 5.1, it follows that for $\rho = \rho(\Lambda)$ the ball $B(0, 2\rho)$ does not contain lattice points in its interior. The proof now follows by the Minkowski Theorem (Theorem 3.2), the formula for the volume of a ball (Lemma 3.4) and Stirling's Formula $\Gamma(x+1) = \sqrt{2\pi x}(x/e)^x(1 + O(1/x))$. □

The inequality of Corollary 5.3 is exactly equivalent to the statement that the packing density of any lattice does not exceed 1. It seems obvious intuitively (and is indeed true) that the packing density of any d-dimensional lattice is strictly less than 1 for any $d > 1$ and that the density approaches zero as the dimension d grows. Thus, Minkowski's Convex Body Theorem (Theorem 3.2) is not optimal for a ball. In fact, the lattices $\mathbb{Z}$, A_2, $A_3 \sim D_3$, D_4, D_5, E_6, E_7 and E_8 are the lattices with the highest packing density in their respective dimensions; see [**CS99**]. Similarly, one can define the packing density of an arbitrary (measurable) set. Blichfeldt's and Minkowski's Theorems are equivalent to stating that the packing density does not exceed 1.

PROBLEMS.

The following simple estimates turn out to be quite useful.

1. For $x = (\xi_1, \dots, \xi_d) \in \mathbb{R}^d$ let $\|x\|_\infty = \max_{i=1,\dots,d} |\xi_i|$. Let $\Lambda \subset \mathbb{R}^d$ be a lattice. Prove that there exists a non-zero point $x \in \Lambda$ such that $\|x\|_\infty \leq \big(\det \Lambda\big)^{1/d}$.

2°. Let $\Lambda \subset \mathbb{R}^d$ be a lattice. Prove that there exists a non-zero point $x \in \Lambda$ such that $\|x\| \leq \sqrt{d}\big(\det \Lambda\big)^{1/d}$.

6. The Minkowski-Hlawka Theorem

Next, we discuss an interesting method to construct higher-dimensional lattices with a reasonably high packing density. As we will see, the packing radius of the d-dimensional unimodular lattice that we construct will be about $\sqrt{d}$ up to a constant factor, which is the best possible by Corollary 5.3. The main idea of the construction is to choose a *random* lattice.

(6.1) Lemma. *Let $M \subset \mathbb{R}^d$ be a Lebesgue measurable set, let $\Lambda \subset \mathbb{R}^d$ be a lattice and let Π be a fundamental parallelepiped of Λ. For $x \in \mathbb{R}^d$ let $\Lambda + x = \big\{u + x : u \in \Lambda\big\}$ be the translation of Λ and let $|M \cap (\Lambda + x)|$ be the number of points from $\Lambda + x$ in M. Then*

$$\int_\Pi |M \cap (\Lambda + x)| \, dx = \operatorname{vol} M.$$

Proof. For a lattice point $u \in \Lambda$ let $[M - u] : \mathbb{R}^d \longrightarrow \mathbb{R}$ be the indicator function of the translation $M - u$; see Section I.7.1. Then $|M \cap (\Lambda + x)| = \sum_{u \in \Lambda} [M - u](x)$ and

$$\int_\Pi |M \cap (\Lambda + x)| \, dx = \sum_{u \in \Lambda} \int_\Pi [M - u] \, dx = \sum_{u \in \Lambda} \operatorname{vol}\Big((\Pi + u) \cap M\Big) = \operatorname{vol} M.$$

The last equality follows since the translates $\Pi + u$ cover M without overlapping by Corollary 2.2. □

PROBLEMS.

1°. Let $\Lambda \subset \mathbb{R}^d$ be a lattice and let $M \subset \mathbb{R}^d$ be a measurable set such that $\operatorname{vol} M < \det \Lambda$. Prove that there exists an $x \in \mathbb{R}^d$ such that $M \cap (\Lambda + x) = \emptyset$.

2°. Let $f(x) = |M \cap (\Lambda + x)|$. Prove that $f(x+u) = f(x)$ for all $u \in \Lambda$.

We are going to apply Lemma 6.1 to very reasonable sets M, such as a ball, and definitely not to a general Lebesgue measurable set. In what follows, we assume that M is "Jordan measurable", which implies that the volume of M can be well approximated by using more and more refined meshes. The reader may always think of M as something familiar, such as a ball or a convex body. Although we don't prove the result in its full power (for so-called "star" bodies M), we still get quite interesting asymptotics. The result was conjectured by H. Minkowski and proved by E. Hlawka in 1944.

(6.2) The Minkowski-Hlawka Theorem. *Let $d > 1$ and let $M \subset \mathbb{R}^d$ be a bounded Jordan measurable set. Let us choose $\delta > \operatorname{vol} M$. Then there exists a lattice $\Lambda \subset \mathbb{R}^d$ such that $\det \Lambda = \delta$ and M does not contain a non-zero point from Λ.*

Proof. Without loss of generality we may assume that $\operatorname{vol} M < 1$ and that $\delta = 1$. Let $e_1, \ldots, e_d$ be the standard basis of $\mathbb{R}^d$ and let $\mathbb{R}^{d-1}$ be the subspace spanned by $e_1, \ldots, e_{d-1}$. Hence we fix a decomposition $\mathbb{R}^d = \mathbb{R}^{d-1} \oplus \mathbb{R}$. Let us fix a small $\alpha > 0$ and let us consider a family of hyperplanes $H_k = \{x : \xi_d = k\alpha\}$, $k \in \mathbb{Z}$. In particular, $H_0 = \mathbb{R}^{d-1}$. Let $M_k = M \cap H_k$ be the $(d-1)$-dimensional slice of M; see Figure 87. We choose α to be small enough so that

(6.2.1) Every point from M_0 lies within the cube $|\xi_i| < \alpha^{-1/(d-1)}$ for $i = 1, \ldots, d-1$ (we can do it since M is bounded).

$$\alpha \sum_{k=-\infty}^{+\infty} \operatorname{vol}_{d-1} M_k < 1. \tag{6.2.2}$$

We can choose such an α because M is Jordan measurable, $\operatorname{vol} M < 1$ and the sum approximates $\operatorname{vol} M$ arbitrarily close if α is sufficiently small.

We construct Λ by choosing a basis. More precisely, we choose the first $d-1$ vectors $u_1, \ldots, u_{d-1}$ of the basis and then choose the remaining vector u_d at random. Let us choose the first $d-1$ basis vectors of Λ:

$$u_i = \alpha^{-1/(d-1)} e_i \quad \text{for} \quad i = 1, \ldots, d-1.$$

Let Π be the fundamental parallelepiped of the basis $u_1, \ldots, u_{d-1}$ in H_0, so

$$\operatorname{vol}_{d-1} \Pi = \alpha^{-1}.$$

Now, for an $x \in \Pi$, let us choose $u_d(x) = x + \alpha e_d$ and let Λ_x be the lattice with the basis $u_1, \ldots, u_{d-1}, u_d(x)$. Since $u_d(x) \in H_1$, $\det \Lambda_x = (\operatorname{vol} \Pi)\alpha = 1$.

We claim that for some $x \in \Pi$, the lattice Λ_x satisfies the desired property. The proof is based on the formula

$$\int_{\Pi} |\Lambda_x \cap M_k| \, dx = \operatorname{vol}_{d-1} M_k \quad \text{for} \quad k \neq 0. \tag{6.2.3}$$

Indeed, let $\Lambda_0 \subset H_0$ be the lattice with the basis $u_1, \ldots, u_{d-1}$. We consider the case of $k \geq 1$ (the case $k \leq -1$ is treated similarly). We can write

$$|\Lambda_x \cap M_k| = |M_k \cap (\Lambda_0 + k u_d(x))| = |M_k \cap (\Lambda_0 + k\alpha e_d + kx)|.$$

Let us think of the hyperplane H_k as a $(d-1)$-dimensional Euclidean space with the origin at $k\alpha e_d$. Then $\Lambda_k = \Lambda_0 + k\alpha e_d$ is a $(d-1)$-dimensional lattice in H_k and $\Lambda_0 + k u_d(x)$ is a translation of Λ_k by the vector kx.

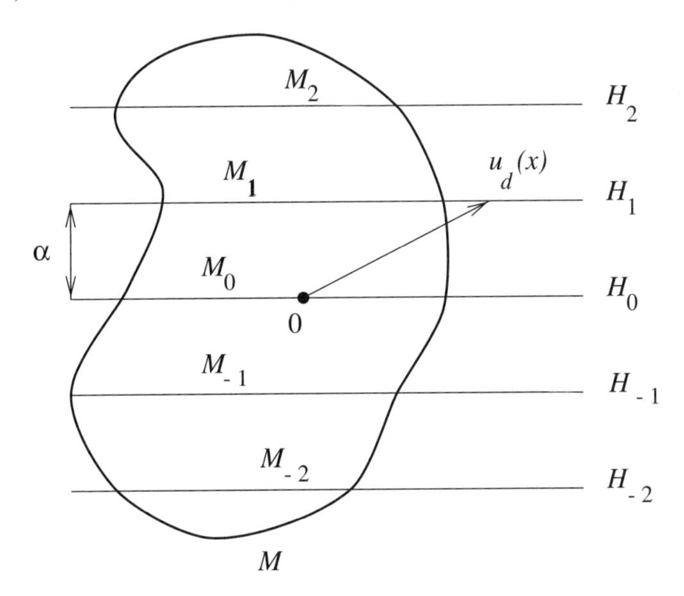

Figure 87

We get

$$\int_{\Pi} |\Lambda_x \cap M_k| \, dx = \int_{\Pi} |M_k \cap (\Lambda_k + kx)| \, dx = k^{-(d-1)} \int_{k\Pi} |M_k \cap (\Lambda_k + y)| \, dy$$

(we substitute $y = kx$). On the other hand, the parallelepiped $k\Pi$ is the union of k^{d-1} non-intersecting lattice shifts $\Pi + u$ for $u \in \Lambda_0$, so by Problem 2, Section 6.1, we get that

$$\int_{k\Pi} |(M_k \cap (\Lambda_0 + y)| \, dy = k^{d-1} \int_{\Pi} |M_k \cap (\Lambda_0 + y)| \, dy.$$

Applying Lemma 6.1, we finally conclude that

$$\int_{\Pi} |M_k \cap (\Lambda_0 + y)| \, dy = \operatorname{vol}_{d-1} M_k$$

and hence (6.2.3) follows. Now we observe that

$$\frac{1}{\operatorname{vol}_{d-1} \Pi} \int_{x \in \Pi} \Big(\sum_{k \neq 0} |M_k \cap \Lambda_x| \Big) \, dx = \alpha \sum_{k \neq 0} \operatorname{vol}_{d-1} M_k < 1$$

by (6.2.2). Therefore, the average value of the number of points in $(M \setminus M_0) \cap \Lambda_x$ is strictly smaller than 1. Therefore, there must be an $x \in \Pi$ such that the intersection $(M \setminus M_0) \cap \Lambda_x$ is empty. By (6.2.1) we conclude that $M_0 \cap \Lambda_x$ consists of at most the zero vector, which completes the proof. □

PROBLEMS.

1. Let ϕ be a Lebesgue integrable function on $\mathbb{R}^d$ and let $\Lambda \subset \mathbb{R}^d$ be a lattice. Prove that there exists a $z \in \mathbb{R}^d$ such that

$$\sum_{u \in \Lambda} \phi(u + z) \leq \frac{1}{\det \Lambda} \int_{\mathbb{R}^d} \phi(x) \, dx.$$

2. Let ϕ be a bounded Riemann integrable function vanishing outside a bounded region in $\mathbb{R}^d$, $d > 1$, and let ϵ be a positive number. Prove that there exists a unimodular lattice $\Lambda \subset \mathbb{R}^d$ such that

$$\sum_{u \in \Lambda \setminus \{0\}} \phi(u) < \epsilon + \int_{\mathbb{R}^d} \phi(x) \, dx.$$

3. Let $M \subset \mathbb{R}^d$, $d > 1$, be a bounded centrally symmetric Jordan measurable set and let $\delta > (1/2)(\operatorname{vol} M)$. Prove that there exists a lattice $\Lambda \subset \mathbb{R}^d$ such that $\det \Lambda = \delta$ and M does not contain any lattice point, except possibly 0.

Hint: Either M does not contain any non-zero lattice point or it contains at least two.

(6.3) Corollary. *For any $\sigma < 2^{-d}$ there exists a d-dimensional lattice $\Lambda \subset \mathbb{R}^d$ whose packing density is at least σ. Similarly, for any*

$$\rho < \frac{\Gamma^{1/d}(d/2+1)}{2\sqrt{\pi}} = \sqrt{\frac{d}{8\pi e}} \Big(1 + O\Big(\frac{1}{d}\Big)\Big)$$

there exists a d-dimensional lattice $\Lambda \subset \mathbb{R}^d$ with packing radius at least ρ and determinant 1.

Proof. Let $B(0,2)$ the ball of radius 2 centered at the origin. By Theorem 6.2, for any $\epsilon > 0$ there exists a d-dimensional lattice $\Lambda \subset \mathbb{R}^d$, such that $\det \Lambda = (1+\epsilon) \operatorname{vol} B(0,2)$ and such that $\Lambda \cap B(0,2) = \{0\}$. Then the packing radius of this lattice is at least 1 and the packing density is at least

$$\frac{\operatorname{vol} B(0,1)}{(1+\epsilon)\operatorname{vol} B(0,2)} = \frac{2^{-d}}{(1+\epsilon)}.$$

Let us choose $\Lambda_1 = \alpha\Lambda$ where $\alpha > 0$ is chosen so that $\det \Lambda_1 = 1$. The packing density of Λ_1 is the same as that of Λ (cf. Problem 4 of Section 5.1), but Λ_1 is now a unimodular lattice. The second part follows from the relationship between the packing density and packing radius; see Problem 2 of Section 5.1 and also Corollary 5.3. □

Lattices whose existence is asserted by Corollary 6.3 are not bad at all when the dimension d is high. For example, the packing radius of such a lattice has the same order of magnitude as the upper bound in Corollary 5.3, that is, of the order of $\sqrt{d}$. It is not easy to present explicitly a sequence of unimodular lattices whose packing radius grows unbounded as the dimension grows. The proof of Theorem 6.2 suggests the idea of how to construct such a lattice: we choose a *random* lattice $\Lambda \subset \mathbb{R}^d$. More precisely, only the last basis vector should be chosen at random, whereas the first $d-1$ basis vectors are very easy to construct deterministically. The exact asymptotics of the best packing radius are not known; note that any improvement by a constant multiplicative factor of the packing radius results in the improvement of the packing density by an exponential (in the dimension) factor.

PROBLEMS.

1. Prove that for any $\sigma < 2^{1-d}$ there exists a lattice in $\mathbb{R}^d$ whose packing density is at least σ.

Hint: Use Problem 3 of Section 6.2.

2*. Prove that there exists a lattice in $\mathbb{R}^d$ whose packing density is at least 2^{1-d}.

Remark: In fact, one can strengthen the bound to $\zeta(d)2^{1-d}$, where $\zeta(d) = \sum_{n=1}^{+\infty} n^{-d}$ and $d > 1$; see Section IX.7 of [**C97**].

7. The Dual Lattice

As is the case with convex duality (polarity), some important information about a lattice can be extracted from a properly defined dual object, which, not surprisingly, turns out to be a lattice.

(7.1) Definition. Let $\Lambda \subset \mathbb{R}^d$ be a lattice. The set

$$\Lambda^* = \Big\{x \in \mathbb{R}^d : \quad \langle x, y\rangle \in \mathbb{Z} \quad \text{for all} \quad y \in \Lambda\Big\}$$

is called the *dual* (or *polar* or *reciprocal*) to Λ.

PROBLEMS.

1°. Let $\Lambda \subset \mathbb{R}^d$ be a lattice with a basis $u_1, \ldots, u_d$. Prove that $\Lambda^* \subset \mathbb{R}^d$ is a lattice with the basis $v_1, \ldots, v_d$, where

$$\langle u_i, v_j \rangle = \begin{cases} 1 & \text{if} \quad i+j=d+1, \\ 0 & \text{otherwise.} \end{cases}$$

Prove that $\det(\Lambda^*) \cdot \det(\Lambda) = 1$.

2°. Prove that $(\Lambda^*)^* = \Lambda$.

3°. Prove that $\left(\mathbb{Z}^d\right)^* = \mathbb{Z}^d$.

4. Prove that $(E_8)^* = E_8$; cf. Example 1.2.4.

5. Prove that D_4 and $(D_4)^*$ are similar; cf. Example 1.2.3 and Problem 4 of Section 5.1.

There is a useful relationship between the packing radius of Λ and the packing radius of Λ^*. Recall that the packing radius $\rho(\Lambda)$ is equal to half the length of a shortest non-zero vector from Λ.

(7.2) Lemma. *Let $\Lambda \subset \mathbb{R}^d$ be a lattice and let $\Lambda^* \subset \mathbb{R}^d$ be the dual lattice. Then, for the packing radii of Λ and Λ^*, we have*

$$\rho(\Lambda) \cdot \rho(\Lambda^*) \leq \frac{d}{4}.$$

Proof. The result follows by Minkowski's Convex Body Theorem. By Problem 2 of Section 5.3, we have

$$\rho(\Lambda) \leq \frac{1}{2}\sqrt{d}\big(\det \Lambda\big)^{1/d} \quad \text{and} \quad \rho(\Lambda^*) \leq \frac{1}{2}\sqrt{d}\big(\det \Lambda^*\big)^{1/d}.$$

The result follows by Problem 1, Section 7.1. □

Using Corollary 5.3, we can replace the upper bound $d/4$ by a better bound of $d/(2\pi e)\big(1 + O(1/d)\big)$, but we will not need this refinement.

PROBLEMS.

1°. Let Λ be a lattice and let $\Lambda_0 \subset \Lambda$ be a sublattice. Prove that

$$\rho(\Lambda) \leq \rho(\Lambda_0) \leq |\Lambda/\Lambda_0| \cdot \rho(\Lambda).$$

2°. Let $\Lambda \subset \mathbb{R}^d$ be a lattice and let $u_1, \ldots, u_d$ be linearly independent vectors from Λ^*. Prove that

$$\max_{i=1,\ldots,d} \|v\| \cdot \|u_i\| \geq 1$$

for each vector $v \in \Lambda \setminus \{0\}$.

3. Let $\Lambda \subset \mathbb{R}^d$ be a lattice and let Λ^* be the dual lattice. Let us construct a basis $v_1, \ldots, v_d$ of Λ^* as follows: let v_1 be a shortest non-zero vector in Λ^* and let $v_k \in \Lambda \setminus \operatorname{span}(v_1, \ldots, v_{k-1})$ be a closest vector to $\operatorname{span}(v_1, \ldots, v_{k-1})$ for $k > 1$; cf. Theorem 1.4. Check that $v_1, \ldots, v_d$ is a basis of Λ^*. Let $u_1, \ldots, u_d$ be a basis of Λ dual to $v_1, \ldots, v_d$:

$$\langle u_i, v_j \rangle = \begin{cases} 1 & \text{if } i + j = d + 1, \\ 0 & \text{otherwise.} \end{cases}$$

Check that $u_1, \ldots, u_d$ is a basis of Λ. Let $L_0 = \{0\}$ and let $L_k = \operatorname{span}(u_1, \ldots, u_k)$ for $k \geq 1$. Prove that there is a vector $u \in \Lambda \setminus \{0\}$ such that

$$\|u\| \leq d \min_{k=1,\ldots,d} \operatorname{dist}(u_k, \ L_{k-1});$$

cf. Problem 2 of Section 1.4.

Remark: This result is due to J.C. Lagarias, H.W. Lenstra and C.-P. Schnorr [**LLS90**].

A quantity which may be viewed as "dual" to the packing radius of a lattice is its covering radius.

(7.3) Definition. Let $\Lambda \subset \mathbb{R}^d$ be a lattice. The largest possible distance from a point in $\mathbb{R}^d$ to the nearest lattice point is called the *covering radius* of Λ and denoted $\mu(\Lambda)$:

$$\mu(\Lambda) = \max_{x \in \mathbb{R}^d} \operatorname{dist}(x, \Lambda).$$

In other words, $\mu(\Lambda)$ is the smallest number α such that the balls of radii α centered at the lattice points cover the whole space $\mathbb{R}^d$.

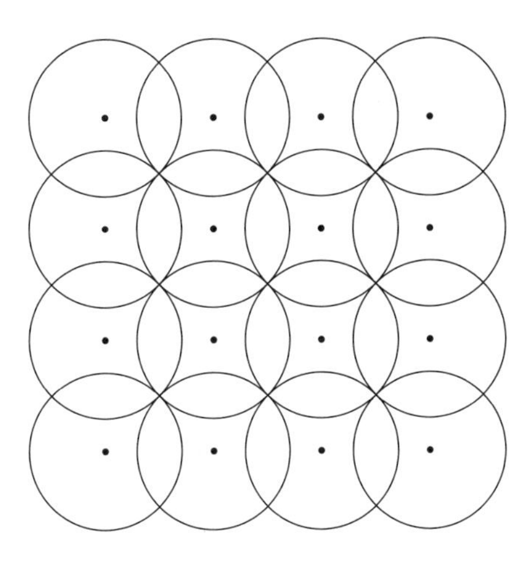

Figure 88. Example: the covering radius of $\mathbb{Z}^2$ is $\frac{\sqrt{2}}{2}$.

PROBLEMS.

1°. Check that the covering radius is well defined and finite.

2. Prove that

$$\mu(\mathbb{Z}^d) = \frac{\sqrt{d}}{2}, \quad \mu(D_3) = 1 \quad \text{and} \quad \mu(D_n) = \frac{\sqrt{n}}{2} \quad \text{for} \quad n \geq 4.$$

3. Prove that $\mu(E_8) = 1$.

There is a remarkable duality result relating the covering radius of a lattice and the packing radius of the dual lattice.

(7.4) Theorem. *Let $\Lambda \subset \mathbb{R}^d$ be a lattice and let $\Lambda^* \subset \mathbb{R}^d$ be the dual lattice. Then the covering radius of Λ and the packing radius of Λ^* are related by the inequalities*

$$\frac{1}{4} \leq \mu(\Lambda) \cdot \rho(\Lambda^*) \leq c(d), \quad \textit{where} \quad c(d) = \frac{1}{4}\sqrt{\sum_{k=1}^{d} k^2} \leq \frac{d^{3/2}}{4}.$$

Proof. We prove the lower bound first (this is the "easy" part and it was known long before the upper bound). Let us choose linearly independent vectors $u_1, \ldots,$ $u_d \in \Lambda$ as follows: u_1 is the shortest non-zero vector in Λ, $u_2 \in \Lambda$ is the shortest vector such that u_1 and u_2 are linearly independent, and so forth, so that $u_d \in \Lambda$ is the shortest vector such that $u_1, \ldots, u_d$ are linearly independent. Then $\|u_1\| \leq \|u_2\| \leq \ldots \leq \|u_d\|$ (the lengths of $\|u_i\|$ are called *successive minima*; cf. Problem 11 of Section 3.2). Let $x = (1/2)u_d$. We claim that

$$\operatorname{dist}(x, \Lambda) = \operatorname{dist}(x, 0) = \operatorname{dist}(x, u_d) = \|u_d\|/2.$$

Indeed, suppose that $v \in \Lambda$ is a point such that $\operatorname{dist}(x, v) < \|u_d\|/2$. Then, by the triangle inequality, $\|v\| \leq \|x\| + \operatorname{dist}(x, v) < \|u_d\|$. If $v \notin \operatorname{span}(u_1, \ldots, u_{d-1})$, we get a contradiction with the definition of u_d. If $v \in \operatorname{span}(u_1, \ldots, u_{d-1})$, then $w = 2v - u_d \in \Lambda$ is linearly independent of $u_1, \ldots, u_{d-1}$ and $\|w\| = \|2(v - x)\| < \|u_d\|$, which is a contradiction.

Thus we conclude that

$$\mu(\Lambda) \geq \frac{\|u_d\|}{2} \geq \frac{\|u_i\|}{2} \quad \text{for} \quad i = 1, \ldots, d.$$

By Problem 2 of Section 7.2, for all $v \in \Lambda^* \setminus \{0\}$ we have

$$\max_{i=1,\ldots,d} \|v\| \cdot \|u_i\| \geq 1 \quad \text{and hence} \quad \mu(\Lambda) \cdot \|v\| \geq \frac{1}{2}.$$

Since $\rho(\Lambda^*) = \|v\|/2$ for a shortest vector $v \in \Lambda^* \setminus \{0\}$, we get the lower bound

$$\mu(\Lambda) \cdot \rho(\Lambda^*) \geq \frac{1}{4}.$$

We prove the upper bound by induction on d. If $d = 1$, then $\Lambda = \{\alpha m : m \in \mathbb{Z}\}$ and $\Lambda^* = \{\alpha^{-1} m : m \in \mathbb{Z}\}$ for some $\alpha > 0$. Hence $\mu(\Lambda) = \alpha/2$ and $\rho(\Lambda^*) = \alpha^{-1}/2$, so $\mu(\Lambda) \cdot \rho(\Lambda^*) = 1/4$.

Suppose that $d > 1$. Let u be a shortest non-zero vector in Λ, so that

$$\|u\| = 2\rho(\Lambda). \tag{7.4.1}$$

Let us identify the orthogonal complement to u with $\mathbb{R}^{d-1}$ and let $pr : \mathbb{R}^d \longrightarrow \mathbb{R}^{d-1}$ be the orthogonal projection. Let $\Lambda_1 = pr(\Lambda)$ be the orthogonal projection of Λ onto $\mathbb{R}^{d-1}$. Then Λ_1 is a lattice in $\mathbb{R}^{d-1}$; see Problem 3 of Section 1.3.

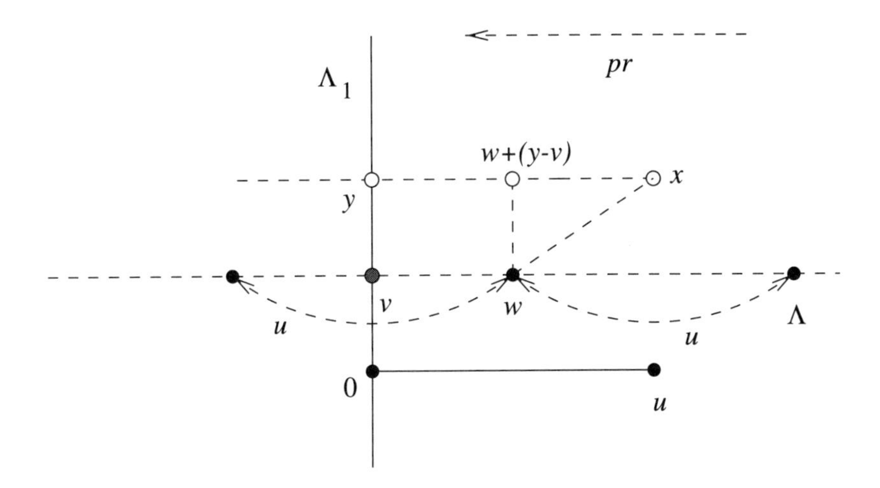

Figure 89. Black points are from Λ, grey points are from Λ_1, white points are from $\mathbb{R}^d$ or $\mathbb{R}^{d-1}$.

Let $\Lambda_1^* \subset \mathbb{R}^{d-1}$ be the lattice dual to Λ_1. For every $a \in \Lambda_1^*$ and every $b \in \Lambda$, we have $\langle a, b\rangle = \langle a, pr(b)\rangle \in \mathbb{Z}$, so $\Lambda_1^* \subset \Lambda^*$. In particular,

$$\rho(\Lambda_1^*) \geq \rho(\Lambda^*). \tag{7.4.2}$$

Let us choose a point $x \in \mathbb{R}^d$ and let us estimate $\operatorname{dist}(x, \Lambda)$.

Let $y = pr(x)$. Let $v \in \Lambda_1$ be a closest point to y in Λ_1, so

$$\operatorname{dist}(y, v) \leq \mu(\Lambda_1). \tag{7.4.3}$$

The line through v parallel to u intersects Λ by a set of equally spaced points, each being of distance $\|u\|$ from the next. Hence we can find a point $w \in \Lambda$ such that $pr(w) = v$ and

$$\operatorname{dist}\bigl(x,\ w + (y - v)\bigr) \leq \frac{\|u\|}{2}; \tag{7.4.4}$$

see Figure 89. By Pythagoras' Theorem,

$$\operatorname{dist}^2(x,w) = \operatorname{dist}^2\big(x,\ w+(y-v)\big) + \operatorname{dist}^2(y,v).$$

Applying (7.4.4) and (7.4.3), we get

$$\operatorname{dist}^2(x,w) \leq \mu^2(\Lambda_1) + \frac{\|u\|^2}{4}.$$

Since x was chosen arbitrary, by using (7.4.1), we conclude that

$$\mu^2(\Lambda) \leq \mu^2(\Lambda_1) + \frac{\|u\|^2}{4} = \mu^2(\Lambda_1) + \rho^2(\Lambda).$$

Applying (7.4.2), the induction hypothesis and Lemma 7.2, we get

$$\begin{aligned}\mu^2(\Lambda)\cdot\rho^2(\Lambda^*) &\leq \mu^2(\Lambda_1)\cdot\rho^2(\Lambda^*) + \rho^2(\Lambda)\cdot\rho^2(\Lambda^*)\\ &\leq \mu^2(\Lambda_1)\cdot\rho^2(\Lambda_1^*) + \rho^2(\Lambda)\cdot\rho^2(\Lambda^*)\\ &\leq c^2(d-1) + \frac{d^2}{16},\end{aligned}$$

which completes the proof. □

It is known that one can choose $c(d)$ in Theorem 7.4 to be γd for some constant $\gamma > 0$. The above proof belongs to J.C. Lagarias, H.W. Lenstra and C.-P. Schnorr [**LLS90**].

PROBLEMS.

1°. Let $u_1, \dots, u_d \in \Lambda$ be linearly independent vectors. Prove that

$$\mu(\Lambda) \leq \frac{1}{2}\sum_{i=1}^{d} \|u_i\|.$$

In Problems 2–4, $\{\{\xi\}\}$ denotes the distance from a number $\xi \in \mathbb{R}$ to the nearest integer, so that $0 \leq \{\{\xi\}\} \leq 1/2$.

2. Let $\theta_1, \dots, \theta_n$ be real numbers such that $m_1\theta_1 + \ldots + m_n\theta_n + m_{n+1} = 0$ for integers $m_1, \dots, m_{n+1}$ implies $m_1 = \ldots = m_{n+1} = 0$. Prove Kronecker's Theorem: for any real vector $a = (\alpha_1, \dots, \alpha_n)$ and for any $\epsilon > 0$ there exists a positive integer m such that $\{\{\alpha_i - m\theta_i\}\} < \epsilon$ for $i = 1, \dots, n$.

Hint: Let $d = n+1$ and let $\tau > 0$ be a number. Consider the set $\Lambda_\tau \subset \mathbb{R}^d$ of all integer linear combinations of the vectors $u_1 = (1, 0, \dots, 0), \dots, u_n = (0, \dots, 0, 1, 0)$, $u_{n+1} = (\theta_1, \dots, \theta_n, \tau^{-1})$. Show that Λ_τ is a lattice and that the packing radius of Λ_τ^* grows to infinity when τ grows. Deduce that the covering radius of Λ_τ approaches 0 as τ grows.

3. Let $\Lambda \subset \mathbb{R}^d$ be a lattice and let $x \in \mathbb{R}^d$ be a point. Prove that for every $v \in \Lambda^* \setminus \{0\}$

$$\frac{\{\{\langle v, x\rangle\}\}}{\|v\|} \leq \operatorname{dist}(x, \Lambda).$$

4*. Prove that for every lattice $\Lambda \subset \mathbb{R}^d$ and every $x \in \mathbb{R}^d$ there is a point $v \in \Lambda^* \setminus \{0\}$ such that

$$\frac{\{\{\langle v, x\rangle\}\}}{\|v\|} \geq \frac{1}{6d^2+1} \operatorname{dist}(x, \Lambda).$$

Remark: This result is due to J. Håstad [**Hå88**].

8. The Flatness Theorem

Minkowski's Convex Body Theorem (Theorem 3.2) asserts that a symmetric convex body of sufficiently large volume contains a lattice point other than the origin. One may ask if a similar statement can be made about a general convex body. It quickly becomes clear that the volume is not an issue here: a convex body without lattice points can have an arbitrarily large volume.

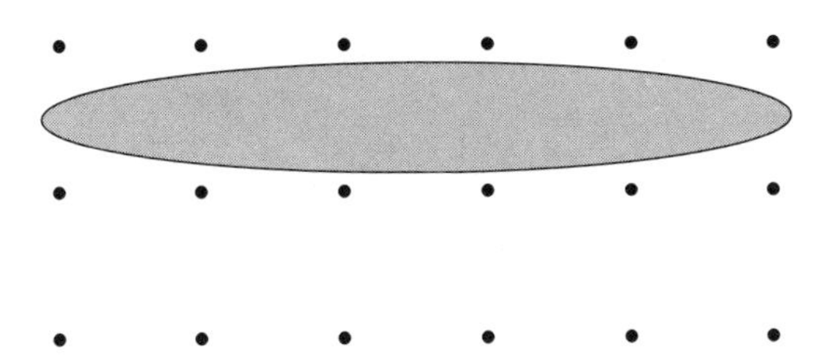

Figure 90. A convex body without lattice points can have an arbitrarily large volume.

Having done some experimenting, we start to feel that a convex body without lattice points must be somewhat "flat".

Let $\Lambda \subset \mathbb{R}^d$ be a lattice and let $\Lambda^* \subset \mathbb{R}^d$ be the dual lattice. Let $v \in \Lambda^*$ be a non-zero lattice point. Then lattice Λ can be "sliced" into "layers" Λ_k of lattice points:

$$\Lambda_k = \Big\{u \in \Lambda: \quad \langle u, v\rangle = k\Big\} \quad \text{for} \quad k \in \mathbb{Z}.$$

It turns out that a convex body without lattice points must be "squeezed" in between some layers that are not very far apart. This is the main content of the "Flatness Theorem" that we prove in this section. In fact, there are many "flatness theorems": they differ from each other by estimating exactly how far apart those layers can be. The best possible bound is not known yet and we don't strive to obtain it. Our objective is to give a simple proof of a reasonably good bound. Our main tool is Theorem 7.4.

(8.1) Lemma. *Let $\Lambda \subset \mathbb{R}^d$ be a lattice and let B be a ball such that $B \cap \Lambda = \emptyset$. Let $v \in \Lambda^*$ be a shortest non-zero vector. Then*

$$\max\bigl\{\langle v, x\rangle : x \in B\bigr\} - \min\bigl\{\langle v, x\rangle : x \in B\bigr\} \leq c(d)$$

where one can choose $c(d) = d^{3/2}$.

Proof. Let a be the center of B and let β be the radius of B, so $B = \bigl\{x : \|x-a\| \leq \beta\bigr\}$. Then

$$\begin{aligned}\max\bigl\{\langle v, x\rangle : x \in B\bigr\} &- \min\bigl\{\langle v, x\rangle : x \in B\bigr\} \\ &= \Bigl(\langle v, a\rangle + \beta\|v\|\Bigr) - \Bigl(\langle v, a\rangle - \beta\|v\|\Bigr) = 2\beta\|v\|.\end{aligned}$$

Since B does not contain lattice points, its radius does not exceed the covering radius of Λ, so $\beta \leq \mu(\Lambda)$. Since v is a shortest non-zero vector from Λ^*, the length of v is twice the packing radius of Λ^*, so $\|v\| = 2\rho(\Lambda^*)$. The proof now follows by Theorem 7.4. □

PROBLEM.

1°. Let $\Lambda_1 \subset \mathbb{R}^d$ be a lattice and let $T : \mathbb{R}^d \longrightarrow \mathbb{R}^d$ be an invertible linear transformation. Prove that $\Lambda = T(\Lambda_1)$ is a lattice and that $\Lambda^* = (T^*)^{-1}(\Lambda_1^*)$.

(8.2) Lemma. *Let $\Lambda \subset \mathbb{R}^d$ be a lattice and let $E \subset \mathbb{R}^d$ be an ellipsoid such that $E \cap \Lambda = \emptyset$. Then there exists a non-zero vector $w \in \Lambda^*$ such that*

$$\max\bigl\{\langle w, x\rangle : x \in E\bigr\} - \min\bigl\{\langle w, x\rangle : x \in E\bigr\} \leq c(d)$$

where one can choose $c(d) = d^{3/2}$.

Proof. Since E is an ellipsoid, there exists an invertible linear transformation T such that $E = T(B)$, where B is a ball. Let $\Lambda_1 = T^{-1}(\Lambda)$. Hence, by Problem 1 of Section 8.1, Λ_1 is a lattice and $\Lambda = T(\Lambda_1)$. Then $B \cap \Lambda_1 = \emptyset$ and hence by Lemma 8.1 there exists a non-zero vector $v \in \Lambda_1^*$ such that

$$\max\bigl\{\langle v, y\rangle : y \in B\bigr\} - \min\bigl\{\langle v, y\rangle : y \in B\bigr\} \leq c(d).$$

Let $w = (T^*)^{-1}v$. By Problem 1 of Section 8.1, we have $w \in \Lambda^* \setminus \{0\}$. Now

$$\begin{aligned}\max\bigl\{\langle w, x\rangle : x \in E\bigr\} &- \min\bigl\{\langle w, x\rangle : x \in E\bigr\} \\ &= \max\bigl\{\langle w,\ T(y)\rangle : y \in B\bigr\} - \min\bigl\{\langle w,\ T(y)\rangle : y \in B\bigr\} \\ &= \max\bigl\{\langle T^*(w),\ y\rangle : y \in B\bigr\} - \min\bigl\{\langle T^*(w),\ y\rangle : y \in B\bigr\} \\ &= \max\bigl\{\langle v, y\rangle : y \in B\bigr\} - \min\bigl\{\langle v, y\rangle : y \in B\bigr\} \leq c(d)\end{aligned}$$

and the result follows. □

Finally, we prove the general "flatness theorem".

(8.3) Theorem. *Let $\Lambda \subset \mathbb{R}^d$ be a lattice and let $K \subset \mathbb{R}^d$ be a convex body such that $K \cap \Lambda = \emptyset$. Then there exists a non-zero vector $w \in \Lambda^*$ such that*

$$\max\bigl\{\langle w, x\rangle : \ x \in K\bigr\} - \min\bigl\{\langle w, x\rangle : \ x \in K\bigr\} \leq c_1(d)$$

where one can choose $c_1(d) = d^{5/2}$.

Proof. Let E be the maximum volume ellipsoid of K; see Section V.2. Assuming that a is the center of E, by Theorem V.2.4 we get

$$E \subset K \subset d(E - a) + a = dE + (1 - d)a.$$

Let $w \in \Lambda^*$ be the vector for the ellipsoid E whose existence is asserted by Lemma 8.2. Then

$$\begin{aligned}\max\bigl\{\langle w, x\rangle : \ x \in K\bigr\} &\leq \max\bigl\{\langle w, x\rangle : \ x \in dE + (1-d)a\bigr\} \\ &= (1-d)\langle w, a\rangle + d\max\bigl\{\langle w, x\rangle : \ x \in E\bigr\}.\end{aligned}$$

Similarly,

$$\begin{aligned}\min\bigl\{\langle w, x\rangle : \ x \in K\bigr\} &\geq \min\bigl\{\langle w, x\rangle : \ x \in dE + (1-d)a\bigr\} \\ &= (1-d)\langle w, a\rangle + d\min\bigl\{\langle w, x\rangle : \ x \in E\bigr\}.\end{aligned}$$

Therefore,

$$\begin{aligned}&\max\bigl\{\langle w, x\rangle : \ x \in K\bigr\} - \min\bigl\{\langle w, x\rangle : \ x \in K\bigr\} \\ &\quad \leq d\max\bigl\{\langle w, x\rangle : \ x \in E\bigr\} - d\min\bigl\{\langle w, x\rangle : \ x \in E\bigr\} \leq dc(d) = c_1(d)\end{aligned}$$

by Lemma 8.2. □

Let $w \in \Lambda^*$ be the vector from Theorem 8.3. Let

$$\Lambda_k = \bigl\{u \in \Lambda : \quad \langle u, w\rangle = k\bigr\} \quad \text{for} \quad k \in \mathbb{Z}$$

be the "layers" of lattice points determined by w. Let H_k be the affine hull of Λ_k. Theorem 8.3 asserts that if a convex body does not contain lattice points, then it may intersect only a small number (polynomial in the dimension d) of affine hyperplanes H_k; cf. Figure 91.

The smallest possible value of $c_1(d)$ is not known. It is conjectured though to be roughly proportional to d. The first (exponential in d) bound for $c_1(d)$ was obtained by A.Ya. Khintchin in 1948. Theorem 8.3 is due to J.C. Lagarias, H.W. Lenstra and C.-P. Schnorr [**LLS90**]. It is known that one can choose $c(d) = \gamma d$ for some $\gamma > 0$ in Lemmas 8.1 and 8.2 [**Ba95**] and that one can choose $c_1(d) = O(d^{3/2})$ in Theorem 8.3 [**BLP99**]. Moreover, if K has a center of symmetry, then one can choose $c_1(d) = O(d \ln d)$ [**Ba95**], [**Ba96**], which is optimal up to a logarithmic factor.

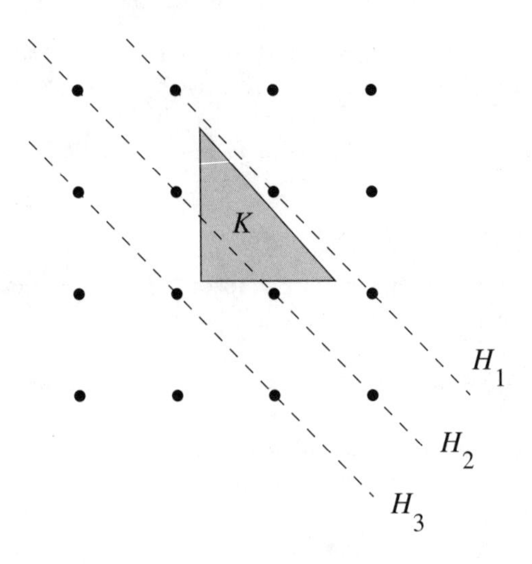

Figure 91

PROBLEMS.

1. Show that we must have $c_1(d) \geq d$ in Theorem 8.3.

2. Let $P \subset \mathbb{R}^d$ be a polytope with the vertices $v_1, \ldots, v_m \in \mathbb{Z}^d$. Suppose that $m > 2^d$. Prove that there is a point $v \in P \cap \mathbb{Z}^d$ different from $v_1, \ldots, v_m$.

3* (J.-P. Doignon, 1972). Let $A_1, \ldots, A_m \subset \mathbb{R}^d$ be convex sets. Prove the following integer version of Helly's Theorem: if

$$\Bigl(A_{i_1} \cap \ldots \cap A_{i_k}\Bigr) \cap \mathbb{Z}^d \neq \emptyset$$

for every collection of $k = 2^d$ sets $A_{i_1}, \ldots, A_{i_k}$, then

$$\Bigl(\bigcap_{i=1}^{m} A_i\Bigr) \cap \mathbb{Z}^d \neq \emptyset.$$

Remark: See [**Do73**].

4. Let $P \subset \mathbb{R}^2$ be a convex polygon with vertices in $\mathbb{Z}^2$. Suppose that P does not contain any point from $\mathbb{Z}^2$ other than its vertices. Prove that there exists a vector $w \in \mathbb{Z}^2 \setminus \{0\}$ such that

$$\max\bigl\{\langle w, x\rangle : \ x \in P\bigr\} - \min\bigl\{\langle w, x\rangle : \ x \in P\bigr\} \leq 1.$$

5* (R. Howe). Let $P \subset \mathbb{R}^3$ be a convex polytope with vertices in $\mathbb{Z}^3$. Suppose that P does not contain any point from $\mathbb{Z}^3$ other than its vertices. Prove that there exists a vector $w \in \mathbb{Z}^3 \setminus \{0\}$ such that

$$\max\bigl\{\langle w, x\rangle : \ x \in P\bigr\} - \min\bigl\{\langle w, x\rangle : \ x \in P\bigr\} \leq 1.$$

Remark: See [**S85**].

9. Constructing a Short Vector and a Reduced Basis

For various reasons we often need to compute efficiently a shortest or a reasonably short non-zero vector in a given lattice. This is the case, for example, for the Flatness Theorem of Section 8 if we want to know a direction in which a convex body without lattice points is flat. In this section, we sketch an efficient algorithm due to A.K. Lenstra, H.W. Lenstra and L. Lovász [**LLL82**]. The procedure is called now the Lenstra-Lenstra-Lovász reduction or just the LLL reduction. Given a basis of a lattice, it produces another basis of the same lattice which has some very useful properties.

(9.1) Reduced basis. Let $\Lambda \subset \mathbb{R}^d$ be a lattice and let $u_1, \ldots, u_d$ be a basis of Λ. Let us describe the properties that we want our basis to satisfy.

Let us define subspaces $\{0\} = L_0 \subset L_1 \subset \ldots \subset L_d = \mathbb{R}^d$ by

$$L_k = \operatorname{span}\big(u_1, \ldots, u_k\big) \quad \text{for} \quad k = 1, \ldots, d.$$

Let $L_k^\perp$ denote the orthogonal complement of L_k and let w_k denote the orthogonal projection of u_k onto $L_{k-1}^\perp$. In other words, $w_1, \ldots, w_d$ is the Gram-Schmidt orthogonalization (without normalization!) of $u_1, \ldots, u_d$.

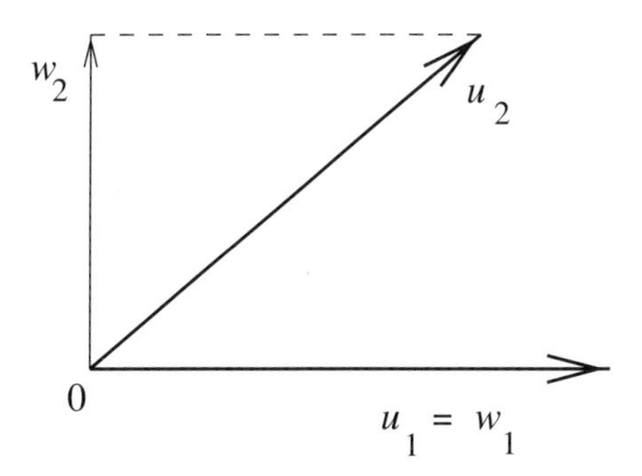

Figure 92

In particular, $w_1 = u_1$, $w_2, \ldots, w_d$ is a basis of $\mathbb{R}^d$ and

$$\|w_k\| = \operatorname{dist}(u_k, \ L_{k-1}) \quad \text{for} \quad k = 1, \ldots, d.$$

Hence we can write

$$u_k = w_k + \sum_{i=1}^{k-1} \alpha_{ki} w_i \quad \text{for} \quad k = 1, \ldots, d. \tag{9.1.1}$$

We say that the basis $u_1, \ldots, u_d$ is *reduced* (or *Lenstra-Lenstra-Lovász reduced* or *LLL reduced*) if the following properties (9.1.2)–(9.1.3) are satisfied:

$$|\alpha_{ki}| \leq \frac{1}{2} \quad \text{for all} \quad 1 \leq i < k \leq d \tag{9.1.2}$$

and

$$\text{dist}^2(u_k,\ L_{k-1}) \leq \tau\, \text{dist}^2(u_{k+1},\ L_{k-1}) \quad \text{for} \quad k = 1, \ldots, d-1 \quad \text{with} \quad \tau = \frac{4}{3}. \tag{9.1.3}$$

The last condition is written in a bizarre way; indeed, the exact value of τ is not really important as long as $1 < \tau < 4$ with the standard choice being $\tau = 4/3$. We will trace the role of the parameter τ throughout the proofs that follow to reveal the mystery of 4/3.

At this point, it is not clear whether a reduced basis exists, let alone how to construct one. But before we show how to construct such a basis, we demonstrate that a reduced basis, if one exists, satisfies some useful properties.

(9.2) Theorem. *Let $\Lambda \subset \mathbb{R}^d$ be a lattice and let $u_1, \ldots, u_d$ be its reduced basis. Then*

$$\|u_1\| \leq 2^{\frac{d-1}{2}} \|v\| \quad \textit{for all} \quad v \in \Lambda \setminus \{0\}.$$

In other words, the first basis vector u_1 is reasonably short.

Proof. By (9.1.1),

$$\text{dist}(u_{k+1},\ L_{k-1}) = \|w_{k+1} + \alpha_{k+1,k} w_k\|.$$

By (9.1.3) and (9.1.2), we get

$$\begin{aligned}\|w_k\|^2 &= \text{dist}^2(u_k,\ L_{k-1}) \leq \tau\, \text{dist}^2(u_{k+1},\ L_{k-1}) = \tau \|w_{k+1}\|^2 + \tau \alpha_{k+1,k}^2 \|w_k\|^2 \\ &\leq \tau \|w_{k+1}\|^2 + (\tau/4)\|w_k\|^2.\end{aligned}$$

Therefore,

$$\|w_{k+1}\|^2 \geq \tau^{-1}(1 - \tau/4)\|w_k\|^2 = \frac{1}{2}\|w_k\|^2$$

(now we see why we should have $\tau < 4$). Iterating the inequality, we get

$$\|w_k\|^2 \geq \frac{1}{2}\|w_{k-1}\|^2 \geq \frac{1}{4}\|w_{k-2}\|^2 \geq \ldots \geq 2^{-k+1}\|w_1\|^2.$$

In particular,

$$\text{dist}(u_k,\ L_{k-1}) = \|w_k\| \geq 2^{\frac{1-d}{2}} \|w_1\| = 2^{\frac{1-d}{2}} \|u_1\|.$$

The proof follows by Problem 2 of Section 1.4. □

PROBLEMS.

1. Let $\Lambda \subset \mathbb{R}^d$ be a lattice and let $u_1, \ldots, u_d$ be its reduced basis. Prove that $\|u_1\| \leq 2^{(d-1)/4} (\det \Lambda)^{1/d}$; cf. Problem 2 of Section 5.3.

2. Let $\Lambda \subset \mathbb{R}^d$ be a lattice and let $u_1, \ldots, u_d$ be its reduced basis. Prove that

$$\prod_{i=1}^{d} \|u_i\| \leq 2^{\frac{d(d-1)}{4}} \det \Lambda;$$

cf. Problem 3 of Section 1.4 and Problem 12 of Section 3.2.

3 (L. Babai). Let $\Lambda \subset \mathbb{R}^d$ be a lattice, let $u_1, \ldots, u_d$ be its reduced basis and let $w_1, \ldots, w_d$ be the Gram-Schmidt orthogonalization of $u_1, \ldots, u_d$. Given a point $b \in \mathbb{R}^d$, show that there exists a point $v \in \Lambda$ such that

$$b - v = \sum_{i=1}^{d} \beta_i w_i \quad \text{where} \quad |\beta_i| \leq \frac{1}{2} \quad \text{for} \quad i = 1, \ldots, d$$

and that for such a point we have

$$\operatorname{dist}(b, v) \leq 2^{d/2-1} \operatorname{dist}(b, \Lambda).$$

4. Let $\Lambda \subset \mathbb{R}^d$ be a lattice, let $u_1, \ldots, u_d$ be its reduced basis and let $u \in \Lambda$ be a shortest non-zero lattice vector. Suppose that $u = \sum_{k=1}^{d} \gamma_k u_k$ for some $\gamma_k \in \mathbb{Z}$. Prove that $|\gamma_k| \leq 3^d$ for $k = 1, \ldots, d$.

Now we discuss why a reduced basis exists and how to construct one.

(9.3) The algorithm to construct a reduced basis. We start with an arbitrary basis $u_1, \ldots, u_d$ of Λ. Indeed, we have to assume that the lattice Λ is given to us somehow; we assume, therefore, that it is given by some basis. The algorithm consists of repeated applications of the following two procedures.

(9.3.1) Enforcing conditions (9.1.2). Given the current basis $u_1, \ldots, u_d$ of Λ, we compute the Gram-Schmidt orthogonalization $w_1, \ldots, w_d$ as in Section 9.1 and compute the expansion (9.1.1). These are problems of linear algebra and can be solved easily. If $|\alpha_{ki}| \leq 1/2$ for all $1 \leq i < k \leq d$, we go to (9.3.2). Otherwise, we locate a pair of indices $i < k$ with $|\alpha_{ki}| > 1/2$ and the largest i. We modify u_k by

$$\text{new } u_k := \text{old } u_k - [\alpha_{ki}] u_i, \quad \text{where} \quad [\alpha_{ki}] \quad \text{is the nearest integer to} \quad \alpha_{ki}$$

(and ties are broken arbitrarily). Clearly, this action does not change the subspaces $L_k = \operatorname{span}(u_1, \ldots, u_k)$ and the vectors $w_1, \ldots, w_d$. It changes, however, some of the coefficients in (9.1.1). The coefficients of (9.1.1) that do change are α_{kj} with $j \leq i$. In particular, we get

$$\text{new } \alpha_{ki} := \text{old } \alpha_{ki} - [\text{old } \alpha_{ki}] = \{\{\text{old } \alpha_{ki}\}\},$$

where $\{\{\xi\}\}$ is the signed distance of a real number ξ to the nearest integer, so $-1/2 \leq \{\{\xi\}\} \leq 1/2$. We repeat this procedure, always "straightening out" the rightmost "wrong" coefficient α_{ki} so that any of $\alpha_{k,i+1}, \ldots, \alpha_{kk}$ do not get "spoiled". After repeating the procedure at most $\binom{d}{2}$ times, (9.1.2) is satisfied.

(9.3.2) Enforcing conditions (9.1.3). We check conditions (9.1.3); they are easy to check having expansions (9.1.1). If the conditions are satisfied, we stop and output the current basis $u_1, \ldots, u_d$. If, in fact,

$$\operatorname{dist}^2(u_k,\ L_{k-1}) > \tau \operatorname{dist}^2(u_{k+1},\ L_{k-1}),$$

we swap u_k and u_{k+1}:

$$\text{new } u_k := \text{old } u_{k+1} \quad \text{and} \quad \text{new } u_{k+1} := \text{old } u_k$$

and go to (9.3.1).

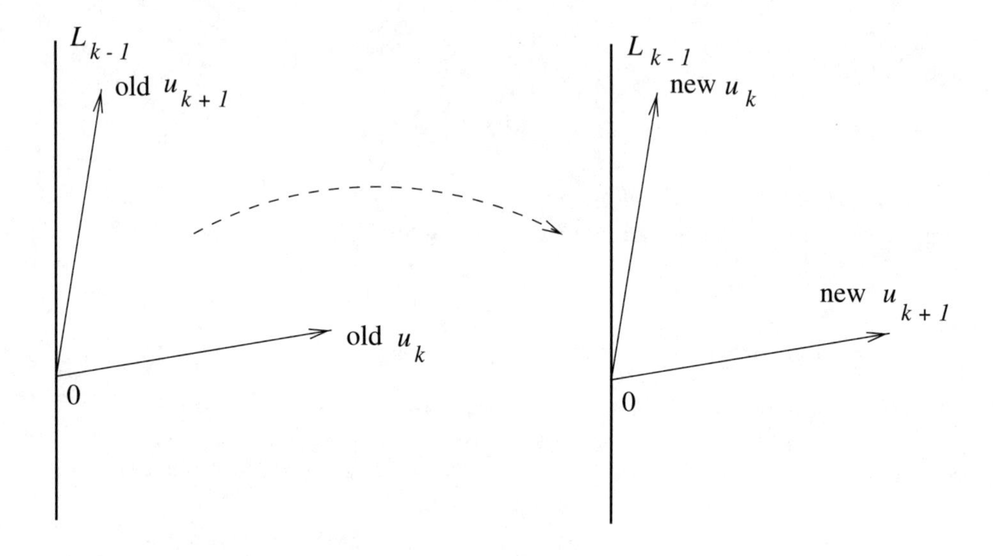

Figure 93. If u_{k+1} is much closer to L_{k-1} than u_k we swap u_k and u_{k+1}.

Clearly, if the algorithm ever stops, it outputs the reduced basis $u_1, \ldots, u_d$. It is not obvious, however, that the algorithm cannot cycle endlessly. In fact, it does stop and, moreover, the running time is quite reasonable. In practice, the algorithm performs quite well. We prove a bound for the running time, which, although not optimal, hints that the algorithm actually runs *in polynomial time* (to prove that the algorithm indeed runs in polynomial time, we also need to verify that the numbers do not grow too wild, which we don't do here).

PROBLEMS.

1°. Let $\Lambda \subset \mathbb{R}^d$ be a lattice and let $u_1, \ldots, u_d \in \Lambda$ be its basis. Let

$$L_0 = \{0\} \quad \text{and} \quad L_k = \operatorname{span}(u_1, \ldots, u_k) \quad \text{for} \quad k = 1, \ldots, d.$$

Let w_k be the orthogonal projection of u_k onto $L_{k-1}^{\perp}$ for $k = 1, \ldots, d$.

For $k = 1, \ldots, d$, let $\Lambda_k = \Lambda \cap L_k$. Considering Λ_k as a lattice in L_k, prove that

$$\det \Lambda_k = \prod_{i=1}^{k} \|w_i\|.$$

2°. Let $\Lambda \subset \mathbb{R}^d$ be a lattice, let $u_1, \ldots, u_k \in \Lambda$ be linearly independent points and let $L_k = \operatorname{span}(u_1, \ldots, u_k)$. Let $\Lambda_k = \Lambda \cap L_k$. Let us consider Λ_k as a lattice in L. Prove that

$$\det \Lambda_k \geq \left(\frac{\lambda}{\sqrt{k}}\right)^k, \quad \text{where} \quad \lambda = \min_{u \in \Lambda \setminus \{0\}} \|u\|.$$

(9.4) Theorem. *Let $\Lambda \subset \mathbb{R}^d$ be a lattice given by its basis $u_1, \ldots, u_d$. Let us define the subspaces $\{0\} = L_0 \subset L_1 \subset \ldots \subset L_d = \mathbb{R}^d$ by*

$$L_k = \operatorname{span}(u_1, \ldots, u_k) \quad \text{for} \quad 1 \leq k \leq d.$$

Let $\Lambda_k = \Lambda \cap L_k$. We consider Λ_k as a lattice in L_k. Let us define

$$D(u_1, \ldots, u_d) = \prod_{k=1}^{d-1} (\det \Lambda_k)$$

and let

$$\lambda = \min_{v \in \Lambda \setminus \{0\}} \|v\|$$

be the length of a shortest non-zero vector of the lattice.

Let m be a positive integer such that

$$\tau^{-m/2} D(u_1, \ldots, u_d) = \left(\frac{\sqrt{3}}{2}\right)^{m/2} D(u_1, \ldots, u_d) < \lambda^{d(d-1)/2} \prod_{k=1}^{d-1} k^{-k/2}.$$

Then Algorithm 9.3 stops after at most $m+1$ applications of (9.3.1) and at most $m+1$ applications of (9.3.2).

In particular, every lattice has a reduced basis.

Proof. Let us see what happens to $D(u_1, \ldots, u_d)$ when we change the basis $u_1, \ldots, u_d$ by performing procedures (9.3.1) and (9.3.2) of the algorithm. Procedure (9.3.1) does not change the spaces L_k and hence does not change $D(u_1, \ldots, u_d)$.

Swapping u_k and u_{k+1} in procedure (9.3.2) changes the space L_k only. Let us see how $\det \Lambda_k$ changes.

Let $w_1, \ldots, w_d$ be the Gram-Schmidt orthogonalization of $u_1, \ldots, u_d$. By Problem 1 of Section 9.3,

$$\det \Lambda_k = \prod_{i=1}^{k} \|w_i\|.$$

In this product, procedure (9.3.2) changes the factor $\|w_k\|$ only. We have

$$\begin{aligned}\|\text{new } w_k\| &= \operatorname{dist}\bigl(\text{new } u_k,\ L_{k-1}\bigr) = \operatorname{dist}\bigl(\text{old } u_{k+1},\ L_{k-1}\bigr)\\ &< \tau^{-1/2} \operatorname{dist}\bigl(\text{old } u_k,\ L_{k-1}\bigr) = \tau^{-1/2}\|\text{old } w_k\|.\end{aligned}$$

Hence each application of swapping in procedure (9.3.2) gets $D(u_1, \ldots, u_d)$ multiplied by $\tau^{-1/2} = \sqrt{3}/2$ or a smaller number (it clear now why we should have $\tau > 1$).

By Problem 2 of Section 9.3, we get

$$D(u_1, \ldots, u_d) = \prod_{k=1}^{d-1} (\det \Lambda_k) \geq \prod_{k=1}^{d-1} \Bigl(\frac{\lambda}{\sqrt{k}}\Bigr)^k = \lambda^{d(d-1)/2} \prod_{k=1}^{d-1} k^{-k/2}$$

for any basis $u_1, \ldots, u_d$ of Λ.

Hence Algorithm 9.3 performs procedure (9.3.2) at most $m+1$ times (the last application checks that conditions (9.1.3) are satisfied and outputs the current basis $u_1, \ldots, u_d$). Since each application of procedure (9.3.2) is accompanied by at most one application of procedure (9.3.1), the result follows. □

C.-P. Schnorr constructed a modification of the algorithm, which, for any fixed $\epsilon > 0$, produces in polynomial time a non-zero lattice vector whose length approximates the length of a shortest non-zero lattice vector within a factor of $(1+\epsilon)^d$ [**Sch87**].

PROBLEM.

1. Suppose that $\Lambda \subset \mathbb{Z}^d$ is a sublattice of the standard integer lattice. Prove that $D(u_1, \ldots, u_d) \geq 1$ for any basis $u_1, \ldots, u_d$ of Λ.

10. Remarks

Our main references are [**C97**], [**GL87**] and [**CS99**]. In particular, [**CS99**] contains a wealth of material on particularly interesting lattices and sphere packings. For the Lenstra-Lenstra-Lovász reduced basis and its numerous applications, see [**Lo86**], [**GLS93**] and the original paper [**LLL82**]. A nice generalization of the reduction for arbitrary norms is given in [**LS92**].

Chapter VIII

Lattice Points and Polyhedra

We discuss the enumeration of lattice points in polyhedra. Our main tools are generating functions, also known as exponential sums, and some identities in the algebra of polyhedra. A parallel theory for exponential integrals is developed in the exercises. Since we are interested in combinatorial rather than metric properties, we consider the case of the standard integer lattice $\mathbb{Z}^d \subset \mathbb{R}^d$ only. The case of a general lattice $\Lambda \subset \mathbb{R}^d$ reduces to that of $\mathbb{Z}^d$ by a change of the coordinates.

1. Generating Functions and Simple Rational Cones

Let $P \subset \mathbb{R}^d$ be a polyhedron and let $\mathbb{Z}^d \subset \mathbb{R}^d$ be the standard integer lattice. For a point $m = (\mu_1, \ldots, \mu_d) \in \mathbb{Z}^d$ we write $\mathbf{x}^m$ for the monomial

$$\mathbf{x}^m = x_1^{\mu_1} \cdots x_d^{\mu_d}$$

in d (complex) variables $(x_1, \ldots, x_d)$. We agree that $x_i^0 = 1$ for all $i = 1, \ldots, d$. The main object of the chapter is the *generating function*

$$f(P, \mathbf{x}) = \sum_{m \in P \cap \mathbb{Z}^d} \mathbf{x}^m;$$

see Figure 94 for an example.

If the sum is infinite, the issue of convergence emerges. Usually, there will be a non-empty open set $U \subset \mathbb{C}^d$ such that the series converges absolutely for all $\mathbf{x} \in U$ and uniformly on compact subsets of U. We don't emphasize analytic rigor here; one can always think that the series behaves "just like the (multiple) geometric series", the basic example being the series $\sum_{m=0}^{\infty} x^m$ which converges to $1/(1-x)$ absolutely for all x with $|x| < 1$ and uniformly on compact subsets of the set $U = \{x \in \mathbb{C} : |x| < 1\}$.

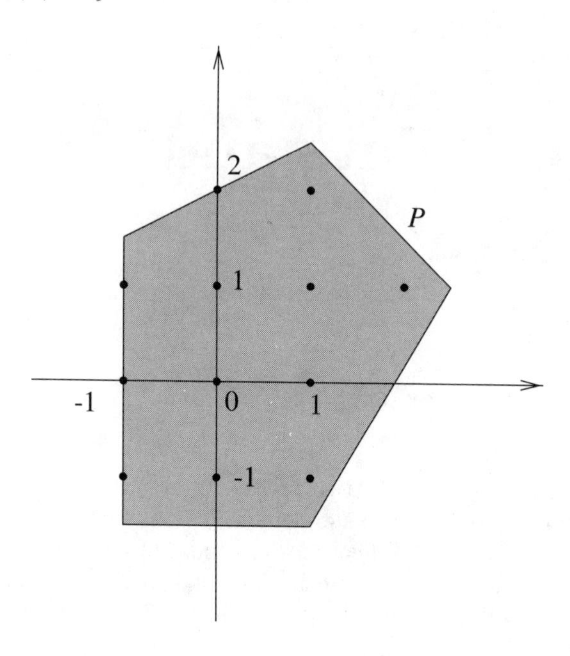

Figure 94. Example: $f(P, \mathbf{x}) = x_1^{-1}x_2^{-1} + x_1^{-1} + x_2^{-1} + 1 + x_1^{-1}x_2 + x_1x_2^{-1} + x_1 + x_2 + x_1x_2 + x_2^2 + x_1x_2^2 + x_1^2x_2$

It is convenient to introduce some notation. For $y = (\eta_1, \ldots, \eta_d) \in \mathbb{C}^d$, let us define

$$\mathbf{e}^y = \big(\exp\{\eta_1\}, \ldots, \exp\{\eta_d\}\big) \in \mathbb{C}^d.$$

Hence for $m = (\mu_1, \ldots, \mu_d) \in \mathbb{Z}^d$ and $\mathbf{x} = \mathbf{e}^y$ we have

$$\mathbf{x}^m = \exp\{\eta_1\}^{\mu_1} \cdots \exp\{\eta_d\}^{\mu_d} = \exp\big\{\eta_1\mu_1 + \ldots + \eta_d\mu_d\big\} = \exp\big\{\langle y, m\rangle\big\}.$$

We denote by $\mathbb{Z}_+^d$ the set of all d-tuples $(\mu_1, \ldots, \mu_d)$ of non-negative integers.

Our immediate goal is to look at the generating function when P is a cone.

(1.1) Definition. Let $u_1, \ldots, u_k \in \mathbb{Z}^d$ be linearly independent lattice vectors and let

$$K = \operatorname{co}\big(u_1, \ldots, u_k\big).$$

The cone K is called a *simple rational cone.*

PROBLEMS.

1°. Check that a simple rational cone in $\mathbb{R}^d$ is a closed convex cone without straight lines.

2°. Let $K = \Big\{(\xi_1, \xi_2) : 0 \leq \xi_2 \leq \xi_1\sqrt{2}\Big\} \subset \mathbb{R}^2$. Prove that K is not a simple rational cone.

3°. Let $K = [0, +\infty) \subset \mathbb{R}^1$. Check that

$$f(K, x) = \sum_{n=0}^{\infty} x^n = \frac{1}{1-x}$$

for all $x \in \mathbb{C}$ such that $|x| < 1$.

4°. Let

$$K = \mathbb{R}^d_+ = \Big\{(\xi_1, \ldots, \xi_d) \in \mathbb{R}^d : \quad \xi_i \geq 0 \quad \text{for all} \quad i = 1, \ldots, d\Big\}.$$

Prove that

$$f(K, x) = \sum_{(\mu_1, \ldots, \mu_d) \in \mathbb{Z}^d_+} x_1^{\mu_1} \cdots x_d^{\mu_d} = \prod_{i=1}^{d} \frac{1}{1 - x_i}$$

for all $(x_1, \ldots, x_d) \in \mathbb{C}$ such that $|x_i| < 1$ for $i = 1, \ldots, d$.

5°. Let $u \in \mathbb{Z}^d$ be an integer vector and let $U = \Big\{\mathbf{x} \in \mathbb{C}^d : |\mathbf{x}^u| < 1\Big\}$. Prove that

$$\sum_{\mu \in \mathbb{Z}_+} \mathbf{x}^{\mu u} = \frac{1}{1 - \mathbf{x}^u}$$

for every $\mathbf{x} \in U$ and that the convergence is absolute and uniform on compact subsets of U.

Here is our first result.

(1.2) Lemma. *Let*

$$K = \operatorname{co}\big(u_1, \ldots, u_k\big),$$

where $u_1, \ldots, u_k \in \mathbb{Z}^d$ are linearly independent vectors. Let

$$\Pi = \Big\{\sum_{i=1}^{k} \alpha_i u_i : \quad 0 \leq \alpha_i < 1\Big\}$$

be the "semi-open" parallelepiped spanned by $u_1, \ldots, u_k$. Let

$$U = \big\{\mathbf{x} \in \mathbb{C}^d : \quad |\mathbf{x}^{u_i}| < 1 \quad \text{for} \quad i = 1, \ldots, k\big\}.$$

Then for all $\mathbf{x} \in U$ the series

$$\sum_{m \in K \cap \mathbb{Z}^d} \mathbf{x}^m$$

converges absolutely and uniformly on compact subsets of U to the rational function

$$f(K, \mathbf{x}) = \Big(\sum_{n \in \Pi \cap \mathbb{Z}^d} \mathbf{x}^n\Big) \prod_{i=1}^{k} \frac{1}{1 - \mathbf{x}^{u_i}}.$$

Proof. The proof resembles that of Lemma VII.2.1. For a real number ξ, let $\lfloor \xi \rfloor$ be the integer part of ξ (the largest integer not exceeding ξ) and let $\{\xi\} = \xi - \lfloor \xi \rfloor$ be the fractional part of ξ.

Let us choose a point $m \in K \cap \mathbb{Z}^d$, so

$$m = \sum_{i=1}^{k} \alpha_i u_i \quad \text{where} \quad \alpha_i \geq 0 \quad \text{for} \quad i = 1, \ldots, d.$$

Let

$$m_1 = \sum_{i=1}^{k} \{\alpha_i\} u_i \quad \text{and} \quad m_2 = \sum_{i=1}^{k} \lfloor \alpha_i \rfloor u_i.$$

Thus $m = m_1 + m_2$, m_1 is an integer point in Π and m_2 is a non-negative integer combination of $u_1, \ldots, u_k$. Hence every point $m \in K \cap \mathbb{Z}^d$ can be represented as the sum of an integer point from Π and a non-negative integer combination of the vectors u_i. As in Lemma VII.2.1, it follows that the representation is unique. It is also clear that the sum of an integer point from Π and a non-negative integer combination of $u_1, \ldots, u_k$ is an integer point from K.

Therefore, we have

$$\sum_{m \in K \cap \mathbb{Z}^d} \mathbf{x}^m = \Big(\sum_{n \in \Pi \cap \mathbb{Z}^d} \mathbf{x}^n \Big) \Big(\sum_{(\nu_1, \ldots, \nu_k) \in \mathbb{Z}_+^k} \mathbf{x}^{\nu_1 u_1 + \ldots + \nu_k u_k} \Big)$$

(as formal power series). The second factor is a multiple geometric series which sums up to

$$\prod_{i=1}^{k} \frac{1}{1 - \mathbf{x}^{u_i}}$$

and the result follows; cf. Problem 5 of Section 1.1. □

PROBLEMS.

1°. Let $K \subset \mathbb{R}^d$ be a simple rational cone as in Lemma 1.2 and let

$$\overline{\Pi} = \Big\{ \sum_{i=1}^{k} \alpha_i u_i : \quad 0 < \alpha_i \leq 1 \quad \text{for} \quad i = 1, \ldots, k \Big\}.$$

Let int K denote the interior of K considered as a convex set in its affine hull. Prove that

$$f(\operatorname{int} K, \mathbf{x}) = \sum_{m \in \operatorname{int} K \cap \mathbb{Z}^d} \mathbf{x}^m = \Big(\sum_{n \in \overline{\Pi} \cap \mathbb{Z}^d} \mathbf{x}^n \Big) \prod_{i=1}^{k} \frac{1}{1 - \mathbf{x}^{u_i}}$$

provided $|\mathbf{x}^{u_i}| < 1$ for $i = 1, \ldots, k$.

2°. Given a simple rational cone $K = \operatorname{co}(u_1, \ldots, u_k)$, let us consider two rational functions

$$f(K, \mathbf{x}) = \Big(\sum_{n \in \Pi \cap \mathbb{Z}^d} \mathbf{x}^n \Big) \prod_{i=1}^{k} \frac{1}{1 - \mathbf{x}^{u_i}} \quad \text{and}$$

$$f(\operatorname{int} K, \mathbf{x}) = \Big(\sum_{n \in \overline{\Pi} \cap \mathbb{Z}^d} \mathbf{x}^n \Big) \prod_{i=1}^{k} \frac{1}{1 - \mathbf{x}^{u_i}}$$

in d complex variables $\mathbf{x} \in \mathbb{C}^d$ as in Lemma 1.2 and Problem 1 above. Let $\mathbf{x}^{-1}$ denote $(x_1^{-1}, \ldots, x_d^{-1})$ for $\mathbf{x} = (x_1, \ldots, x_d)$. Prove the *reciprocity relation*

$$f(\operatorname{int} K, \mathbf{x}^{-1}) = (-1)^k f(K, \mathbf{x}).$$

Hint: Consider the transformation $m \longmapsto u - m$, where $u = u_1 + \ldots + u_k$. Show that it establishes a bijection between the sets $\Pi \cap \mathbb{Z}^d$ and $\overline{\Pi} \cap \mathbb{Z}^d$.

3°. Let $u_1, \ldots, u_k \in \mathbb{Z}^d$ be linearly independent vectors, let $L_k = \operatorname{span}\big(u_1, \ldots, u_k\big)$ and let $\Lambda_k = \mathbb{Z}^d \cap L_k$. Let us consider Λ_k as a lattice in L_k. Suppose that $u_1, \ldots, u_k$ is a basis of Λ_k. Prove that under the conditions of Lemma 1.2, we have

$$f(K, \mathbf{x}) = \prod_{i=1}^{k} \frac{1}{1 - \mathbf{x}^{u_i}}.$$

Here is a continuous version of Lemma 1.2.

4. Let $u_1, \ldots, u_d \in \mathbb{R}^d$ be linearly independent vectors and let $K = \operatorname{co}\big(u_1, \ldots, u_d\big)$ be the cone spanned by $u_1, \ldots, u_d$. Let

$$\Pi = \Big\{ \sum_{i=1}^{d} \alpha_i u_i : \quad 0 \leq \alpha_i \leq 1 \quad \text{for} \quad i = 1, \ldots, d \Big\}$$

be the parallelepiped spanned by $u_1, \ldots, u_d$.

Prove that for $c = a + ib$ where $a \in \operatorname{int} K^\circ$ (recall that K° is a polar of K) and $b \in \mathbb{R}^d$, we have

$$\int_K \exp\{\langle c, x \rangle\} \, dx = (\operatorname{vol} \Pi) \prod_{i=1}^{d} \langle -c, u_i \rangle^{-1}$$

(we let $\langle a + ib, x \rangle = \langle a, x \rangle + i \langle b, x \rangle$).

Hint: Applying a linear transformation, we may assume that $K = \mathbb{R}^d_+$ is the standard non-negative orthant.

Here are some other interesting generating functions.

5. Let a and b be coprime positive integers. Let

$$S = \Big\{ \mu_1 a + \mu_2 b : \quad \mu_1, \mu_2 \in \mathbb{Z} \quad \text{and} \quad \mu_1, \mu_2 \geq 0 \Big\}$$

be the set of all non-negative integer combinations of a and b (in other words, $S \subset \mathbb{Z}_+$ is a semigroup generated by a and b; cf. Problem 11 of Section VII.1.2). Prove that

$$\sum_{m \in S} x^m = \frac{1 - x^{ab}}{(1 - x^a)(1 - x^b)}$$

provided $|x| < 1$.

6*. Let a, b and c be coprime positive integers. Let

$$S = \Big\{ \mu_1 a + \mu_2 b + \mu_3 c : \quad \mu_1, \mu_2, \mu_3 \in \mathbb{Z} \quad \text{and} \quad \mu_1, \mu_2, \mu_3 \geq 0 \Big\}$$

be the set of all non-negative integer combinations of a, b and c. Hence $S \subset \mathbb{Z}_+$ is a semigroup generated by a, b and c. Prove that there exist positive integers δ_i, $i = 1, \ldots, 5$, and numbers $\epsilon_i \in \{-1, 1\}$, $i = 1, \ldots, 5$, such that

$$\sum_{m \in S} x^m = \frac{1 + \epsilon_1 x^{\delta_1} + \epsilon_2 x^{\delta_2} + \epsilon_3 x^{\delta_3} + \epsilon_4 x^{\delta_4} + \epsilon_5 x^{\delta_5}}{(1 - x^a)(1 - x^b)(1 - x^c)}$$

provided $|x| < 1$.

Remark: See [**BP99**] for discussion and some references.

2. Generating Functions and Rational Cones

Our next goal is to extend Lemma 1.2 to a larger class of sets.

(2.1) Definitions. Let $c_i \in \mathbb{Z}^d$, $i = 1, \ldots, n$, be integer vectors. The set

$$K = \Big\{ x \in \mathbb{R}^d : \quad \langle c_i, x \rangle \leq 0 \quad \text{for} \quad i = 1, \ldots, n \Big\}$$

is called a *rational cone.*

Let $c_i \in \mathbb{Z}^d$ be integer vectors and let $\alpha_i \in \mathbb{Z}$ be integer numbers for $i = 1, \ldots, n$. The set

$$P = \Big\{ x \in \mathbb{R}^d : \quad \langle c_i, x \rangle \leq \alpha_i \quad \text{for} \quad i = 1, \ldots, n \Big\}$$

is called a *rational polyhedron.*

We denote by $\mathbb{Q}^d$ the set of all points with rational coordinates in $\mathbb{R}^d$.

A polytope $P \subset \mathbb{R}^d$ is called an *integer* (resp. *rational*) polytope provided the vertices of P are points from $\mathbb{Z}^d$ (resp. $\mathbb{Q}^d$).

We will need "rational" versions of the Weyl-Minkowski Theorem; see Corollary II.4.3 and Corollary IV.1.3.

PROBLEMS.

1°. Let $P \subset \mathbb{R}^d$ be a rational polyhedron and let v be a vertex of P. Prove that v has rational coordinates.

Hint: Cf. Theorem II.4.2.

2°. Let $P \subset \mathbb{R}^d$ be a bounded rational polyhedron. Prove that P is a rational polytope.

Hint: Use Problem 1 above and Corollary II.4.3.

3° Let $P \subset \mathbb{R}^d$ be a rational polytope. Prove that the polar $P^\circ \subset \mathbb{R}^d$ is a rational polyhedron.

Hint: Cf. Problem 7 of Section IV.1.1

4°. Let $P \subset \mathbb{R}^d$ be a rational polytope. Prove that P is a rational polyhedron.

Hint: Use Problem 3 above and Corollary IV.1.3.

5°. Let $P \subset \mathbb{R}^d$ be rational polytope. Prove that there exists a positive integer δ such that δP is an integer polytope.

Next, we prove that a rational cone without straight lines has an integer polytope as a base; see Definition II.8.3.

(2.2) Lemma. *Let $K \subset \mathbb{R}^d$, $K \neq \{0\}$, be a rational cone without straight lines. Then there exists an integer polytope $Q \subset \mathbb{R}^d$ which is a base of K. In other words, there exist points $v_1, \ldots, v_n \in \mathbb{Z}^d$ such that every point $x \in K \setminus \{0\}$ has a unique representation $x = \lambda y$ for $y \in Q = \operatorname{conv}\big(v_1, \ldots, v_n\big)$ and $\lambda > 0$.*

Proof. Suppose that

$$K = \big\{x : \langle c_i, x \rangle \leq 0 \quad \text{for} \quad i = 1, \ldots, m\big\},$$

where $c_i \in \mathbb{Z}^d$. Let $c = c_1 + \ldots + c_m$, so c is an integer vector. Let us prove that $\langle c, x \rangle < 0$ for all $x \in K \setminus \{0\}$.

Clearly, $\langle c, x \rangle \leq 0$ for every $x \in K$. On the other hand, if $\langle c, x \rangle = 0$ for some $x \in K$, then we must have $\langle c_i, x \rangle = 0$ for $i = 1, \ldots, m$ (if the sum of non-positive numbers is 0, each number should be equal to 0). Since we assumed that K does not contain straight lines, we must have $x = 0$.

In particular, since $K \neq \{0\}$, we have $c \neq 0$.

Let us define an affine hyperplane

$$H = \big\{x \in \mathbb{R}^d : \quad \langle c, x \rangle = -1\big\}$$

and let $P = K \cap H$. Hence for every $x \in K \setminus \{0\}$ there is a $\lambda > 0$ such that $\lambda x \in P$. Thus P is a base of K.

Clearly, P is a rational polyhedron. We claim that P is a polytope. To demonstrate this, we prove that P does not contain rays (see Section II.16). Indeed, suppose that P contains a ray $a + \tau b$ for $\tau \geq 0$. Then $b \neq 0$ and we must have $\langle c_i, b \rangle \leq 0$ for $i = 1, \ldots, m$ and hence $b \in K$. On the other hand, we must have $\langle c, b \rangle = 0$, which is a contradiction. By Lemma II.16.3, P must be a convex hull of the set of its extreme points and hence, by Theorem II.4.2, P must be a polytope. Finally, by Problem 2 of Section 2.1, P is a rational polytope.

Choosing $Q = \delta P$ for some appropriate positive integer δ (cf. Problem 5 of Section 2.1), we obtain an integer polytope Q which is a base of K. □

PROBLEM.

1. Prove that $K \subset \mathbb{R}^d$ is a rational cone if and only if K can be written as $K = \operatorname{co}\bigl(u_1, \ldots, u_n\bigr)$ for some $u_1, \ldots, u_n \in \mathbb{Z}^d$.

To reduce the case of a rational cone to the case of a simple rational cone, we need an intuitively obvious, although not-so-easy-to-prove, fact that every polytope adopts a *triangulation*, that is, it can be represented as a union of simplices such that every two simplices can intersect only at a common face.

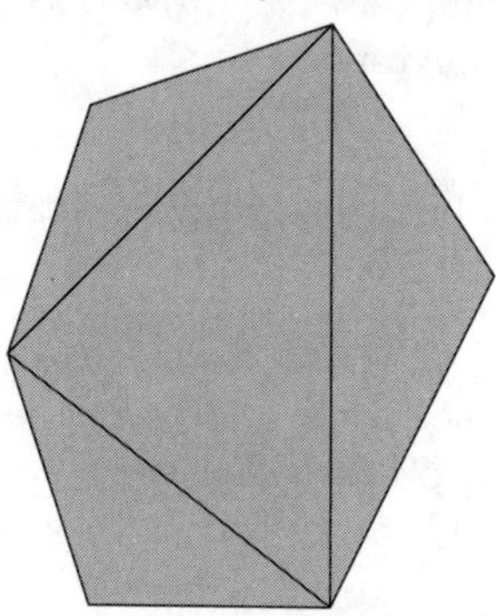

Figure 95. A triangulation of a polygon

Since a rigorous proof may require considerable effort, we sketch only a possible approach below; see Chapter 9 of [**Z95**].

(2.3) Lemma. *Let $P \subset \mathbb{R}^d$ be a polytope with the vertices $v_1, \ldots, v_n$. There exists a partition $I_1 \cup \ldots \cup I_m = \{1, \ldots, n\}$ such that for the polytopes*

$$\Delta_j = \operatorname{conv}\bigl(v_i : i \in I_j\bigr), \quad j = 1, \ldots, m,$$

we have

1. *the points $\{v_i : i \in I_j\}$ are affinely independent for all $j = 1, \ldots, m$ and $\dim \Delta_j = \dim P$ for $j = 1, \ldots, m$;*
2. $$P = \bigcup_{j=1}^{m} \Delta_j;$$
3. *the intersection $\Delta_j \cap \Delta_k$, if non-empty, is a proper common face of Δ_i and Δ_j.*

Sketch of Proof. Without loss of generality, we may assume that $\dim P = d$. Let us consider $\mathbb{R}^d$ as the hyperplane $\xi_{d+1} = 0$ in $\mathbb{R}^{d+1}$. Thus we think of vertices v_i of P as points $(v_i, 0)$ in $\mathbb{R}^{d+1}$. Let us "lift" v_i slightly into $\mathbb{R}^{d+1}$. Namely, we let $u_i = (v_i, \tau_i)$, where $\tau_i > 0$ are "generic" numbers for $i = 1, \ldots, n$. Let $Q = \operatorname{conv}\big(u_i : i = 1, \ldots, n\big)$ be the "lifted" polytope, so $Q \subset \mathbb{R}^{d+1}$. One can show that if the τ_i are sufficiently generic, then Q is a *simplicial* $(d+1)$-dimensional polytope, that is, every facet of Q is a d-dimensional simplex.

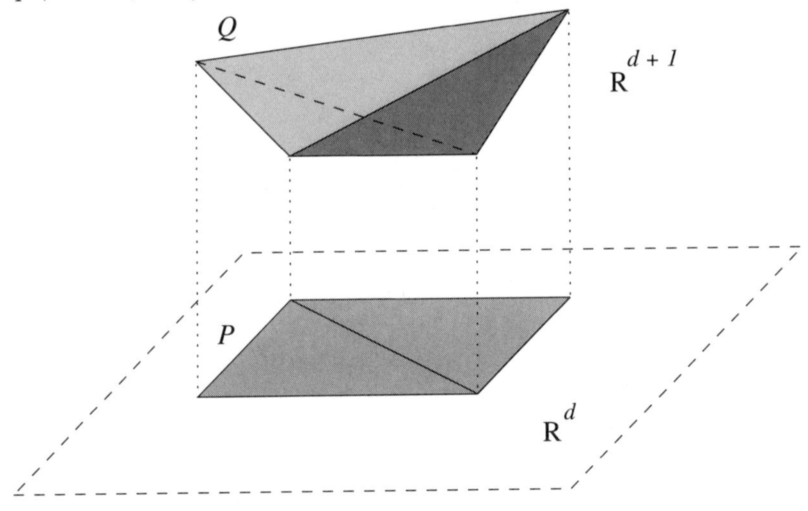

Figure 96. Lifting a polytope to obtain its triangulation

For a generic point $x \in P$, the straight line (x, τ) intersects ∂Q at two points: one belonging to the "lower" facet and the other belonging to the "upper" facet. The projections of the lower facets induce a triangulation of P. $\square$

Now we are ready to prove the main result of this section (it is still a lemma, not a theorem though).

(2.4) Lemma. *Let $K \subset \mathbb{R}^d$ be a rational cone without straight lines and let K° be the polar of K. Let us define a subset $U \subset \mathbb{C}^d$ by*

$$U = \Big\{\mathbf{e}^{x+iz}: \quad x \in \operatorname{int} K^\circ \quad \text{and} \quad z \in \mathbb{R}^d\Big\}.$$

Then $U \subset \mathbb{C}^d$ is a non-empty open set and for every $x \in U$ the series

$$\sum_{m \in K \cap \mathbb{Z}^d} \mathbf{x}^m$$

converges absolutely and uniformly on compact subsets of U to a rational function

$$f(K, \mathbf{x}) = \sum_{i=1}^{n} \frac{p_i(\mathbf{x})}{(1 - \mathbf{x}^{u_{i1}}) \cdots (1 - \mathbf{x}^{u_{id}})},$$

where $p_i(\mathbf{x})$ are Laurent polynomials and $u_{ij} \in \mathbb{Z}^d$ are integer vectors for $i = 1, \ldots, n$ and $j = 1, \ldots, d$.

Proof. Without loss of generality, we assume that $K \neq \{0\}$. By Lemma 2.2, there exists an integer polytope Q which is a base of K. Triangulating Q (Lemma 2.3), we represent K as a union of simple rational cones (see Definition 1.1) K_i, $i \in I$, such that the intersection of every two cones is a common face of the cones, which, if not $\{0\}$, must be a simple rational cone. Using the Inclusion-Exclusion Formula (see Section I.7), we can represent the indicator function of K as a linear combination of the indicator functions of K_i:

$$[K] = \sum_{i \in I} \epsilon_i [K_i] \quad \text{where} \quad \epsilon_i \in \{-1, 1\}.$$

Hence the same relation holds for the generating functions (considered as an identity between formal power series):

$$\sum_{m \in K \cap \mathbb{Z}^d} \mathbf{x}^m = \sum_{i \in I} \epsilon_i \Big(\sum_{m \in K_i \cap \mathbb{Z}^d} \mathbf{x}^m \Big).$$

Clearly, for all $\mathbf{x} \in U$ and all $m \in K \cap \mathbb{Z}^d$ we have $|\mathbf{x}^m| < 1$, so by Lemma 1.2 the series in the right-hand side of the identity converges absolutely and uniformly on compact subsets of U to rational functions in $\mathbf{x}$ of the required type (multiplying the numerator and denominator of each fraction by some binomials $(1 - \mathbf{x}^u)$, we can ensure that each denominator is the product of exactly d binomials).

It remains to show that the set U is open. If $\dim K^\circ < d$, then by Theorem II.2.4 the cone K° is contained in a hyperplane and hence $K = (K^\circ)^\circ$ contains a straight line, which is a contradiction. Thus $\dim K^\circ = d$ and hence the result follows. □

PROBLEMS.

1. In Lemma 2.4, let $k = \dim K$ and let $\operatorname{int} K$ denote the interior of K considered as a convex set in its affine hull. Prove that for every $\mathbf{x} \in U$ the series

$$\sum_{m \in \operatorname{int} K \cap \mathbb{Z}^d} \mathbf{x}^m$$

converges absolutely to a rational function $f(\operatorname{int} K, \mathbf{x})$ and that

$$f(\operatorname{int} K, \mathbf{x}^{-1}) = (-1)^k f(K, \mathbf{x})$$

(the *reciprocity relation*).

Hint: Use Problems 1 and 2 of Section 1.2 and Problems 7 and 8 of Section VI.3.3.

Here is a continuous version of Lemma 2.4.

2. Let $K \subset \mathbb{R}^d$ be a polyhedral cone without straight lines. Prove that for $c = x + iy$, where $x \in \operatorname{int} K^\circ$ and $y \in \mathbb{R}^d$, the integral

$$\int_K \exp\{\langle c, x \rangle\} \, dx$$

converges to a rational function

$$\sum_{i=1}^{m} \alpha_i \prod_{j=1}^{d} \langle -c, u_{ij} \rangle^{-1},$$

where $u_{ij} \in \mathbb{R}^d$ are some vectors and α_i are some real numbers.

Hint: Use Problem 4 of Section 1.2.

3. Generating Functions and Rational Polyhedra

In this section, we prove the main result of this chapter. But first we need one more lemma.

(3.1) Lemma. *Let $P \subset \mathbb{R}^d$ be a rational polyhedron without straight lines. Then there exists a non-empty open set $U \subset \mathbb{C}^d$ such that for all $\mathbf{x} \in U$ the series*

$$\sum_{m \in P \cap \mathbb{Z}^d} \mathbf{x}^m$$

converges absolutely and uniformly on compact subsets of U to a rational function $f(P, \mathbf{x})$ of $\mathbf{x}$.

Proof. Let us identify $\mathbb{R}^d$ with the affine hyperplane H defined by the equation $\xi_{d+1} = 1$ in $\mathbb{R}^{d+1}$. Suppose that P is defined by a system of linear inequalities:

$$P = \Big\{x \in \mathbb{R}^d : \quad \langle c_i, x \rangle \leq \alpha_i, \quad i = 1, \ldots, n\Big\}, \quad \text{where} \quad c_i \in \mathbb{Z}^d \quad \text{and} \quad \alpha_i \in \mathbb{Z}$$

for $i = 1, \ldots, n$. Let us define $K \subset \mathbb{R}^{d+1}$ by

$$K = \Big\{(x, \ \xi_{d+1}) : \quad \langle c_i, x \rangle - \alpha_i \xi_{d+1} \leq 0 \quad \text{for} \quad i = 1, \ldots, n \quad \text{and} \quad \xi_{d+1} \geq 0\Big\}.$$

Clearly, $K \subset \mathbb{R}^{d+1}$ is a rational cone. If P is bounded, then $K = \operatorname{co}(P)$ and if P is unbounded, then $K = \operatorname{cl}\big(\operatorname{co}(P)\big)$. Note that $P = K \cap H$; see Figure 97.

Moreover, K does not contain straight lines. Indeed, suppose that K contains a straight line in the direction of $y = (\eta_1, \ldots, \eta_{d+1})$. Then we must have $\eta_{d+1} = 0$ since the last coordinate must stay non-negative and then $\eta_1 = \ldots = \eta_d = 0$ since P does not contain straight lines.

By Lemma 2.4, there exists a non-empty open set $U_1 \subset \mathbb{C}^{d+1}$ such that for all $\mathbf{y} = (\mathbf{x}, x_{d+1}) \in U_1$ the series

$$\sum_{m \in K \cap \mathbb{Z}^{d+1}} \mathbf{y}^m = \sum_{(m_1, \mu) \in K \cap \mathbb{Z}^{d+1}} \mathbf{x}^{m_1} x_{d+1}^{\mu}$$

converges absolutely and uniformly on compact subsets of U_1 to a rational function $f\big(K, (\mathbf{x}, x_{d+1})\big)$.

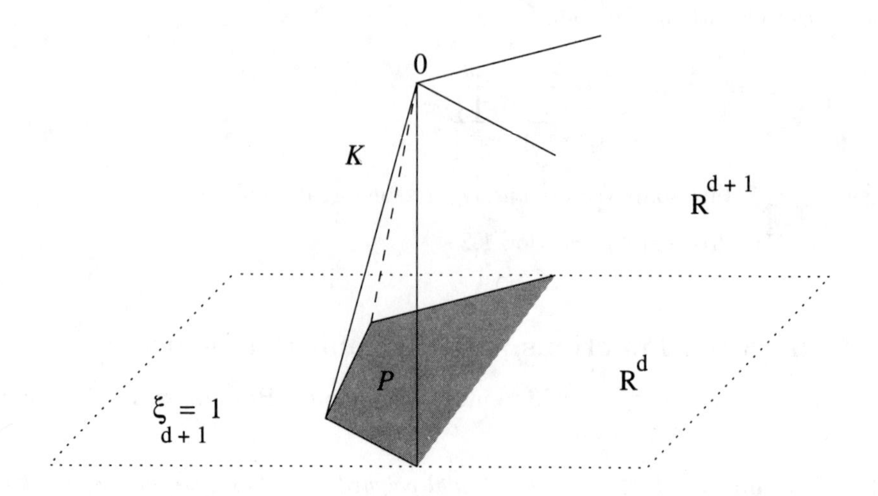

Figure 97. The reader who imagines the positive direction as upward may need to view this picture upside down.

We obtain $f(P, \mathbf{x})$ by differentiating $f\big(K, (\mathbf{x}, x_{d+1})\big)$ with respect to x_{d+1} and substituting $x_{d+1} = 0$ into the derivative.

Indeed, we observe that for every lattice point $(m_1, \mu) \in K$ the last coordinate μ is non-negative. By a standard result in complex analysis, we can differentiate the series and conclude that the series

$$\sum_{\substack{(m_1,\mu)\in K\cap\mathbb{Z}^{d+1}\\ m_1\in\mathbb{Z}^d,\mu\geq 1}} \mu\mathbf{x}^{m_1}x_{d+1}^{\mu-1} = \sum_{m_1\in P\cap\mathbb{Z}^d} \mathbf{x}^{m_1} + \sum_{\substack{(m_1,\mu)\in K\cap\mathbb{Z}^{d+1}\\ m_1\in\mathbb{Z}^d,\mu\geq 2}} \mu\mathbf{x}^{m_1}x_{d+1}^{\mu-1} \tag{3.1.1}$$

converges absolutely and uniformly on compact sets in U_1 to a rational function

$$\frac{\partial}{\partial x_{d+1}} f\big(K, (\mathbf{x}, x_{d+1})\big).$$

Let $U \subset \mathbb{C}^d$ be the projection of U_1: $(\mathbf{x}, x_{d+1}) \longmapsto \mathbf{x}$. Substituting $x_{d+1} = 0$ in (3.1.1), we conclude that for every $\mathbf{x} \in U$ the series

$$\sum_{m_1\in P\cap\mathbb{Z}^d} \mathbf{x}^{m_1}$$

converges absolutely and uniformly on compact subsets of U_1 to the rational function

$$f(P, \mathbf{x}) = \frac{\partial}{\partial x_{d+1}} f\big(K, (\mathbf{x}, x_{d+1})\big)\Big|_{x_{d+1}=0}.$$

$\square$

PROBLEMS.

1°. In the situation of Lemma 3.1, let $m \in \mathbb{Z}^d$ be a lattice vector and let $P+m$ be the translation of P. Prove that

$$f(P+m, \mathbf{x}) = \mathbf{x}^m f(P, \mathbf{x}).$$

Here is a continuous version of Lemma 3.1.

2. Let $P \subset \mathbb{R}^d$ be a polyhedron without straight lines. Prove that there exists a non-empty open set $U \subset \mathbb{C}^d$ and a rational function $\phi : \mathbb{C}^d \longrightarrow \mathbb{C}$ such that for all $c \in U$ we have

$$\int_P \exp\{\langle c, x\rangle\}\ dx = \phi(c)$$

and the integral converges absolutely. Again, we let $\langle c, x\rangle = \langle a, x\rangle + i\langle b, x\rangle$ for a complex vector $c = a + ib$.

Hint: Use Problem 2 of Section 2.4 and the trick of Lemma 3.1. Instead of differentiating, use the Laplace transform.

3*. Let $u_1, \ldots, u_n \in \mathbb{Z}^d$ be vectors such that the cone $K = \operatorname{co}(u_1, \ldots, u_n)$ does not contain straight lines. Let

$$S = \Big\{\sum_{i=1}^n \alpha_i u_i \quad \text{where} \quad \alpha_1, \ldots, \alpha_n \quad \text{are non-negative integers}\Big\}$$

be the semigroup generated by $u_1, \ldots, u_n$; cf. Problem 6 of Section 1.2. Prove that there exists a non-empty open set $U \subset \mathbb{C}^d$ such that for all $\mathbf{x} \in U$ the series $\sum_{m \in S} \mathbf{x}^m$ converges absolutely and uniformly on compact subsets of U to a rational function in $\mathbf{x}$.

Hint: Let $\mathbb{R}^n_+$ be the non-negative orthant in $\mathbb{R}^n$. Construct a linear transformation $T : \mathbb{R}^n \longrightarrow \mathbb{R}^d$ such that $T(\mathbb{Z}^n_+) = S$. Construct a set $Q \subset \mathbb{R}^n_+$ which is a finite union of rational polyhedra and such that the restriction $T : Q \cap \mathbb{Z}^n \longrightarrow S$ is a bijection. Apply Lemma 3.1.

We are getting ready to prove the central result of this chapter. We state it in the form of the existence theorem for some particular valuation; cf. Sections I.7 and I.8. We need the rational analogue of the algebra $\mathcal{P}(\mathbb{R}^d)$ of polyhedra; see Definition I.9.3.

(3.2) Definitions. The real vector space spanned by the indicator functions $[P]$ of rational polyhedra $P \subset \mathbb{R}^d$ is called the *algebra of rational polyhedra* in $\mathbb{R}^d$ and denoted $\mathcal{P}(\mathbb{Q}^d)$. Let $\mathbb{C}(x_1, \ldots, x_d)$ denote the complex vector space of all rational functions in d variables.

PROBLEMS.

1°. Check that the intersection of rational polyhedra is a rational polyhedron.

2°. Prove that $\mathcal{P}(\mathbb{R}^d)$ (resp. $\mathcal{P}(\mathbb{Q}^d)$) is spanned by the indicator functions $[P]$ of polyhedra $P \subset \mathbb{R}^d$ (resp. rational polyhedra $P \subset \mathbb{R}^d$) without straight lines.

3°. Let $P \subset \mathbb{R}^d$ be a rational polyhedron which contains a straight line. Prove that there exists a vector $m \in \mathbb{Z}^d \setminus \{0\}$ such that $P + m = P$.

Hint: Assume that $P = \{x : \langle c_i, x\rangle \leq \alpha_i \text{ for } i = 1, \ldots, n\}$ for some vectors $c_i \in \mathbb{Z}^d$ and some numbers $\alpha_i \in \mathbb{Z}$. Prove that there exists a vector $u \in \mathbb{R}^d \setminus \{0\}$ such that $\langle c_i, u\rangle = 0$ for all $i \in I$. Deduce that one can choose $u \in \mathbb{Q}^d \setminus \{0\}$. Conclude that one can choose $u \in \mathbb{Z}^d \setminus \{0\}$.

The theorem below was proved independently by A.V. Pukhlikov and A.G. Khovanskii [**PK92**] and J. Lawrence [**L91b**].

(3.3) Theorem. *There exists a map*

$$\mathcal{F} : \mathcal{P}(\mathbb{Q}^d) \longrightarrow \mathbb{C}(x_1, \ldots, x_d)$$

from the algebra of rational polyhedra in $\mathbb{R}^d$ to the space of rational functions in d complex variables $\mathbf{x} = (x_1, \ldots, x_d)$ such that the following hold:

1. *The map $\mathcal{F}$ is a valuation, that is, a linear transformation.*
2. *If $P \subset \mathbb{R}^d$ is a rational polyhedron without straight lines, then $\mathcal{F}[P] = f(P, \mathbf{x})$ is the rational function such that*

$$f(P, \mathbf{x}) = \sum_{m \in P \cap \mathbb{Z}^d} \mathbf{x}^m$$

provided the series converges absolutely.

3. *For a function $g \in \mathcal{P}(\mathbb{Q}^d)$ and an integer vector $m \in \mathbb{Z}^d$, let $h(x) = g(x-m)$ be the shift of g. Then $\mathcal{F}(h) = \mathbf{x}^m \mathcal{F}(g)$.*
4. *If $P \subset \mathbb{R}^d$ is a rational polyhedron containing a straight line, then $\mathcal{F}([P]) \equiv 0$ (the rational function that is identically zero).*

Proof. We know how to define $\mathcal{F}[P]$ for a rational polyhedron $P \subset \mathbb{R}^d$ without straight lines. By Lemma 3.1 there is a non-empty open set $U \subset \mathbb{C}^d$ such that the series

$$\sum_{m \in P \cap \mathbb{Z}^d} \mathbf{x}^m$$

converges for all $\mathbf{x} \in U$ to a rational function $f(P, \mathbf{x})$. Hence we let $\mathcal{F}[P] = f(P, \mathbf{x})$. By Problem 2 of Section 3.2, the algebra $\mathcal{P}(\mathbb{Q}^d)$ is spanned by the indicator functions of rational polyhedra $[P]$ without straight lines, so we may try to extend $\mathcal{F}$ by linearity. To be able to do that, we should show that whenever

$$\sum_{i=1}^{n} \alpha_i [P_i] = 0 \tag{3.3.1}$$

for rational polyhedra P_i without straight lines and real numbers α_i, we must have

$$\sum_{i=1}^{n} \alpha_i f(P_i, \mathbf{x}) \equiv 0. \tag{3.3.2}$$

Indeed, suppose that (3.3.1) holds. For a non-empty subset $I \subset \{1, \ldots, n\}$, let

$$P_I = \bigcap_{i \in I} P_i.$$

Using the Inclusion-Exclusion Formula (see Lemma I.7.2), we obtain that

$$\Big[\bigcup_{i=1} P_i\Big] = \sum_{\substack{I \subset \{1, \ldots, n\} \\ I \neq \emptyset}} (-1)^{|I|-1} [P_I].$$

Multiplying the above identity by $[P_i]$, we obtain

$$[P_i] = \sum_{\substack{I \subset \{1, \ldots, n\} \\ I \neq \emptyset}} (-1)^{|I|-1} [P_{I \cup \{i\}}].$$

Hence

$$\sum_{m \in P_i \cap \mathbb{Z}^d} \mathbf{x}^m = \sum_{\substack{I \subset \{1, \ldots, n\} \\ I \neq \emptyset}} (-1)^{|I|-1} \sum_{m \in P_{I \cup \{i\}} \cap \mathbb{Z}^d} \mathbf{x}^m$$

as formal power series. Now P_i is a rational polyhedron without straight lines and $P_{I \cup \{i\}} \subset P_i$ are rational polyhedra as well, so by Lemma 3.1 there is a non-empty open set $U \subset \mathbb{C}^d$ where all involved series converge absolutely. Therefore,

$$f(P_i, \mathbf{x}) = \sum_{\substack{I \subset \{1, \ldots, n\} \\ I \neq \emptyset}} (-1)^{|I|-1} f(P_{I \cup \{i\}}, \mathbf{x}). \tag{3.3.3}$$

Let us choose a non-empty $I \subset \{1, \ldots, n\}$. Multiplying (3.3.1) by $[P_I]$, we get

$$\sum_{i=1}^{n} \alpha_i [P_{I \cup \{i\}}] = 0.$$

Therefore,

$$\sum_{i=1}^{n} \alpha_i \Big(\sum_{m \in P_{I \cup \{i\}} \cap \mathbb{Z}^d} \mathbf{x}^m \Big) = 0$$

as formal power series. Since P_I is a rational polyhedron without straight lines and $P_{I \cup \{i\}} \subset P_I$, there is a non-empty open subset $U \subset \mathbb{C}^d$ where all series involved converge absolutely. Therefore, for every non-empty $I \subset \{1, \ldots, n\}$, we have

$$\sum_{i=1}^{n} \alpha_i f(P_{I \cup \{i\}}, \mathbf{x}) \equiv 0. \tag{3.3.4}$$

From (3.3.3) and (3.3.4) we get (3.3.2). Thus (3.3.1) implies (3.3.2) and we can extend $\mathcal{F}$ by linearity to the whole algebra $\mathcal{P}(\mathbb{Q}^d)$.

Since by Problem 1 of Section 3.1 $\mathcal{F}([P+m]) = \mathbf{x}^m \mathcal{F}([P])$ for every rational polyhedron P without straight lines and every $m \in \mathbb{Z}$ and the indicator functions $[P]$ span $\mathcal{P}(\mathbb{Q}^d)$, Part 3 follows.

It remains to prove Part 4. Let $P \subset \mathbb{R}^d$ be a rational polyhedron with a straight line. By Problem 3 of Section 3.2 we have $P + m = P$ for some non-zero integer vector m. Then by Part 3 we must have

$$\mathcal{F}([P]) = \mathcal{F}([P+m]) = \mathbf{x}^m \mathcal{F}([P]).$$

Hence $\mathcal{F}([P]) \equiv 0$ and the proof is completed. □

(3.4) Example. Let $d=1$, let $P_+ = [0, +\infty)$ and let $P_- = (-\infty, 0]$. Thus P_+ and P_- are rational polyhedra in $\mathbb{R}^1$. Let $P_0 = P_- \cap P_+ = \{0\}$ and let $P = P_- \cup P_+ = \mathbb{R}$. We have

$$\sum_{m \in P_+ \cap \mathbb{Z}} x^m = \sum_{m \geq 0} x^m = \frac{1}{1-x},$$

where the series converges absolutely for all x such that $|x| < 1$. Thus by Part 2 of Theorem 3.3, we must have

$$\mathcal{F}[P_+] = \frac{1}{1-x}.$$

Similarly,

$$\sum_{m \in P_- \cap \mathbb{Z}} x^m = \sum_{m \leq 0} x^m = \frac{1}{1-x^{-1}},$$

where the series converges absolutely for all x such that $|x| > 1$. Thus by Part 2 of Theorem 3.3 we must have

$$\mathcal{F}[P_-] = \frac{1}{1-x^{-1}}.$$

Now, $P_0 = \{0\}$, so we must have

$$\mathcal{F}[P_0] = 1.$$

The polyhedron P is the whole line $\mathbb{R}$ and hence by Part 4

$$\mathcal{F}[P] = 0.$$

By the Inclusion-Exclusion Formula,

$$[P] = [P_-] + [P_+] - [P_0].$$

Then, by Part 1 we must have

$$0 = \mathcal{F}[P] = \mathcal{F}[P_-] + \mathcal{F}[P_+] - \mathcal{F}[P_0] = \frac{1}{1-x} + \frac{1}{1-x^{-1}} - 1$$
$$= \frac{1}{1-x} - \frac{x}{1-x} - 1,$$

which is indeed the case. Note that there are no x for which both series for $\mathcal{F}[P_-]$ and $\mathcal{F}[P_+]$ converge.

PROBLEMS.

Here is a continuous version of Theorem 3.3.

1. Prove that there exists a valuation $\Phi : \mathcal{P}(\mathbb{R}^d) \longrightarrow \mathbb{C}(\gamma_1, \ldots, \gamma_d)$ such that the following hold:

(1) If $P \subset \mathbb{R}^d$ is a polyhedron without straight lines, then $\Phi([P])$ is the rational function $\phi(P, c)$ in $c = (\gamma_1, \ldots, \gamma_d)$ such that

$$\phi(P, c) = \int_P \exp\{\langle c, x\rangle\}\, dx \quad \text{for} \quad c = (\gamma_1, \ldots, \gamma_d)$$

provided the integral converges absolutely (cf. Problem 2 of Section 3.1).

(2) For a function $g \in \mathcal{P}(\mathbb{R}^d)$ and a vector $a \in \mathbb{R}^d$, let $h(x) = g(x - a)$. Then $\Phi(h) = \exp\{\langle c, a\rangle\}\Phi(g)$.

(3) If $P \subset \mathbb{R}^d$ is a polyhedron with straight lines, then $\Phi(P) \equiv 0$.

2. Let $P \subset \mathbb{R}^d$ be a d-dimensional polytope defined by a system of linear inequalities,

$$P = \Big\{x : \quad \langle u_i, x\rangle \leq \alpha_i \quad \text{for} \quad i = 1, \ldots, n\Big\},$$

where $\alpha_i > 0$ and $\|u_i\| = 1$ for $i = 1, \ldots, n$. Suppose that

$$F_i = \Big\{x \in P : \quad \langle u_i, x\rangle = \alpha_i\Big\}$$

is a facet of P for $i = 1, \ldots, n$ and let μ_i be the Lebesgue measure on the affine hull of F_i. Prove that for every $c \in \mathbb{C}^d$ and every $v \in \mathbb{R}^d$ one has

$$\langle c, v\rangle \int_P \exp\{\langle c, x\rangle\}\, dx = \sum_{i=1}^n \langle v, u_i\rangle \int_{F_i} \exp\{\langle c, x\rangle\}\, d\mu_i.$$

Hint: Use Stokes' Formula; see [**Barv93**].

4. Brion's Theorem

In this section, we discuss the structure of the algebra of (rational) polyhedra in more detail and obtain an important formula for the valuation $\mathcal{F}$.

(4.1) Definition. Let $P \subset \mathbb{R}^d$ be a polyhedron and let $v \in P$ be a point. We define the *support cone* of P at v as

$$\operatorname{cone}(P, v) = \Big\{x \in \mathbb{R}^d : \quad \epsilon x + (1 - \epsilon)v \in P \quad \text{for some} \quad 0 < \epsilon < 1\Big\}.$$

Strictly speaking, $\operatorname{cone}(P, v)$ is not a cone since it has its vertex at v.

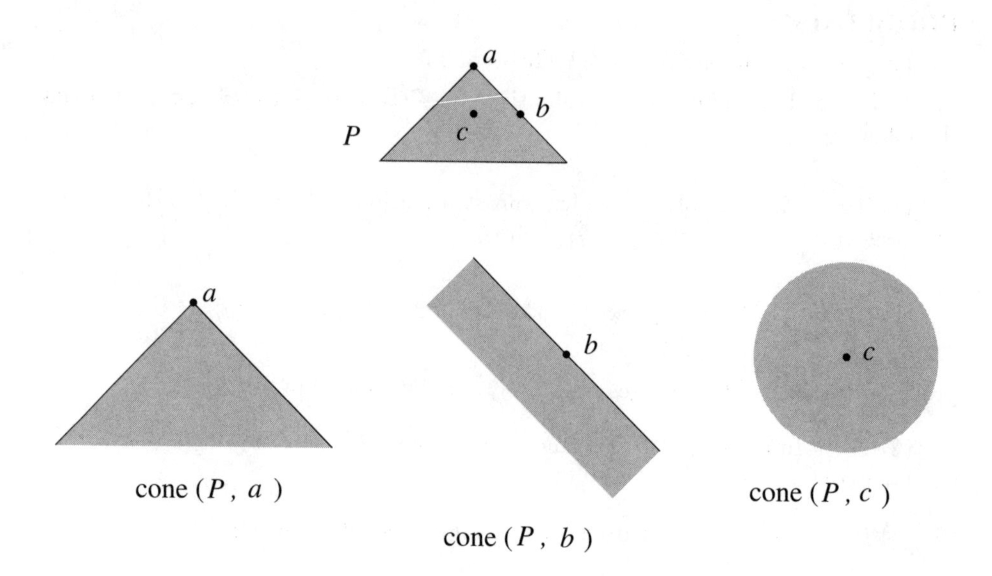

Figure 98. Example: a polyhedron P and its support cones

PROBLEMS.

1°. Check that the support cone at v is the translation of the cone of feasible directions at v (see Problem 2 of Section II.16.1) by the vector v.

2°. Let $P \subset \mathbb{R}^d$ be a polyhedron defined by a system of linear inequalities

$$P = \{x \in \mathbb{R}^d: \quad \langle c_i, x\rangle \leq \beta_i \quad \text{for} \quad i = 1, \ldots, n\}$$

and let $v \in P$ be a point. Let $I = \{i : \langle c_i, v\rangle = \beta_i\}$ be the set of inequalities active on v. Prove that

$$\operatorname{cone}(P, v) = \{x \in \mathbb{R}^d: \quad \langle c_i, x\rangle \leq \beta_i \quad \text{for} \quad i \in I\}$$

(if $I = \emptyset$, we let $\operatorname{cone}(P, v) = \mathbb{R}^d$). In particular, the support cone of a rational polyhedron is a rational polyhedron.

3°. Let $P \subset \mathbb{R}^d$ be a polyhedron and let $u, v \in P$ be points which lie in the interior of the same face F of P. Prove that $\operatorname{cone}(P, v) = \operatorname{cone}(P, u)$. Hence we can talk about the support cone, denoted $\operatorname{cone}(P, F)$, of a face F of P.

Our goal is to prove that "up to straight lines", every polyhedron is the sum of its support cones.

(4.2) Definition. Let $\mathcal{P}_0(\mathbb{R}^d)$ (resp. $\mathcal{P}_0(\mathbb{Q}^d)$) denote the subspace of the algebra $\mathcal{P}(\mathbb{R}^d)$ (resp. $\mathcal{P}(\mathbb{Q}^d)$) spanned by the indicator functions $[P]$ of polyhedra (resp. rational polyhedra) with straight lines.

We start with the case of a standard simplex; cf. Problem 1 of Section I.2.2.

(4.3) Lemma. *Let $\Delta \subset \mathbb{R}^{d+1}$ be the standard d-dimensional simplex, $\Delta = \operatorname{conv}\big(e_i : i = 1, \ldots, d+1\big)$, where $e_1, \ldots, e_{d+1}$ are the standard basis vectors. Then one can find rational polyhedra $P_k \subset \mathbb{R}^d$, $k = 1, \ldots, N$, such that*

1. *each polyhedron P_k contains a straight line parallel to $e_i - e_j$ for some pair $1 \leq i < j \leq d+1$;*
2. *we have*

$$[\Delta] - \sum_{i=1}^{d+1} \big[\operatorname{cone}(\Delta, e_i)\big] = \sum_{k=1}^{N} \alpha_k [P_k] \quad \textit{for some} \quad \alpha_k \in \{-1, 1\}.$$

In particular, modulo $\mathcal{P}_0(\mathbb{Q}^d)$, the indicator function of the standard simplex is the sum of the indicator functions of the support cones at its vertices.

Proof. Let us identify $\mathbb{R}^d$ with the affine hull of $e_1, \ldots, e_{d+1}$. Let $\overline{H_i^+}$ be the closed halfspace $\xi_i \geq 0$ in $\mathbb{R}^d$. Then

$$\Delta = \bigcap_{i=1}^{d+1} \overline{H_i^+} \quad \text{and} \quad \operatorname{cone}(\Delta, e_j) = \bigcap_{i \neq j} \overline{H_i^+}.$$

By the Inclusion-Exclusion Formula (see Lemma I.7.2), we have

$$[\mathbb{R}^d] = \Big[\bigcup_{i=1}^{d+1} \overline{H_i^+}\Big] = \sum_{\substack{I \subset \{1, \ldots, d+1\} \\ I \neq \emptyset}} (-1)^{|I|-1} \bigcap_{i \in I} \overline{H_i^+}.$$

Let

$$P_I = \bigcap_{i \in I} \overline{H_i^+}.$$

If $I = \{1, \ldots, d+1\}$, we have $P_I = \Delta$. If $I = \{1, \ldots, d+1\} \setminus \{i\}$, we have $P = \operatorname{cone}(\Delta, e_i)$. All other polyhedra P_I contain straight lines. In particular, if $i, j \notin I$, then P_I contains a straight line in the direction of $e_i - e_j$. □

PROBLEMS.

1°. Let $\operatorname{cone}(P, F)$ be the support cone of P at a face $F \subset P$; see Problem 3 of Section 4.1. Let us fix a $0 < k < d$. Prove that for the standard simplex Δ

$$[\Delta] - \sum_{\substack{F \text{ is a face of } \Delta \\ \dim F \leq k}} (-1)^{\dim F} \big[\operatorname{cone}(\Delta, F)\big]$$

is a linear combination of the indicator functions of polyhedra P_j each of which contains a $(k+1)$-dimensional affine subspace.

2°. Let $P \subset \mathbb{R}^n$ be a rational polyhedron and let $T : \mathbb{R}^n \longrightarrow \mathbb{R}^d$ be a linear transformation with a rational matrix. Prove that $T(P)$ is a rational polyhedron in $\mathbb{R}^d$.

Hint: "Rationalize" the reasoning of Section I.9.

3°. Let $P \subset \mathbb{R}^n$ be a polyhedron, let $T: \mathbb{R}^n \longrightarrow \mathbb{R}^d$ be a linear transformation and let $Q = T(P)$. Let $v \in P$ be a point and let $u = T(v)$. Check that $\operatorname{cone}(Q, u) = T\big(\operatorname{cone}(P, v)\big)$.

Next, we generalize Lemma 4.3 to (rational) polytopes.

(4.4) Lemma. *Let $P \subset \mathbb{R}^d$ be a polytope (resp. rational polytope) with the vertices $v_1, \ldots, v_m$. Then we can write*

$$[P] = g + \sum_{i=1}^{m} \big[\operatorname{cone}(P, v_i)\big]$$

for some function $g \in \mathcal{P}_0(\mathbb{R}^d)$ (resp. for some $g \in \mathcal{P}_0(\mathbb{Q}^d)$).

Proof. Let us consider the standard simplex $\Delta \subset \mathbb{R}^{m+1}$ and let T be the linear transformation, $T: \mathbb{R}^{m+1} \longrightarrow \mathbb{R}^d$ such that $T(e_i) = v_i$. Then $T(\Delta) = P$. Using Lemma 4.3, we can write

$$[\Delta] = \sum_{i=1}^{m+1} \big[\operatorname{cone}(\Delta, e_i)\big] + \sum_{k=1}^{N} \alpha_k [Q_k],$$

where each Q_k is a rational polyhedron containing a straight line in some direction $e_i - e_j$. Linear relations among indicator functions of polyhedra are preserved under linear transformations; see Problem 1 of Section I.9.3. Applying the transformation T to the above identity and using Problem 3 of Section 4.3, we get

$$[P] = [T(\Delta)] = \sum_{i=1}^{m} \big[\operatorname{cone}(P, v_i)\big] + \sum_{k=1}^{N} \alpha_k [T(Q_k)].$$

By Problem 2 of Section 4.3, each $T(Q_k)$ is a (rational) polyhedron. Moreover, since $v_1, \ldots, v_m$ are distinct, $e_i - e_j \notin \ker T$ for every pair of indices $i \neq j$. Therefore, each polyhedron $T(Q_k)$ contains some straight line in the direction of $T(e_i - e_j)$. The proof now follows. □

PROBLEMS.

1*. For a face F of a polytope $P \subset \mathbb{R}^d$, let $\operatorname{cone}(P, F)$ be defined as in Problem 3, Section 4.1. By convention, P is a face of itself (so $\operatorname{cone}(P, P) = \mathbb{R}^d$). Prove *Gram's relation*, also known as the *Brianchon-Gram Theorem:*

$$[P] = \sum_{F} (-1)^{\dim F} \big[\operatorname{cone}(P, F)\big],$$

where the sum is taken over all non-empty faces F of P, including $F = P$.

Remark: The following proof was suggested by J. Lawrence. Let us choose the origin in the interior of P and let $Q = P^\circ$ be the polar of P. Using Problem 8 of Section VI.3.3, write

$$(-1)^{d-1}[Q] = \sum_F (-1)^{\dim F}[\operatorname{conv}(F \cup \{0\})],$$

where the sum is taken over all faces $F \neq Q$ of Q, including the empty face. Apply the polarity valuation $\mathcal{D}$ of Theorem IV.1.5 to both sides of the identity and use Theorem VI.1.3 to interpret the resulting identity as Gram's relation for P.

2°. Let $P \subset \mathbb{R}^d$ be a polyhedron with a straight line. Prove that P does not have a vertex.

3°. Let $P_1, P_2 \subset \mathbb{R}^d$ be polyhedra and let $v_1 \in P_1$ and $v_2 \in P_2$ be points. Let us define $P_1 \times P_2 \subset \mathbb{R}^d \oplus \mathbb{R}^d = \mathbb{R}^{2d}$ by

$$P_1 \times P_2 = \big\{(x, y) : \quad x \in P_1, y \in P_2\big\}.$$

Prove that $P = P_1 \times P_2$ is a polyhedron and that for $v = (v_1, v_2)$, we have

$$\operatorname{cone}(P_1 \times P_2, v) = \operatorname{cone}(P_1, v_1) \times \operatorname{cone}(P_2, v_2).$$

Finally, we extend Lemma 4.4. to (rational) polyhedra.

(4.5) Theorem. *Let $P \subset \mathbb{R}^d$ be a polyhedron (resp. rational polyhedron). Then*

$$[P] = g + \sum_{v \text{ is a vertex of } P} \big[\operatorname{cone}(P, v)\big]$$

for some function $g \in \mathcal{P}_0(\mathbb{R}^d)$ (resp. for some $g \in \mathcal{P}_0(\mathbb{Q}^d)$).

In words: modulo the indicator functions of (rational) polyhedra with straight lines, the indicator function of every (rational) polyhedron P is equal to the sum of the indicator functions of the support cones of P at the vertices of P.

Proof. Suppose that P is defined by a system of linear inequalities

$$P = \big\{x : \langle c_i, x\rangle \leq \beta_i, \quad i = 1, \ldots, n\big\},$$

where $c_i \in \mathbb{Z}^d$ and $\beta_i \in \mathbb{Z}$. First, we observe that if P does not have vertices, then by Lemma II.3.5 the polyhedron is either empty or contains a straight line; in both cases the result is immediate. Suppose, therefore, that P has vertices and let $Q = \operatorname{conv}\big(\operatorname{ex} P\big)$ be their convex hull. Then Q is a (rational) polytope; cf. Problem 2 of Section 2.1. Let

$$K = \big\{x : \langle c_i, x\rangle \leq 0 : \quad i = 1, \ldots, n\big\}$$

be the recession cone of P; see Section II.16 and, in particular, Problem 3 of Section II.16.1. Thus K is a (rational) cone without straight lines and Lemma II.16.3 implies that

$$P = Q + K.$$

By Lemma 4.4, we can write

$$[Q] = \sum_{v \in \text{ex}(P)} \big[\text{cone}(Q, v)\big] + \sum_{i \in I} \alpha_i [Q_i],$$

where the Q_i are (rational) polyhedra with straight lines and the α_i are numbers.

Let us consider $\mathbb{R}^{2d}$ as a direct sum of two copies of $\mathbb{R}^d$: $\mathbb{R}^{2d} = \mathbb{R}^d \oplus \mathbb{R}^d$. For sets $X, Y \subset \mathbb{R}^d$, let $X \times Y = \big\{(x, y) : x \in X, y \in Y\big\} \subset \mathbb{R}^{2d}$ be its direct product. Multiplying the last identity by $[K]$, we get

$$\big[Q \times K\big] = \sum_{v \in \text{ex}(P)} \big[\text{cone}(Q, v) \times K\big] + \sum_{i \in I} \alpha_i \big[Q_i \times K\big].$$

For $v \in \mathbb{R}^d$, let $\overline{v} = (v, 0) \in \mathbb{R}^{2d}$. Using Problem 3 of Section 4.4, we can write

$$\big[Q \times K\big] = \sum_{v \in \text{ex}(P)} \big[\text{cone}\left(Q \times K,\ \overline{v}\right)\big] + \sum_{i \in I} \alpha_i \big[Q_i \times K\big].$$

Clearly, $Q_i \times K$ are (rational) polyhedra.

Let $T : \mathbb{R}^{2d} \longrightarrow \mathbb{R}^d$, $(x, y) \longmapsto x + y$ be the projection. Applying T to both parts of the identity (cf. Problem 1 of Section I.9.3 and Problem 3 of Section 4.3), we get

$$\begin{aligned}[P] = \big[P + K\big] &= \sum_{v \in \text{ex}(P)} \big[\text{cone}\left(Q + K, v\right)\big] + \sum_{i \in I} \alpha_i \big[Q_i + K\big] \\ &= \sum_{v \in \text{ex}(P)} \big[\text{cone}\left(P, v\right)\big] + \sum_{i \in I} \alpha_i \big[Q_i + K\big].\end{aligned}$$

Since $Q_i + K$ are (rational) polyhedra with straight lines, the result follows. □

Now we are ready to prove the main result of this section. The theorem below was first obtained by M. Brion in 1988 [**Bri88**] (see also [**Bri92**]) by methods of algebraic geometry. Since then many elementary proofs appeared; see, for example, [**PK92**], [**L91b**], [**BV97**] and [**Barv93**]. We obtain the result as a corollary of Theorem 4.5 (this is the approach of [**PK92**] and [**L91b**]).

(4.6) Corollary (Brion's Theorem). *Let $\mathcal{F} : \mathcal{P}(\mathbb{Q}^d) \longrightarrow \mathbb{C}(x_1, \ldots, x_d)$ be the valuation of Theorem 3.3. Then, for every rational polyhedron $P \subset \mathbb{R}^d$, one has*

$$\mathcal{F}\big[P\big] = \sum_{v \text{ is a vertex of } P} \mathcal{F}\big[\text{cone}(P, v)\big].$$

Proof. Follows by Theorem 3.3 and Theorem 4.5. □

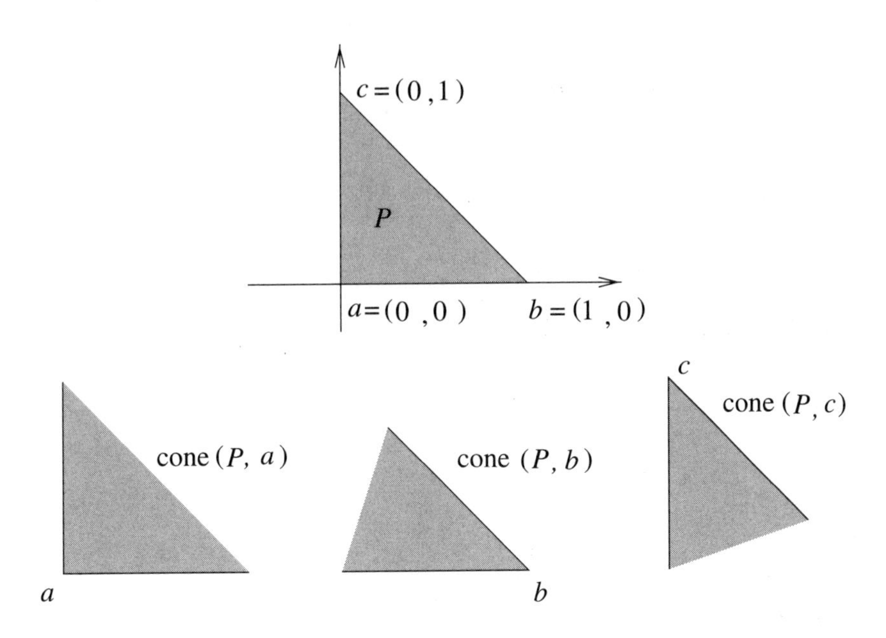

Figure 99

(4.7) Example. For the triangle P in Figure 99, there are three vertices a, b and c and three support cones. For cone(P, a), we have

$$\sum_{m \in \operatorname{cone}(P,a) \cap \mathbb{Z}^2} \mathbf{x}^m = \sum_{(\mu_1,\mu_2) \in \mathbb{Z}^2_+} x_1^{\mu_1} x_2^{\mu_2} = \frac{1}{(1-x_1)(1-x_2)}$$

provided $|x_1|, |x_2| < 1$. Therefore, by Part 2 of Theorem 3.3, we have

$$\mathcal{F}\big[\operatorname{cone}(P,a)\big] = \frac{1}{(1-x_1)(1-x_2)}.$$

The support cone at b, translated to the origin, is spanned by the vectors $a - b = (-1, 0)$ and $c - b = (-1, 1)$. Since we have

$$\left| \det \begin{pmatrix} -1 & -1 \\ 0 & 1 \end{pmatrix} \right| = 1,$$

by Corollary VII.2.6 the fundamental parallelepiped of $a - b$ and $c - b$ does not contain any lattice point other than the origin. Therefore, by Lemma 1.2 we have

$$\sum_{m \in \operatorname{cone}(P,b)} \mathbf{x}^m = \mathbf{x}^b \sum_{(\mu_1,\mu_2) \in \mathbb{Z}^2_+} \mathbf{x}^{\mu_1(a-b)+\mu_2(c-b)} = \frac{x_1}{(1-x_1^{-1})(1-x_1^{-1}x_2)}$$

provided $x_1 < 1$ and $|x_2/x_1| < 1$. Hence by Part 2 of Theorem 3.3,

$$\mathcal{F}\big[\operatorname{cone}(P, b)\big] = \frac{x_1}{(1 - x_1^{-1})(1 - x_1^{-1}x_2)}.$$

Similarly, we get

$$\mathcal{F}\big[\operatorname{cone}(P, c)\big] = \frac{x_2}{(1 - x_2^{-1})(1 - x_1 x_2^{-1})}.$$

Now

$$\begin{aligned}
&\mathcal{F}\big[\operatorname{cone}(P, a)\big] + \mathcal{F}\big[\operatorname{cone}(P, b)\big] + \mathcal{F}\big[\operatorname{cone}(P, c)\big] \\
&= \frac{1}{(1 - x_1)(1 - x_2)} + \frac{x_1}{(1 - x_1^{-1})(1 - x_1^{-1}x_2)} + \frac{x_2}{(1 - x_2^{-1})(1 - x_1 x_2^{-1})} \\
&= \frac{(x_1 - x_2) - x_1^3(1 - x_2) + x_2^3(1 - x_1)}{(1 - x_1)(1 - x_2)(x_1 - x_2)} = \frac{(x_1 - x_2) - (x_1^3 - x_2^3) + x_1x_2(x_1^2 - x_2^2)}{(1 - x_1)(1 - x_2)(x_1 - x_2)} \\
&= \frac{1 - (x_1^2 + x_1x_2 + x_2^2) + x_1x_2(x_1 + x_2)}{(1 - x_1)(1 - x_2)} \\
&= \frac{(1 - x_1^2) - (x_1x_2 - x_1^2x_2) - (x_2^2 - x_1x_2^2)}{(1 - x_1)(1 - x_2)} \\
&= \frac{1 + x_1 - x_1x_2 - x_2^2}{1 - x_2} = \frac{(1 - x_2^2) + x_1(1 - x_2)}{1 - x_2} = 1 + x_1 + x_2.
\end{aligned}$$

On the other hand,

$$\sum_{m \in P \cap \mathbb{Z}^2} \mathbf{x}^m = 1 + x_1 + x_2,$$

hence we must have

$$\mathcal{F}\big[P\big] = 1 + x_1 + x_2$$

and Brion's identity indeed holds.

It is interesting to note that while each of the three rational functions corresponding to the support cones of P have singularities, all the singularities cancel each other out in the sum. We observe also that there are no values x_1 and x_2 for which all the three series defining $\mathcal{F}[\operatorname{cone}(P, a)]$, $\mathcal{F}[\operatorname{cone}(P, b)]$ and $\mathcal{F}[\operatorname{cone}(P, c)]$ simultaneously converge.

PROBLEMS.

1 (M. Brion). Let $P \subset \mathbb{R}^d$ be a rational polyhedron. Prove that

$$\mathcal{F}[\operatorname{int} P] = \sum_{v \text{ is a vertex of } P} \mathcal{F}\big[\operatorname{int} \operatorname{cone}(P, v)\big],$$

where the interiors are taken in the affine hull of the polyhedron.

Remark: See [**Bri88**].

2°. Let Φ be the valuation of Problem 1, Section 3.4. Prove that

$$\Phi[P] = \sum_{v \text{ is a vertex of } P} \Phi\big[\operatorname{cone}(P, v)\big]$$

for any polyhedron $P \subset \mathbb{R}^d$.

5. The Ehrhart Polynomial of a Polytope

As an application of Brion's Theorem (Corollary 4.6), we describe how the number of lattice points in a polytope changes when the polytope is subjected to dilation by an integer.

Suppose that $P \subset \mathbb{R}^d$ is a (rational) polytope. Obviously, we can compute the number $|P \cap \mathbb{Z}^d|$ of integer points in P by substituting $\mathbf{x} = (1, \ldots, 1)$ into the generating function

$$f(P, \mathbf{x}) = \sum_{m \in P \cap \mathbb{Z}^d} \mathbf{x}^m.$$

If, however, we are to use Brion's Theorem to evaluate $f(\mathbf{x})$, we should exercise some care since $\mathbf{x} = (1, \ldots, 1)$ is a pole of every generating function

$$f\big(\operatorname{cone}(P, v), \mathbf{x}\big) = \sum_{m \in \operatorname{cone}(P,v) \cap \mathbb{Z}^d} \mathbf{x}^m;$$

see Example 4.7. A way to resolve this problem is to approach the point $\mathbf{x} = 1$ via some curve and compute an appropriate limit. To see how this works, we prove the existence of the *Ehrhart polynomial*, named after E. Ehrhart who first studied them.

(5.1) Theorem. *Let $P \subset \mathbb{R}^d$ be an integer polytope. Then there exists a univariate polynomial poly, called the Ehrhart polynomial of P, of degree at most d such that for every positive integer k we have*

$$|kP \cap \mathbb{Z}^d| = poly(k).$$

In words: the number of integer points in the dilated polytope is a polynomial in the coefficient of dilation.

Proof. Let $v_1, \ldots, v_n$ be the vertices of P, hence $v_i \in \mathbb{Z}^d$ for $i = 1, \ldots, n$. Then the vertices of kP are $kv_1, \ldots, kv_n$. Let

$$K_i = \operatorname{cone}(P, v_i) - v_i$$

be the support cone of P at v_i translated to the origin. Thus K_i is a rational cone (cf. Problems 1 and 2 of Section 4.1). It is not hard to see that

$$\operatorname{cone}(kP, v_i) = kv_i + K_i.$$

Applying Part 3 of Theorem 3.3, we conclude that

$$\mathcal{F}\big[\operatorname{cone}(kP, v_i)\big] = \mathbf{x}^{kv_i} \mathcal{F}[K_i].$$

By Corollary 4.6 and Part 2 of Theorem 3.3,

$$\sum_{m \in kP \cap \mathbb{Z}^d} \mathbf{x}^m = \mathcal{F}\big[kP\big] = \sum_{i=1}^{n} \mathbf{x}^{kv_i} \mathcal{F}\big[K_i\big].$$

We observe that as k changes, the $\mathcal{F}[K_i]$ remain the same and only the $\mathbf{x}^{kv_i}$ change. Applying Lemma 2.4, we conclude that

$$\mathcal{F}[K_i] = \frac{p_i(\mathbf{x})}{(1-\mathbf{x}^{u_{i1}})\cdots(1-\mathbf{x}^{u_{id}})},$$

where $u_{i1}, \ldots, u_{id}$ are integer vectors and p_i are Laurent polynomials in $\mathbf{x} = (x_1, \ldots, x_d)$. Thus we can write

$$\sum_{m \in kP\cap\mathbb{Z}^d} \mathbf{x}^m = \sum_{i=1}^{n} \frac{\mathbf{x}^{kv_i} p_i(\mathbf{x})}{(1-\mathbf{x}^{u_{i1}})\cdots(1-\mathbf{x}^{u_{id}})}. \tag{5.1.1}$$

Let us choose some very special $\mathbf{x}$. Let $c \in \mathbb{R}^d$ be a vector such that $\langle c, u_{ij}\rangle \neq 0$ for all i, j and let τ be a number. We substitute

$$\mathbf{x} = \mathbf{e}^{\tau c}$$

in (5.1.1) and observe what happens as $\tau \longrightarrow 0$, so that $\mathbf{x}$ approaches $(1, \ldots, 1)$. In the left-hand side, we get

$$\sum_{m \in kP\cap\mathbb{Z}^d} \mathbf{x}^m = \sum_{m \in kP\cap\mathbb{Z}^d} \exp\{\tau\langle c, m\rangle\}.$$

The sum is an analytic function of τ. Expanding it in the neighborhood of $\tau = 0$, we see that the constant term is the number $|kP \cap \mathbb{Z}^d|$ of lattice points in kP. Let us see what we get in each fraction of the right-hand side.

We have

$$\begin{aligned}
&\frac{\mathbf{x}^{kv_i} p_i(\mathbf{x})}{(1-\mathbf{x}^{u_{i1}})\cdots(1-\mathbf{x}^{u_{id}})} \\
&\qquad = \frac{\exp\{\tau\langle c, kv_i\rangle\} \cdot p_i(\mathbf{e}^{\tau c})}{\big(1-\exp\{\tau\langle c, u_{i1}\rangle\}\big)\cdots\big(1-\exp\{\tau\langle c, u_{id}\rangle\}\big)} \\
&\qquad = \tau^{-d} \exp\{\tau\langle c, kv_i\rangle\} p_i(\mathbf{e}^{\tau c}) \prod_{j=1}^{d} \frac{\tau}{1-\exp\{\tau\langle c, u_{ij}\rangle\}}.
\end{aligned} \tag{5.1.2}$$

Now, we observe that the part

$$p_i(\mathbf{e}^{\tau c}) \prod_{j=1}^{d} \frac{\tau}{1-\exp\{\tau\langle c, u_{ij}\rangle\}} = \sum_{l=0}^{\infty} \alpha_{il}\tau^l$$

is an analytic function of τ which does not depend on k at all. On the other hand, the Laurent expansion of

$$\tau^{-d} \exp\{\tau\langle c, kv_i\rangle\} = \tau^{-d} \sum_{l=0}^{\infty} k^l \frac{\langle c, v_i\rangle^l}{l!} \tau^l$$

contains some negative terms and does depend on k. We are interested in the constant term of the expansion of (5.1.2). Collecting the terms, we conclude that the constant term of (5.1.2) is

$$\sum_{l=0}^{d} k^l \frac{\langle c, v_i \rangle^l}{l!} \alpha_{i,d-l},$$

which is a polynomial in k. Equating the constant terms in (5.1.1), we get

$$|kP \cap \mathbb{Z}^d| = \sum_{i=1}^{n} \sum_{l=0}^{d} k^l \frac{\langle c, v_i \rangle^l}{l!} \alpha_{i,d-l} \equiv poly(k),$$

which completes the proof. □

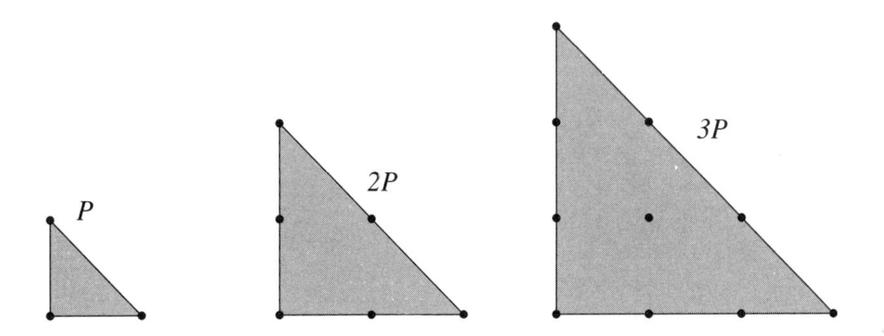

Figure 100. Example: a triangle P and its dilations $2P$ and $3P$. One can observe that $|kP \cap \mathbb{Z}^2| = k^2/2 + 3k/2 + 1$.

PROBLEMS.

1. Let $P \subset \mathbb{R}^d$ be an integer polytope and let $p(k) = |kP \cap \mathbb{Z}^d|$ be its Ehrhart polynomial. Prove that $\deg p = \dim P$.

2. Deduce the existence of the Ehrhart polynomial for integer polygons from Pick's Formula; see Problem 6 of Section VII.2.6. Prove that in the case of integer polygons the coefficients of the Ehrhart polynomial are non-negative.

3. Let $P \subset \mathbb{R}^d$ be an integer polytope and let p be its Ehrhart polynomial. Prove that for any positive integer k we have

$$p(-k) = (-1)^{\dim P} \left| \operatorname{int}(kP) \cap \mathbb{Z}^d \right|,$$

where the interior of a polytope is considered with respect to its affine hull (the *reciprocity relation*).

Hint: Use Problem 1 of Section 4.7 and Problem 1 of Lemma 2.4.

4. Let $a_d(k)$ denote the number of $d \times d$ matrices with non-negative integer entries such that all row and column sums are equal to k.

a) Prove that for each d, $a_d(k)$ is a polynomial $p_d(k)$ in k of degree $(d-1)^2$.

Hint: See Section II.5.

b) Deduce from Problem 3 that $p_d(-1) = \ldots = p_d(-d+1) = 0$.

5. Let $P \subset \mathbb{R}^d$ be an integer polytope such that $\dim P = d$ and let p be its Ehrhart polynomial. Prove that the highest coefficient of p is equal to the volume of P.

6*. Prove that the constant term of the Ehrhart polynomial is 1.

7. Let $P \subset \mathbb{R}^3$ be the tetrahedron with the vertices $(0,0,0)$, $(1,0,0)$, $(0,1,0)$ and $(1,1,n)$, where $n > 0$ is an integer parameter. Prove that the Ehrhart polynomial of P is

$$p(k) = \frac{n}{6}k^3 + k^2 + \frac{12-n}{6}k + 1.$$

8. Let $P \subset \mathbb{R}^d$ be a rational polytope and let m be a positive integer such that mP is an integer polytope. Let $f(k) = |kP \cap \mathbb{Z}^d|$. Let us fix a positive integer k_0. Prove that $f(k_0 + nm)$ is a polynomial in n, where n is a positive integer.

9. Let $c_1, \ldots, c_m \in \mathbb{Z}^d$ be integer vectors and let $\Gamma \subset \mathbb{Z}^m$ be a set of m-tuples $a = (\alpha_1, \ldots, \alpha_m)$ such that the polyhedron

$$P(a) = \Big\{x \in \mathbb{R}^d : \quad \langle c_i, x \rangle \leq \alpha_i \quad \text{for} \quad i = 1, \ldots, m\Big\}$$

is an integer polytope for all $a \in \Gamma$ and any two polytopes $P(a)$ and $P(b)$ for $a, b \in \Gamma$ have the same combinatorial structure (that is, there is an inclusion-preserving bijection between the set of faces of $P(a)$ and the set of faces of $P(b)$). Prove that there is an m-variant polynomial p such that

$$|P(a) \cap \mathbb{Z}^d| = p(a) \quad \text{for all} \quad a \in \Gamma.$$

10. Let $c_1, \ldots, c_m \in \mathbb{R}^d$ be vectors and let $\Gamma \subset \mathbb{R}^m$ be a set of m-tuples $a = (\alpha_1, \ldots, \alpha_m)$ such that the polyhedron

$$P(a) = \Big\{x \in \mathbb{R}^d : \quad \langle c_i, x \rangle \leq \alpha_i \quad \text{for} \quad i = 1, \ldots, m\Big\}$$

is a polytope for all $a \in \Gamma$ and any two polytopes $P(a)$ and $P(b)$ for $a, b \in \Gamma$ have the same combinatorial structure. Prove that there exists an m-variant polynomial $p(a)$ such that

$$\operatorname{vol} P(a) = p(a) \quad \text{for all} \quad a \in \Gamma.$$

Hint: Cf. Problem 2 of Section 4.7, Problem 1 of Section 3.4 and Problem 2 of Section 2.4.

6. Example: Totally Unimodular Polytopes

In one special case, Brion's Theorem (Corollary 4.6) gives a particularly succinct representation of the generating function.

(6.1) Definition. Let $u_1, \ldots, u_k \in \mathbb{Z}^d$ be linearly independent vectors and let $K = \operatorname{co}(u_1, \ldots, u_k)$. Let $L_k = \operatorname{span}(u_1, \ldots, u_k)$ and let $\Lambda_k = \mathbb{Z}^d \cap L_k$. We say that K is a *unimodular cone* provided $u_1, \ldots, u_k$ is a basis of Λ_k, considered as a lattice in L_k. We call $u_1, \ldots, u_k$ *generators* of K.

Let $P \subset \mathbb{R}^d$ be an integer polytope. We say that P is *totally unimodular* provided the support cone at every vertex of P is a translation of a unimodular cone.

Some important polytopes are totally unimodular.

PROBLEMS.

1. Let Δ be the standard $(d-1)$-dimensional simplex in $\mathbb{R}^d$:

$$\Delta = \Big\{(\xi_1, \ldots, \xi_d) : \quad \sum_{i=1}^{d} \xi_i = 1 \quad \text{and} \quad \xi_i \geq 0 \quad \text{for} \quad i = 1, \ldots, d\Big\}.$$

Prove that Δ is a totally unimodular polytope.

2. Let us fix positive integers m and n and let us identify $\mathbb{R}^d$, $d = mn$, with the space of $m \times n$ real matrices (ξ_{ij}). Thus $\mathbb{Z}^d$ is identified with the space of all $m \times n$ integer matrices. Let us fix positive integers $\alpha_1, \ldots, \alpha_m$ and $\beta_1, \ldots, \beta_n$ and let $P \subset \mathbb{R}^d$ be the polyhedron of all non-negative $m \times n$ matrices with row sums $\alpha_1, \ldots, \alpha_m$ and column sums $\beta_1, \ldots, \beta_n$. Suppose that $\dim P = (m-1)(n-1)$ and that P is a simple polytope; see Definition VI.5.1 (which means that $\alpha_1, \ldots, \alpha_m$ and $\beta_1, \ldots, \beta_n$ are chosen in a sufficiently generic way). Prove that P is a totally unimodular polytope.

Remark: More generally, a sufficiently generic transportation polytope (see Section II.7) is totally unimodular. Non-negative integer matrices with prescribed row and column sums are called *contingency tables.*

3. Let $u_1, \ldots, u_d \in \mathbb{Z}^d$ be linearly independent lattice points and let $K = \operatorname{co}(u_1, \ldots, u_d)$. Prove that K can be dissected into the union of unimodular cones, that is, there is a decomposition

$$K = \bigcup_{i=1}^{n} K_i,$$

where each cone K_i is unimodular and the intersection $K_i \cap K_j$ of every two distinct cones K_i and K_j is a proper face of both.

4. Let $K \subset \mathbb{R}^d$ be a unimodular cone such that $\dim K = d$. Prove that the polar $K^\circ \subset \mathbb{R}^d$ is a unimodular cone.

For totally unimodular polytopes, Brion's Theorem (Corollary 4.6) gives a particularly nice identity.

(6.2) Corollary. *Let $P \subset \mathbb{R}^d$ be a totally unimodular polytope with the vertices $v_1, \ldots, v_n$. Suppose that $\dim P = k$ and that $\operatorname{cone}(P, v_i) = v_i + K_i$, where K_i is a unimodular cone with the generators $u_{i1}, \ldots, u_{ik}$ for $i = 1, \ldots, n$. Then*

$$\sum_{m \in P \cap \mathbb{Z}^d} \mathbf{x}^m = \sum_{i=1}^{n} \frac{\mathbf{x}^{v_i}}{(1 - \mathbf{x}^{u_{i1}}) \cdots (1 - \mathbf{x}^{u_{ik}})}.$$

Proof. Follows by Brion's Theorem and Problem 3 of Section 1.2. □

PROBLEMS.

1. Let $K \subset \mathbb{R}^d$ be a unimodular cone with the generators $u_1, \ldots, u_d$. Let us define vectors $u_1^*, \ldots, u_d^*$ by

$$\langle u_i, u_j^* \rangle = \begin{cases} 1 & \text{if } i = j, \\ 0 & \text{otherwise.} \end{cases}$$

Let $v \in \mathbb{Q}^d$ be a rational vector. Prove that

$$\mathcal{F}\big[K + v\big] = \frac{\mathbf{x}^w}{(1 - \mathbf{x}^{u_1}) \cdots (1 - \mathbf{x}^{u_d})}, \quad \text{where} \quad w = \sum_{i=1}^{d} \lceil \langle v, u_i^* \rangle \rceil u_i.$$

Here $\lceil \xi \rceil$ denotes the smallest integer greater than or equal to ξ.

2. Let $P \subset \mathbb{R}^d$ be a rational polytope with the vertices $v_1, \ldots, v_n$ such that the support cone of P at v_i is a translation of a unimodular cone with the generators $u_{i1}, \ldots, u_{id}$. Let us define $u_{i1}^*, \ldots, u_{id}^*$ by

$$\langle u_{ij}, u_{ik}^* \rangle = \begin{cases} 1 & \text{if } k = j, \\ 0 & \text{otherwise.} \end{cases}$$

Prove that

$$\sum_{m \in P \cap \mathbb{Z}^d} \mathbf{x}^m = \sum_{i=1}^{n} \frac{\mathbf{x}^{w_i}}{(1 - \mathbf{x}^{u_{i1}}) \cdots (1 - \mathbf{x}^{u_{id}})} \quad \text{where} \quad w_i = \sum_{j=1}^{d} \lceil \langle v_i, u_{ij}^* \rangle \rceil u_{ij}.$$

Here is a continuous version of Corollary 6.2.

3. Let $P \subset \mathbb{R}^d$ be a simple polytope with the vertices $v_1, \ldots, v_n$. Suppose that $\dim P = d$ and that $\operatorname{cone}(P, v_i) = v_i + K_i$, where $K_i = \operatorname{co}\big(u_{i1}, \ldots, u_{id}\big)$ for some linearly independent vectors $u_{i1}, \ldots, u_{id}$. Let α_i be the volume of the parallelepiped spanned by $u_{i1}, \ldots, u_{id}$. Prove that

$$\int_P \exp\big\{\langle c, x \rangle\big\} \, dx = \sum_{i=1}^{n} \exp\big\{\langle c, v_i \rangle\big\} \alpha_i \prod_{j=1}^{d} \frac{1}{\langle -c, u_{ij} \rangle}.$$

The formula of Corollary 6.2 provides a way to compute the number of points in a totally unimodular polytope P by specializing $\mathbf{x} = (1, \ldots, 1)$. Of course, one should do it carefully because of the singularities in the right-hand side. One way to do it is to choose a vector $c \in \mathbb{R}^d$ such that $\langle c, u_{ij} \rangle \neq 0$ for all i and j, substitute $\mathbf{x} = \mathbf{e}^{\tau c}$, where τ is a real parameter, and compute the constant term of the Laurent expansion of the left-hand side in the neighborhood of $\tau = 0$, as we did in the proof of Theorem 5.1. Another possibility is to substitute $\mathbf{x}$ sufficiently close to $(1, \ldots, 1)$ and round the result to the nearest integer.

(6.3) The duality trick. It happens sometimes that a polytope P is rather close to being totally unimodular and yet is not totally unimodular. For example, if the transportation polytope P in Problem 2 of Section 6.1 is not simple, it cannot be totally unimodular. Then it becomes a problem to compute $\mathcal{F}[K]$, where K is the translation to the origin of the support cone of P at some integer vertex v.

A computationally efficient way to handle $\mathcal{F}[K]$ is as follows. Suppose, for simplicity, that $K \subset \mathbb{R}^d$ is a d-dimensional rational cone. Let $K^\circ \subset \mathbb{R}^d$ be the polar of K. Let us dissect (or otherwise decompose) K° into the union of unimodular cones; see Problem 3 of Section 6.1. Thus, by the Inclusion-Exclusion Formula, we can write

$$[K^\circ] = \sum_{i=1}^{n} [K_i] \quad \pm \quad \text{indicators of lower-dimensional rational cones,}$$

where $K_i \subset \mathbb{R}^d$ are d-dimensional unimodular cones. Polarity preserves linear relations between closed cones; see Corollary IV.1.6. By the Bipolar Theorem (Theorem IV.1.2), we have $(K^\circ)^\circ = K$. Moreover, polars of lower-dimensional rational cones are rational cones containing straight lines. Hence we can write

$$[K] = \sum_{i=1}^{n} [K_i^\circ] \quad \pm \quad \text{indicators of rational cones containing straight lines.}$$

By Theorem 3.3, Part 4, valuation $\mathcal{F}$ ignores rational polyhedra with straight lines and hence

$$\mathcal{F}[K] = \sum_{i=1}^{n} \mathcal{F}[K_i^\circ].$$

By Problem 4 of Section 6.1, K_i° are unimodular cones. Hence we get a closed formula for $\mathcal{F}[K]$.

The computational savings for passing to the polar cones and then going back compared to the direct decomposition of K into unimodular cones are twofold. First, we are able to ignore the lower-dimensional cones completely. Second, in many problems it is much easier to decompose the polar cone K° into unimodular cones. Suppose, for example, that P is a transportation polytope of Problem 2, Section 6.1. Let v be a vertex of P and let $K = \operatorname{cone}(P, v) - v$. Let us consider K as a cone in the subspace $L = \operatorname{aff}(P) - v$ and let $K^\circ \subset L$ be its polar in that subspace. Then *any* triangulation of K° produces totally unimodular cones. Thus if P is sufficiently close to being a simple polytope, we get a closed formula for the generating function

$$\mathcal{F}[P] = \sum_{m \in P \cap \mathbb{Z}^d} \mathbf{x}^m.$$

7. Remarks

Gram's relation or the Brianchon-Gram Theorem expresses the indicator function of a polyhedron P as an alternating sum of the indicators of the support cones at the bounded faces of P (cf. Problem 1 of Section 4.4 where the formula is stated for polytopes). It is discussed, for example, in [**L91a**].

Triangulations, which we discussed very briefly (see Lemma 2.3), are discussed in detail in [**Z95**].

A survey of results pertaining to this chapter (with an algorithmic slant) can be found in [**BP99**]. An implementation of the algorithm for counting contingency tables based on Brion's Theorem (see Problem 2 of Section 6.1) is found in [**DS+**].

For Ehrhart polynomials and their interesting properties, see Chapter IV of [**St97**]. An analytical approach leading to interesting closed formulas is found in [**DR97**]. For extension to lattice semigroups (cf. Problem 3 of Section 3.1), see [**Kho95**] and [**BP99**]. Lattice points in irrational polytopes exhibit a very interesting behavior [**Sk98**]. That topic, however, is beyond the scope of this book.

Bibliography

[AEZ93] G.E. Andrews, S.B. Ekhad and D. Zeilberger, *A short proof of Jacobi's formula for the number of representations of an integer as a sum of four squares*, Amer. Math. Monthly **100** (1993), 274–276.

[AN87] E.J. Anderson and P. Nash, *Linear Programming in Infinite-Dimensional Spaces. Theory and Applications*, Wiley, Interscience, 1987.

[AP79] Y.H. Au-Yeung and Y.T. Poon, *A remark on the convexity and positive definiteness concerning Hermitian matrices*, Southeast Asian Bull. Math. **3** (1979), 85–92.

[B97] K. Ball, *An elementary introduction to modern convex geometry*, Flavors of Geometry, Math. Sci. Res. Inst. Publ., vol. 31, Cambridge Univ. Press, Cambridge, 1997, pp. 1–58.

[Ba95] W. Banaszczyk, *Inequalities for convex bodies and polar reciprocal lattices in* $\mathbb{R}^n$, Discrete Comput. Geom. **13** (1995), 217–231.

[Ba96] W. Banaszczyk, *Inequalities for convex bodies and polar reciprocal lattices in* $\mathbb{R}^n$*. II. Application of* K*-convexity*, Discrete Comput. Geom. **16** (1996), 305–311.

[Bar82] I. Bárány, *A generalization of Carathéodory's theorem*, Discrete Math. **40** (1982), 141–152.

[Barv92] A. Barvinok, *Combinatorial complexity of orbits in representations of the symmetric group*, Representation Theory and Dynamical Systems, Adv. Soviet Math., vol. 9, Amer. Math. Soc., Providence, RI, 1992, pp. 161–182.

[Barv93] A. Barvinok, *Computing the volume, counting integral points, and exponential sums*, Discrete Comput. Geom. **10** (1993), 123–141.

[Barv95] A. Barvinok, *Problems of distance geometry and convex properties of quadratic maps*, Discrete Comput. Geom. **13** (1995), 189–202.

[Barv01] A. Barvinok, *A remark on the rank of positive semidefinite matrices subject to affine constraints*, Discrete Comput. Geom. **25** (2001), 23–31.

[BC98] J. Bochnak, M. Coste and M.-F. Roy, *Real Algebraic Geometry*, A Series of Modern Surveys in Mathematics 36, Springer-Verlag, Berlin, 1998.

[BH75] A.E. Bryson, Jr. and Y.C. Ho, *Applied Optimal Control. Optimization, Estimation, and Control. Revised printing*, Hemisphere Publishing Corp. Distributed by Halsted Press [John Wiley & Sons], 1975.

[Bl02] G. Blekherman, *Convexity properties of the cone of non-negative polynomials*, preprint (2002), Univ. of Michigan.

[BL81] L.J. Billera and C.W. Lee, *A proof of the sufficiency of McMullen's conditions for f-vectors of simplicial convex polytopes*, J. Combin. Theory Ser. A **31** (1981), 237–255.

[BLP99] W. Banaszczyk, A.E. Litvak, A. Pajor and S.J. Szarek, *The flatness theorem for nonsymmetric convex bodies via the local theory of Banach spaces*, Math. Oper. Res. **24** (1999), 728–750.

[Bo98] V.I. Bogachev, *Gaussian Measures*, Mathematical Surveys and Monographs, vol. 62, American Mathematical Society, Providence, RI, 1998.

[Bou87] N. Bourbaki, *Topological Vector Spaces, Chapters 1–5*, Elements of Mathematics, Springer-Verlag, 1987.

[BP99] A. Barvinok and J.E. Pommersheim, *An algorithmic theory of lattice points in polyhedra*, New Perspectives in Algebraic Combinatorics (Berkeley, CA, 1996–97), Math. Sci. Res. Inst. Publ., vol. 38, Cambridge Univ. Press, Cambridge, 1999, pp. 91–147.

[Br61] L. Brickman, *On the field of values of a matrix*, Proc. Amer. Math. Soc. **12** (1961), 61–66.

[Bri88] M. Brion, *Points entiers dans les polyèdres convexes*, Ann. Sci. École Norm. Sup. (4) **21** (1988), 653–663.

[Bri92] M. Brion, *Polyèdres et réseaux*, Enseign. Math. (2) **38** (1992), 71–88.

[Brø83] A. Brøndsted, *An Introduction to Convex Polytopes*, Graduate Texts in Mathematics, vol. 90, Springer-Verlag, New York-Berlin, 1983.

[BS96] L.J. Billera and A. Sarangarajan, *The combinatorics of permutation polytopes*, Formal Power Series and Algebraic Combinatorics (New Brunswick, NJ, 1994), DIMACS Ser. Discrete Math. Theoret. Comput. Sci., vol. 24, Amer. Math. Soc., Providence, RI, 1996, pp. 1–23.

[BV97] M. Brion and M. Vergne, *Residue formulae, vector partition functions and lattice points in rational polytopes*, J. Amer. Math. Soc. **10** (1997), 797–833.

[C97] J.W.S. Cassels, *An Introduction to the Geometry of Numbers*, Classics in Mathematics, Springer-Verlag, Berlin, 1997.

[CH88] G.M. Crippen and T.F. Havel, *Distance Geometry and Molecular Conformation*, Chemometrics Series 15, Wiley, New York, 1988.

[Co90] J.B. Conway, *A Course in Functional Analysis. Second edition*, Graduate Texts in Mathematics, vol. 96, Springer-Verlag, New York, 1990.

[CS99] J.H. Conway and N.J.A. Sloane, *Sphere Packings, Lattices and Groups. Third edition. With additional contributions by E. Bannai, R. E. Borcherds, J. Leech, S. P. Norton, A. M. Odlyzko, R. A. Parker, L. Queen and B. B. Venkov*, Grundlehren der Mathematischen Wissenschaften, vol. 290, Springer-Verlag, New York, 1999.

[Da71] Ch. Davis, *The Toeplitz-Hausdorff theorem explained*, Canad. Math. Bull. **14** (1971), 245–246.

[DG63] L. Danzer, B. Grünbaum and V. Klee, *Helly's theorem and its relatives*, Proc. Sympos. Pure Math., vol. VII, Amer. Math. Soc., Providence, R.I., 1963, pp. 101–180.

[DL97] M.M. Deza and M. Laurent, *Geometry of Cuts and Metrics*, Algorithms and Combinatorics 15, Springer, 1997.

[Do73] J.-P. Doignon, *Convexity in cristallographical lattices*, J. Geometry **3** (1973), 71–85.

[DR97] R. Diaz and S. Robins, *The Ehrhart polynomial of a lattice polytope*, Ann. of Math. (2) **145** (1997), 503–518.

[DS+] J.A. De Loera and B. Sturmfels, *Algebraic unimodular counting*, to appear, Math. Programming.

[E93] J. Eckhoff, *Helly, Radon, and Carathéodory type theorems*, Handbook of Convex Geometry (P.M. Gruber and J.M. Wills, eds.), vol. A, Elsevier, North-Holland, 1993, pp. 389–448.

[F93] W. Fulton, *Introduction to Toric Varieties*, Annals of Mathematics Studies, vol. 131, Princeton University Press, Princeton, NJ, 1993.

[FK99] A. Frieze and R. Kannan, *Quick approximation to matrices and applications*, Combinatorica **19** (1999), 175–220.

[FL76] S. Friedland and R. Loewy, *Subspaces of symmetric matrices containing matrices with a multiple first eigenvalue*, Pacific J. Math. **62** (1976), 389–399.

[FLM77] T. Figiel, J. Lindenstrauss and V.D. Milman, *The dimension of almost spherical sections of convex bodies*, Acta Math **139** (1977), 53–94.

[G92] M.B. Gromova, *The Birkhoff-von Neumann theorem for polystochastic matrices*; translation of Operations research and statistical simulation, No. 2 (Russian), 3–15, Leningrad. Univ., Leningrad, 1974. Selecta Math. Soviet. **11** (1992), 145–158.

[GL87] P.M. Gruber and C.G. Lekkerkerker, *Geometry of Numbers. Second Edition*, North-Holland Mathematical Library, vol. 37, North-Holland Publishing Co., Amsterdam, 1987.

[GLS93] M. Grötschel, L. Lovász and A. Schrijver, *Geometric Algorithms and Combinatorial Optimization. Second Edition*, Algorithms and Combinatorics, vol. 2, Springer-Verlag, Berlin, 1993.

[Gr67] B. Grünbaum, *Convex Polytopes. With the cooperation of Victor Klee, M. A. Perles and G. C. Shephard*, Pure and Applied Mathematics, vol. 16, Interscience Publishers, John Wiley & Sons, Inc., New York, 1967.

[H95] B. Hendrickson, *The molecule problem: exploiting structure in global optimization*, SIAM J. Optim. **5** (1995), 835–857.

[Hå88] J. Håstad, *Dual vectors and lower bounds for the nearest lattice point problem*, Combinatorica **8** (1988), 75–81.

[JL84] W.B. Johnson and J. Lindenstrauss, *Extensions of Lipschitz mappings into a Hilbert space*, Conference in modern analysis and probability (New Haven, Conn., 1982), Contemp. Math., vol. 26, Amer. Math. Soc., Providence, RI, 1984, pp. 189–206.

[K95] G. Kalai, *Combinatorics and convexity*, Proceedings of the International Congress of Mathematicians Vol. 1, 2 (Zürich, 1994), Birkhäuser, Basel, 1995, pp. 1363–1374.

[Kh97] A. Ya. Khinchin, *Continued Fractions*, Dover Publications, Inc., Mineola, NY, 1997.

[Kho95] A.G. Khovanskii, *Sums of finite sets, orbits of commutative semigroups and Hilbert functions. (Russian)*, Funktsional. Anal. i Prilozhen. **29** (1995), 36–50; translation in Funct. Anal. Appl. **29** (1995), 102–112.

[Kl63] V. Klee, *The Euler characteristic in combinatorial geometry*, Amer. Math. Monthly **70** (1963), 119–127.

[KN77] M.G. Krein and A.A. Nudelman, *The Markov Moment Problem and Extremal Problems. Ideas and Problems of P. L. Chebyshev and A. A. Markov and their Further Development*, Translations of Mathematical Monographs, vol. 50, American Mathematical Society, Providence, R.I., 1977.

[KR97] D.A. Klain and G.-C. Rota, *Introduction to Geometric Probability*, Cambridge University Press, 1997.

[KS66] S. Karlin and W.J. Studden, *Tchebycheff Systems: with Applications in Analysis and Statistics*, Wiley, Interscience, 1966.

[L88] J. Lawrence, *Valuations and polarity*, Discrete Comput. Geom. **3** (1988), 307–324.

[L91a] J. Lawrence, *Polytope volume computation*, Math. Comp. **57** (1991), 259–271.

[L91b] J. Lawrence, *Rational-function-valued valuations on polyhedra*, Discrete and computational geometry (New Brunswick, NJ, 1989/1990), DIMACS Ser. Discrete Math. Theoret. Comput. Sci., vol. 6, Amer. Math. Soc., Providence, RI, 1991, pp. 199–208.

[L97] J. Lawrence, *A short proof of Euler's relation for convex polytopes*, Canad. Math. Bull. **40** (1997), 471–474.

[Le01] M. Ledoux, *The Concentration of Measure Phenomenon*, Mathematical Surveys and Monographs, vol. 89, Amer. Math. Soc., Providence, RI, 2001.

[Li66] J. Lindenstrauss, *A short proof of Liapounoff's convexity theorem*, J. Math. Mech. **15** (1966), 971–972.

[LLL82] A.K. Lenstra, H.W. Lenstra, Jr. and L. Lovász, *Factoring polynomials with rational coefficients*, Math. Ann. **261** (1982), 515–534.

[LLS90] J.C. Lagarias, H.W. Lenstra, Jr. and C.-P. Schnorr, *Korkin-Zolotarev bases and successive minima of a lattice and its reciprocal lattice*, Combinatorica **10** (1990), 333–348.

[Lo79] L. Lovász, *On the Shannon capacity of a graph*, IEEE Trans. Inform. Theory **25** (1979), 1–7.

[Lo86] L. Lovász, *An Algorithmic Theory of Numbers, Graphs and Convexity*, CBMS-NSF Regional Conference Series in Applied Mathematics, vol. 50, SIAM, Philadelphia, PA, 1986.

[LS92] L. Lovász and H.E. Scarf, The generalized basis reduction algorithm, Math. Oper. Res. **17** (1992), 751–764.

[M56] J.C. Mairhuber, *On Haar's theorem concerning Chebychev approximation problems having unique solutions*, Proc. Amer. Math. Soc. **7** (1956), 609–615.

[Ma80] W.S. Massey, *Singular Homology Theory*, Graduate Texts in Mathematics, vol. 70, Springer-Verlag, 1980.

[Mat02] J. Matoušek, *Lectures on Discrete Geometry*, Graduate Texts in Mathematics, vol. 212, Springer-Verlag, 2002.

[Mc93a] P. McMullen, *Valuations and dissections*, Handbook of Convex Geometry (P.M. Gruber and J.M. Wills, eds.), vol. B, Elsevier, North-Holland, 1993, pp. 933–990.

[Mc93b] P. McMullen, *On simple polytopes*, Invent. Math. **113** (1993), 419–444.

[Mc96] P. McMullen, *Weights on polytopes*, Discrete Comput. Geom. **15** (1996), 363–388.

[MiS86] V.D. Milman and G. Schechtman, *Asymptotic Theory of Finite-Dimensional Normed Spaces. With an Appendix by M. Gromov*, Lecture Notes in Mathematics, vol. 1200, Springer-Verlag, Berlin, 1986.

[MO79] A.W. Marshall and I. Olkin, *Inequalities: Theory of Majorization and its Applications*, Academic Press, New York, 1979.

[MR95] R. Motwani and P. Raghavan, *Randomized Algorithms*, Cambridge University Press, Cambridge, 1995.

[MS83] P. McMullen and R. Schneider, *Valuations on convex bodies*, Convexity and Applications (P.M. Gruber and J.M. Wills, eds.), Birkhäuser, Basel, 1983, pp. 170–247.

[MSh71] P. McMullen and G.C. Shephard, *Convex Polytopes and the Upper Bound Conjecture. Prepared in collaboration with J. E. Reeve and A. A. Ball*, London Mathematical Society Lecture Note Series, vol. 3, Cambridge University Press, London-New York, 1971.

[MT93] A. Megretsky and S. Treil, *Power distribution inequalities in optimization and robustness of uncertain systems*, J. Math. Systems Estim. Control **3** (1993), 301–319.

[N96] M.B. Nathanson, *Additive Number Theory. The Classical Bases*, Graduate Texts in Mathematics, vol. 164, Springer-Verlag, New York, 1996.

[NN94] Yu. Nesterov and A. Nemirovskii, *Interior-Point Polynomial Algorithms in Convex Programming*, SIAM Studies in Applied Mathematics, vol. 13, Society for Industrial and Applied Mathematics (SIAM), Philadelphia, PA, 1994.

[P94] G. Pisier, *The Volume of Convex Bodies and Banach Space Geometry*, Cambridge Tracts in Mathematics, vol. 94, Cambridge University Press, Cambridge, 1989.

[PK92] A.V. Pukhlikov and A.G. Khovanskii, *The Riemann-Roch theorem for integrals and sums of quasipolynomials on virtual polytopes. (Russian)*, Algebra i Analiz **4** (1992), 188–216; translation in St. Petersburg Math. J. **4** (1993), 789–812.

[PS98] C.H. Papadimitriou and K. Steiglitz, *Combinatorial Optimization: Algorithms and Complexity*, Dover, Mineola, New York, 1998.

[R95] B. Reznick, *Uniform denominators in Hilbert's seventeenth problem*, Math. Z. **220** (1995), 75–97.

[R00] B. Reznick, *Some concrete aspects of Hilbert's 17th Problem*, Real Algebraic Geometry and Ordered Structures (Baton Rouge, LA, 1996), Contemp. Math., vol. 253, Amer. Math. Soc., Providence, RI, 2000, pp. 251–272.

[Ru91] W. Rudin, *Functional Analysis. Second Edition*, International Series in Pure and Applied Mathematics, McGraw-Hill, Inc., New York, 1991.

[S85] H.E. Scarf, *Integral polyhedra in three space*, Math. Oper. Res. **10** (1985), 403–438.

[Sc93] R. Schneider, *Convex Bodies: the Brunn-Minkowski Theory*, Encyclopedia of Mathematics and its Applications, vol. 44, Cambridge University Press, Cambridge, 1993.

[Sch87] C.-P. Schnorr, *A hierarchy of polynomial time lattice basis reduction algorithms*, Theoret. Comput. Sci. **53** (1987), 201–224.

[Schr86] A. Schrijver, *Theory of Linear and Integer Programming*, Wiley-Interscience Series in Discrete Mathematics, John Wiley & Sons, Chichester, 1986.

[Se95] R. Seidel, *The upper bound theorem for polytopes: an easy proof of its asymptotic version*, Comput. Geom. **5** (1995), 115–116.

[Sk98] M.M. Skriganov, *Ergodic theory on $SL(n)$, Diophantine approximations and anomalies in the lattice point problem*, Invent. Math. **132** (1998), 1–72.

[St80] R.P. Stanley, *The number of faces of a simplicial convex polytope*, Adv. in Math. **35** (1980), 236–238.

[St97] R.P. Stanley, *Enumerative Combinatorics. Vol. 1. With a foreword by Gian-Carlo Rota. Corrected reprint of the 1986 original*, Cambridge Studies in Advanced Mathematics, vol. 49, Cambridge University Press, Cambridge, 1997.

[V01] R.J. Vanderbei, *Linear Programming. Foundations and Extensions. Second Edition*, International Series in Operations Research & Management Science, vol. 37, Kluwer Academic Publishers, Boston, MA, 2001.

[VB96] L. Vandenberghe and S. Boyd, *Semidefinite programming*, SIAM Rev. **38** (1996), 49–95.

[Ve70] A.M. Vershik, *Some remarks on infinite-dimensional problems in linear programming (Russian)*, Uspehi Mat. Nauk **25** (1970), 117–124.

[Ve84] A.M. Vershik, *Quadratic forms that are positive on the cone and quadratic duality*, Zap. Nauchn. Sem. Leningrad. Otdel. Mat. Inst. Steklov. (LOMI) **134** (1984), 59–83 (Russian); English transl. in Journal of Soviet Mathematics **36**, 39–56.

[VT68] A.M. Vershik and V. Temel't, *Certain questions on approximation of the optimal value of infinite-dimensional linear programming problems. (Russian)*, Sibirsk. Mat. Z. **9** (1968), 790–803.

[W94] R. Webster, *Convexity*, Oxford University Press, New York, 1994.

[We97] R. Wenger, *Helly-type theorems and geometric transversals*, Handbook of Discrete and Computational Geometry (J.E. Goodman and J. O'Rourke, eds.), CRC Press, Boca Raton, New York, 1997, pp. 63–82.

[YKK84] V.A. Yemelichev, M.M. Kovalëv and M.K. Kravtsov, *Polytopes, Graphs and Optimization*, Cambridge University Press, Cambridge, 1984.

[Z95] G.M. Ziegler, *Lectures on Polytopes*, Graduate Texts in Mathematics, vol. 152, Springer-Verlag, Berlin, 1995.

[Ž97] R. Živaljević, *Topological methods*, Handbook of Discrete and Computational Geometry (J.E. Goodman and J. O'Rourke, eds.), CRC Press, Boca Raton, New York, 1997, pp. 209–224.

Index